Plant Biotechnology: Recent Advancements and Developments

Suresh Kumar Gahlawat • Raj Kumar Salar
Priyanka Siwach • Joginder Singh Duhan
Suresh Kumar • Pawan Kaur
Editors

Plant Biotechnology: Recent Advancements and Developments

Springer

Editors
Suresh Kumar Gahlawat
Department of Biotechnology
Chaudhary Devi Lal University
Sirsa, Haryana, India

Raj Kumar Salar
Department of Biotechnology
Chaudhary Devi Lal University
Sirsa, Haryana, India

Priyanka Siwach
Department of Biotechnology
Chaudhary Devi Lal University
Sirsa, Haryana, India

Joginder Singh Duhan
Department of Biotechnology
Chaudhary Devi Lal University
Sirsa, Haryana, India

Suresh Kumar
Department of Biotechnology
Chaudhary Devi Lal University
Sirsa, Haryana, India

Pawan Kaur
Department of Biotechnology
Chaudhary Devi Lal University
Sirsa, Haryana, India

ISBN 978-981-10-4731-2 ISBN 978-981-10-4732-9 (eBook)
DOI 10.1007/978-981-10-4732-9

Library of Congress Control Number: 2017942816

Printed on acid-free paper

This Springer imprint is published by Springer Nature
The registered company is Springer Nature Singapore Pte Ltd.
The registered company address is: 152 Beach Road, #21-01/04 Gateway East, Singapore 189721, Singapore

Preface

The present century is a century of knowledge, indeed a century of the mind, and it belongs to biotechnology, which has brought unprecedented advances in all walks of life. The origin of biotechnology can be traced back to prehistoric times, when microorganisms were used for processes like fermentation. The advent of biotechnology has opened up a wide horizon in the field of biological research. It has developed into an important force in the creation of employment, production, and trade, and affects national economy and the quality of human life throughout the world. The range and the significance of biotechnological activities are expanding at an ever-increasing rate so that the impact of biotechnology on the world economy and human welfare is likely to be far greater than that of any other discipline. Our effort is to introduce students and researchers to cutting-edge techniques and applications of biotechnology. Therefore, chapters were selected on climate change, global warming, genetic engineering, crop improvements, RNAi technology and nanoparticles, etc., for this book.

The book entitled *Plant Biotechnology: Recent Advancements and Developments* will be highly useful for students, teachers, and researchers in all disciplines of life sciences, agricultural sciences, and biotechnology in universities, research institutions, and biotechnology companies. It will provide the readers a comprehensive knowledge of topics on genomics, plant biotechnology, and bionanotechnology. The chapters have been written with special reference to the latest developments in the above broader areas of biotechnology that impact the biotechnology industry. A list of references at the end of each chapter is provided for the readers to learn more about a particular topic. Typically, these references include basic research, research papers, review articles, and articles from the popular literature.

We as editors are grateful to the contributing authors for providing necessary expertise and commitment. Without their contribution, this book would not have been possible. Our sincere thanks go to the University Grants Commission, New Delhi, for sponsoring the National Conference on Biotechnology: Emerging Trends, which was organized by the Department of Biotechnology, Chaudhary Devi Lal University, Sirsa, India, during February 2016 and attended by most contributing authors. Last but not the least, we also thank our family members for their constant support and encouragement while we were working on this project.

Sirsa, Haryana, India

Suresh Kumar Gahlawat
Raj Kumar Salar
Priyanka Siwach
Joginder Singh Duhan
Suresh Kumar
Pawan Kaur

Contents

About the Editors

Suresh Kumar Gahlawat, Ph.D. is the chairperson, Department of Biotechnology, and dean, Faculty of Life Sciences, Chaudhary Devi Lal University (CDLU), Sirsa, India. He received postdoctoral BOYSCAST fellowship and DBT Overseas Associateship from the Ministry of Science & Technology, Government of India, for carrying out research at FRS Marine Laboratory, Aberdeen, the UK. He has completed four R&D projects from UGC, ICAR, and Government of Haryana. His research interests include the development of molecular diagnostic methods for bacterial and viral diseases. He published more than 70 research papers in journals of national and international repute, written five books, and supervised M.Phil. and Ph.D. research work of 12 students.

Raj Kumar Salar, Ph.D. is a professor at the Department of Biotechnology, CDLU, Sirsa, India. He was awarded a postdoctoral fellowship from the Ministry of Education, Slovakia, to pursue postdoctoral research at Slovak University of Technology, Bratislava, and has also visited Japan, Norway, Austria, and Hungary for academic pursuits. Dr. Salar received several R&D projects from UGC, New Delhi, HSCST, Chandigarh, and DST, New Delhi. He was felicitated with King Abdulaziz City for Science and Technology (KACST) award for the best paper published in 3 Biotech. He has supervised several Ph.D. and M.Phil. students for their research. Dr. Salar is also a reviewer of several international journals.

Priyanka Siwach, Ph.D. is a professor at the Department of Biotechnology, CDLU, Sirsa. She was awarded with Silver Jubilee Gold Medal for her Ph.D. thesis by CCS Haryana Agricultural University, Hisar. She has received R&D projects from UGC, New Delhi. Her current research areas include genome annotations, identification and functional annotation of miRNAs, and molecular marker analysis for improvement of quality traits. She has also worked on micropropagation and biochemical characterization of medicinal plants and filed two patents in this field. She is a reviewer of several national and international journals.

Joginder Singh Duhan, Ph.D. (Microbiology) is a senior assistant professor at the Department of Biotechnology, CDLU, Sirsa. He is the recipient of Young Scientist Award given by the Association of Microbiologists of India and K.K. Nanda award for best research and several other best paper awards. He has more than 70 research papers and edited two books published by Springer and Narosa. He has received R&D projects from UGC and CSIR, New Delhi. He has supervised the research work of 15 M.Phil. and 5 Ph.D. students. His specialization includes microbial biotechnology and fermentation technology. He is a reviewer of several national and international journals.

Suresh Kumar, Ph.D. is presently working in the Department of Biotechnology, CDLU, Sirsa. His specialization includes microbial and environmental biotechnology. He has published more than 15 research papers in journals of national and international repute.

Pawan Kaur, Ph.D. is presently working in the Department of Biotechnology, CDLU, Sirsa. She has post-Ph.D. research experience in the area of drug delivery. Her area of interests includes synthesis of nanoparticles, nanocomposites, and antimicrobial activity and cytotoxicity of nanoformulations in biotechnology. She has published more than 15 research papers in journals of national and international repute.

1 Advances in Computational Tools for Plant microRNA Identification

Megha Sihag, Khushboo Sethi, S.K. Gahlawat, and Priyanka Siwach

Abstract

Continuous development in biotechnological techniques has led to major breakthrough discoveries in life sciences, one of which is the finding of a particular class of small RNA molecules known as microRNAs (miRNA). miRNAs have been found to regulate various biological activities in all life forms. Though plant miRNAs were discovered later than animal miRNAs, these have been found to reveal remarkable importance in gene regulation during plant development as well as toward responding to any stimuli. Unraveling the entire mystery of these small molecules is the first step to gaining a better understanding of their function. Initially, miRNAs were identified largely by experimental techniques like forward genetic screening, cloning, and microarray, but these techniques were expensive, more time-consuming, and unable to reveal poorly expressed RNA molecules. Advances in computational tools, made during the last decade, have done wonders in miRNA research work. Many novel miRNAs were discovered using these approaches in plant systems, though confirmation requires experimental validation. In this chapter, an effort has been made to understand the characteristic features of plant miRNA compared to animal miRNAs and to review the recent advances in various computational tools made for identification of plant miRNAs.

M. Sihag • K. Sethi • S.K. Gahlawat • P. Siwach (✉)
Department of Biotechnology, Chaudhary Devi Lal University, Sirsa 125 055, Haryana, India
e-mail: psiwach29@gmail.com

S.K. Gahlawat et al. (eds.), *Plant Biotechnology: Recent Advancements and Developments*, DOI 10.1007/978-981-10-4732-9_1

1.1 Introduction

MicroRNAs (miRNA) are small noncoding RNAs that play an important role in regulation of gene expression at posttranscriptional stage (Ambros 2001; Carrington and Ambros 2003; Bartel 2004). The first miRNA (lin-4) was discovered by Ambros and colleagues in 1993 while working on developmental regulation in *C. elegans* (Lee et al. 1993). For seven years, lin-4 was considered as an unusual thing, until the discovery of a second miRNA in *C. elegans*, called let-7 (Reinhart et al. 2000). Following this, let-7 homologs were identified in many vertebrate species, including humans (Pasaquinelli et al. 2000), and were later on found to be evolutionarily conserved molecules. The potential of miRNA in regulation of gene expression was not recognized until 2001 (Ambros 2001; Lau et al. 2001; Lagos-Quintana et al. 2001). Extensive follow-up work by many laboratories revealed the concepts of miRNA biogenesis and function (Kurihara and Watanabe 2004; Lee et al. 2002; Bartel 2004; Yu et al. 2005).

Plant miRNAs were identified nearly a decade later than animal miRNAs (Llave et al. 2002; Reinhart et al. 2002; Park et al. 2002). Since then, different approaches have been undertaken to identify new miRNAs. A public repository for all published miRNA sequences and associated annotation was established in 2002, named as MicroRNA Registry, initially, and as miRbase, later on. The first release of miRBase in 2002 contained 218 miRNA loci from five species, while the latest miRbase release (v21, June 2014) contains 28,645 entries from 206 species, of which 6150 miRNAs are from different plant species as on June 2016, indicating the intensity and speed of miRNA research projects. A separate database for plant miRNA, plant microRNA database (PMRD) has also been developed to integrate the data pertaining to plant miRNAs available from public resources (Zhang et al. 2010). The database was established in 2008 and has been updated regularly. One update was made on February 14, 2014, with a total of 10,594 miRNA entries from 128 plant species in the PMRD (http://bioinformatics.cau.edu.cn/PMRD). In November 2014, PMRD was changed to PNRD (Plant Noncoding RNA Database), to help biologists with their research in the RNA world. Noncoding RNA (ncRNA) is a functional RNA molecule that is not translated into a protein and is highly abundant and functionally important. The latest update of PNRD was made on May 2, 2016, with a total of 28,214 ncRNA entries from 166 plant species in the PNRD in which 16,390 were miRNA entries from 154 species. (http://structuralbiology.cau.edu.cn/PNRD/). Some other plant-based databases are MicroPC (http://www.biotec.or.th/isl/micropc) and PmiRKB (http://bis.zju.edu.cn/pmirkb/), which are basically annotation databases and have greatly enhanced plant miRNA research. Since the identification of first miRNA in plants, a lot of advancement has taken place in the miRNA identification approach and different tools have been developed for this. Various efforts have been made from time to time to discuss and compare various strategies adopted for miRNA identification (Meyers et al. 2008; Unver et al. 2009; Mendes et al. 2009; Hu et al. 2012). Since the computational tools and laboratory techniques, as well as the understanding of miRNA genes, are continuously evolving, this chapter is another effort in this direction to review various advances made for plant miRNA identification.

1.2 Characteristic Features of Plant miRNAs Versus Animal miRNAs

In both animal and plant cells, miRNA genes are transcribed by polymerase II to capped and polyadenylated primary miRNA (pri-miRNA) transcript (Lee et al. 2004). These pri-miRNAs are processed to form stem-loop precursor miRNA (pre-miRNA) (Kurihara and Watanabe 2004), with a minimal folding free energy index (MFEI) which is further processed to mature miRNAs. The latter enter a ribonucleoprotein complex known as the RNA-induced silencing complex (RISC), which regulates targeted gene expression (Bartel 2004). However, the biogenesis of plant and animal miRNAs differ in some aspects (Fig. 1.1). Studies have revealed larger and more variable precursor miRNA molecules for plants than for animals (Reinhart et al. 2002). In animals, processing of pri-miRNA to the pre-miRNA is mediated by

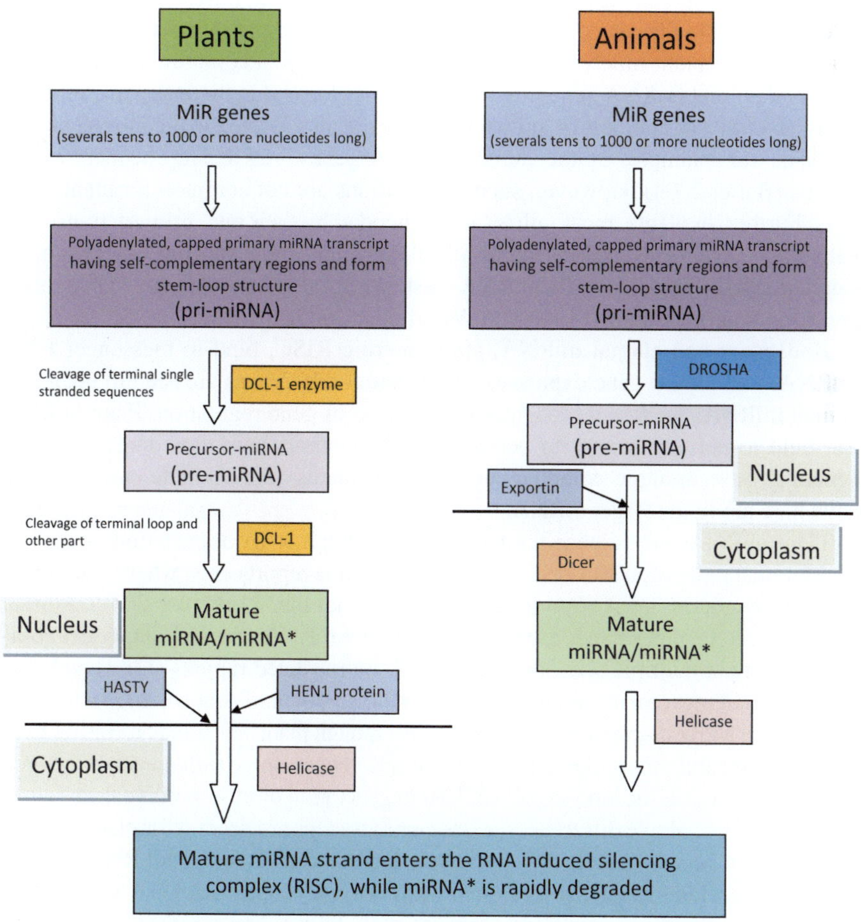

Fig. 1.1 Comparison of miRNA biogenesis mechanism in plants and animals

two enzymes Drosha and Dicer, in nucleus and cytoplasm respectively (Kim et al. 2009). On the other hand, in plants, a single enzyme (Dicer like-1) carries out both the processing steps in the nucleus only (Fahlgren et al. 2007), the same enzyme carrying out both cleavage steps, indicating that the stem-loop intermediate is a very transient structure in plants, compared to in animals. A stronger preference for a U at the 5′ terminus of the plant miRNAs has also been observed (Lau et al. 2001; Reinhart et al. 2002; Bartel and Bartel 2003). The differences between plant and animal miRNAs do not pertain to biogenesis only, but are also visible in their conservation, genomic organization, and mode of action (Millar and Waterhouse 2005; Zhou and Guan 2012).

Plant miRNAs are less conserved than animal miRNAs; usually only the mature miRNAs are conserved in plants instead of miRNA precursors that are usually conserved in animals (Bartel 2004). Phylogenetic studies indicate that plant miRNAs have a very deep origin in plant phylogeny, since the last common ancestor of bryophytes and seed plants (Zhang et al. 2005). The age of plant miRNA is suggested to be comparable to the age of miRNA regulation in metazoans; however, no evidence yet shows that plant miRNAs and animal miRNAs have a common ancestor. Both plant and animal miRNA genes are predominantly located in the intergenic regions, though significant numbers of animal miRNAs are also located in the introns of pre-mRNAs; for example, of the human miRNA genes, ~25% are encoded within introns (Bartel 2004). However, such observations are not common for plant miRNAs. Further, in plants, most miRNAs are encoded by their own primary transcript, with a few exceptional cases such as miR395 (which is present four times within a single transcript in rice), while miRNA clusters appear to have developed to a much greater extent in animals (Millar and Waterhouse 2005).

Both plant and animal miRNA, after entering RISC, bind to messenger RNA (mRNA) and inhibit gene expression. Substantial differences are seen in plant and animal miRNAs for target recognition and mode of gene regulation. Plant miRNA can bind to perfect or near to perfect complementary sites; most target mRNAs generally have a single complementary site in plants, and this site can be found anywhere along the target mRNA.

Plant miRNAs can silence the target gene through RNA degradation as well as translational repression (Brodersen et al. 2008). Some reports exist where plant miRNAs cause reduced level of proteins in plants but not mRNA (Cullen 2011). Animal miRNAs, on the other hand, usually bind to target mRNAs through imperfect complementarity at multiple sites located at the 3′ untranslated regions (UTR) and stop the ribosomal movements along the mRNA (Carrington and Ambros 2003).

From the above discussions, it is obvious that though plant and animal miRNAs have considerable similarities, they do have slight differences; these differences, if ignored during miRNA identification steps, will lead to generation of false-positive data.

In addition to above differences, extra cautious steps are needed for plant miRNA identification because plants have relatively large and complex small RNA populations within which miRNAs are often a minority. By contrast, most of the small RNAs in the animals are miRNAs (Meyers et al. 2008.). The complex small RNAs population in plants is largely due to the plant-specific RNA polymerase IV/RNA polymerase V (PolIV/PolV)–dependent short interfering RNAs (siRNAs) as well as secondary siRNAs, some of which are trans-acting. The presence of these diverse, endogenous siRNA populations in plants makes the precise identification of miRNAs more challenging (Meyers et al. 2008).

1.3 Methods of Plant miRNA Identification

Presently a number of methods are available which have not only speeded up miRNA identification but have also enhanced the efficiency and accuracy. Although miRNAs were first discovered through forward genetic screens in roundworms (Lee et al. 1993; Reinhart et al. 2000), this method has not been successful for miRNA gene identification in plants. The first plant miRNA was discovered by cloning experiments (Llave et al. 2002; Reinhart et al. 2002). Intensive refinement in cloning techniques for small RNA molecules has helped to successfully identify many new plant miRNA (Lu et al. 2005). However, cloning is biased toward RNAs that are expressed highly and broadly. MicroRNAs expressed at low levels or only in specific cell types or in response to certain environmental stimuli are more difficult to clone. Sequence-based biases in cloning procedures might also cause certain miRNAs to be missed (Ambros et al. 2003; Mendes et al. 2009).

Other experimental approaches have also been used to investigate new miRNAs such as Northern blot, microarray, and in situ hybridization, but these are also tedious and time-consuming. The advent of next-generation sequencing (NGS) technology reduced the cost for discovery and offers significant advantage of identifying even the less abundant miRNAs. It also provides a more reliable and sensitive method to quantify known miRNAs (Table 1.1). However, to discover new miRNAs from NGS data, some kinds of miRNA prediction algorithms with proper computational infrastructure are required (Friedlander et al. 2008). Several computational tools have been developed to complement experimental approaches to identify and validate novel miRNAs from high-throughput platforms such as NGS (Huang et al. 2011).

The main miRNA characteristics used by different computational tools are their length, high sequence conservation among species, and structural features like hairpin and minimal folding free energy (Li et al. 2010). Presently, various computational plant miRNA identification methods exist, which can be broadly divided into two main strategies: comparative-genomics-based methods and ab initio methods (Table 1.2). However, more specific and further subclassification can also be applied.

Comparative-genomics-based approach is one of the most extensively developed methods for miRNA detection, both in plants and in animals (Bonnet et al. 2004; Jones-Rhoades and Bartel 2004). A comparative genomics study across taxa has shown that many mature miRNAs are evolutionarily conserved from species to species in plants, though pre-miRNAs are less conserved. Because the mature miRNA sequences are short (~22 nt), sequence alignment tools like BLAST (Altschul et al. 1990) can only find the nearly perfect matches due to the large number of irrelevant hits. Sequence alignment alone may fail to detect the distant homologs that diverge in sequence yet are conserved in structure. So, more sensitive approaches that can

Table 1.1 Experimental tools applied for discovery of plant miRNA

Techniques used	Constraints	Some examples of work reference
Cloning	Time-consuming	Lu et al. (2005), Sunkar et al. (2005), Xie et al. (2005) and Zhao et al. (2012)
	Data is noisy (may clone and sequence a large number of degraded RNA fragments from the samples).	
	May not detect miRNAs that have low expression levels	
Microarray	Time-consuming	Chambers and Shuai (2009), Ding et al. (2011) and Liu et al. (2008)
	Low chances of detection of miRNA with low expression level	
In situ hybridization	Sensitive	Zhao et al. (2012)
	Time-consuming	
	Tedious	
Next-generation sequencing	To discover new miRNAs from NGS data, some kinds of miRNA prediction algorithms with proper computational infrastructure are required	He et al. (2013) and Jiao et al. (2011)

consider both sequence and structure conservation are needed. On the other hand, ab initio methods predict miRNAs in a single genome without using conservation of structure or comparative sequence analysis. The number of nonconserved miRNAs is estimated to be relatively large, which enables the identification of completely novel miRNAs for which no close homologs are known. Unlike comparative genomics approaches, ab initio approaches can discover species-specific miRNAs without known homology sequences (Li et al. 2010). Ab initio methods, in general, do not depend entirely on known genetic sequence information but use it as a training set to extract features and then make use of a machine-learning classification algorithm to identify candidate clips. Previously, an attempt has been made to compare and discuss various ab initio methods for miRNA discovery in metazoans as well as plants (Wu et al. 2012). At present, many tools belong to each category. Some of the popular ones are discussed below.

1.3.1 microHARVESTER

This is a popular homology-based tool introduced for plant miRNA identification (Dezulian et al. 2006). It takes advantage of the conservation of mature miRNA. It uses a BLAST sequence similarity search to first generate a set of candidates which is then rigorously refined by a series of filters exploiting the structural features specific to plant miRNAs to achieve specificity. In the first filtration step, those sequences of the candidate set whose aligned segments do not span most of the mature segment of the query are discarded. In the second filtration step, a modified Smith–Waterman pairwise alignment algorithm (Smith and Waterman 1981) is

Table 1.2 Some popular computational tools for plant miRNA identification

Tools	General features	References	Web sources
microHARVESTER	Identifies candidate miRNA homologs in any set of sequences, given a query miRNA. The candidate genes are subjected to various filters before final conclusion.	Dezulian et al. (2006)	http://ab.inf.uni-tuebingen.de/software/microHARVESTER/
MIRcheck	Uses a sequence/structure specification and 20mer coordinates. A strong conservation criteria is applied.	Jones-Rhoades and Bartel (2004)	http://web.wi.mit.edu/bartel/pub/software.html
Semrina	Uses a putative target sequence as input and allows to search for miRNAs that target this sequence.	Munoz-Merida et al. (2012)	http://www.bioinfocabd.upo.es/semirna
PsRobot	Web-based tool for identification of smRNAs with stem-loop precursors and their target genes.	Wu et al. (2012)	http://omicslab.genetics.ac.cn/psRobot/
MaturePred	A machine-learning method based on support vector machine which predicts positions of plant miRNAs for new plant pre-miRNA candidates.	Xuan et al. (2011)	http://nclab.hit.edu.cn/maturepred/
MiRDeep-P	Enables the expression patterns of annotated miRNA genes from small RNA library generated by NGS and explores the expression pattern of annotated miRNA genes.	Yang and Li (2011)	http://faculty.virginia.edu/lilab/miRDP/

(continued)

Table 1.2 (continued)

Tools	General features	References	Web sources
miRDeepFinder	A software package developed to identify and functionally analyze plant microRNAs (miRNAs) and their targets from small RNA data sets obtained from deep sequencing.	Xie et al. (2012)	http://www.leonxie.com/DeepFinder.php
miRCat	miRCat is a part of a tool kit developed for analysis of large-scale plant sRNA data sets. It identifies mature miRNAs and their precursors.	Moxon et al. (2008)	http://srna-tools.cmp.uea.ac.uk/plant/cgi-bin/srna-tools.cgi?rm=input_form&tool=mircat
C-mii	C-mii pipeline is based on combined steps and criteria from previous studies and also incorporates several tools.	Numnark et al. (2012)	http://www.biotec.or.th/isl/c-mii

applied to precisely determine the mature sequence in the candidate precursor from the optimal alignment of the query mature sequence against the corresponding segment of the BLAST hit. A candidate is discarded if the length of the mature sequences differs by >2 nt. In the third filtration step, the minimal free energy structure of the candidate sequence using RNAfold (Hofacker et al. 1994) is predicted and its putative miRNA* sequence is determined. A candidate is discarded if more than six nucleotides of its miRNA* are not predicted to form bonds with its mature miRNA (keeping in mind the 2 nt offset between miRNA and miRNA*). From a selection of all candidates that pass each filter, a multiple sequence alignment is constructed, using T-Coffee (Notredame et al. 2000), of a region that includes the miRNA, the miRNA*, and the "loop" sequence in between the miRNA and the miRNA*.

Of late, such a strategy has been extensively used for identification of miRNA genes from expressed sequence tag (EST) and a genome survey sequence (GSS) data sets of plants (Zhang et al. 2005, 2007). To further minimize the presence of false-positives, the concept of calculation of minimal free energy index (MFEI) was also introduced (Zhang et al. 2006a, b, c). The miRNA precursor sequences were reported to have higher MEFI than other coding or noncoding small RNA molecules (Zhang et al. 2006b). Using this approach, many miRNAs have been successfully identified in different plant species, such as soybean (Chen et al. 2009), maize (Zhang et al. 2006c), tomato (Yin et al. 2008), tobacco (Frazier et al. 2010), potato

(Xie et al. 2011), wheat (Han et al. 2009), mustard (Xie et al. 2007), citrus (Song et al. 2009), switch grass (Xie et al. 2010), apple (Gleave et al. 2008), cotton (Wang et al. 2012), and coffee (Akter et al. 2014).

1.3.2 MIRcheck

MIRcheck (Jones-Rhoades and Bartel 2004) uses a sequence/structure specification and the coordinates of the 20mer within the hairpin. It then uses a series of metrices concerning the number of unpaired nucleotides and bulges in the miRNA mature regions and the length of the hairpin. Sequences overlapping repetitive elements are eliminated, and a strong conservation criterion is applied, retaining only those stem-loops where the mature miRNA appears in both the genomes and exhibit high conservation in both miRNA and miRNA* sequence. Additionally, stem-loops are tested for robust folding, indicating that their secondary structures do not change substantially in the presence of flanking sequences. With this tool, Jones-Rhoades and Bartel (2004) identified 379 good plant miRNA candidates in 228 unique loci, of which 23 had their expression experimentally verified.

1.3.3 Semrina

This tool takes a putative target sequence such as a messenger RNA (mRNA) as input, and allows users to search for miRNAs that target this sequence (Munoz-Merida et al. 2012). It can also be used to determine whether small RNA sequences from massive sequencing analysis represent true miRNAs and to search for miRNAs in new genomes using homology. Semrina has shown a high level of accuracy using various test sets, and gives users the ability to search for miRNAs with several different adjustable parameters. Semrina is a user-friendly and intuitive web server. It is useful in searching for miRNAs involved in particular pathways as well as those in newly sequenced genomes.

1.3.4 PsRobot

PsRobot is a web-based easy-to-use tool dedicated to the identification of smRNAs with stem-loop-shaped precursors (such as microRNAs and short hairpin RNAs) and their target genes or transcripts (Wu et al. 2012). It performs fast analysis to identify smRNAs with stem-loop-shaped precursors among batch input data and predicts their targets using a modified Smith–Waterman algorithm. The multiple user-adjustable parameters enable the software to meet different needs of users. To facilitate better classification and functional analysis of input sequences, PsRobot integrates the expression information of input sequences in reported plant smRNA binding protein pull-down assay or mutants of major smRNA biogenesis pathway genes. For example, strong association with ARGONAUTE1 (AGO1) protein and expression depletion in dcl1 mutant will strengthen the confidence of a smRNA with stem-loop precursor as a miRNA. It also incorporates the available mRNA degradome data for users to evaluate the reliability of miRNA target prediction results. The multiplicity of miRNA binding sites on a single target as well as the cross-species conservation status of the target sites are also analyzed and provided. PsRobot can be either used online or downloaded and installed locally. The local version offers a larger capacity for input data size and has the function to incorporate user-uploaded degradome data. PsRobot has been found to outperform many of the tools in plant miRNA and target prediction.

1.3.5 MaturePred

MaturePred, a machine-learning method based on support vector machine, has been developed to predict the positions of plant miRNAs for the new plant pre-miRNA candidates (Xuan et al. 2011). Since the plant pre-miRNAs are cleaved into the miRNA:miRNA* duplexes, the prediction model considers the position-specific features for the whole miRNA:miRNA* duplexes. A set of informative features are selected to improve the prediction accuracy. Two-stage sample selection algorithm is proposed to combat the serious imbalance problem between real and pseudo miRNA:miRNA* duplexes. Extraction of the informative features is the key for improved performance of our SVM-based prediction model. MaturePred considers not only the position-specific features of a single nucleotide but also the structure-related, energy-related, and stability-related features, totaling 160 features. Some of the selected features are discussed below-.

A miRNA* is defined to have the same size as the miRNA candidate but lies on the opposite strand with its 3′ end starting two nucleotides before the matching position of the miRNA candidate's 5′ end. In order to obtain the miRNA:miRNA* candidates, two windows slide with step 1 in a pre-miRNA. If the sequence in the sliding window 1 is regarded as a miRNA candidate, the sequence in the sliding window 2 is regarded as the corresponding miRNA* candidate. The combination of windows 1 and 2 is a miRNA:miRNA* candidate. When the starting position of the miRNA candidate is coincident with the starting position of the actual miRNA, the miRNA:miRNA* candidate is a real miRNA:miRNA* duplex. Otherwise, the candidate is a pseudo miRNA:miRNA* duplex. The other information used is the stability of 5′ end of miRNA. According to miRNA biogenesis, the 5′ end of a miRNA is usually less stable than that of the corresponding miRNA*. It is useful for determining the functional strands where the miRNAs locate. Therefore, the stability of the first nucleotide at the 5′ end of miRNA/miRNA* is considered and denoted as miRNA_5′end and miRNA*_5′end, respectively. When the first position is (A, L), (G, L), (C, L), or (U, L), the feature (miRNA_5′end/miRNA*_5′end) value is assigned to 0. When it is (G, M) or (U, M), and there is a G-U or U-G wobble pair, the feature value is assigned to 1. When it is (A, M) or (U, M), and there is an A-U or U-A pair, the feature value is assigned to 2. When it is (G, M) or (C, M), and there is a G-C or C-G pair, the feature value is assigned to 3. The real miRNA:miRNA* duplexes typically are of greater binding stability and are less likely to be broken, and MaturePred uses this information as well during the training part.

Efficacy of MaturePred has been demonstrated by Xuan et al. (2011) while working with the real miRNA:bmiRNA* duplexes and the pseudo miRNA:miRNA* duplexes in the experimentally verified pre-miRNAs (1366) from database miRBase 14 (http://www.mirbase.org/). They demonstrated the importance of careful feature extraction, feature selection, and training sample selection in achieving effective prediction performance. Particularly, according to the characteristics of plant miRNAs, 160 features were extracted and 86 informative features were selected. Each negative sample (pseudo miRNA:miRNA* duplex) was mapped into the 86-dimensional space. About 17,803 representative negative samples were selected as the training samples to combat the class imbalance problem between the positive and negative samples. The proposed two-stage sample selection method can also be applied to other class imbalance problem in bioinformatics, such as identifying the SNP sites in the EST sequences.

1.3.6 miRDeep-P

Based on ultradeep sampling of small RNA libraries by next-generation sequencing, miRDeep-P enables users to explore expression patterns of annotated miRNA genes and discover novel ones (Yang and Li 2011). To run this application, the reads should be preprocessed by removing adapters, discarding reads shorter than 15 nucleotides and parsing them into FASTA format with their copy number recorded. With correctly formatted input files, miRDeep-P maps the reads to the reference (either genomic or transcriptomic) sequences using Bowtie (Langmead et al. 2009). For a given mapped read, the optimal size of the window from which to extract reference sequences for predicting RNA secondary structure is needed. However, miRDeep-P contains a module for users to empirically determine what window sizes to use in case a set of validated miRNA genes is available. The secondary structures of the extracted reference sequences along with all reads mapped to such sequences are processed by the miRDeep core algorithm (Friedlander et al. 2008), with a plant-specific scoring system. The output from the core algorithm is then filtered with additional plant-specific criteria based on known characteristics of plant miRNA genes (Meyers et al. 2008). The overall process quantifies the signature distribution of small RNA reads and thereby provides reliable information.

1.3.7 miRDeepFinder

miRDeepFinder is a software package developed to identify and functionally analyze plant microRNAs (miRNAs) and their targets from small RNA data sets obtained from deep sequencing (Xie et al. 2012). The functions available in miRDeepFinder include preprocessing of raw data, identifying conserved miRNAs, mining and classifying novel miRNAs, miRNA expression profiling, predicting miRNA targets, and gene pathway and gene network analysis involving miRNAs. The fundamental design of miRDeepFinder is based on miRNA biogenesis, miRNA-mediated gene regulation and target recognition, such as perfect or near-perfect hairpin structures, different read abundances of miRNA and miRNA*, and targeting patterns of plant miRNAs. To test the accuracy and robustness of miRDeepFinder, a small RNA deep sequencing data set of *Arabidopsis thaliana* published in the GEO database of NCBI was explored (Xie et al. 2012). The test retrieved 128 of 131 (97.7%) known miRNAs that have a more than 3 read count in *Arabidopsis*. Because many known miRNAs are not associated with miRNA*s in small RNA data sets, miRDeepFinder was also designed to recover miRNA candidates without the presence of miRNA*. To mine as many miRNAs as possible, miRDeepFinder allows users to compare mature miRNAs and their miRNA*s with other small RNA data sets from the same species. Cleaveland software package was also incorporated into miRDeepFinder for miRNA target identification using degradome sequencing analysis. Using this new computational tool, Xie et al. (2012) identified 13 novel miRNA candidates with miRNA*s from *Arabidopsis* and validated 12 of them experimentally. Interestingly, of the 12 verified novel miRNAs, a miRNA named AC1 spans the exons of two genes (UTG71C4 and UGT71C3). Both the mature AC1 miRNA and its miRNA* were also found in four other small RNA data sets.

1.3.8 miRCat

miRCat is a part of tool kit developed for analysis of large-scale plant sRNA data sets (Moxon et al. 2008). It identifies mature miRNAs and their precursors. Users upload a FASTA file of sRNA sequences mapped to a plant genome using PatMaN (Prüfer et al. 2008) and grouped into loci. The program screens a number of empirical criteria for bona fide miRNA loci. Briefly, it searches for a two-peak alignment pattern of sRNAs on one strand of the locus and assesses the secondary structures of a series of putative precursor transcripts using the RNAfold and randfold. miRCat has been tested on several high-throughput plant sRNA data sets and shows a high level of sensitivity and specificity.

1.3.9 C-mii

This software tool is developed for plant miRNA and target identification (Numnark et al. 2012). C-mii pipelines are based on combined steps and criteria from previous studies and also incorporated with several tools such as standalone BLAST and UNAFold and preinstalled databases including miRBase, UniProt, and Rfam. C-mii provides the following distinguished features. First, it comes with graphical user interfaces of well-defined pipelines for both miRNA and target identifications with reliable results. Second, it provides a set of filters allowing users to reduce the number of results corresponding to the constraints in plant miRNA and target biogenesis. Third, it extends the standard computational steps of miRNA target identification with miRNA-target folding module and GO annotation. Fourth, it supplies the bird's-eye views of the identification results with infographics and grouping information. Fifth, it provides helper functions for database update and autorecovery to ease system usage and maintenance. Finally, it supports multiproject and multithread management to improve the computational speed.

1.4 Conclusion

Biocomputational tools have enabled the discovery of novel biomolecules and their interactions in complex data sets. Cracking the complete mystery of miRNA is critical to understanding the developmental and regulatory aspects of plant systems. A number of methods, as summarized in this review, have been developed in an attempt to identify plant miRNAs. It is very difficult to accurately compare the methods because they use different data sets and may be based on different parameters to evaluate their performance. Selection of a methodology to use for a study probably depends on the information available. Further, some tools generate huge false-positive data, while the majority do not provide insights into the function or regulatory role of the predicted candidates. The lack of a clear and simpler pipeline to predict and validate miRNA candidates also makes the task of predicting miRNA transcripts and their encoded miRNAs complicated. The presence of various small RNA populations in plants aggravates the challenge. Ongoing improvements in computational tools as well as laboratory techniques are quite promising for achieving breakthrough levels in miRNA research in the near future.

References

Akter A, Islam MM, Mondal SI, Mahmud Z (2014) Computational identification of miRNA and targets from expressed sequence tags of coffee (*Coffea arabica*). Saudi Arabia J Biol Sci 21:3–12

Altschul SF, Gish W, Miller W et al (1990) Basic local alignment search tool. J Mol Biol 215(3):403–410

Ambros V (2001) microRNAs: tiny regulators with great potential. Cell 107(7):823–826

Ambros V, Bartel B, Bartel DP et al (2003) A uniform system for microRNA annotation. RNA 9:277–279

Bartel B, Bartel DP (2003) MicroRNAs: at the root of plant development? Pl Physiol 132:709–717

Bartel DP (2004) MicroRNAs: genomics, biogenesis, mechanism and function. Cell 116:281–297

Bonnet E, Wuyts J, Rouze P, Van de Peer Y (2004) Detection of 91 potential conserved plant microRNAs in *Arabidopsis thaliana* and *Oryza sativa* and identification of important target genes. Proc Natl Acad Sci U S A 101:11511–11516

Brodersen P, Sakvarelidze-Achard L, Bruun-Rasmussen M et al (2008) Widespread translational inhibition by plant miRNAs and siRNAs. Sci 320(5880):1185–1190

Carrington JC, Ambros V (2003) Role of microRNAs in plant and animal development. Sci 301:336–338

Chambers C, Shuai B (2009) Profiling microRNA expression in Arabidopsis pollen using microRNA array and real time PCR. BMC Pl Biol 9(87):1–11

Chen R, Hu Z, Zhang H (2009) Identification of microRNA in wild soybean (*Glycine soja*). J Integr Plant Biol 51(12):1071–1079

Cullen BR (2011) Viruses and microRNAs: RISCy interactions with serious consequences. Genes Dev 25:1881–1894

Dezulian T, Remmert M, Palatnik JF et al (2006) Identification of plant homologs. Bioinformatics 22:359–360

Ding Y, Chen Z, Zhu C (2011) Microarray-based analysis of cadmium-responsoive microRNAs in rice. J Exp Biol 62(10):1–11

Fahlgren N, Howell MD, Kasschau KD et al (2007) High-throughput sequencing of Arabidopsis microRNAs: evidence for frequent birth and death of miRNA genes. PLoS One 2:e219

Frazier TP, Xie F, Freistaedter A et al (2010) Identification and characterization of microRNAs and their target genes in tobacco (*Nicotiana tabacum*). Planta 232:1289–1308

Friedlander MR, Chen W, Adamidi C, Maaskola J et al (2008) Discovering microRNAs from deep sequencing data using miRDeep. Nat Biotechnol 26:407–415

Gleave AP, Ampomah-Dwamena C, Berthold S, Dejnoprat S et al (2008) Identification and characterisation of primary microRNAs from apple (*Malus domestica* cv. Royal Gala) expressed sequence tags. Tree Genet Genomes 4:343–358

Han Y, Luan F, Zhu H, Shao Y et al (2009) Computational identification of microRNAs and their targets in wheat (*Triticum aestivum* L.) Sci China C Life Sci 52:1091–1100

He C, Cui K, Zhang J, Duan A, Zeng Y (2013) Next-generation sequencing-based mRNA and microRNA expression profiling analysis revealed pathways involved in the rapid growth of developing culms in *Moso bamboo*. BMC Pl Biol 13(119):2–14

Hofacker IL, Fontana W, Stadler PF, Bonhoeffer LS et al (1994) Fast folding and comparison of RNA secondary structures. Monatsh Chem 125:167–188

Hu LL, Huang Y, Wang QC, Zou Q, Jiang Y (2012) Benchmark comparison of ab initio microRNA identification methods and software. Genet Mol Res 11(4):4525–4538

Huang Y, Zou Q, Wang SP, Tang SM et al (2011) The discovery approaches and detection methods of micro-RNAs. Mol Biol Rep 38:4125–4135

Jiao Y, Song W, Zhang M, Lai J (2011) Identification of novel maize miRNAs by measuring the precision of precursor processing. BMC Pl Biol 11:141. doi:10.1186/1471-2229-11-141

Jones-Rhoades MW, Bartel DP (2004) Computational identification of plant MicroRNAs and their targets, including a stress-induced miRNA. Mol Cell 14:787–799

Kim J, Jung JH, Reyes JL, Kim YS et al (2009) microRNA directed cleavage of ATHB15 mRNA regulated vascular development in Arabidopsis inflorescence stems. Pl J 42:84–94

Kurihara Y, Watanabe Y (2004) *Arabidopsis* Micro-RNA biogenesis through dicer like 1 protein functions. Proc Natl Acad Sci U S A 101:12753–12758

Lagos-Quintana M, Rauhat R, Lendeckel W, Tueshi T (2001) Identification of novel genes coding for small expressed RNAs. Sci 294:853–858

Langmead B, Trapnell C, Pop M, Salzberg SL (2009) Ultrafast and memory-efficient alignment of short DNA sequences to the human genome. Genome Biol 10:R25

Lau NC, Lim LP, Weinstein EG, Bartel DP (2001) The abundant class of tiny RNAs with probable regulatory roles in *Caenorhabditis elegans*. Sci 294:858–862

Lee RC, Feinbaum RL, Ambros V (1993) The *C. elegans* heterochronic gene lin-4 encodes small RNAs with antisense complementary to lin-14. Cell 75:843–854

Lee Y, Jeaon K, Lee JT, Kim S, Kim VN (2002) MicroRNA maturation: stepwise processing and subcellular localization. Eur Mol Biol J 21:4663–4670

Lee Y, Kim M, Han J, Yeom KM, Lee S, Baek SH, Kim VN (2004) MicroRNA genes are transcribed by RNA polymerase II. The EMBP J 23:4051–4060

Li L, Xu J, Yang D, Tan X, Wang H (2010) Computational approaches for microRNA studies: a review. Mamm Genome 21:1–12

Liu H, Tian X, Li Y, Wu C, Zheng C (2008) Microarray-based analysis of stress-regulated microRNAs in *Arabidopsis thaliana*. RNA 14:836–843

Llave C, Kasschau KD, Rector MA, Carrington JC (2002) Endogenous and silencing-associated small RNAs in plants. The Pl Cell 14:1605–1619

Lu S, Sun YH, Shi R, Clark C et al (2005) Novel and mechanical stress-responsive microRNAs in *Pouplas trichocarpa* that are absent from *Arabidopsis*. Plant Cell 17:2186–2203

Mendes ND, Freitas AT, Sagot MF (2009) Current tools for the identification of miRNA genes and their targets. Nucleic Acids Res 37:2419–2433

Meyers BC, Axtell MJ, Bartel B, Bartel DP et al (2008) Criteria for annotation of plant microRNAs. The Pl Cell 20:3186–3190

Millar AA, Waterhouse PM (2005) Plant and animal microRNAs: similarities and differences. Funct Integ Genomics 5(3):129–135

Moxon S, Schwach F, Dalmay T, MacLean D et al (2008) A toolkit for analyzing large scale plant small RNA datasets. Bioinformatics 24(19):2252–2253

Munoz-Merida A, Perkins JR, Viguera E et al (2012) Semirna: searching for plant miRNAs using target sequences. OMICS 16(4):168–177

Notredame C, Higgins DG, Heringa J (2000) T-coffee: a novel method for fast and accurate multiple sequence alignment. J Mol Biol 302:205–217

Numnark S, Mhuantong W, Ingsris S, Wang X, Wichadakul D (2012) C-mii: a tool for plant miRNA and target identification. BMC Genomics 13(7):S16

Park W, Li J, Song R, Messing J, Chen X (2002) CARPEL FACTORYU, a dicer homolog and HEN1, a novel protein, act in microRNA metabolism in *Arabidopsis thaliana*. Curr Biol 12:1484–1495

Pasaquinelli AE, Reinhart BJ, Slack F et al (2000) Conservation of the sequence and temporal expression of let-7 heterochromic regulatory RNA. Nature 408:86–89

Prüfer K, Stenzel U, Dannemann M, Green RE et al (2008) PatMaN: rapid alignment of short sequences to large databases. Bioinformatics 24:1530–1531

Reinhart BJ, Slack FJ, Basson M, Pasquinelli AE et al (2000) The 21-nucleotide let-7 RNA regulates developmental timing in *Caernorhabditis elegans*. Nature 403:901–906

Reinhart BJ, Weinstein EG, Rhoades MW, Bartel B, Bartel DP (2002) MicroRNAs in plants. Genes Dev 16:1616–1626

Smith TF, Waterman MS (1981) Identification of common molecular subsequences. J Mol Biol 147:195–197

Song C, Fang J, Li X, Liu H, Thomas C (2009) Identification and characterization of 27 conserved microRNAs in citrus. Planta 230:671–685

Sunkar R, Girke T, Jain PK, Zhu J (2005) Cloning and characterization of microRNAs from Rice. Pl Cell 17(5):1397–1411

Unver T, Namuth-Covert DM, Budak H (2009) Review of current methodological approaches for characterizing microRNAs in plants. Int J Pl Genomics. doi:10.1155/2009/262463

Wang M, Wang Q, Wang B (2012) Identification and characterization of MicroRNAs in Asiatic cotton (*Gossypium arboreum* L.) PLoS One 7(4):e33696

Wu H, Ma Y, Chen T, Wang M, Wang X (2012) PsRobot: a web-based plant small RNA meta-analysis toolbox. Nucleic Acids Res 40:W22–W28. doi:10.1093/nar/gks554

Xie F, Frazier TP, Zhang B (2010) Identification and characterization of microRNAs and their targets in the bioenergy plant switchgrass (*Panicum virgatum*). Planta 232:417–434

Xie F, Frazier TP, Zhang B (2011) Identification, characterization and expression analysis of MicroRNAs and their targets in the potato (*Solanum tuberosum*). Gene 473:8–22

Xie F, Huang SQ, Guo K, Xiang AL et al (2007) Computational identification of novel microRNAs and targets in *Brassica napus*. FEBS Lett 581:1464–1474

Xie F, Xiao P, Chen D, Xu L, Zhang B (2012) miRDeepFinder: a miRNA analysis tool for deep sequencing of plant small RNAs. Pl Mol Biol 80(1):75–84

Xie Z, Allen E, Fahlgren N, Calamar A, Givan SA, Carrington JC (2005) Expression of Arabidopsis MIRNA genes. Pl Physiol 138(4):2145–2154

Xuan P, Guo M, Huang Y, Li W, Huang Y (2011) MaturePred: efficient identification of MicroRNAs within novel plant pre-MiRNAs. PLoS One 6(11):e27422. doi:10.1371/journal.pone.0027422
Yang X, Li L (2011) miRDeep-P: a computational tool for analyzing the microRNA Transcriptome in plants. Bioinformatics 1:1–2
Yin Z, Li C, Han X, Shen F (2008) Identification of conserved microRNAs and their target genes in tomato (*Lycopersicon esculentum*). Gene 414:60–66
Yu B, Yang Z, Li J, Minakhina S, Yang M et al (2005) Methylation as a crucial step in plant microRNA biogenesis. Sci 307(5711):932
Zhang BH, Pan XP, Anderson TA (2006a) Identification of 188 conserved maize microRNAs and their targets. FEBS Lett 580:3753–3762
Zhang BH, Pan XP, Cox SB, Cobb GP, Anderson TA (2006b) Evidence that miRNAs are different from other RNAs. Cell Mol Life Sci 63:246–254
Zhang BH, Pan XP, Wang QL et al (2005) Identification and characterization of new plant microRNAs using EST analysis. Cell Res 15(5):336–360
Zhang BH, Pan X, Cobb GP, Anderson TA (2006c) Plant microRNA: a small regulatory molecule with big impact. Dev Biol 289:3–16
Zhang BH, Wang Q, Wang K, Pan X et al (2007) Identification of cotton microRNAs and their targets. Gene 397:26–37
Zhang Z, Yu J, Li D, Zhang Z et al (2010) PMRD: plant microRNA database. Nucleic Acids Res 38(Database issue):D806–D813
Zhao M, Tai H, Sun S, Zhang F et al (2012) Cloning and characterization of maize miRNAs involved in response to nitrogen deficiency. PLoS One 7(1):e29669
Zhou JS, Guan J (2012) Finding MicroRNA targets in plants: current status and perspectives. Genomics Proteomics Bioinformatics 10:264–275

Control of Gene Expression by RNAi: A Revolution in Functional Genomics

2

Suresh Kumar and Raj Kumar Salar

Abstract

RNA interference is an evolutionarily preserved gene-silencing phenomenon in eukaryotic organisms, which is triggered by double-stranded RNA (dsRNA) derived from viral replication intermediates or transposable elements or an introduced synthetic oligonucleotide. Dicer, an endonuclease, processes dsRNA into 21- to 23-base-pair (bp) small interfering RNA (siRNA) with 2-nucleotide 3′ overhangs, which are further linked with the RNA-induced silencing complex (RISC). Argonaute, an active (AGO)-family protein of RISC remove the sense strand of siRNA, guided by the antisense strand, mediates the destruction of cognate target RNA with exquisite potency and selectivity. RNAi is employed to downregulate the activity of a gene with hereditability, and other significant advantages have revolutionized functional genomics to discover genes involved in disease progression and validate gene function, holding promise for the expansion of new therapeutics to treat several diseases. RNAi has enormous exciting applications in plant biotechnology, generating new hopes to produce flowers with desired colors, and to enhance crop yield and shelf life, while in humans it is used to treat various forms of cancer, infectious diseases, and neurodegenerative disorders, though the stability, off-target activity, and in vivo delivery of small interfering RNAs remain significant obstacles. This chapter reviews the mechanism of RNA interference, its applications, and challenges encountered while using this technology for research purposes.

S. Kumar • R.K. Salar (✉)
Department of Biotechnology, Chaudhary Devi Lal University, Sirsa 125 055, Haryana, India
e-mail: rajsalar@rediffmail.com

S.K. Gahlawat et al. (eds.), *Plant Biotechnology: Recent Advancements and Developments*, DOI 10.1007/978-981-10-4732-9_2

2.1 Introduction

The central dogma of molecular biology describes the process by which the genetic information in DNA is converted into a functional product. In eukaryotic cells, there are several steps in the pathway leading from DNA to protein, and all of them can be controlled (Fig. 2.1). Eukaryotic cells can control the expression of a gene through (1) control of when and how often a given gene is transcribed; (2) processing of primary RNA transcripts; (3) transport of selected mRNAs from the nucleoplasm to the cytoplasm; (4) systemic degradation of selective mRNAs; (5) selection of which mRNAs are translated by ribosomes; and (6) selective activation or inactivation of posttranslational proteins. Gene expression can be regulated at six mentioned steps; however, for most genes, the control of transcription, posttranscriptional processing, translation, and posttranslational modification is paramount.

The regulation of transcription and translation has long been considered the "Holy Grail" of molecular biology (Smith 2009). Researchers have made exciting progress in developing a novel and effective approach to regulate transcription and translation activity of any gene. They have accomplished this by introducing a synthetic oligonucleotide molecule, which triggers a cellular process called *RNA interference* (RNAi). The basic goal of RNAi or antisense is to find or design small complementary oligonucleotides and get them inside a cell where they can bind to a segment of mRNA, prevent translation, or initiate cleavage of mRNA to interrupt protein synthesis. Probably, RNAi is the earliest immune protecting mechanism that enables organisms to shield their genome from foreign nucleic acids such as transposons and viruses (Waterhouse et al. 2001). It also refers to the accurate influence of DNA or RNA so that they transcribe to form a complimentary copy of normal—or *sense*—messenger RNA (mRNA). The antisense strand will target and bind with a specific mRNA of a desired gene to inhibit its phenotypic expression via different mechanisms (Fig. 2.2). This inhibition is sometime called *knockdown* or *knockout* subject to the gene function being either partially or completely abolished; it also permits researchers to set the usefulness of the designed sequence.

RNAi was initially reported in transgenic plants as cosuppression, a mechanism of gene suppression similar to that of transcriptional gene silencing (TGS) or

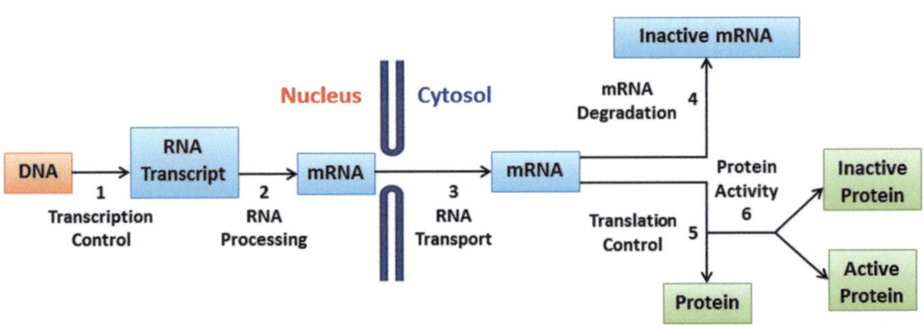

Fig. 2.1 Eukaryotic gene expression can be controlled at six different steps (Adapted from Bruce et al. 2013)

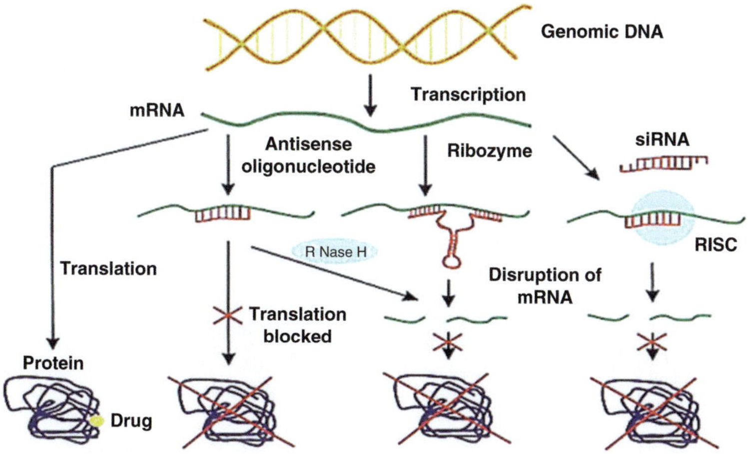

Fig. 2.2 Different gene expression–controlling techniques. Most conventional drugs bind to particular sites of a protein to inhibit their expression, whereas antisense strands bind with their complementary target mRNA and block translation of that mRNA (antisense oligonucleotides) or may induce mRNA degradation by RNase H/ribozymes. The RNAi approach is accomplished with siRNA/miRNA having 2-nucleotide 3′ overhangs that are associated with RISC and trigger degradation of the target mRNA (Adapted from Kurreck 2003)

posttranscriptional gene silencing (PTGS). This technique refers to several distinct but related approaches such as classical antisense, ribozyme, quelling (Lindbo 2012), and aptamer technology (Zhang and Wang 2005) (Table 2.1).

RNAi is a posttranscriptional method triggered with the introduction of double-stranded RNA (dsRNA) either through a virus or transgenes homologous to the target gene facilitating sequence-specific gene silencing. Dicer, a ribonuclease III enzyme processes long double-stranded RNAs to generate duplexes of ≈21 nt with 2-nucleotide 3′- overhangs called *small interfering RNA* (siRNA) that induce sequence-specific messenger RNA (mRNA) degradation. RNAi has been reported to naturally occur in organisms as diverse as bacteria, fungi, algae, plants, and animals (invertebrates and vertebrates) (Table 2.2).

The first evidence of gene silencing via dsRNA was reported successfully in roundworm, *Caenorhabditis elegans*, and this phenomenon has been termed *RNA interference* (RNAi) (Fire et al. 1998). In 2006, Fire and Mello shared the Nobel Prize in Physiology and Medicine for their innovative work on RNAi. Further studies in fruit fly, *Drosophila melanogaster*, have contributed greatly toward comprehension of the biochemical nature of the RNAi pathway (Elbashir et al. 2001). RNAi has become an invaluable research tool in both in vitro and in vivo studies because synthetic dsRNA introduced into cells can selectively and robustly induce suppression of specific genes of interest. This chapter reviews the RNAi mechanism, its applications, and challenges with the use of this technology for research purposes.

Table 2.1 Characteristics of gene expression–controlling technologies (After Zhang and Wang 2005)

Technology	Active molecule	Molecular target	Cellular site of action[a]	Suggested mechanism of action	Status of drug discovery and development
Antisense	DNA or RNA	RNA[b]	Cytoplasm	Translation blockage, RNase H activation, inhibition of splicing, disruption of RNA structure	Clinical use[c], clinical trials
RNAi	RNA	RNA	Cytoplasm	Promotion of mRNA degradation	In vivo and in vitro study
Ribozyme	RNA	RNA	Cytoplasm	Translation inhibition, destruction of RNA structure	Clinical use
Triplex forming oligonucleotide	DNA	DNA	Nucleus	Transcription inhibition	Preclinical
Aptamer	DNA or RNA	Protein	Nucleus, cytoplasm	Interference with protein function	Clinical phase I trial
Riboswitch	RNA	RNA	Nucleus, cytoplasm	Inhibition or promotion of gene expression at the level of transcription, translation, or RNA splicing	Clinical use, clinical trials
Quelling	RNA	RNA	Nucleus, cytoplasm	Induction of silencing of transgenes and cognate endogenous mRNAs at the posttranscriptional level	Clinical use, clinical trials

[a]Site of action refers to the initial site of interaction of the active molecule with its target molecule and may not be the same as the site where biological activity occurs
[b]Any kind of RNA can be targeted: pre-RNA, mRNA, or viral RNA
[c]The first antisense drug, Vitravene, has been approved for the treatment of patients with cytomegalovirus-induced retinitis

2.2 Historical Perspectives

RNAi was initially discovered in transgenic plants where antisense RNAs prevent transcription of a peculiar gene and indirectly from observing unpredicted results in experiments performed to obtain transgenic petunias with a higher content of anthocyanin pigments by introducing additional copies of chalcone synthase transgene (Napoli et al. 1990; Jorgensen et al. 1996). To the researchers' surprise, they obtain variegated or chimeric flowers instead of intense purple coloration (Fig. 2.3).

Table 2.2 RNAi and related gene-silencing mechanisms in different organisms

Phylum	Organism	Mechanism	Effector	References
Monera	*Escherichia coli*	RNAi	dsRNA	Tchurikov et al. (2000)
Fungi	*Neurospora crassa*	Quelling	Transgenes	Ruvkun (2001)
	Schizosaccharomyces pombe	Quelling	dsRNA	Raponi and Arndt (2003), Verdel et al. 2004
Algae	*Chlamydomonas reinhardtii*	Quelling	dsRNA	Winston et al. (2002)
Plants	*Arabidopsis thaliana*	PTGS	Transgenes	Elmayan et al. (1998)
	Petunia	PTGS	Transgenes	Elmayan et al. (1998)
	Nicotiana benthamiana	PTGS	Transgenes	Voinnet and Baulcombe (1997)
	Lycopersicon esculentum	PTGS	Transgenes	Hamilton et al. (1998)
Invertebrates	*Caenorhabditis elegans*	RNAi	dsRNA	Kelly and Fire (1998)
	C. briggsae	RNAi	dsRNA	Fjose et al. (2001)
	Brugia malayi	RNAi	dsRNA	Aboobaker and Blaxter (2003)
	Drosophila melanogaster	RNAi Cosuppression	dsRNA Transgenes	Misquitta and Patterson (1999) and Pal Bhadra et al. (1998)
	Hydra magnipapillata	RNAi	dsRNA	Lohmann et al. (1999)
	Trypanosoma brucei	RNAi	dsRNA	Wang et al. (2000a, b)
Vertebrates	*Mus musculus*	RNAi	dsRNA	Wianny and Goetz (2000)
	Xenopus laevis	RNAi	dsRNA	Nakano et al. (2000)
	Danio rerio	RNAi	dsRNA	Wargelius et al. (1999)
	Homo sapiens	RNAi	dsRNA	Chiu and Rana (2002)

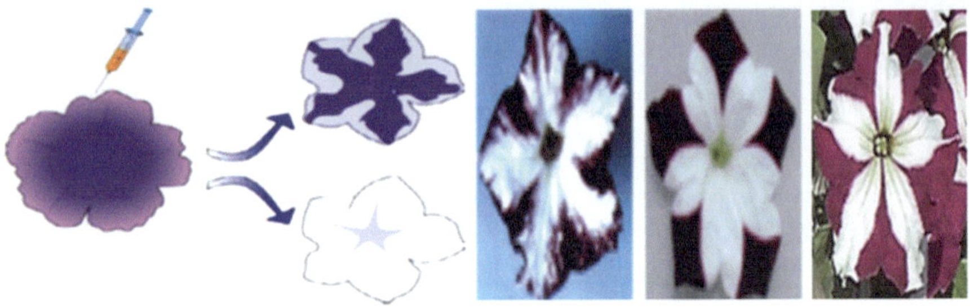

Fig. 2.3 Introduction of the transgene to enhance *purple* color in *Petunia* (Adapted from Matzke and Matzke 2004; Kumar et al. 2012)

They proposed that both the transgene and the plants' endogenous purple pigment gene may be suppressed (Napoli et al. 1990; Hannon 2002). Further, they coined the term *cosuppression* for this phenomenon (Napoli et al. 1990), as the expression of the endogenous gene (the original purple color) and the transgene (to deepen the purple color) were suppressed, but the molecular mechanism behind this remained unknown. It was subsequently described that suppression of gene expression may occur at the transcriptional level (TGS) or at the posttranscriptional level (PTGS).

The unknown mechanism responsible for gene suppression is now known as *RNA interference* (RNAi), discovered in 1998 by researchers working on a nematode, *Caenorhabditis elegans* (Fire et al. 1998). A similar phenomenon was seen in *Neurospora crassa*, where it was named *quelling* (Ruvkun 2001). It is evident that PTGS is activated by double-stranded RNA (dsRNA) and also provides many amazing manifestations with profound implications. The dsRNA comes from viruses and transposon activity in the natural RNAi process, while it can be injected into the cells in experimental processes (Elbashir et al. 2001). One strand of the dsRNA is identical in sequence to a region of the target mRNA molecule called the *sense strand*, and the complimentary is termed an *antisense strand*. An enzyme complex called *Dicer* is similar to RNase III, which recognizes dsRNA, and cleaves it into approximately 22-nucleotide-long fragments. These fragments are termed *small interfering RNAs* (siRNAs), which remain in double-stranded form with very short 3′ overhangs (Elbashir et al. 2002) and act as a template for RNAi-inducing silencing complex (RISC) to silence specifically the expression of a targeted gene. A similar form of RNAi is termed *PTGS* and is supposed to operate either at the transcriptional or translational level in diverse organisms. Various findings and benchmarks in RNAi expansion are listed in Table 2.3.

2.3 Components of RNAi

Several key enzymes and genes encoding them have been identified in several model organisms by analysis of mutants. A few components of the RNAi machinery were identified and served as initiators and effectors while others served as

Table 2.3 Benchmarks in the discovery and expansion of RNA interference

Discovery	Reference(s)
Double-stranded RNA activation of RNAi more efficiently than antisense RNA	Fire et al. (1998)
Nourishing of *C. elegans* with bacteria expressing dsRNA	Timmons and Fire (1998)
siRNAs mediating the process of RNAi	Hamilton and Baulcombe (1999)
Genes involved in RNAi and RNAi inheritance	Grishok et al. (2000)
MicroRNAs	Reinhart et al. (2002), Pareek et al. (2015)
Mechanism of RNAi beginning	Zamore et al. (2000)
Cloning of Dicer-encoding gene	Bernstein et al. (2001)
Sequence-specific silencing of a gene in in vitro mammalian cell culture	Elbashir et al. (2001)
Suggestion of siRNA design rules	Elbashir et al. (2002)
RNAi delivery by plasmids (short hairpin RNA, shRNA)	Brummelkamp et al. (2002a)
Viral delivery of shRNA	Brummelkamp et al. (2002b)
Successful use of therapeutic siRNAs in animals	Song et al. (2003)
Viral shRNA–induced animal knockdowns	Rubinson et al. (2003)
C. elegans genome screen using RNAi	Kamath et al. (2003)
Demonstration of RNAi effect on heterochromatin formation	Schramke and Allshire (2003)
Demonstration of siRNAs triggering interferon response	Sledz et al. (2003), Bridge et al. (2003)
Human genome screen using RNAi	Kiger et al. (2003), Berns et al. (2004)
Enzyme-mediated RNAi library synthesis	Sen et al. (2004)
Large-scale mammalian RNAi screens	Paddison et al. (2004)
Suppression of DET1 gene expression under fruit-specific promoters to enhance carotenoid and flavonoid levels in tomato	Davuluri et al. (2005)
Successful use of nanosecond pulsed laser–induced stress wave to deliver siRNA in plants for posttranscriptional gene silencing	Tang et al. (2006)
Reduction of lysine catabolism specifically during seed development to produce a white-flowered transgenic variety by silencing the chalcone synthase	Zhu and Galili (2003, 2004), Tang and Galili (2004), Tang et al. (2007)
Systemic administration of siRNA via targeted nanoparticles	Davis et al. (2010)
Use of fluorescent conjugated polymer nanoparticles (CPNs) to deliver siRNAs and knockdown the cellulose biosynthesis pathway (*NtCesA-1a* and *NtCesA-1b*)	Silva et al. (2010)
Development of an efficient reporter-based siRNA validation system for reverse functional genomics and molecular therapeutics	Tsai and Chang (2013)
Stable RNA interference rules for silencing	Fellmann and Lowe (2014)
Lipopeptide nanoparticles for siRNA delivery in rodents	Dong et al. (2014)

Table 2.4 Genes involved in RNAi-mediated gene silencing

Genes	Plants	Fungi	Animal
RdRp	SGS2/SDE1	QDE1	EGO1
elf2C	AGO1/AGO2/AGO7	QDE2	RDE1
RNase D	–	–	MUT7
RNA helicase	MUT6/SDE3	–	–
RecQ DNA helicase	–	QDE3	–
NMD proteins	–	–	SMG2/SMG5/SMG7
Coiled coil protein	SGS3	–	–

enhancers and transmitters of the gene-silencing process. These are described in the following sections.

2.3.1 Silencing Genes

Several eukaryotic genes governing gene silencing have been identified through genetic or biochemical screening of mutants reduced in transgene-induced RNA silencing (Table 2.4). These genes encode many proteins and enzymes in different eukaryotic organisms, for example a coiled coil protein (SGS3), a presumed RNA-dependent RNA polymerase (SGS2/SDE1), proteins containing PAZ/PIWI domain (AGO), and RNA helicase (SDE3). The coiled coil protein–encoding gene SGS3 was reported only in plants and it helps in silencing mediates by transgenes containing inverted repeats (Muangsan et al. 2004). SGS2/SDE1 of plants is related to QDE1 and EGO1 of *Neurospora crassa* and *C. elegans*. Similarly, AGO1 and AGO2 are related to QDE2 and RDE1 of *Neurospora* and *C. elegans* (Tabara et al. 1999). AGO7 of plants has no role in transgene silencing; rather, its primary function is in the regulation of developmental timing (Hunter et al. 2003). QDE3 protein of *Neurospora crassa* is probably responsible for the interaction between injected transgenes or an endogenous gene desired to trigger gene silencing (Cogoni and Macino 1999; Pickford et al. 2003). SMG2/SMG5/SMG7 of *C. elegans* facilitates nonsense-mediated mRNA decay (NMD), i.e., selective degradation of mRNA that contains premature stop codons (Anders et al. 2003). MUT6/SDE3 protein of *Chlamydomonas reinhardtii* degrades certain aberrant RNAs, i.e., improperly processed transcripts, which are often produced by transposons and some transgenes (Wu-Scharf et al. 2000). MUT7 protein of *C. elegans* is similar to RNase D enabling degradation of transposon transcripts, preventing transposase production and transposition (Ketting et al. 1999).

2.3.2 Dicer

The endonuclease Dicer is an RNase III family enzyme that acts as an essential component in RNAi technology involving dsRNA processing and target mRNA degradation. It is commonly known as the "fear of every dsRNA" as it cleaves large dsRNA into uniformly sized, 21 to 25 bp–nucleotide small RNAs (siRNAs) with 2-nucleotide 3′ overhangs and 5′ phosphorylated ends (Bernstein et al. 2001). Dicer encoded by more than one gene in several organisms has been reported with each Dicer specifically processing dsRNAs that comes from a definite source such as transcripts containing inverted repeats, dsRNA formed during RNA virus replication, and an introduced synthetic dsRNA (Margis et al. 2006). In *Arabidopsis thaliana*, DCL1 (Dicer like protein 1) processes miRNA precursor (transcript containing inverted repeats) exported from the nucleoplasm (Xie et al. 2004). DCL2, DCL3, and DCL4 (Dicer like proteins 2–4) preferentially process long dsRNA to produce siRNA establishing and retaining DNA methylation (Meister and Tuschl 2004; Henderson et al. 2006; de Souza et al. 2007) and are involved in virus defense mechanisms. In *Drosophila melanogaster*, Dcr-1/Loquacious create miRNA (Jiang et al. 2005; Saito et al. 2005; Lim et al. 2016) and Dcr-2/R2D2 produces siRNA. Dcr-1 has structural homology with Dcr-2, though they show differences in ATP dependence and substrate specifications (Liu et al. 2003; Jiang et al. 2005). DCL1 shows ATP-independent activity with high affinity toward the stem–loop structure of RNA (miRNA), whereas DCL2, DCL3, and DCL4 have ATP-dependent activity with substrate specificity to dsRNA. Further, DCL2 functions in association with dsRNA binding protein R2D2, forming a heterodimeric complex.

Dicer is a multidomain enzyme comprising an aminoterminal helicase domain, dual RNase III motifs, a dsRNA binding domain, and a PAZ/PIWI domain similar to the protein of the Argonaute family (Hammond et al. 2001a). The slicer activity of Dicer is probably catalyzed by its PAZ/PIWI domain. Complete digestion due to Dicer generates dsRNA fragments of 12–15 bp, i.e., half the size of siRNAs. Dicer, a specific endonuclease, acts as a dimmer, which processes dsRNA with the help of two catalytic domains of each monomer encoded by a consensus sequence (Agarwal et al. 2003). In the initial step of RNAi technology, Dicer works as a catalyst and generates small RNAs (siRNA or miRNA) as precursors recognized by RISC. Later, Argonaute protein, a catalytic component of RISC, has the ability to cleave mRNA complementary to the antisense strand (Jaronczyk et al. 2005).

2.3.3 RNA-Induced Silencing Complex (RISC) and Guide RNAs

RISC is an siRNA/miRNA-directed endonuclease, which recognizes the complementary mRNA and activates RNase to cleave the mRNA. About 20 to 23 bp small dsRNAs with 2-nucleotide overhangs at 3′ ends are associated with RISC then target and disrupt the cognate mRNA to reduce the levels of translation and knockdown the function of the specific gene (Hammond et al. 2000, 2001b). RISC acts as a catalyst to cut a single phosphodiester bond of mRNA (Schwarz et al. 2004). Though the components of RISC are not completely known, it includes members of the Argonaute family (Hammond et al. 2001a) that have been involved in processes directing posttranscriptional silencing (Stevenson 2004).

The noncoding RNAs (ncRNAs) are thought to be associated with sequence-specific nuclease and work as guides to target specific mRNA by following the Watson–Crick base-pairing rule. Studies of a few components of RISC belonging to Argonaute family proteins reported in *C. elegans*, *Arabidopsis thaliana*, *Drosophila melanogaster*, and other eukaryotic organisms have shown that some act as endonucleases and are referred to as *slicers*. Most of the AGO proteins contain an N-terminal domain, PAZ (PIWI-Argonaute-Zwille) and PIWI (P-element Induced Wimpy) domain, which adopts an RNase H conformation with endonuclease activity (Song et al. 2003).

In humans, Ago2 protein has slicer activity (Meister et al. 2004), whereas in *Drosophila*, both Ago1 and Ago2 act as slicers. Ago1 has miRNA-directed cleavage of RNA, while Ago2 cleavage activity is guided by siRNAs. *Arabidopsis thaliana* has ten Argonaute genes, but only Ago1 has been identified as a slicer that uses both miRNAs and siRNAs as guides (Baumberger and Baulcombe 2005). The members of the Argonaute family have been related to a gene-silencing mechanism at both the transcriptional or posttranscriptional and translational levels either by disrupting the target mRNA or by inhibiting translation (Hammond et al. 2000, 2001a; Martinez et al. 2002; Martinez and Tuschl 2004).

2.3.4 RNA-Dependent RNA Polymerase (RdRp)

RNA-dependent RNA polymerase has a vital role in both initiating and amplifying the silencing effect of RNAi or the posttranscriptional gene-silencing mechanism. The effects of RNAi are more effective and systemic in nature than the antisense approach (Dalmay et al. 2000) due to the activity of RdRp that results in transcription and replication of the RNA viral genome. RdRp was isolated and purified from tomato leaves then its catalytic properties were characterized by biochemical analysis (Schiebel et al. 1993a, b). An excess amount of dsRNA is considered better for the production of more siRNA to survive dilution by cell division. The sense strand or antisense strand is probably used as a template by RdRp to convert it into dsRNA or repeatedly copies the dsRNAs to produce a population of siRNA that could then interact with the target mRNA and amplify the effect of RNAi silencing.

2.3.5 Translation Initiation Factors

The RNAi mechanism is carried out in two steps such as the initiation and effector steps (Hutvagner and Zamore 2002). In the initiation step, dsRNA is processed and produces 21- to 25-nucleotide fragments known as *siRNAs* (Naykanen et al. 2001). It was apparent that siRNAs and miRNAs are produced by the slicer activity of Dicer on the dsRNA strand. In the effector step, siRNA/miRNA is associated with RISC. The complex is activated with the hydrolysis of ATP resulting in unwinding of siRNA (passenger strand and guide strand) and further directs the cleavage of the target mRNA into ~12-nucleotide-long fragments from the 5′ upstream of the siRNA (Naykanen et al. 2001; Hutvagner and Zamore 2002).

2.3.6 RNA and DNA Helicases

Various proteins from eukaryotic organisms seem to carry out eradication of aberrant RNA and they are clustered in biochemically similar groups of RNA–DNA helicases. A strain of *C. reinhardtii* carrying mut- 6 mutation was found to diminish silencing by a transgene, and by also activating the transposons (Wu-Scharf et al. 2000). MUT6 protein is involved in cleavage of unusual RNAs and thus could be a part of the RNAi-related mechanism. In *Neurospora crassa*, three quelling-defective mutants (qde1, qde2, and qde3) have been isolated (Cogoni and Macino 1997) and only qde3 gene encodes a 1955-amino-acid protein, which shows homology with other proteins belonging to the family of RecQ DNA helicases that specifically unwind DNA–DNA interaction between transgenes or an endogenous gene required for gene-silencing activation by producing an aberrant RNA, which is related to the human proteins for Bloom syndrome and Werner syndrome (Yu et al. 1996; Cogoni and Macino 1999). In addition, QDE3 protein is thought to be involved in the activation step of transcription gene silencing (Saxena 2014).

2.3.7 Transmembrane Protein (Channel or Receptor)

It is well established that the diffusion of gene silencing from one cell to another is systemic in nature (Winston et al. 2002); however, a transgenic strain of *C. elegans* HC57 has failed to show systemic RNAi. It produces a membrane protein, SID1, composed of 776 amino acids with a signal peptide and 11 putative transmembrane domains. The transmembrane protein may either act as a channel/pore for the transport of the siRNA signal or be required for endocytosis of the systemic RNAi signal (Saxena 2014).

2.4 Types of Small RNA Molecules

Recent discoveries of small regulatory RNA molecules have identified various small noncoding RNA molecules as listed in Table 2.5. The small noncoding dsRNA molecules control the expression of a gene via either PTGS or RNAi as discussed below.

2.4.1 Small Interfering RNA (siRNA)

Small interfering RNAs (siRNAs) are short dsRNA molecules that have a significant role in the RNA interference mechanism. Their role was first discovered in the 1990s by David Baulcombe's group in plants. The siRNAs have a definite

Table 2.5 Sizes and functions of various small noncoding RNAs involved in gene expression (After Cech and Steiz 2014)

Type of RNA	Definition	Size[a]	Function
mRNA	Messenger RNA	2–5 kb	Contains a coding region that encodes the sequence of amino acids within a polypeptide. mRNAs are polycistronic in prokaryotes and monocistronic in eukaryotes, having 5′- and 3′-untranslated sequences
tRNA	Transfer RNA	70–90 nt	RNA adaptor connecting an mRNA codon and the activated form of the cognate amino acid during translation on the ribosome
rRNA	Ribosomal RNA	120–5025 nt	RNA component of the small or large ribosomal subunit; the largest is a ribozyme
snRNA	Small nuclear RNA	100–300 nt	RNA localized in the eukaryotic cell nucleus
snoRNA	Small nucleolar RNA	70 nt	Essential for pre-rRNA processing by serving as a guide RNA
miRNA	MicroRNA	22 nt	RNA that (in complex with AGO protein) uses seed sequences near its 5′ end to base pair with a target mRNA to induce deadenylation and decay or translational regulation
siRNA	Small interfering RNA	22 nt	Product of Dicer cleavage of dsRNA; when it complexes with an AGO protein, it induces cleavage of a perfectly complementary target RNA
gRNA	Guide RNA	40–80 nt	Base pairs with an RNA target, orienting bound proteins to carry out a site-specific cleavage, ligation, or modification reaction
Riboswitch	RNA element within mRNA	40–140 nt	Toggles between two conformations upon exposure to a small-molecule ligand or another stimulus to inhibit or promote gene expression at the level of transcription, posttranscriptional processing, or translation
RNA of RNase P	RNA component of ribonuclease P	400 nt	Catalytic subunit of the enzyme that removes 5′-leaders from pre-tRNA
Telomerase RNA	Telomerase RNA	160–1.2 kb	Provides a template for telomeric DNA synthesis and scaffold protein assembly
Scan RNA	Small conjugation-specific RNA	28 nt	dsRNAs produced by an RNAi-related mechanism that recognize genomic internal eliminated sequences in the developing ciliates and target them for destruction

[a]Sizes of ncRNAs are typical values or ranges; most are approximate

configuration, i.e., two hybridized strands of 21–24 bases (usually 22 bp), each with phosphorylated 5′ ends and hydroxylated 3′ ends with 2-nucleotide overhangs at the 3′ terminus. In RNAi, long dsRNAs introduced in certain organisms or cells are processed into ~22 nt fragments by Dicer. The siRNA fragments coupled with RISC then target and cleave the complementary mRNA, thus inhibiting translation of targeted genes. This offers an easy and efficient way to silence a specific gene post-transcriptionally and also provides a potent tool to study gene function, target validation, and signal transduction.

2.4.2 MicroRNA (miRNA)

The miRNA noncoding genes were discovered in *Drosophila*, *C. elegans*, *Arabidopsis thaliana*, *Oryza sativa*, and land plants (Lee and Ambrose 2001; Lee et al. 2003; Axtell and Bartel 2005; Amor et al. 2009; Axtell et al. 2011; Liu et al. 2013). The initial transcription involved 70-nucleotide stem–loop structure primary transcripts called *pri-miRNAs*. In the nucleus, Drosha and DGCR8 processed the base of the hairpin loop structure to generate precursor miRNA (pre-miRNA), then exported to the cytosol through a carrier protein, exportin 5. In the cytoplasm, Dicer cleaves the pre-miRNAs to yield mature and functional double-stranded miRNA approximately 21–22 nucleotides long. The strands whose sequence is complementary to the target mRNA are associated with RISC leading to either mRNA degradation or translational repression (Lee et al. 1993; Hutvagner and Zamore 2002; Pareek et al. 2015). However, miRNAs have also been reported to enhance translation (Vasudevan 2012). Since its discovery, miRNA has uncovered its promising role both in negative regulation (transcript degradation and translational suppression) and positive regulation (transcriptional and translational activation) (Grosshans and Slack 2002; Kim et al. 2005).

2.4.3 Small Hairpin RNA or Short Hairpin RNA (shRNA)

Small hairpin RNAs made in the cell from an engineered DNA composed of a promoter and terminator separated by an inversely repeated sequence of the target gene are inserted (with a spacer region between the repeats) allowing the RNA transcribed from such a transgene to fold back with itself to form a hairpin loop structure. This is an advantageous feature of shRNA over siRNA in that it is continuously produced by the target cell's own machinery. The transgene constructs have been used in eukaryotes and have been expressed under the control of strong constitutive promoters such as U6, H1, tRNA (RNA polymerase III-dependent promoter), and cauliflower mosaic virus (CaMV) 35S promoter (RNA polymerase II-dependent promoter) (Miyagishi and Taira 2002; Kawasaki and Taira 2003). RNA polymerase III is ideal for transcribing small RNAs as its transcripts do not require further processing. Plasmid and viruses have been used to produce functional shRNA, resulting in persistent and robust gene silencing in vitro and in vivo (Rubinson et al. 2003).

The structure, length, composition, and origin of hairpins seem to be crucial in initiating RNAi-mediated gene silencing (Waterhouse et al. 1998). RNAi has provided a rapid and effective means to silence genes in stem cells and transgenic animals (Kunath et al. 2003; Reich et al. 2003; Tiscornia et al. 2003). The target cell can be prompted to produce shRNA by altered DNA sequences injected into the cell through a small gene cassette which is transported to the nucleus. The inserted DNA either becomes a part of the cell's own DNA or may persist in the nucleus, and starts the cell producing the specific shRNA, which follows a route virtually similar to that of miRNAs to suppress the expression of a targeted gene. RNAi could be triggered through either siRNA or miRNA or shRNA with certain differences in their properties as described in Table 2.6.

2.5 Mechanism of RNAi-Mediated Gene Silencing

Gene expression can be tailored by antisense oligonucleotides, or by ribozyme and RNA interference (Zhang et al. 2005), by means of either posttranscriptional inhibition or a splicing shift. Posttranscriptional inhibition is accomplished by several mechanisms including steric blockade of ribosomal access to the target mRNA, induction of RNase H cleavage of mRNA, and inhibition of ribosomal assembly as they bind a target RNA through Watson–Crick base pairing (Fig. 2.2). The net outcome of these processes is the diminished translation of target proteins. A comparison of different gene-silencing approaches is provided in Table 2.7.

The process of RNAi uses small double-stranded RNAs to knockdown the function of a gene at either the posttranscriptional or transcriptional level. The PTGS mechanism is supposed to trigger translational repression due to partial complimentarity with the targeted sequence or degradation of sequence-specific mRNAs having a perfectly complementary sequence. The endogenous microRNAs trigger either translational repression or degradation of mRNA when the antisense strand has partial homology with the target mRNA. The sequence-specific cleavage is demonstrated by exogenous small interfering RNAs (siRNAs) or endogenous short hairpin RNAs (shRNAs) with perfect or partial Watson–Crick base pairing to the desired mRNA target. The construction and processing of miRNAs need host machinery guided by one of the two miRNA strands to the target mRNA (Fig. 2.4). Likewise, siRNAs and shRNAs exploit almost similar elements and may compete with the function of natural miRNAs.

Primary miRNAs (pri-miRNAs) process to produce 70-nucleotide stem–loop structures by a microprocessor complex comprising Drosha and Pasha/DiGeorge syndrome critical region gene 8 (DGCR8). Pasha/DGCR8 are obligatory for recognition and specific cleavage of the hairpin loop structure (Martin et al. 2009; Nyugen et al. 2015; Herbert et al. 2016; Kwon et al. 2016). The pre-miRNAs are exported to the cytoplasm through a transporter protein, exportin 5, with further processing into 22-nucleotide miRNA mature duplexes by the Dicer/TRBP complex (Lee et al. 2006; Fukunaga et al. 2012; Lee and Doudna 2012). The partial complementary miRNA duplexes are linked with the Argonaute protein component of RISC that

Table 2.6 Differences between general properties of siRNA, miRNA, and shRNA

Criterion	shRNA	siRNA	miRNA
Nomenclature	Short hairpin RNA	Small interfering RNA	MicroRNA
Source	Nuclear expression	Laboratory synthesis, encoded by virus and transposons	Encoded by eukaryotic nuclear DNA and by viral DNA
Prior to Dicer processing	70–100 nucleotides with interspersed and hairpin structure	Double-stranded RNA containing from 30 to over 100 nucleotides	Pre-miRNA containing 70–100 nucleotide with interspersed and hairpin structure
Structure	21 to 23-nucleotide RNA duplex with 2-nucleotide 3′ overhang	21 to 24-nucleotide RNA duplex with 2-nucleotide 3′ overhang	19 to 25-nucleotide RNA duplex with 2-nucleotide 3′ overhang
Complimentary	Partially complimentary to mRNA	Fully complimentary to mRNA	Partial homology to mRNA, typically targeting the 3′ untranslated region of mRNA
Delivery methods	Viral and nonviral gene therapy vectors to the nucleus	Synthetic/natural polymers and lipids to the cytoplasm	Vectors and other gene transfer techniques to the nucleus
mRNA target	Multiple (could be over 100 at the same time)	One	Multiple (could be over 100 at the same time)
Mechanism of gene regulation	Translational repression	Cleavage of mRNA	Translational repression, cleavage of mRNA (rare, only when there is a high level of homology between miRNA and mRNA)
Persistence	Expressed for up to 3 years	99% degraded after 48 hours	Average half-life of miRNA ~5 days but up to 10 times more stable than mRNA
Administration	Local and systemic injection	Local or inadequate systemic injection	Local and systemic injection
Possibility of specific "off-target" effects	Lower than siRNA	Higher than shRNA	Higher than shRNA
Possibility of nonspecific "off-target" effects	Lower immune activation, inflammation, and toxicity	Higher immune activation, inflammation, and toxicity	Lower immune activation, inflammation, and toxicity
Application	Chronic, life-threatening diseases or disorders; where low doses are desirable	Acute disease conditions; where high doses are tolerable; drug target; diagnostic and biomarker tool	Acute disease conditions; where high doses are tolerable; drug target; diagnostic and biomarker tool

Table 2.7 Comparison between different gene-silencing technologies

Feature	RNAi	Ribozymes	Antisense technology
Structure	21 to 23 nt–long molecules with a 3′ overhang inducing posttranscriptional gene silencing	RNA molecules that act as an enzyme, even in complete absence of proteins	15–20 nt, which are complimentary to their target mRNA sequence
Mechanism of action	Degradation of dsRNA precursors into short 21 to 23 bp siRNA/miRNA by the enzymes Dicer and Dicer like proteins	Catalytic activity of breaking or forming covalent bonds with extraordinary specificity and simple catalytic domains	Either activating RNase H, which cleaves RNA moiety of DNA–RNA hybrid, or by steric blockade of ribosomes
Choice of target regions	Not effective against intron targets and RNAs that are exclusively nuclear like spliceosome RNAs	Can target both introns and subcellular components	Can be designed and targeted against intron regions
Design parameters	Set of rules has been proposed for designing siRNA/miRNA against target mRNA	GUC triplets limit choice of target finding in secondary and tertiary structures of target mRNA	ONs containing contiguous G quartet as well as one containing CpG islands must be avoided
Alterations	Can be chemically altered as reported in the literature so as to increase specificity and half-life with lower toxicity	Sequences can be added to change target specificity. But stabilization of ribozyme is even more difficult as introduction of modified nucleotides very often leads to conformational changes that abolish catalytic activity	Easily altered to improve selectivity and efficacy
Off-target activity and aptamer activity	Off-target effects may also occur in cases of high concentrations of siRNA	Show aptamer activity	Off-target effects are significant and also show aptamer activity
SNP detection	Existence of a SNP reduces efficiency	Can discriminate single base polymorphism	–

removes the passenger strand and the guide strand mediates silencing of the targeted homologous mRNA. The activated RISC can silence gene expression either by preventing translation or through transporting the complex to cytoplasmic processing bodies (p-bodies) where the mRNA is deadenylated and degraded (Han et al. 2006).

Similarly, shRNAs are transcribed from an engineered DNA construct and undertake similar processing to that of miRNA. The perfect Watson–Crick base pairing between the antisense strand and the target mRNA triggers RISC-mediated

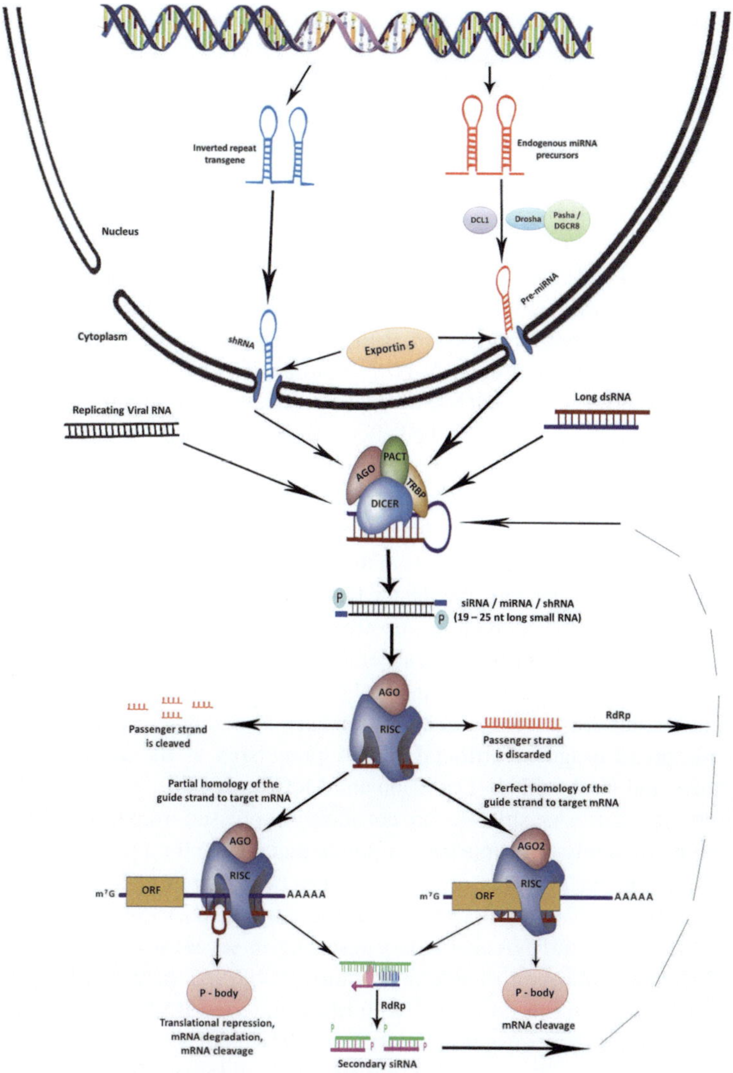

Fig. 2.4 Mechanism of RNAi-mediated gene silencing (Modified and adapted from de Souza et al. 2007; Burnett and Rossi 2012)

cleavage of the target mRNA (Tolia and Joshua-Tor 2006) while siRNA, a synthetic double-stranded RNA, is introduced in the cytoplasm of a cell, where it is diced by Dicer/TRBP complex to approximately 21 to 23 nt fragments with 2 nt overhangs at both 3′ ends. The guide strand of siRNA is associated with the Argonaute protein component of RISC and leads to sequence-specific cleavage of mRNA by perfect Watson–Crick base pairing (Martinez et al. 2002).

The Argonaute component of RISC has two conserved regions such as the PAZ and PIWI domains. The PAZ is attached to the 3′-OH group of the guide strand and the PIWI domain contains the mRNA cleavage site (Nykanen et al. 2001). Further, the sense strand after unwinding may be degraded or used as a template by the RNA-dependent-RNA-polymerase (RdRp) enzyme, which probably amplifies the dsRNA thus spreading the silencing signal to improve the efficiency of RNAi mechanisms (Matzke et al. 2001; de Souza et al. 2007). It has been established in plants that RdRp enzyme mediates primer-dependent (siRNA) and primer-independent (aberrant RNA) synthesis of dsRNA (Baulcombe 2007). The production of dsRNA from a single-stranded template (either sense or antisense strand) may be linked to the primer-independent progression, hence silencing can be initiated in virus-infected plants or with modified RNAs. The initial siRNA is generated by the Dicer enzyme, whereas secondary siRNA would be produced from a unique pathway having indispensable involvement of the RdRp enzyme (Pak and Fire 2007).

The last step comprises specific targeting of the mRNA sequence. During this step, RISC loaded with guide strand search for the complimentary sequences in the transcript mRNA pool. Once the complementary sequence is recognized, the guide strand binds to the PAZ domain site that stimulates the PIWI domain of the Argonaute to break one phosphodiester bond on the targeted mRNA strand at a distance of about ten nucleotides upstream from the 5′ end of the targeted mRNA. The mRNA is hydrolyzed at several locations, depending on siRNA homology, producing the cleaved mRNA untranslatable (Nykanen et al. 2001) or destroyed by cellular exonucleases (Rand et al. 2005). This remarkable property has encouraged widespread usage of artificial siRNA precursors as therapeutics to silence endogenous and viral mRNA (Davidson and McCray 2011).

In contrast to siRNAs, shRNAs are constantly expressed from promoters and can knockdown efficiently the expression of the targeted gene for a prolonged duration. Although prominent amounts of shRNAs may be required to achieve maximum silencing of the mRNA target, expression from strong RNA Pol III promoters can improve the natural miRNA machinery resulting in severe toxicity (Grimm et al. 2010). To avoid saturation of RNAi machinery, shRNAs may be expressed as a polycistronic transcript or combined with ribozymes and RNA decoys (Zhang and Rossi 2010). Since their inception, interest in siRNA/shRNA technologies has surpassed other antisense strategies due to their ability to knockdown specific mRNA, ease of siRNA synthesis, and getting persistent silencing as the siRNA can self-proliferate by RdRp activity as well as persistent catalytic activity in RISC for longer periods (de Souza et al. 2007; Castanotto and Rossi 2009).

Finally, in comparison with different antisense technologies that function stoichiometrically on the mRNA target, siRNAs is persistently recycled causing disruption of mRNA. It may be anticipated that siRNA-mediated cleavage of the mRNA target occurs at ten nucleotides upstream to the 5′ end of the guide strand; 5′ rapid amplification of cDNA ends (5′- RACE) PCR is employed to precisely ensure sequence-specific cleavage by AGO2 (Soutschek et al. 2004; de Souza et al. 2007), an important thought in validating that the siRNA is working through the RNAi pathway (Davis et al. 2010; Neff et al. 2011; Burnett and Rossi 2012).

2.6 RNAi and Functional Genomic Studies

The classical functional genomics techniques assessing a peculiar phenotype needs proximity of a specific sequence of marker genes and several generations of crosses are needed for identifying a specific mutant allele for a specific genotype. Microarray technology currently allows determining the level of expression of each gene in the genome to be assessed under different physiological conditions. This has led to vast generation and addition of information regarding genes whose expression is significantly studied in different physiological states. For example, enormous database features have been established for genes that are awkwardly regulated in diseases. In a very few cases this has resulted in the finding of key genes involved in the progression of the diseases and provided novel therapeutic targets. However, many times the pattern of gene expression is far too advanced for appropriate determination of the relatively smallest unchanged bystanders.

RNAi is used as a high-throughput method, in which a single gene in the genome is knocked down at a time and the cells are observed for a desired phenotype, i.e., death of a cultured cancer cell but not a normal cell. Since RNAi is a sequence-specific process this requires selection of a unique or conserved region of the target gene ensuring that a target gene of a multiples gene family can be silenced. Organisms that contain multiple gene families have required numerous assays to look at the involvement of a single gene as well as a whole family of genes; this approach is very tedious and labor intensive. However, this approach has been used successfully in the laboratory to investigate cell death signaling by TRAIL (tumor necrosis factor related apoptosis inducing ligand), an agent that might have therapeutic potential against various cancers (Aza-Blanc et al. 2003; Tewari and Vidal 2003). Further, this approach was also used to recognize the familial cylindromatosis tumor suppressor gene (CYLD) as a deubiquitinating enzyme in the nuclear factor-κB pathway (Brummelkamp et al. 2003). It has been discovered that aspirin can target this pathway and thus could be employed as a novel therapeutic approach to this rare inherited cancer. Similarly, in tobacco and rice, RNAi can knockdown specific target sequences when the 3′ UTR region is used as a trigger sequence (Ifuku et al. 2003; Miki et al. 2005). In addition, RNAi enables the generation of a dominant loss-of-function mutation in polyploidy plants (Yamamoto et al. 2005).

RNAi is also used as large pools of constructed viral vectors and applies a selection pressure such that only cells with the expected amendment remain viable and proliferate. The uniqueness of the genes knocked down in the surviving cells can then be identified by sequencing the vectors. This method is being used to investigate genes involved in agriculture for improvement of traits such as decreasing the levels of natural toxins, improvement of nutritional value, developing pest resistant varieties, etc., (Yogindran and Ranjam 2015) and in animal health specifically for neurodegenerative diseases, HIV, diabetes, cancer, etc. It has been used successfully to identify some novel components of the p53 tumor suppressor gene signaling pathway (Berns et al. 2004).

Cotton seeds are rich in dietary proteins but unpalatable to humans as they contain a natural toxic terpenoid called *gossypol*. The RNAi mechanism has been used successfully to downregulate the expression of δ-cadinene synthase enzyme, which is essential for the synthesis of gossypol (Kumar et al. 2006). RNAi technology has also been employed for developing barley varieties that are resistant to BYDV (barley yellow dwarf virus) (Wang et al. 2000a, b). In rice, RNAi has been used to decrease the level of glutenin protein and produce rice varieties with low glutenin content (Kusaba et al. 2003).

Both methods demonstrate substantial promises in identifying unique genes that could be targeted to design important therapeutic agents for inhibition either by conventional drug discovery methods or by RNA interference itself. Currently, several research projects are sanctioned to produce or design siRNAs that will silence a specific gene in model organisms using an approach called *host-delivered RNAi* (hdRNAi) (Gheysen and Vanholme 2007; Gordon and Waterhouse 2007; Auer and Frederick 2009) using partial sequence homology between plant and animal genes.

Nowadays, RNAi is a significant tool to alter gene function as well as providing a possible strategy for genetic rectification and genetic therapy in organisms where hitherto a genetic approach was not developed. Further, RNAi has revolutionized genomics permitting understanding of gene expression in eukaryotic cells and certainly a possible answer for many other challenges in functional genomics, and it will replace the classical genetic technologies and modifications.

2.6.1 Advantages and Disadvantages of RNAi in Functional Genomics

RNAi has both advantages and drawbacks when used in functional genomics to regulate the function of a gene. The foremost advantage of RNAi is its ability to specifically target a gene and silence it in a sequence-specific manner. It can be used to accomplish varying levels of gene silencing, using the same intron-containing hairpin RNA (ihpRNA) construct in an organism under different physiological conditions, as well as in different organisms, and permits selection of organisms with varying rates of gene silencing. In addition to this, the efficacy and level of knockdown can be controlled, which mediate targeted gene silencing at selected stages of growth in a chosen tissue. Thus, RNAi provides us with a great degree of flexibility in the field of functional genomics.

There are certain disadvantages of RNAi in that it requires the exact sequence of the target gene whose function is intended to be controlled. Once this sequence information is available, the remaining process is, however, relatively fast. This problem can be overcome by selecting a sequence homologous to a genome that has already been sequenced. Secondly, safe and efficient delivery methods for dsRNA are a limiting step for RNAi-based approaches. Due to this, further research is needed to design vectors that can be used safely and reliably. Further, it has been problematic to detect a mutant in which there has been little change in gene expression; however, in plants, several marker genes are being developed that will indicate if a change in gene expression occurs (Matthew 2004).

2.7 RNAi Delivery Methods

RNAi has rapidly advanced from a bench-scale observation to a fascinating area of research in biotechnology, genomics, and medicine. However, expansion of RNAi technology has been stalled since siRNAs are unstable in cells and delivery across the cell membrane is highly inefficient. Several methods have been attempted for direct or indirect delivery of exogenous dsRNA to animal and plant cells with different degrees of gene silencing. To improve the efficacy of RNAi technology the enzymatic stability and cellular uptake of siRNA should be significantly improved while their off-target effects and immunoactivation should be decreased.

2.7.1 RNAi Delivery in Animals

In animals, several methods have been established to facilitate delivery of RNAi to target cells in a cell culture or whole organisms including physical methods (electroporation, hydrodynamic injection, and nanoparticle delivery) (Plasterk and Ketting 2000; Akhtar and Benter 2007; Lewis and Wolff 2007; Li and Szoka 2007; Al-Dosari and Gao 2009; Gao and Huang 2009; Zhao and Feng 2015), chemical methods (liposome-mediated gene delivery, peptide-based gene delivery, chemical modification, and conjugation with lipids) (Shuey et al. 2002; Giladi et al. 2003; Simeoni et al. 2004; Akhtar and Benter 2007; Nguyen et al. 2008), and other methods (intraocular delivery, intratumoral delivery, local electroporation into muscle, local delivery to the central nervous system, and intranasal delivery to the airway) (Gao and Huang 2009). For direct delivery of exogenous dsRNA in target cells and organs, electroporation is frequently used in primitive organisms, whereas microinjection is ideal to deliver dsRNA into germ lines or early embryos of multicellular organisms. However, there is the problem of dsRNA dilution due to cell proliferation resulting in early genes being more easily silenced than late genes.

The siRNA is covalently linked to a ligand through cleavable spacers that improve its cellular uptake as well as conferring cell-specific targeting. In addition, chemical modifications to sugar phosphate backbones or bases of siRNAs have been revealed to improve their stability without affecting their biological activity with a reduced dosing frequency to enhance gene silencing, besides preventing eliciting of an immune response and decreasing off-target effects (Saxena 2014). The systemic delivery of siRNA is necessary to downregulate gene functions; however, it is challenging to introduce short dsRNAs directly into cells using physical and chemical transfection techniques.

Alternatively, animal cells can be transfected naturally using viral vectors including adenovirus, adeno-associated virus, lentivirus, retrovirus, herpesvirus, baculovirus, etc. The viruses used for siRNA/miRNA delivery are attenuated by using either culture techniques or genetic engineering techniques. The host range of viral vectors can be increased by pseudotyping with envelope proteins of other viruses. The main advantage of using a virus-mediated dsRNA delivery system is that it can infect both dividing and nondividing cells with high efficiency compared with nonviral methods (Rubinson et al. 2003; Shen et al. 2003; Hosono et al. 2004; Cronin et al. 2005; Sabbioni et al. 2007; Lambeth et al. 2009). Despite initial promise as a vector to deliver siRNA into animals, viral vectors are hindered due to adverse effects observed such as induction of toxic immune responses and unintentional gene expression changes following random integration into the host genome (Akhtar and Benter 2007).

2.7.2 RNAi Delivery in Plants

Nowadays, RNAi is used as a tool in plant transformation technologies to regulate gene functions by presenting dsRNA in a way similar to sense and antisense RNAs, with higher efficiency and bringing new traits such as disease resistance, stress resistance (biotic and abiotic), and improved nutritional quality and yield in different crops. The advanced RNAi delivery methods allow speeding up of the study of gene loss functions in targeted plant cells or tissues by using transient assay systems. The most reliable and frequently used transient delivery systems to deliver dsRNA into plant cells are virus infections, direct gene bombardment, and agroinfiltration.

2.7.2.1 Viral Vectors

The altered RNA and DNA viruses serve as vectors for gene silencing in different plant species such as *Arabidopsis* (Burch-Smith et al. 2004), *Nicotiana benthamiana* (Velásquez et al. 2009), pea (*Pisum sativum*) (Grønlund et al. 2010), tomato (*Lycopersicon lycopersicum*) (Fei et al. 2011), and barley (*Hordeum vulgare*) (Liang et al. 2012). A few viruses—viz., tobacco mosaic virus (TMV), potato virus X (PVX), and tobacco rattle virus (TRV)—can be used for both phenotypic expression and gene silencing (Kumagai et al. 1995; Angell and Baulcombe 1999; MacFarlane and Popovich 2000; Mallory et al. 2002). The efficacy of viral vectors varies according to the plant species being used; biosafety regulations for using some viruses are a limiting factor for using virus-induced gene silencing (VIGS) (Shang et al. 2007). However, modified viruses have been used for endogenous dsRNA production in both transient and stable transformation protocols targeting transcripts and controlling pests during herbivorous insect–plant interactions (Kumar and Mysore 2011).

Further, use of viral vectors to silence endogenous plant genes requires cloning of homologous gene fragments into the virus without compromising viral replication and movement. This was first demonstrated in RNA viruses by inserting sequences into the TMV and later in DNA viruses by replacing the coat protein gene with a homologous sequence (Kjemtrup et al. 1998). Phenotypic markers were used for investigating gene silencing of phytoene desaturase and chalcone synthase, and suggesting a new measure of tissue-specific silencing as both genes are involved in the carotenoid metabolic pathway. By suppressing the phytoene desaturase gene, a dramatic reduction in leaf carotene content results in variegated appearance of the leaf (Turnage et al. 2002). Similarly, overexpression of the chalcone synthase gene caused a chimeric phenotype instead of producing the expected intense orange color (Cogoni et al. 1994).

2.7.2.2 Biolistic Delivery

Particle bombardments (biolistic) are employed as a direct delivery technique for RNA or DNA constructs in stable transformations; the system is being used additionally in useful transient assays particularly in finding out plant pigmentation patterns (Shang et al. 2007). Particle bombardment on mature plants is sometimes random and restricted to the targeted tissue in showing a phenotypical amendment. Therefore, biolistic is employed on specific tissues like flower petals, young leaves and developing seeds.

Synthetic siRNAs delivered into plants by a biolistic technique results in reduced green fluorescent fusion protein (GFP) expression (Liu et al. 2002a, b). Bombarding of cells with particles coated with dsRNA or DNA that code hairpin constructs or antisense RNA triggers the RNAi pathway. The silencing effect of RNAi in rice is sometimes detected quickly after a day of bombardment for GFP, and it continues for up to a few days postbombardment. Systemic spreading of the GFP gene silencing occurs after 2 weeks of bombardment and northern hybridization with systemic leaves indicates that the introduced siRNAs induce de novo synthesis of siRNAs that accumulate to cause silencing (Klahre et al. 2002)). Additionally, microprojectile bombardment is employed on monocot plants wherever *Agrobacterium*-mediated transformation is usually restricted (Schweizer et al. 2000).

2.7.2.3 Agroinfiltration

The commonly used method in RNAi transient assays is agroinfiltration, like a technique for transient expression of T-DNA vectors with delivery by *Agrobacterium tumefaciens*. The introduction of *Agrobacterium* transports a DNA construct into the intracellular areas of leaves for eliciting RNA silencing and is known as *agroinoculation* or *agroinfiltration* (Hily and Liu 2007). The injected DNA is expressed transiently and encodes as either sense or antisense RNA, which is naturally a hairpin RNA. The hairpin constructs are particularly effective; as a result their dsRNA can be processed directly to siRNAs, whereas the constructs expressing as ssRNA can be helpful to induce silencing (Johansen and Carrington 2001; Voinnet 2001; Mlotshwas et al. 2002; Tenllado et al. 2003) with its suppressors and for easy protein purification (Johansen and Carrington 2001; Voinnet 2001; Mlotshwas et al. 2002; Tenllado et al. 2003). It is a fast, versatile, and convenient method for delivering silencing constructs to proliferative cells or tissues and getting a high level of native phenotypic effect, which may be biochemically analyzed, before continuing on to totally transgenic plants. Agroinfiltration transient assays are optimized for several plant species including monocots such as rice (Andrieu et al. 2012).

2.8 Application of RNAi

With the innovation of RNAi, the possibility of antisense therapeutics has blown into an innovative and highly competitive field that has attracted researchers all over the world and indeed it is now one of the most invested areas in the field of biotechnology. Besides being an area of intense, upfront basic research, the RNAi process holds the key to future technological applications in various sectors including agriculture, biotechnology, animal health, therapeutic intervention, etc. Some of the selected applications of RNAi are listed in the following sections.

2.8.1 Gene Knockdown

RNA inference is a highly efficient sequence-specific interaction of siRNA with targeted mRNA to downregulate gene expressions in virtually all eukaryotic organisms. The siRNAs can be processed to cause a drastic reduction in the expression of a targeted gene with investigation of the effects of this reduction genetically or biochemically to confirm the physiological role of the gene product. As RNAi could not totally eliminate expression of the gene, it is more appropriate to call it *knockdown* to discriminate it from *knockout* processes in which expression of a gene is totally eliminated.

2.8.2 RNAi as a Novel and Potential Therapeutic Agent

The systemic nature of RNAi plays an important role in many therapeutic applications. This approach depends on designing and processing of dsRNA that can specifically and efficiently silence any gene. The effectiveness of the approach depends on the design and process of dsRNA to silence any gene in a highly specific and efficient manner. The therapeutic approach is problematic when dsRNAs with ≥30 nucleotides may prompt nonsequence-specific interferon/immune responses in mammalian cells. However, this problem could be solved by introduction of siRNAs/miRNAs into cultured mammalian cells. Currently, throughout the world, researchers are focusing on this vital research theme for the curing of different diseases. Interestingly, for human diseases the altered expression of one gene or a few genes could be amenable to RNAi mechanisms. The diseases mainly include cancer, autoimmune diseases, hepatitis, malaria, genetic disorders, metabolic diseases, viral infections, cardiovascular and cerebrovascular diseases, neurodegenerative disorders, etc.

- RNAi has the capability to cure cancer or prevent tumor growth by inhibiting cancer cells through suppression of specific gene expression, cell cycle regulatory genes or antiapoptotic genes within cancer cells. RNAi is effectively used to target a gene specifically involved in the growth or survival of the cancer cell to kill cancer cells without damaging normal cells (Saxena 2014).
- A research report suggested that there is a possibility of blocking entry and replication of HIV by suppressing the chemokine receptor gene of HIV-1 with RNAi. However, despite the possibility of RNAi-mediated inhibition of HIV, the virus itself is a difficult target to inhibit due to a higher mutation rate (Martinez et al. 2002). This leads to it escaping from being targeted. Therefore, further research is required to prevent escaping or avoiding of the viral mutation.

- RNAi seems to be a good treatment for genetic diseases because of its sequence-specific nature and ability to target a single mutation in the genome, which is causing the disease. Once the target has been located, the siRNA is effectively delivered, and it has the potential to downregulate the single mutation, which could restore proper health (Saxena 2014).
- RNAi holds great promise in therapeutics to suppress disease-causing endogenous genes, especially metabolic diseases that code so-called nondruggable targets. Moreover, this technology is nontoxic as well, as no adverse effects are observed due to the high potency, specificity, and chemical structure of RNAi. It can be used to target specific genes involved in gluconeogenesis synthesis and also provides an in vivo proof of principle for the development of RNAi therapy for diabetes (Koo et al. 2004; Taniguchi et al. 2005).
- Malaria is a mosquito-borne infectious disease, which is the primary cause of morbidity and mortality in the developing world. A recent report has shown that RNAi technology is a vital technique for the identification of genetic features of vector-borne diseases. Female anopheles mosquitoes were transfected with a transgene that contained two copies of the target gene arranged as an inverted repeat configuration. The expression of dsRNA could be monitored by placing it under the control of a tissue- and time-specific promoter; ultimately dsRNA expression could be tailored to coincide spatially and temporally with the journey of the parasite through the mosquito. Both parasite receptors and immune components that protect the parasite could be used as putative targets for inhibiting parasite resistance through RNAi and the mosquito strains that have been rendered recalcitrant to malaria transmission could be released in the field to replace wild-type, permissive populations and achieve malaria eradication (Angaji et al. 2010).
- RNAi has the potential to reduce the impairment of heart and brain tissues by preventing the process of atherosclerosis. It can also reduce the expression of a cell adhesion molecule–producing gene in vascular endothelial cells, as well as preventing these cells from recruiting neutrophils and macrophages to the damaged site (Jarad et al. 2002).
- RNAi is capable of inhibiting hepatitis B virus replication as well as its expression in vitro and in vivo, which will continue to provide a research direction for developing an innovative therapeutic approach for hepatitis B virus infection (Ying et al. 2006).
- RNAi could be used to develop a newer therapy to cure neurodegenerative disorders through suppression of the expression of a particular gene. Currently many researchers throughout the world are focusing on Alzheimer disease, amyotrophic lateral sclerosis, Parkinson disease, Huntington disease, and spinocerebellar ataxia due to the age-linked nature of these diseases (Saxena 2014).

2.8.3 RNAi Role in Transgenic Plants

RNAi is now a widely used technique for the development of genetically modified food. An earlier study demonstrated that the caffeine content in *Coffea* plants has been markedly reduced by suppression of the expression of the caffeine synthase gene by RNAi technology (Ogita et al. 2003). This technology has enabled the creation of varieties of *Coffea* that produce natural coffee with very low caffeine content, subsequently bypassing the need for extraction (Van Uyen 2006). By decreasing the caffeine, its side effects (which include insomnia, restlessness, and palpitations) can be reduced.

- In plants, it provides a defense mechanism to protect against infection by viruses, transposons, and other insertional elements. Two examples are a barley variety resistant to barley yellow dwarf virus (BYDV) (Wang et al. 2000a, b) and a banana variety resistant to banana bract mosaic virus (BBrMV) (Rodoni and Dale 1999).
- Flavr Savr, a transgenic tomato exhibiting slow ripening and rot resistant traits, was produced by a Californian company, Calgene. The tomato was made more resistant to rotting by adding an antisense gene that interferes with the production of polygalacturonase, which normally degrades pectin in the cell wall and cause softening of fruit, making them more susceptible to being damaged by microbial infections. The RNAi construct was used to downregulate aminocyclopropane-carboxylic acid oxidase (ACC oxidase), a key enzyme of the ethylene biosynthetic pathway, resulting in delayed ripening of the tomato and improved shelf life as well (Bruening and Lyons 2000; Martineau 2001).
- The production of forest trees with low lignin content by suppression of the lignin biosynthesis pathway may significantly reduce processing costs and pollution by the paper and alcohol industries (Hu et al. 1999).
- White-flowered chimeric gentians are produced by suppressing the chalcone synthase gene (Nishihara et al. 2005). A similar technique is used to produce blue roses by disrupting the cyanidin gene, while carnation roses are produced by introducing the delphinidin gene (Van Uyen 2006).
- RNAi has been used to improve apple quality and shelf life by reducing allergenicity and sorbitol distribution (Gilissen et al. 2005), which affects starch accumulation and the sugar–acid balance (Teo et al. 2006), through specific gene silencing (Dandekari et al. 2004).
- Combining Bt technology with RNAi will probably augment product performance and it may provide protection against pests as well as hopefully restricting the development of resistance to Bt proteins (Goldstein et al. 2009).
- RNAi can be used to silence the gene responsible for the production of a neurotoxin, β-oxalylaminoalanine-L-alanine (BOAA), in a leguminous crop, *Lathyrus sativus*. Reduction of the levels of this antinutritional factor to a safer concentration may protect people from a paralytic disease, lathyrism (Williams et al. 2004).

2.8.4 Improvement of Crop Nutritional Value

Foods rich in carbohydrates, such as fiber, are considered to be health promoting (Williams 1995). A plant-derived carbohydrate is starch, which is composed of amylopectin (branched) and amylose (unbranched) as major nutritional sources found in cereals, tubers, and vegetables. RNAi is used to enhance the relative content of amylose in wheat grains using a designed antisense gene construct to specifically silence the genes within a seed-specific promoter (Regina et al. 2006), resulting in an increase in amylose content to over 70% of the total starch content (Tang et al. 2007).

- Production of a dominant high-lysine maize variety by suppression of the expression of the 22 kD maize zein storage protein results in a decrease in lysine catabolism and improved seed germination (Zhu and Galili 2003, 2004; Tang and Galili 2004; Angaji et al. 2010).
- RNAi has been employed to improve rice plants with reduced levels of glutelin and to create a rice variety called LGC-1 (*low glutelin content 1*) by downregulating the expression of the glutelin-producing gene (Kusaba et al. 2003); further, this can offer relief to kidney patients who are unable to digest glutelin.
- RNAi is used in cotton to silence two key fatty acid desaturase genes encoding stearoyl acyl career protein D9 desaturase and oleoyl phosphatidylcholine u6 desaturase. Knockdown of these genes in cotton led to nutritionally improved high oleic and stearic acid cottonseed oil that is more suitable for human consumption (Liu et al. 2002a, b).
- RNAi-mediated silencing of DET1 expression has been employed in tomato using specific promoters to enhance carotenoid and flavonoid levels without affecting plant growth (Williams et al. 2004).
- RNAi has been employed to increase the content of a health-promoting carotenoid, β-carotene, in potato by silencing the β-carotene hydroxylase gene, which convert β-carotene to zeaxanthin. Increased consumption of this will mitigate the incidence of vitamin A deficiency in populations as well as decreasing the frequency of heart diseases, skin cancer, and lung cancer (Eck et al. 2007).

2.9 RNAi Prospects for the Future

RNAi is an imperative area of research all over the world for refining the currently available protocols of RNAi in diverse fields. A comprehensive understanding of RNAi mechanisms is necessary to alter gene expression and advance this technology. It is now in a more cutting-edge state but still has not come out of its infancy. The current scenario of RNAi as an experimental tool is such that many investigators are now aware of the technology, but most have not hitherto implemented it in their own studies. The development of RNAi kits by several companies will facilitate the implementation of RNAi procedures by essentially any researcher, regardless of their understanding of RNAi mechanisms.

RNAi has a diverse array of applications in agriculture, genomics, and pharmaceutical biotechnology. One of the earliest uses of RNAi has been identification of novel targets in complex pathways and their relevance to various drug discoveries. This approach is advantageous as it permits fast target identification and processing without the need for pre-existing knowledge of the target biology. Using bioinformatics, libraries of designed siRNAs can be used to elucidate novel targets for any biological pathway. RNAi is highly reproducible and allows for high-throughput testing of potential targets with high specificity and sensitivity in functional analysis of thousands of genes simultaneously, as it requires small amounts of siRNA oligonucleotides. Viral infection is one of the foremost problems existing worldwide for flora and fauna. RNAi could represent the next generation of antiviral therapeutics, and DNA encoding siRNAs should be useful in various forms of gene therapy.

With regard to the use of RNAi as a therapeutic agent, its potential remains to be established. The possible outcomes of RNAi in eukaryotes will be combating of deadly diseases such as cancer, neurodegenerative disorders, and parasite infections, as well as in the analysis of functions and interactions of genes in cell developmental biology. An endogenous or exogenous dsRNA is expected to be developed as a nucleic acid–based medicine to treat various forms of cancer. Because targeted gene silencing is an important parameter in the potential success of RNAi cancer therapies, each gene associated with cell proliferation (mitosis), metastasis, angiogenesis, and drug resistance could be used as a possible target to downregulate that particular gene function. Importantly, silencing of such genes must not affect the functions of normal cells. Ongoing research and future preclinical trials in model animals will certainly help to optimize RNAi protocols and also design a new class of therapeutics for applications in humans.

RNAi could be used to produce crop plants that confer immunity against biotic stresses (viruses and transposons), abiotic stresses (heat, droughts, and floods), and various pests. It can also be employed to produce high nutritional crop varieties with higher yields, reduced levels of toxins, and reduced lignin content. The nutritional value of foods has an increasing role in the prevention of various human diseases associated with malnutrition. In another way, an RNAi approach would involve utilization of an inducible promoter system in order to produce dsRNA only upon infection and not constitutively to downregulate the key enzymes in biosynthetic pathways.

Thus, we can say that RNAi is gradually becoming an exciting molecular biology tool in exploration of the future of functional genomics. Its use in silencing of viral genes, development of siRNA-based therapeutics, and systemic silencing to generate plant varieties with improved nutritional content and maximum possible yields (to feed extra mouths, which will increase by two billion over the next 30 years) and low-lignin-content plants allowing easier, more environmentally friendly processing of fiber for the production of good-quality paper and cloth, etc., will be the major outcomes in the near future.

2.10 Challenges for RNAi as a Therapy

siRNA has promising therapeutic applications as it is efficient for knockdown of specific genes, as demonstrated in previous selected in vitro and in vivo studies. Certain challenges encountered while developing RNAi technology are siRNA stability and targeting, off-target silencing, activation of immune responses, and efficient siRNA delivery, requiring more advancement to exploit the full potential of this technology.

2.10.1 siRNA Stability and Targeting

Extracellular siRNAs are sensitive to degradation by enzymes found in serum and tissues. The short half-life of siRNAs (Behlke 2006) is a challenge for target-site accumulation to therapeutically appropriate levels (Guo et al. 2010). To be effectively used as therapeutic agents, siRNAs must not only survive in vitro and in vivo, but also reach their target cells in the specific tissue or organ that expresses the aberrant gene(s) of interest. After reaching their target cells, siRNAs still face a number of hurdles before they can exert their gene-silencing activity. The large size and negative charge of siRNAs prevents their transport across the plasma membrane and inhibits intracellular accumulation. Further, intracellular siRNAs remain vulnerable to degradation by intracellular RNases and still need to be recognized by and incorporated into RISC with high efficiency.

2.10.2 Off-Target Silencing

Microarray technology has demonstrated that siRNA treatment may result in suppression of genes other than the desired gene targets, referred to as *off-target gene silencing* (Jackson et al. 2003). Usually off-target silencing is not desirable because it can result in modification of gene expression and unexpected cell transformation. Previous studies have suggested that most off-target silencing occurs due to partial homology of nucleotides in the "seed region" of the siRNA sequence (Birmingham et al. 2006; Jackson et al. 2006).

Similarly, a few siRNAs result in altered gene expression because of "seed region" homology with endogenous miRNAs, which themselves regulate the expression of a gene (Aagaard and Rossi 2007). Weaker selection of the guide strand over the passenger strand by RISC can lead to a higher probability of matching undesired targets for a siRNA duplex. Off-target silencing can be significantly reduced or eradicated by implementing a predictive bioinformatics approach at the stage of siRNA design.

The potential risk of side effects appears from the feature that separates RNAi from other antisense technologies, i.e., the use of host cellular components for directing sequence-specific silencing. Employment of siRNAs/miRNAs to target particular cellular or viral transcripts fundamentally takes over the endogenous RNAi machinery, and little is known regarding the potential for saturating the RNAi mechanism in primary cells, while saturation of RISC is demonstrable in cultured cells (Hutvagner et al. 2004).

This problem may also be overcome by using symmetrical (19–23 base pairs with 2 nt overhangs at both 3′ ends) and asymmetrical (a 2 nt 3′ overhang at one end and a blunt end at the other) synthetic siRNAs (Kim et al. 2005) to improve the efficacy of the siRNA and decreasing off-target effects (Amarzguioui and Rossi 2008). It is important to strengthen the basic research studies on off-target effects of siRNAs and on the design of effective iRNAs (Saxena et al. 2003; Scacheri et al. 2004). Clear understanding of RNAi leading to nonspecific effects of short dsRNAs is necessary before the use of siRNAs or miRNAs in patient trials.

2.10.3 Activation of Immune Response

A previous study reveals that siRNA duplexes ≥30 nucleotides long can elicit nonsequence-specific interferon/immune responses in mammalian cells and cause cell death (Reynolds et al. 2006). Certain siRNAs can bind to and activate Toll-like receptor 7 (TLR7) if they contain the supposed "danger motif" (5′-GUCCUUCAA-3′) or similar GU-rich sequences that also can be recognized by TLR7 (Hornung et al. 2005). Different immune reactions are provoked among different cell types, thus it is difficult to predict in vivo responses on the basis of in vitro work. Obviously, immunogenicity and toxicity are grounds for concern that must be addressed in developing RNAi for therapeutic use. But this could be overcome by delivering siRNAs/miRNAs into cultured mammalian cells. Impressive work in this direction in different diseases is going on throughout the world. Basically, every human disease caused by activity from one or a few genes could be amenable to RNAi mechanisms.

2.10.4 siRNA Delivery

The problem of delivery has delimited RNAi technology for about two decades. Targeted delivery to particular cells or tissue types is still in its infancy for oligonucleotide-based therapeutics. On the other hand, virus-mediated delivery of therapeutic oligonucleotides has the usual problems related to safety and systemic delivery of viral vectors. However, RNAi-based research has opened the door to effective and probable therapeutic utility. More careful studies are desirable to test whether backbone-modified or nuclease-resistant siRNAs will move into clinical trials more rapidly than synthetic deoxyoligonucleotides.

2.11 Salient Features of RNAi

A few significant features of RNAi are as follows:

- Only double-stranded RNA is the interfering agent.
- There is a high degree of sequence-specific gene silencing with less effort.
- It is highly potent and effective (only some double-stranded RNA molecules per cell are required for effective interference).
- Silencing could be introduced in different developmental stages.
- Systemic silencing is possible.
- It avoids problems with abnormalities caused by a knocked down gene in early stages (which could mask desired observations).
- Silencing effects are heritable.

2.12 Conclusion

RNAi is advancing at an impressive pace for analyzing the activity of a gene in disease development and progression, and provides a gene-specific therapeutic development tool for reverse genetics to silence the expression of genes as desired, with further use for making transgenic organisms and rationalizing drug designing. However, more improvement is desperately needed to enhance the stability of siRNA, with development of efficient delivery systems and reduced off-target activity to exploit the full potential of RNAi in terms of its extraordinary efficacy and specificity. Further, RNAi is simply another tool in the biotech toolkit, along with genetic engineering and marker-assisted breeding, which can be used to develop novel and sustainable approaches to improve crop quality and productivity, meeting the food, clothing, and energy needs of an ever-increasing population. Thus, current classical genetic technologies need more and more molecular tools to provide a possible answer for many other challenges in functional genomics.

References

Aagaard L, Rossi JJ (2007) RNAi therapeutics: principles, prospects and challenges. Adv Drug Deliv Rev 59(2–3):75–86. doi:10.1016/j.addr.2007.03.005

Aboobaker AA, Blaxter ML (2003) Use of RNA interference to investigate gene function in the human filarial nematode parasite *Brugia malayi*. Mol Biochem Parasitol 129:41–51

Agarwal A, Dasaradhi PVN, Mohmmed A, Malhotra P et al (2003) RNA interference: biology, mechanism, and applications. Microbiol Mol Biol Rev 67(4):657–685

Akhtar S, Benter I (2007) Toxicogenomics of non-viral drug delivery systems for RNAi: potential impact on siRNA-mediated gene silencing activity and specificity. Adv Drug Deliv Rev 59:164–182. doi:10.1016/j.addr.2007.03.010

Al-Dosari MS, Gao X (2009) Non-viral gene delivery: principle, limitations, and recent progress. AAPS J 11:671–681. doi:10.1208/s12248-009-9143-y

Amarzguioui M, Rossi JJ (2008) Principles of Dicer substrate (D-siRNA) design and function. Methods Mol Biol 442:3–10. doi:10.1007/978-1-59745-191-8_1

Amor BB, Wirth S, Merchan F, Laporte P et al (2009) Novel long non-protein coding RNAs involved in *Arabidopsis* differentiation and stress responses. Genome Res 19:57–69. doi:10.1101/gr.080275.108

Anders KR, Grimson A, Anderson P (2003) SMG-5, required for *C. elegans* nonsense-mediated mRNA decay, associates with SMG-2 and protein phosphatase 2A. EMBO J 22(3):641–650. doi:10.1093/emboj/cdg056

Andrieu A, Breitler J, Sire C et al (2012) An in planta, *Agrobacterium*-mediated transient gene expression method for inducing gene silencing in rice (*Oryza sativa* L.) leaves. Rice 5(1):23. doi:10.1186/1939-8433-5-23

Angaji SA, Hedayati SS, Poor RH, Poor SS, Shiravi S, Madani S (2010) Application of RNA interference in plants. Plant Omics J 3:77–84

Angell SM, Baulcombe DC (1999) Technical advance: potato virus X amplicon mediated silencing of nuclear genes. Plant J 20:357–362

Auer C, Frederick R (2009) Crop improvement using small RNAs: applications and predictive ecological risk assessments. Trends Biotechnol 27:644–651. doi:10.1016/j.tibtech.2009. 08.005

Axtell MJ, Bartel DP (2005) Antiquity of microRNAs and their targets in land plants. Plant Cell 17(6):1658–1673. doi:10.1105/tpc.105.032185

Axtell MJ, Westholm JO, Lai EC (2011) Vive la différence: biogenesis and evolution of microRNAs in plants and animals. Genome Biol 12(4):221–233. doi:10.1186/gb-2011-12-4-221

Baulcombe DC (2007) Amplified silencing. Science 315:199–200. doi:10.1126/science.1138030

Baumberger N, Baulcombe DC (2005) Arabidopsis ARGONAUTE1 is an RNA slicer that selectively recruits microRNAs and short interfering RNAs. Proc Natl Acad Sci U S A 102:11928–11933. doi:10.1073/pnas.0505461102

Behlke MA (2006) Progress towards *in vivo* use of siRNAs. Mol Ther 13(4):644–670. doi:10.1016/j.ymthe.2006.01.001

Berns K, Hijmans EM, Mullenders J et al (2004) A large-scale RNAi screen in human cells identifies new components of the p53 pathway. Nature 428:431–437. doi:10.1038/nature 02371

Bernstein E, Caudy AA, Hammond SM, Hannon GJ (2001) Role for a bidentate ribonuclease in the initiation step of RNA interference. Nature 409:363–366. doi:10.1038/35053110

Birmingham A, Anderson EM, Reynolds A et al (2006) 3′ UTR seed matches, but not overall identity, are associated with RNAi off-targets. Nat Methods 3(3):199–204. doi:10.1038/nmeth854

Bridge AJ, Pebernard S, Ducraux A, Nicoulaz AL, Iggo R (2003) Induction of an interferon response by RNAi vectors in mammalian cells. Nat Genet 34:263–264. doi:10.1038/ng 1173

Bruce A, Dennis B, Karen H, Johnson AD et al (2013) Essential cell biology, 4th edn, Garland Science, Taylor & Francis Group, New York, p 864. ISBN 978-0-8153-4454-4

Bruening G, Lyons J (2000) The case of the Flavr Savr tomato. Calif Agric 54(4):6–7. doi:10.3733/ca.v054n04p6

Brummelkamp TR, Bernards R, Agami R (2002a) A system for stable expression of short interfering RNAs in mammalian cells. Science 296:550–553. doi:10.1126/science.1068 999

Brummelkamp TR, Bernards R, Agami R (2002b) Stable suppression of tumorigenicity by virus-mediated RNA interference. Cancer Cell 2(3):243–247

Burch-Smith TM, Anderson JC, Martin GB, Dinesh Kumar SP (2004) Applications and advantages of virus-induced gene silencing for gene function studies in plants. Plant J 39:734–746. doi:10.1111/j.1365-313X.2004.02158.x

Burnett JC, Rossi JJ (2012) RNA-based therapeutics: current progress and future prospects. Chem Biol 19(1):60–71. doi:10.1016/j.chembiol.2011.12.008

Castanotto D, Rossi JJ (2009) The promises and pitfalls of RNA interference based therapeutics. Nature 457:426–433. doi:10.1038/nature07758

Cech TR, Steitz JA (2014) The noncoding RNA revolution—trashing old rules to forge new ones. Cell 157(1):77–94. doi:10.1016/j.cell.2014.03.008

Chiu YL, Rana TM (2002) RNAi in human cells: basic structural and functional features of small interfering RNA. Mol Cell 10(3):549–561

Cogoni C, Macino G (1997) Isolation of quelling-defective (qde) mutants impaired in post-transcriptional transgene-induced gene silencing in *Neurospora crassa*. Proc Natl Acad Sci U S A 94:10233–10238

Cogoni C, Macino G (1999) Posttranscriptional gene silencing in *Neurospora* by a RecQ DNA helicase. Science 286(5448):2342–2344

Cogoni C, Romano N, Macino G (1994) Suppression of gene expression by homologous transgenes. Antonie Van Leeuwenhoek 65(3):205–209

Dalmay T, Hamilton A, Rudd S, Angell S, Baulcombe DC (2000) An RNA-dependent RNA polymerase gene in *Arabidopsis* is required for posttranscriptional gene silencing mediated by a transgene but not by a virus. Cell 101(5):543–553

Dandekari AM, Teo G, Defilippi BG et al (2004) Effect of down-regulation of ethylene biosynthesis on fruit flavor complex in apple fruit. Transgenic Res 13(4):373–384

Davidson BL, McCray PB Jr (2011) Current prospects for RNA interference based therapies. Nat Rev Genet 12:329–340. doi:10.1038/nrg2968

Davis ME, Zuckerman JE, Choi CH, Seligson D et al (2010) Evidence of RNAi in humans from systemically administered siRNA via targeted nanoparticles. Nature 464:1067–1070. doi:10.1038/nature08956

Davuluri GR, van Tuinen A, Fraser PD et al (2005) Fruit-specific RNAi-mediated suppression of DET1 enhances carotenoid and flavonoid content in tomatoes. Nat Biotechnol 23(7):890–895. doi:10.1038/nbt1108

Dong Y, Love KT, Dorkin JR et al (2014) Lipopeptide nanoparticles for potent and selective siRNA delivery in rodents and nonhuman primates. Proc Natl Acad Sci U S A 111(15):3955–3960. doi:10.1073/pnas.1322937111

Eck VJ, Conlin B, Garvin D, Mason H et al (2007) Enhancing beta-carotene content in potato by RNAi-mediated silencing of the beta-carotene hydroxylase gene. Am J Potato Res 84:331–342. doi:10.1007/BF02986245

Elbashir SM, Harborth J, Lendeckel W, Yalcin A, Weber K, Tuschl T (2001) Duplexes of 21 nucleotide RNAs mediate RNA interference in cultured mammalian cells. Nature 411(6836):494–498. doi:10.1038/35078107

Elbashir SM, Harborth J, Weber K, Tuschl T (2002) Analysis of gene function in somatic mammalian cells using small interfering RNAs. Methods 26(2):199–213. doi:10.1016/S1046-2023(02)00023-3

Elmayan T, Balzergue S, Beon F, Bourdon V et al (1998) Arabidopsis mutants impaired in co-suppression. Plant Cell 10(10):1747–1758

Fei Z, Joung JG, Tang X, Zheng Y et al (2011) Tomato functional genomics database: a comprehensive resource and analysis package for tomato functional genomics. Nucleic Acids Res 39(Database issue):D1156–D1163. doi:10.1093/nar/gkq991

Fellmann C, Lowe SW (2014) Stable RNA interference rules for silencing. Nat Cell Biol 16(1):10–18. doi:10.1038/ncb2895

Fire A, Xu S, Montgomery MK, Kostas SA, Driver SE, Mello CC (1998) Potent and specific genetic interference by double-stranded RNA in *Caenorhabditis elegans*. Nature 391(6669):806–811. doi:10.1038/35888

Fjose A, Ellingsen S, Wargelius A, Seo HC (2001) RNA interference: mechanisms and applications. Biotechnol Annu Rev 7:31–57

Fukunaga R, Han BW, Hung JH et al (2012) Dicer partner proteins tune the length of mature miRNAs in flies and mammals. Cell 151:533–546. doi:10.1016/j.cell.2012.09.027

Gao K, Huang L (2009) Non-viral methods for siRNA delivery. Mol Pharm 6(3):651–658. doi:10.1021/mp800134q

Gheysen G, Vanholme B (2007) RNAi from plants to nematodes. Trends Biotechnol 25(3):89–92. doi:10.1016/j.tibtech.2007.01.007

Giladi H, Ketzinel-Gilad M, Rivkin L, Felig Y et al (2003) Small interfering RNA inhibits hepatitis B virus replication in mice. Mol Ther 8(5):769–776

Giliseen LJWJ, Bolhaar STH, Matos CI et al (2005) Silencing of the major apple allergen mal 1 by using RNA interference approach. J Allergy Clin Immunol 115:369–384

Goldstein DA, Songstad D, Sachs E, Petrick J (2009) RNA interference in plants. Monsanto http://www.monsanto.com

Gordon KHJ, Waterhouse PM (2007) RNAi for insect-proof plants. Nat Biotechnol 25(11):1231–1232. doi:10.1038/nbt1107-1231

Grimm D, Wang L, Lee JS, Schurmann N et al (2010) Argonaute proteins are key determinants of RNAi efficacy, toxicity, and persistence in the adult mouse liver. J Clin Invest 120(9):3106–3119. doi:10.1172/JCI43565

Grishok A, Tabara H, Mello CC (2000) Genetic requirements for inheritance of RNAi in *C. elegans*. Science 287(5462):2494–2497

Grønlund M, Olsen A, Johansen E, Jakobsen I (2010) Protocol: using virus-induced gene silencing to study the arbuscular mycorrhizal symbiosis in *Pisum sativum*. Plant Methods 6:28. doi:10.1186/1746-4811-6-28

Grosshans H, Slack FJ (2002) Micro RNAs: small is plentiful. J Cell Biol 156(1):17–21. doi:10.1083/jcb.200111033

Guo P, Coban O, Snead NM, Trebley J, Hoeprich S, Guo S (2010) Engineering RNA for targeted siRNA delivery and medical application. Adv Drug Deliv Rev 62(6):650–666. doi:10.1016/j.addr.2010.03.008

Hamilton AJ, Baulcombe DC (1999) A species of small antisense RNA in posttranscriptional gene silencing in plants. Science 286(5441):950–952

Hamilton AJ, Brown S, Yuanhai H, Ishizuka M, Lowe A (1998) A transgene with repeated DNA causes high frequency, post-transcriptional suppression of ACC-oxidase gene expression in tomato. Plant J 15(6):737–746

Hammond SM, Bernstein E, Beach D, Hannon GJ (2000) An RNA-directed nuclease mediates post-transcriptional gene silencing in *Drosophila* cells. Nature 404:293–296. doi:10.1038/35005107

Hammond SM, Boettcher S, Caudy AA, Kobayashi R, Hannon GJ (2001a) Argonaute2, a link between genetic and biochemical analyses of RNAi. Science 293(5532):1146–1150. doi:10.1126/science.1064023

Hammond SM, Caudy AA, Hannon GJ (2001b) Post-transcriptional gene silencing by double stranded RNA. Nat Rev Genet 2(2):110–119. doi:10.1038/35052556

Han J, Lee Y, Yeom KH, Nam JW et al (2006) Molecular basis for the recognition of primary microRNAs by the Drosha-DGCR8 complex. Cell 125(5):887–901. doi:10.1016/j.cell.2006.03.043

Hannon GJ (2002) RNA interference. Nature 418:244–251. doi:10.1038/418244a

Herbert KM, Sarkar SK, Mills M et al (2016) A heterotrimer model of the complete microprocessor complex revealed by single-molecule subunit counting. RNA 22:175–183. doi:10.1261/rna.054684.115

Hilly JM, Liu Z (2007) An overview of small RNAs. In: Bassett CL (ed) Regulation of gene expression in plants. Springer, pp 123–147

Hornung V, Guenthner-Biller M, Bourquin C, Ablasser A et al (2005) Sequence-specific potent induction of IFN-alpha by short interfering RNA in plasmacytoid dendritic cells through TLR7. Nat Med 11(3):263–270. doi:10.1038/nm1191

Hu W, Harding SA, Lung J, Popko JL et al (1999) Repression of lignin biosynthesis promotes cellulose accumulation and growth in transgenic trees. Nat Biotechnol 17(8):808–812. doi:10.1038/11758

Hunter C, Sun H, Poethig RS (2003) The *Arabidopsis* heterochronic gene ZIPPY is an ARGONAUTE family member. Curr Biol 13(19):1734–1739

Hutvagner G, Zamore PD (2002) A microRNA in a multiple turn-over RNAi enzyme complex. Science 297(5589):2056–2060. doi:10.1126/science.1073827

Hutvagner G, Simard MJ, Mello CC, Zamore PD (2004) Sequence-specific inhibition of small RNA function. PLoS Biol 2(4):E98. doi:10.1371/journal.pbio.0020098

Ifuku K, Yamamoto Y, Sato F (2003) Specific RNA interference in psbP genes encoded by multigene family in *Nicotiana tabacum* with a short 3′-untranslated sequence. Biosci Biotechnol Biochem 67(1):107–113

Jackson AL, Bartz SR, Schelter J et al (2003) Expression profiling reveals off-target gene regulation by RNAi. Nat Biotechnol 21(6):635–637. doi:10.1038/nbt831

Jackson AL, Burchard J, Schelter J et al (2006) Widespread siRNA "off-target" transcript silencing mediated by seed region sequence complementarity. RNA 12(7):1179–1187. doi:10.1261/rna.25706

Jarad G, Wang B, Khan S, DeVore J, Miao H et al (2002) Fas activation induces renal tubular epithelial cell beta 8 integrin expression and function in the absence of apoptosis. J Biol Chem 277(49):47,826–47,833. doi:10.1074/jbc.M204901200

Jaronczyk K, Carmichael J, Hobman T (2005) Exploring the functions of RNA interference pathway proteins: some functions are more RISCy than others. Biochem J 387(3):561–571. doi:10.1042/BJ20041822

Jiang F, Ye X, Liu X, Fincher L, McKearin D, Liu Q (2005) Dicer-1 and R3D1-L catalyze microRNA maturation in *Drosophila*. Genes Dev 19(14):1674–1679. doi:10.1101/gad. 1334005

Johansen LK, Carrington JC (2001) Silencing on the spot. Induction and suppression of RNA silencing in the *Agrobacterium*-mediated transient expression system. Plant Physiol 126(3):930–938

de Jonge J, Holtrop M, Wilschut J, Huckriede A (2006) Reconstituted influenza virus envelope as an efficient carrier system for the cellular delivery of small interfering RNAs. Gene Ther 13(5):400–411. doi:10.1038/sj.gt.3302673

Jorgensen RA, Cluster PD, English J, Que Q, Napoli CA (1996) Chalcone synthase co-suppression phenotypes in petunia flowers: comparison of sense vs. antisense constructs and single-copy vs. complex T-DNA sequences. Plant Mol Biol 31(5):957–973

Kamath RS, Fraser AG, Dong Y, Poulin G, Durbin R, Gotta M (2003) Systematic functional analysis of the *Caenorhabditis elegans* genome using RNAi. Nature 421(6920):231–237. doi:10.1038/nature01278

Kawasaki H, Taira K (2003) Short hairpin type of dsRNAs that are controlled by tRNA (Val) promoter significantly induce RNAi-mediated gene silencing in the cytoplasm of human cells. Nucleic Acids Res 31(2):700–707

Kelly WG, Fire A (1998) Chromatin silencing and the maintenance of a functional germline in *Caenorhabdtis elegans*. Development 125(13):2451–2456

Ketting RF, Haverkamp TH, van Luenen HG, Plasterk RH (1999) Mut-7 of *C. elegans*, required for transposon silencing and RNA interference, is a homolog of Werner syndrome helicase and RNase D. Cell 99:133–141

Kiger AA, Baum B, Jones S et al (2003) A functional genomic analysis of cell morphology using RNA interference. J Biol 2:27–41. doi:10.1186/1475-4924-2-27

Kim DH, Behlke MA, Rose SD et al (2005) Synthetic dsRNA Dicer substrates enhance RNAi potency and efficacy. Nat Biotechnol 23(2):222–226. doi:10.1038/nbt1051

Kjemtrup S, Sampson KS, Peele CG et al (1998) Gene silencing from plant DNA carried by a gemini virus. Plant J 14(1):91–100. doi:10.1046/j.1365-313X.1998.00101.x

Klahre U, Crete P, Leuenberger SA, Iglesias VA, Meins F (2002) High molecular weight RNAs and small interfering RNAs induce systemic post transcriptional gene silencing in plants. Proc Natl Acad Sci U S A 99(18):11981–11986. doi:10.1073/pnas.182204199

Koo SH, Satoh H, Herzig S, Lee CH, Hedrick S et al (2004) PGC-1 promotes insulin resistance in liver through PPAR-alpha-dependent induction of TRB-3. Nat Med 10(5):530–534. doi:10.1038/nm1044

Kumagai MH, Donson J, della-Cioppa G et al (1995) Cytoplasmic inhibition of carotenoid biosynthesis with virus-derived RNA. Proc Natl Acad Sci U S A 92(5):1679–1683

Kumar SM, Mysore KS (2011) New dimensions for VIGS in plant functional genomics. Trends Plant Sci 16(12):656–665. doi:10.1016/j.tplants.2011.08.006

Kumar SG, Campbell LM, Puckhaber L et al (2006) Engineering cottonseed for use in human nutrition by tissue-specific reduction of toxic gossypol. Proc Natl Acad Sci U S A 103(48):18054–18059. doi:10.1073/pnas.0605389103

Kumar P, Kamle M, Pandey A (2012) RNAi: new era of functional genomics for crop improvement. In: Goyal A, Maheshwari P (eds) Frontiers on recent developments in plant science. Springer 1:24–38. doi:10.2174/97816080540391120101

Kunath T, Gish G, Lickert H, Jones N et al (2003) Transgenic RNA interference in ES cell-derived embryos recapitulates a genetic null phenotype. Nat Biotechnol 21(5):559–561. doi:10.1038/nbt813

Kurreck J (2003) Antisense technologies: improvement through novel chemical modifications. Eur J Biochem 270(8):1628–1644. doi:10.1046/j.1432-1033.2003.03555.x

Kusaba M, Miyahara K, Lida S, Fukuoka H et al (2003) Low-glutelin content 1: a dominant mutation that suppresses the glutelin multigene family via RNA silencing in rice. Plant Cell 15(6):1455–1467

Kwon SC, Nguyen TA, Choi YG et al (2016) Structure of human DROSHA. Cell 164:81–90. doi:10.1016/j.cell.2015.12.019

Lee RC, Ambrose V (2001) An extensive class of small RNAs in *C. elegans*. Science 294(5545):862–864. doi:10.1126/science.1065329

Lee HY, Doudna JA (2012) TRBP alters human precursor microRNA processing *in vitro*. RNA 18:2012–2019. doi:10.1261/rna.035501.112

Lee RC, Feinbaum RL, Ambrose V (1993) The *C. elegans* heterochronic gene lin-4 encodes small RNAs with antisense complementarity to lin-14. Cell 75(5):843–854

Lee NS, Dohjima T, Bauer G et al (2002) Expression of small interfering RNAs targeted against HIV-1 rev transcripts in human cells. Nat Biotechnol 20(5):500–505. doi:10.1038/nbt0502-500

Lee SS, Lee RY, Fraser AG, Kamath RS et al (2003) A systematic RNAi screen identifies a critical role for mitochondria in *C. elegans* longevity. Nat Genet 33(1):40–48. doi:10.1038/ng1056

Lee Y, Hur I, Park SY, Kim YK et al (2006) The role of PACT in the RNA silencing pathway. EMBO J 25(3):522–532. doi:10.1038/sj.emboj.7600942

Lewis DL, Wolff JA (2007) Systemic siRNA delivery via hydrodynamic intravascular injection. Adv Drug Deliv Rev 59(2–3):115–123. doi:10.1016/j.addr.2007.03.002

Li W, Szoka FC Jr (2007) Lipid-based nanoparticles for nucleic acid delivery. Pharm Res 24(3):438–449. doi:10.1007/s11095-006-9180-5

Liang J, Deng G, Long H, Pan Z et al (2012) Virus-induced silencing of genes encoding LEA protein in Tibetan hulless barley (*Hordeum vulgare* ssp. *vulgare*) and their relationship to drought tolerance. Mol Breed 30(1):441–451. doi:10.1007/s11032-011-9633-3

Lim MYT, Alvin WTN, Chou Y, Lim TP et al (2016) The *Drosophila* Dicer-1 partner Loquacious enhances miRNA processing from hairpins with unstable structures at the dicing site. Cell Rep 15(8):1795–1808. doi:10.1016/j.celrep.2016.04.059

Lindbo JA (2012) A historical overview of RNAi in plants. Methods Mol Biol 894:1–16. doi:10.1007/978-1-61779-882-5_1

Liu YL, Schiff M, Kumar DSP (2002a) Virus induced gene silencing in tomato. Plant J 31(6):777–786

Liu Q, Singh S, Green A (2002b) High-oleic and high-stearic cottonseed oils: nutritionally improved cooking oils developed using gene silencing. J Am Coll Nutr 21(3):205–211

Liu Q, Rand TA, Kalidas S, Du F et al (2003) R2D2, a bridge between the initiation and effector steps of the *Drosophila* RNAi pathway. Science 301(5641):1921–1925. doi:10.1126/science.1088710

Liu TT, Zhu D, Chen W, Deng W et al (2013) A global identification and analysis of small nucleolar RNAs and possible intermediate-sized non-coding RNAs in *Oryza sativa*. Mol Plant 6(3):830–846. doi:10.1093/mp/sss087

Lohmann JU, Endl I, Bosch TCG (1999) Silencing of developmental genes in *Hydra*. Dev Biol 214(1):211–214. doi:10.1006/dbio.1999.9407

MacFarlane SA, Popovich AH (2000) Efficient expression of foreign proteins in roots from tobravirus vectors. Virology 267(1):29–35. doi:10.1006/viro.1999.0098

Mallory AC, Parks G, Endres MW et al (2002) The amplicon-plus system for high-level expression of transgenes in plants. Nat Biotechnol 20(6):622–625. doi:10.1038/nbt0602-622

Martin R, Smibert P, Yalcin A, Tyler DM et al (2009) A *Drosophila pasha* mutant distinguishes the canonical miRNA and mirtron pathways. Mol Cell Biol 29:861–870. doi:10.1128/MCB.01524-08

Martineau B (2001) First fruit: the creation of the Flavr Savr tomato and the birth of genetically engineered food. Tata McGraw-Hill, New York, p 269

Martinez J, Tuschl T (2004) RISC is a 5′ phosphomonoester producing RNA endonuclease. Genes Dev 18:975–980. doi:10.1101/gad.1187904

Martinez J, Patkaniowska A, Urlaub H, Lührmann R, Tuschl T (2002) Single-stranded antisense siRNAs guide target RNA cleavage in RNAi. Cell 110(5):563–574

Martínez MA, Gutiérrez A, Armand-Ugón M et al (2002) Suppression of chemokine receptor expression by RNA interference allows for inhibition of HIV-1 replication. AIDS 16(18):2385–2390

Matthew L (2004) RNAi for plant functional genomics. Comp Funct Genom 5:240–244. doi:10.1002/cfg.396

Matzke MA, Matzke AJM (2004) Planting the seeds of a new paradigm. PLoS Biol 2(5):e133. doi:10.1371/journal.pbio.0020133

Matzke M, Matzke AJM, Kooter JM (2001) RNA: guiding gene silencing. Science 293:1080–1083. doi:10.1126/science.1063051

Meister G, Tuschl T (2004) Mechanisms of gene silencing by double-stranded RNA. Nature 431:343–349. doi:10.1038/nature02873

Meister G, Landthaler M, Patkaniowska A, Dorsett Y et al (2004) Human Argonaute 2 mediates RNA cleavage targeted by miRNAs and siRNAs. Mol Cell 15(2):185–197. doi:10.1016/j.molcel.2004.07.007

Miki D, Itoh R, Shimamoto K (2005) RNA silencing of single and multiple members in a gene family of rice. Plant Physiol 138(4):1903–1913. doi:10.1104/pp.105.063933

Misquitta L, Paterson BM (1999) Targeted disruption of gene function in *Drosophila* by RNA interference (RNAi): a role for nautilus in embryonic somatic muscle formation. Proc Natl Acad Sci U S A 96(4):1451–1456

Miyagishi M, Taira K (2002) U6 promoter driven siRNAs with four uridine 3′ overhangs efficiently suppress targeted gene expression in mammalian cells. Nat Biotechnol 20:497–500

Mlotshwas S, Voinnet O, Mette MF, Matzke M et al (2002) RNA silencing and its mobile silencing signal. Plant Cell 14:289–301

Muangsan N, Beclin C, Vaucheret H, Robertson D (2004) Geminivirus VIGS of endogenous genes requires SGS2/SDE1 and SGS3 and defines a new branch in the genetic pathway for silencing in plants. Plant J 38(6):1004–1014. doi:10.1111/j.1365-313X.2004.02103.x

Nakano H, Amemiya S, Shiokawa K, Taira M (2000) RNA interference for the organizer specific gene Xlim-1 in *Xenopus* embryos. Biochem Biophys Res Commun 274(2):434–439. doi:10.1006/bbrc.2000.3178

Napoli C, Lemieux C, Jorgensen R (1990) Introduction of a chimeric chalcone synthase gene into petunia results in reversible co-suppression of homologous gene in trans. Plant Cell 2(4):279–289. doi:10.1105/tpc.2.4.279

Neff CP, Zhou J, Remling L, Kuruvilla J, Zhang J et al (2011) An aptamer-siRNA chimera suppresses HIV-1 viral loads and protects from helper CD4(+) T cell decline in humanized mice. Sci Transl Med 3(66):66ra6. doi:10.1126/scitranslmed.3001581

Nguyen T, Menocal EM, Harborth J, Fruehauf JH (2008) RNAi therapeutics: an update on delivery. Curr Opin Mol Thera 10(2):158–167

Nguyen TA, Jo MH, Choi YG, Park J et al (2015) Functional anatomy of the human microprocessor. Cell 161:1374–1387. doi:10.1016/j.cell.2015.05.010

Nishihara M, Nakatsuka T, Yamamura S (2005) Flavonoid components and flower color change in transgenic tobacco plants by suppression of chalcone isomerase gene. FEBS Lett 579(27):6074–6078. doi:10.1016/j.febslet.2005.09.073

Nykanen A, Haley B, Zamore PD (2001) ATP requirements and small interfering RNA structure in the RNA interference pathway. Cell 107:309–321

Ogita S, Uefuji H, Yamaguchi Y, Koizumi N, Sano H (2003) Producing decaffeinated coffee plants. Nature 423(6942):823. doi:10.1038/423823a

Paddison P, Silva JM, Conklin DS et al (2004) A resource for large-scale RNA-interference-based screens in mammals. Nature 428(6981):427–431. doi:10.1038/nature02370

Pak J, Fire A (2007) Distinct populations of primary and secondary effectors during RNAi in *C. elegans*. Science 315:241–244. doi:10.1126/science.1132839

Pal Bhadra M, Bhadra U, Birchler JA (1998) Role of multiple transacting regulators in modifying the effect of the retrotransposon copia on host gene expression in *Drosophila*. Mol Gen Genet 259(2):198–206

Pareek M, Yogindran S, Mukherjee SK, Rajam MV (2015) In: Bahadur B, Ranjam MV, Sahijram L, Krishnamurthy KV (eds) Plant biology and biotechnology: vol II: plant genomics and biotechnology. Springer, pp 639–661. doi:10.1007/978–81–322-2283-5_32

Pickford A, Braccini L, Macino G, Cogoni C (2003) The QDE-3 homologue RecQ-2 co-operates with QDE-3 in DNA repair in *Neurospora crassa*. Curr Genet 42(4):220–227. doi:10.1007/s00294-002-0351-6

Plasterk RH (2002) RNA silencing: the genome's immune system. Science 296(5571):1263–1265. doi:10.1126/science.1072148

Plasterk RH, Ketting RF (2000) The silence of the genes. Curr Opin Genet Dev 10(5):62–567

Rand TA, Petersen S, Du F, Wang X (2005) Argonaute2 cleaves the anti-guide strand of siRNA during RISC activation. Cell 123(4):621–629. doi:10.1016/j.cell.2005.10.020

Raponi M, Arndt GM (2003) Double-stranded RNA-mediated gene silencing in fission yeast. Nucleic Acids Res 31(15):4481–4489

Regina A, Bird A, Topping D, Bowden S et al (2006) High-amylose wheat generated by RNA interference improves indices of large-bowel health in rats. Proc Natl Acad Sci U S A 103(10):3546–3551. doi:10.1073/pnas.0510737103

Reich SJ, Fosnot J, Kuroki A et al (2003) Small interfering RNA (siRNA) targeting VEGF effectively inhibits ocular neovascularization in a mouse model. Mol Vis 9:210–216

Reinhart BJ, Weinstein EG, Rhodes MW, Bartel B, Bartel DP (2002) MicroRNAs in plants. Genes Dev 16(13):1616–1626. doi:10.1101/gad.1004402

Reynolds A, Anderson EM, Vermeulen A, Fedorov Y et al (2006) Induction of the interferon response by siRNA is cell type– and duplex length–dependent. RNA 12(6):988–993. doi:10.1261/rna.2340906

Rodoni BC, Dale JL (1999) Harding RM characterization and expression of the coat protein-coding region of the banana bract mosaic potyvirus, development of diagnostic assays and detection of the virus in banana plants from five countries in Southeast Asia. Arch Virol 144(9):1725–1737

Rubinson DA, Dillon CP, Dwiatkowski AU et al (2003) A lentivirus based system to functionally silence genes in primary mammalian cells, stem cells and transgenic mice by RNA interference. Nat Genet 33(3):401–406. doi:10.1038/ng1117

Ruvkun G (2001) Glimpses of a tiny RNA world. Science 294(5543):797–799. doi:10.1126/science.1066315

Saito K, Ishizuka A, Siomi H, Siomi MC (2005) Processing of pre-microRNAs by the Dicer-1-Loquacious complex in *Drosophila* cells. PLoS Biol 3(7):1202–1212. doi:10.1371/journal.pbio.0030235

Saxena J (2014) RNA interference and its applications. In: Ravi I, Baunthiyal M, Saxena J (eds) Advances in biotechnology. Springer, Dordretch, p 259. doi:10.1007/978-81-322-1554-7. isbn:978-81-322-1554-7

Saxena S, Jonsson ZO, Dutta A (2003) Small RNAs with imperfect match to endogenous mRNA repress translation, implications for off-target activity of small inhibitory RNA in mammalian cells. J Biol Chem 278(45):44,312–44,319. doi:10.1074/jbc.M307089200

Scacheri PC, Rozenblatt-Rosen O, Caplen NJ et al (2004) Short interfering RNAs can induce unexpected and divergent changes in the levels of untargeted proteins in mammalian cells. Proc Natl Acad Sci U S A 101(7):1892–1897. doi:10.1073/pnas.0308698100

Schiebel W, Haer B, Marinkovic S, Kianner A, Sanger HL (1993a) RNA-directed RNA polymerase from tomato leaves, purification and physical properties. J Biol Chem 268(16):11851–11857

Schiebel W, Haer B, Marinkovic S, Kianner A, Sanger HL (1993b) RNA directed RNA polymerase from tomato leaves. II. Catalytic *in vitro* properties. J Biol Chem 268(16):11858–11867

Schramke V, Allshire R (2003) Hairpin RNAs and retrotransposon LTRs effect RNAi and chromatin based gene silencing. Science 301(5636):1069–1074. doi:10.1126/science. 1086870

Schwarz D, Tomari Y, Zamore P (2004) The RNA-induced silencing complex is a mg-dependent endonuclease. Curr Biol 14(9):787–791. doi:10.1016/j.cub.2004.03.008

Schweizer P, Pokorny J, Schulze-Lefert P, Dudler R (2000) Double stranded RNA interference with gene functions at the single cell in cereals. Plant J 24(6):895–903

Sen G, Wehrman TS, Myers JW (2004) Restriction enzyme generated siRNA (REGS) vector and libraries. Nat Genet 36(2):183–189. doi:10.1038/ng1288

Shang Y, Schwinn KE, Bennett MJ, Hunter DA et al (2007) Methods for transient assay of gene function in floral tissues. Plant Methods 3:1–12. doi:10.1186/1746-4811-3-1

Shen C, Buck AK, Liu X, Winkler M, Reske SW (2003) Gene silencing by adenovirus-delivered siRNA. FEBS Lett 539(1–3):111–114

Shuey DJ, McCallus DE, Giordano T (2002) RNAi: gene silencing in therapeutic intervention. Drug Discov Today 7:1040–1046

Silva AT, Nguyen A, Ye C, Verchot E, Moon JH (2010) Conjugated polymer nanoparticles for effective siRNA delivery to tobacco BY-2 protoplasts. BMC Plant Biol 10:291–304. doi:10.1186/1471-2229-10-291

Simeoni F, Morris MC, Heitz F, Divita G (2004) Insight into the mechanism of the peptide-based delivery system MPG: implications for delivery of siRNA into mammalian cells. Nucleic Acids Res 31:2717–2724. doi:10.1093/nar/gkg385

Sledz CA, Holko M, de Veer MJ, Silverman RH, Williams BR (2003) Activation of the interferon system by short-interfering RNAs. Nat Cell Biol 5:834–839. doi:10.1038/ncb1038

Smith JE (2009) Biotechnology, 5th edn. Cambridge University Press, Cambridge, p 280. ISBN:978–0–511-46394-5

Song E, Lee SK, Wang J, Ince N et al (2003) RNA interference targeting Fas protects mice from fulminant hepatitis. Nat Med 9:347–351. doi:10.1038/nm828

Soutschek J, Akinc A, Bramlage B, Charisse K et al (2004) Therapeutic silencing of an endogenous gene by systemic administration of modified siRNAs. Nature 432:173–178. doi:10.1038/nature03121

de Souza AJ, Mendes BMJ, de Assis AF, Filho M (2007) Gene silencing: concepts, applications and perspectives in woody plants. Sci Agric 64(6):645–656. doi:10.1590/S0103-90162007000600014

Stevenson M (2004) Therapeutic potential of RNA interference. New Eng J Med 351:1772–1777. doi:10.1056/NEJMra045004

Tabara H, Sarkissian M, Kelly WG et al (1999) The rde-1 gene, RNA interference, and transposon silencing in *C. elegans*. Cell 99(2):123–132

Tang G, Galili G (2004) Using RNAi to improve plant nutritional value: from mechanism to application. Trends Biotechnol 22(9):463–469. doi:10.1016/j.tibtech.2004.07.009

Tang W, Weidner DA, Hu BY et al (2006) Efficient delivery of small interfering RNA to plant cells by a nanosecond pulsed laser-induced stress wave for posttranscriptional gene silencing. Plant Sci 171(3):375–381. doi:10.1016/j.plantsci.2006.04.005

Tang G, Galili G, Zhuang X (2007) RNAi and microRNA: breakthrough technologies for the improvement of plant nutritional value and metabolic engineering. Metabolomics 3:357–369. doi:10.1007/s11306-007-0073-3

Taniguchi CM, Ueki K, Kahn CR (2005) Complementary roles of IRS-1 and IRS-2 in the hepatic regulation of metabolism. J Clin Invest 115:718–727. doi:10.1172/JCI200523187

Tchurikov NA, Chistykva LG, Zavilgelsky GB et al (2000) Gene specific silencing by expression of parallel complementary RNA in *Escherichia coli*. J Biol Chem 275(34):26523–26529. doi:10.1074/jbc.M002833200

Tenllado F, Martinez-Garcia B, Vargas M, Diaz-Ruiz JR (2003) Crude extracts of bacterially expressed dsRNA can be used to protect plants against virus infection. BMC Biotechnol 3:3–14. doi:10.1186/1472-6750-3-3

Teo G, Suziki Y, Uratsu SL, Lampien B et al (2006) Silencing leaf sorbitol synthesis alters long-distance portioning and apple fruit quality. Proc Natl Acad Sci U S A 103:18,842–18,847. doi:10.1073/pnas.0605873103

Thakur A (2003) RNA interference revolution. Electron J Biotechnol 6(1):39–49. doi:10.2225/vol6-issue1-fulltext-1

Timmons L, Fire A (1998) Specific interference by ingested dsRNA. Nature 395:854. doi:10.1038/27579

Tiscornia G, Singer O, Ikawa M, Verma IM (2003) A general method for gene knockdown in mice by using lentiviral vectors expressing small interfering RNA. Proc Natl Acad Sci U S A 100:1844–1848. doi:10.1073/pnas.0437912100

Tolia NH, Joshua-Tor L (2006) Slicer and the Argonautes. Nat Chem Biol 3:36–43. doi:10.1038/nchembio848

Tsai WH, Chang WT (2013) Construction of simple and efficient siRNA validation systems for screening and identification of effective RNAi-targeted sequences from mammalian genes. In: Ochs MF (ed) Gene function analysis, Method Mol Biol 1101:321–338. doi:10.1007/978-1-62703-721-1_15

Turnage MA, Muangsan N, Peele CG, Robertson D (2002) Geminivirus based vectors for gene silencing in *Arabidopsis*. Plant J 30(1):107–117

Van Uyen N (2006) Novel approaches in plants breeding RNAi technology. In: Proceedings of International Workshop on Biotechnology in Agriculture, pp 12–16

Vasudevan S (2012) Posttranscriptional upregulation by microRNAs. Wiley Interdisciplinary Rev RNA 3:311–330. doi:10.1002/wrna.121

Velásquez AC, Chakravarthy S, Martin GB (2009) Virus-induced gene silencing (VIGS) in *Nicotiana benthamiana* and tomato. J Vis Exp 28:e1292. doi:10.3791/1292

Verdel A, Jia S, Gerber S et al (2004) RNAi-mediated targeting of heterochromatin by the RITS complex. Science 303:672–676. doi:10.1126/science.1093686

Voinnet O (2001) RNA silencing as a plant immune system against viruses. Trends Genet 17(8):449–459

Voinnet O, Baulcombe DC (1997) Systemic signaling in gene silencing. Nature 389:553. doi:10.1038/39215

Wang Z, Morris JC, Drew ME, Englund PT (2000a) Inhibition of *Trypanosoma brucei* gene expression by RNA interference using an integratable vector with opposing T7 promoters. J Biol Chem 275(51):40174–40179. doi:10.1074/jbc.M008405200

Wang M, Abbott D, Waterhouse PM (2000b) A single copy of a virus derived transgene encoding hairpin RNA gives immunity to barley yellow dwarf virus. Mol Plant Path 1:401–410

Wargelius A, Ellingsen S, Jose FA (1999) Double stranded RNA induces specific developmental defects in zebrafish embryos. Biochem Biophys Res Commun 263(1):156–161. doi:10.1006/bbrc.1999.1343

Waterhouse PM, Graham MW, Wang MB (1998) Virus resistance in gene silencing in plants can be induced by simultaneous expression of sense and antisense RNA. Proc Natl Acad Sci U S A 95(23):13959–13964

Waterhouse PM, Wang MB, Lough T (2001) Gene silencing as an adaptative defense against viruses. Nature 411:834–842. doi:10.1038/35081168

Wianny F, Zernicka-Goetz M (2000) Specific interference with gene function by double stranded RNA in early mouse development. Nat Cell Biol 2:70–75. doi:10.1038/35000016

Williams ME (1995) The American Geriatrics Society's complete guide to aging & health. Harmony Books, New York, p 494. isbn:9780517595398

Williams M, Clark G, Sathasivan K, Islam AS (2004) RNA interference and its application in crop improvement. Plant Tissue Culture 14(1):1–18

Wilson JA, Jayasena S, Khvorova A, Sabatinos S et al (2003) RNA interference blocks gene expression and RNA synthesis from hepatitis C replicons propagated in human liver cells. Proc Natl Acad Sci U S A 100:2783–2788. doi:10.1073/pnas.252758799

Winston WM, Molodowitch C, Hunter CP (2002) Systemic RNAi in *C. elegans* requires the putative transmembrane protein SID-1. Science 295:2456–2459. doi:10.1126/science. 1068836

Wu-Scharf D, Jeong B, Zhang C, Cerutti H (2000) Transgene and transposon silencing in *Chlamydomonas reinhardtii* by a DEAH-box RNA helicase. Science 290(5494):1159–1162

Xie Z, Johansen LK, Gustafson AM, Kasschau KD et al (2004) Genetic and functional diversification of small RNA pathways in plants. Public Library Sci Biol 2:642–652. doi:10.1371/journal.pbio.0020104

Yamamoto Y, Ifuku K, Sato F (2005) Suppression of psbP and psbQ genes in *Nicotiana tabacum* by RNA interference technique. In: van der Est A, Bruce D (eds) Photosynthesis: fundamental aspects to global perspectives. Kluwers Academic Publisher, Dordrecht, pp 798–799

Ying RS, Fan XG, Zhu C, Li N, Zhang BX (2006) Inhibition of hepatitis B virus replication and expression by RNA interference *in vivo*. Zhonghua Ganzangbing Zazhi 14(1):15–18

Yogindran S, Ranjam MV (2015) RNAi for crop improvement. In: Bahadur B, Ranjam MV, Sahijram L, Krishnamurthy KV (eds) Plant biology and biotechnology: vol II: plant genomics and biotechnology. Springer, pp 623–637. doi:10.1007/978–81–322-2283-5_31

Yu CE, Oshima J, Fu YH, Wijsman EM, Hisama F et al (1996) Positional cloning of the Werners syndrome gene. Science 272:258–262

Zamore PD, Tuschl T, Sharp PA, Bartel DP (2000) RNAi: double-stranded RNA directs the ATP dependent cleavage of mRNA at 21–23 nucleotide intervals. Cell 101:25–33. doi:10.1016/S0092-8674(00)80620-0

Zhang J, Rossi JJ (2010) Strategies in designing multigene expression units to down regulate HIV-1. Methods Mol Biol 623:123–136. doi:10.1007/978-1-60761-588-0_8

Zhang R, Wang H (2005) Antisense technology. In: Curiel DT, Douglas JT (eds) Cancer gene therapy. Humana Press, p 502. eISBN:1–59259–785-8

Zhang YC, Taylor MM, Samson WK, Phillips MI (2005) Antisense inhibition oligonucleotides, ribozymes and siRNAs. From: Methods in Molecular Medicine 106:334. In: Phillips I (ed) Antisense therapeutics, 2nd edn. Humana Press Inc, Totowa 11–34, p 334

Zhao J, Feng SS (2015) Nanocarriers for delivery of siRNA and co-delivery of siRNA and other therapeutic agents. Nanomedicine 10(14):2199–2228. doi:10.2217/nnm.15.61

Zhu X, Galili G (2003) Increased lysine synthesis coupled with a knockout of its catabolism synergistically boosts lysine content and also trans-regulates the metabolism of other amino acids in *Arabidopsis* seeds. Plant Cell 15:845–853. doi:10.1105/tpc.009647

Zhu X, Galili G (2004) Lysine metabolism is concurrently regulated by synthesis and catabolism in both reproductive and vegetative tissues. Plant Physiol 135:129–136. doi:10.1104/pp.103.037168

Engineering Abiotic Stress Tolerance Traits for Mitigating Climate Change

3

Sumit Jangra, Aakash Mishra, Disha Kamboj, Neelam R. Yadav, and Ram C. Yadav

Abstract

Crop yield, survival, and biomass production are negatively influenced by abiotic stresses. Due to multigenic nature, the molecular basis of abiotic stress tolerance is difficult to understand. Modern agriculture faces various challenges which include global climate change, complex field environment, and the combination of abiotic stress. To improve abiotic tolerance in crop plants, a combined effect of various molecular and biochemical approaches will be needed. Advanced molecular biology tools are used to enlighten the regulation mechanism of abiotic stress tolerance based on expression analysis of various stress-linked genes. The data collected from high-throughput transcription profiling, identification of specific protein network on large scale, molecular modeling and their association with environmental changes in plants all reveal information about plant system which is used for engineering plants against various environmental changes. Various genes for abiotic stress tolerance in crop plants have been identified and cloned to develop stress tolerant plants. In spite of various advancements in the technology, the success in developing stress tolerant plants is limited. This review enlightens the recent advancement in transgenic technology for the betterment of crop plants against abiotic stress.

S. Jangra • D. Kamboj • N.R. Yadav • R.C. Yadav (✉)
Department of Molecular Biology, Biotechnology and Bioinformatics, CCS Haryana Agricultural University, Hisar 125 004, Haryana, India
e-mail: rcyadavbiotech@yahoo.com

A. Mishra
Department of Plant Sciences, University of California, Davis, CA 95616, USA

S.K. Gahlawat et al. (eds.), *Plant Biotechnology: Recent Advancements and Developments*, DOI 10.1007/978-981-10-4732-9_3

3.1 Introduction

Stress can be outlined as a state that adversely affects growth, development, and/or productivity. The productivity of crop plants is affected by various biotic and abiotic stresses. Abiotic stress includes water scarcity, fluctuation in temperature, chemical toxicity, and oxidative stress which cause a serious damage to the crop plants. Various signaling molecules and molecular cascades are activated on stress application which leads to activation of various stress-related genes, which is the molecular basis of plant adaptation to different environmental stress. Abiotic stress tolerance, which is multigenic in nature, is much more unmanageable than biotic stress tolerance, which is monogenic in nature. Efforts made to improve plants against various stresses by genetic engineering have resulted in various achievements; genetically complex nature of abiotic stress tolerance and various side effects of altering the control mechanism have made this task extremely difficult. Keeping this in mind, breeding and classical physiology should be supplemented with modern biotechnological tools. When compared with normal conditions, there is a huge decline in the crop productivity under stressed conditions. Improving tolerance of crop plants to abiotic stress could, therefore, enable them to maintain growth and development during the normal fluctuations of adverse conditions, and buffer crops against the large swings in yield experienced from one year to the next.

3.2 Drought

Drought can be defined as water scarcity stress because of reduced water potential of soil or increased water potential of leaves due to dry and hot weather conditions. Plants undergo various morphological, metabolic, and physiological changes under water deficit conditions, which include the closure of stomata, reduction in photosynthesis, cell growth, biomass production, and increased respiration. Plants respond to water scarcity both at the molecular and cellular levels, for example, by the accumulation of stress-related proteins and osmolytes. Drought leads to enhanced production of various phytohormones such as abscisic acid (ABA), which leads to closure of stomata and triggers stress-related gene expression. Drought inducible gene expression is governed by both ABA-dependent and ABA-independent regulatory system. Drought stress causes a huge reduction in crop yield and productivity, making it a great threat to agriculture; due to this there is a great demand for crop plants with enhanced drought tolerance.

3.3 Production of Drought Tolerant Transgenic Plants

3.3.1 Engineering Glycine Betaine Synthesis

A quaternary ammonium compound known as glycine betaine is found in ten angiosperm plant families and in some marine algae. Glycine betaine is highly water soluble, electrically neutral over a broad pH range, and protects plant macromolecules, such as proteins and cell membrane, under extreme environmental conditions. A large number of experiments conducted in vitro has confirmed osmoprotectant nature in glycine betaine by stabilizing functional proteins and membrane structures under osmotic stress. Glycine betaine synthesis is mainly studied in *Chenopodaciae* (spinach, beet, etc.). In these plant species, glycine betaine is formed by two-stage oxidation of choline, CMO (choline monooxygenase) catalyzes the first step and BADH (betaine aldehyde dehydrogenase) catalyzes the second step. CMO and BADH both are induced by stress. Glycine betaine is also found in *E. coli* and *Arthrobacter globoformis* but the biosynthetic pathway is different. In *E. coli* production is catalyzed by CDH (choline dehydrogenase) followed by BADH (betaine aldehyde dehydrogenase) and in *Arthrobacter globoformis* single enzyme encoded by *cod A* (choline oxidase) catalyzes the production.

Attempts have been made to manipulate glycine betaine production by transformation with the relevant plant gene, *E. coli* or *A. globoformis*. Holmstrom et al. (1994) transformed tobacco with *bet B* gene for BADH resulted in the very low amount of glycine betaine accumulation in leaves under the control of a strong light inducible promoter. Similarly, transgenic rice was generated with COD in chloroplast or cytosol, and transgenic plants accumulated betaine at a concentration of 1 μmole/g of fresh weight (FW) in leaves (Sakamoto and Murata 1998). Transgenic poplar expressing bacterial *cod A* showed enhanced growth and stress tolerance under the control of SWPA2 (oxidative stress inducible) promoter (Ke et al. 2016).

3.3.2 Engineering Amino Acid (Proline) Synthesis

Proline is an amino acid present in almost all living organism including plants. Different experiments conducted by various scientists have confirmed the role of proline in stress tolerance. Proline protects plants from various abiotic stress by preventing membrane damage, maintaining enzymes and other proteins quaternary structure, scavenging free radicals, buffering cytosolic pH, balancing cell redox potential, activating stress-related gene expression, and signaling molecules. Two different precursors, that is, glutamate and ornithine, are involved in proline biosynthesis. Pyrroline-5-carboxilate synthetase (P5CS) catalyzes the conversion of glutamate to proline and the conversion of ornithine to proline is catalyzed by an ornithine-oxo-acid transaminase. Proline level was elevated in transgenic tobacco and rice plants by overexpressing P5CS and it was found that transgenic plants showed improved tolerance to drought and salinity (Kishor et al. 1995; Zhu et al. 1998). Yamada et al. (2005) overexpressed P5CS gene under the strong consecutive promoter CaMv 35S, in Petunia from *Arabidopsis*, and it was found that transgenic petunia showed better growth under drought stress. Similarly, improved drought tolerance was observed in transgenic wheat, overexpressing *Vigna aconitifolia,* P5CS gene under the control of stress-induced complex-AIPC promoter (Verduscolo et al. 2007).

3.3.3 Engineering Sugar Accumulation

Different carbohydrate sugars (fructan, trehalose) and sugar alcohols (Sorbitol, Mannitol) are involved in osmoregulation, membrane protein stabilization, and scavenging reactive oxygen species. Pilon-Smits et al. (1999) incorporated a cDNA encoding 1-sucrose: sucrose fructosyl transferase (1-SST) in stomatal guard cell protoplast of sugar beet. It was found that the plants regenerated from the engineered protoplast cells accumulated low molecular weight fructans in their tap roots. When dehydration stress was applied, it was found that plants engineered with 1-SST gene performed better than nontransformed plants. Similarly, Holmstrom et al. (1996) engineered tobacco plants with a gene encoding for trehalose-6-phosphate synthase (TPSI) under the control of *Arabidopsis* promoter gene, *rbcs,* and it was found that the plants transformed with TPSI gene accumulated trehalose and showed improved drought tolerance. Abebe et al. (2003) transformed wheat with *E. coli*, *mtlD* gene, and it was found that the transformed plants accumulated mannitol and showed improved salinity and drought tolerance. Overexpression of bacterial *mtlD* gene in peanut causes accumulation of mannitol at a concentration of 3.02–4.74 μg/g FW and the transformed plants exhibited improved drought tolerance (Bhauso et al. 2014).

3.3.4 Engineering Polyamines Synthesis

Polyamines (putrescine, spermidine, and spermine) are aliphatic amines like phytohormones with aliphatic nitrogen structure and are important biomolecules of every living organism including plants. Polyamines protect plants from different biotic and abiotic stress by scavenging free radicals and are important components of various metabolic and physiological processes, such as cell growth and development. Therefore, genetic transformation of crop plants with genes involved in polyamine biosynthesis may provide a path for enhanced stress tolerance (Table 3.1).

3.3.5 Engineering Transcription Factors

Basic knowledge of molecular, biochemical, physiological, and regulatory pathway is mandatory for the development of stress tolerant crop plants. Among these, signal transduction pathway has a major role in the expression of stress-related gene and activation of transcription factors that regulate multiple pathways and have emerged as a key factor in editing complex metabolic and physiological pathways. Based on the structure of binding domain transcription factors have been classified into several families, that is, DREB, MYB, MYC, ERF, bZIP, HSF, CBF, WRKY, and Zinc figure; all these transcription factors play a crucial role in abiotic stress tolerance. Enhanced drought tolerance was observed in potato by overexpressing hot pepper bZIP like transcription factor, without any loss in yield (Seok-Jun Moon et al. 2015).

Table 3.1 Engineered plants with enhanced polyamines to mitigate abiotic stress

Introduced gene	Function of introduced gene	Gene origin	Engineered plant	Behavior of engineered plant
SAMDC (S-adenosyl methionine decarboxylase)	*SAMDC* plays a key role in the biosynthesis of polyamines, viz. Spd & Spm	*Homo sapiens sapiens*	*Nicotiana tabacum* var. *xanthi*	Resistance to drought, salinity, and biotic stress
SPDS (Spermidine synthase)	SPDS converts substrate into spermidine	*Corymbiaficifolia*	*Ipomoea batatas*	Resistance to drought and salinity
SPDS (Spermidine synthase)	SPDS converts substrate into spermidine	*Corymbiaficifolia*	*Arabidopsis thaliana*	Effective against multiple stress
ADC (Arginine decarboxylase)	ADC converts arginine to diamine	*D. stramonium*	*Oryza sativa*	Drought tolerance
SAMDC1 (S-adenosyl methionine decarboxylase)	*SAMDC* plays a key role in the biosynthesis of polyamines, viz. Spd & Spm	*Arabidopsis thaliana*	*Arabidopsis thaliana*	Broad spectrum stress tolerance
ADC (Arginine decarboxylase)	ADC converts arginine to diamine	*Arabidopsis thaliana*	*Arabidopsis thaliana*	Drought tolerance
ADC (Arginine decarboxylase)	ADC converts arginine to diamine	*Arabidopsis thaliana*	*Solanum melongena*	Effective against multiple stress
SPDS (Spermidine synthase)	SPDS converts substrate into spermidine	*Malus domestica*	Pyrus	Multiple abiotic stress tolerance

Gill and Tuteja (2010)

Transgenic tobacco showed improved drought and salinity tolerance upon expression of new AP2/ERF transcription factor (Wang et al. 2015). Transgenic perennial ryegrass showed increased drought tolerance upon overexpression of ubiquitin-like LpHUB1 transcription factor (Patel et al. 2015). Overexpression of DREB transcription factor, ATDREB1A confers stress tolerance (drought) in transgenic *Salvia miltiorrhiza* (Tao Wei et al. 2016).

3.4 Salinity

Saline conditions in soil and water are one of the major limiting factors that retard plant growth and productivity worldwide. Approximately 800 million hectares of land worldwide are affected by salinity (including both saline and sodic soils), that is, nearly 6% of world's total area. Low water potential in the roots, ion toxicity (Na^+ and Cl^-), and nutrient imbalance caused by saline conditions lead to deleterious effects on plant growth and productivity. With the rise in population, there is a demand for increased food production using sustainable agricultural practices, so there is a need to develop salinity tolerant crop plants which can survive high salt concentration in soil and water without any reduction in yield. Modern molecular biology approaches in combination with conventional breeding is a way toward sustainable crop improvement which is the need of today's world.

3.4.1 Production of Salinity Tolerant Transgenic Plants

3.4.1.1 Increasing Compatible Solute Production

A class of osmoprotective compounds produced by the salt tolerant organism under salinity stress of varying degree is known as compatible solutes. These are small and nontoxic compounds, which can maintain the osmotic potential of the cell and can stabilize the protein and cellular structures. Hayashi et al. (1997) engineered *Arabidopsis thaliana* with *cod A* gene that encodes COD with a signal for expressing the gene in chloroplast and it was found that the accumulation of betaine in leaves was 1 μmole/g FW and the transgenic plants showed enhanced abiotic stress tolerance. Transgenic pigeon pea, overexpressing *Vigna aconitifolia* PCSF129A gene showed an accumulation of proline and improved abiotic stress tolerance. It was found that transgenic plants were able to grow better than nontransformed plants at 200 mM NaCl (Surekha et al. 2014).

3.4.1.2 Antioxidant Protection

Higher salt concentration in soil and water leads to the production of reactive oxygen species (ROS) and free radicals in plants which includes superoxide radicals (O^{2-}), hydrogen peroxide (H_2O_2), and hydroxyl radicals ($OH^•$). These free radicals and reactive oxygen species leads to oxidative damage in the cell. To overcome the deleterious effects of free radicals and reactive oxygen species plant uses various low molecular weight antioxidants like ascorbic acid and reduced glutathione and utilizes various enzymes like superoxide dismutase, catalase, ascorbate peroxidase, glutathione peroxidase and glutathione S-transferase to combat free radicals and reactive oxygen species. Prashanth et al. (2008) overexpressed Cu/ZnSOD from *Avicennia marina* in rice and it was found that transgenic rice showed enhanced salt tolerance. Overexpression of Mn-SOD in Arabidopsis resulted in enhanced salt stress tolerance (Wang et al. 2004).

3.4.1.3 Ion Balancing

Ion ratio is balanced in the cell by sodium-potassium pump. Any imbalance in the ion ratio in plants results in higher sodium ion concentration inside the cell through the tract that functions in the acquirement of potassium. Na^+ toxicity results in alteration of various biochemical pathways. Various strategies are utilized by plants to maintain low Na^+ concentration inside the cell, which includes the exclusion of Na^+ through plasma membrane Na^+/H^+ antiporter (AtSOS1) or by concentrating excess Na^+ in vacuoles through vacuolar Na^+/H^+ antiporter. Transporter proteins can be manipulated with the help of genetic engineering to develop salt tolerant transgenic plants. AtNHX1 overexpression in *Arabidopsis* resulted in raised salt tolerance (Apse et al. 1999). Transgenic wheat lines overexpressing AtNHX1showed better growth and development in terms of higher grain yield and heavier and larger grain under salt stress (Xue et al. 2004). Vacuolar H^+ pyrophosphate gene AVP1 of *Arabidopsis* was expressed in peanut. Transgenic peanut showed improved drought and salinity tolerance than nontransformed plants. The *AVP1*-expressing plants generated higher biomass than nontransformed plants (Hua Qin et al. 2013).

3.5 Heat

When the temperature goes beyond a threshold value of time period that is enough to cause permanent harm to plant growth and development is called heat stress. Continuous fluctuations in temperature cause several molecular, biochemical, morphological, physiological, and metabolic changes in plants which lead to retardation of plant growth and development and ultimately reduces yield, which is undesirable. The ability to plants to grow well without reducing yield under high temperature is called thermotolerance. Modern biotechnological tools in combination with conventional plant breeding methods can be employed to mitigate inauspicious effects of heat stress.

3.5.1 Production of Heat Tolerant Transgenic Plants

3.5.1.1 Engineering Chaperone Activity

High temperature denatures proteins and causes a loss in the structural and functional attributes of proteins. Proteins lose their biological activity at high temperature because proteins get aggregated and misfolded. Heat stress leads to the synthesis of heat shock proteins (HSPs). The main function of heat shock protein is to act as molecular chaperones and protect the cell from heat stress by stabilizing and refolding denatured proteins. HSPs are low molecular weight compounds ranging from 15 to 105 kDa, and therefore can be grouped into five major classes. Different heat shock proteins are found in different organelles, namely, mitochondria, chloroplast, endoplasmic reticulum, nucleus, and cytosol.

HSP class	Size (kDa)	Cellular location
HSP100	100–114	Cytosol, mitochondria, chloroplast
HSP90	80–94	Cytosol, endoplasmic reticulum
HSP70	69–71	Cytosol/nucleus, mitochondria, chloroplast
HSP60	57–60	Mitochondria, chloroplast
smHSP	15–30	Cytosol, mitochondria, chloroplasts, ER

Boston et al. (1996)

Attempts were made to transform plants with various HSP genes to enhance thermotolerance. *Nicotiana tabacum* was transformed with *OsHsp101* gene under the control of strong consecutive promoter CaMV 35S. Transgenic plants showed better growth and development than nontransgenic plants under heat stress (Chang et al. 2007). *Pusa basmati 1* was transformed with *AtHsp101* gene from *Arabidopsis thaliana* under the control of strong consecutive promoter CaMV 35S; transgenic plants showed enhanced thermotolerance (Katiyar-Agarwal et al. 2003). Similarly, Qi et al. (2011) demonstrated that overexpression of mitochondrial *mtHsp70* inhibits programmed cell death and prevents reactive oxygen species formation in rice protoplast.

3.5.1.2 Change in Level of Osmolyte Accumulation

The raised levels of osmolytes impart increase in reducing the state of the environment of the cell and exert the energy status to an extreme level thereby preventing impairment to the proteins present in the cell. Enhanced production of osmolytes in the transgenic plants leads to an elevation in the level of tolerance. At higher temperature, BADH gene of spinach was expressed overly in transgenic tobacco plants, leading to a reduction in the association activity of the rubisco activase enzyme with the membranes of the thylakoids thereby defending the rubisco activity of the enzyme and absorption of CO_2 (Yang et al. 2005). Similarly, BADH protein from *Atriplex hortensis* was overexpressed in the transgenic wheat at elevated temperature because the membranes of thylakoid with lipid content got stabilized by the temperature hike as demonstrated by Wang et al. 2010.

3.5.1.3 Antioxidation and Detoxification Pathway Engineering

Plant cells experience hike in the temperature against normal response by elevating the production levels of H_2O_2 and ROS. When the ROS level is less in plant cells, it is believed to be vital in maintaining stress-related events in signal transduction. The genes that produce antioxidants and enzymes to detoxify ROS are utilized for tolerance to a higher temperature in the engineered plants. The enzyme glutathione peroxidase taken from tomato was expressed in tobacco caused enhancement in the level of thermotolerance (Chen et al. 2004). The plants can be prevented by high-temperature stress if supplied with the higher dose of Cu/Zn superoxide dismutase (Tang et al. 2006).

3.5.1.4 Engineering Transcription Factor Genes

All biological processes involve the use of transcription factors (TFs) for the production of crops with the traits which can tolerate stress. TFs fall into many different families like HSF, MYB, MYC, NFY, ERF/AP2, COR proteins, WRKY, etc. that govern the expression of genes in response to stress. When WRKY11 gene was overexpressed in the rice, it raised the level of drought and thermotolerance in the transformed plants as proved by Wu et al. (2009). Also, Sakuma et al. (2006) demonstrated that TF like DREB2A has a significant role in *Arabidopsis* against high temperature.

3.6 Cold

Modification in physiological and physical change due to low temperature is termed as chilling injury. This is one of the important factors in reducing the production of the crops and causing a reduction in the economy for the producers. Some of the species from the regions of the tropical and subtropical area are liable to low temperature and leads to a reduction in their growth, productivity, and causes necrosis, chlorosis, and sometimes even death. So, by making some crop plants tolerant to chilling can be a good way for the betterment of all.

3.6.1 Production of Chilling Tolerant Transgenic Plants

3.6.1.1 Stabilizing the Membrane

Freezing causes a very vicious effect that causes the membrane of the plant to get destroyed or damaged. This destruction occurs due to the keen dehydration. Membranes are made up of lipids and proteins, and lipids mainly consist of saturated and unsaturated fatty acids. Lipids having saturated fatty acids change their integrity at a higher temperature and become solid as compared to lipids having unsaturated fatty acids. Thus, unsaturated fatty acids are responsible for the fluidity state of the membrane. Many trials have been conducted to modify the composition of fatty acids in the membrane to increase the stability. Transplastomic tobacco plants were produced by Craig et al. (2008) using PEG-mediated method of DNA uptake by protoplasts that expresses D9 desaturase gene. The transplastomic plants had altered fatty acid composition as compared to control plants and thus they were more resistant to cold.

3.6.1.2 Osmoprotectant

As plants are sessile in nature, they have developed various mechanisms to combat the adverse conditions around them. They protect themselves by producing toxic compounds or by synthesizing different kinds of osmolytes. So, engineering the osmolytes in some of the crop species can make them tolerant to cold stress. Park et al. (2004) took a tomato and inserted *cod A* gene from *Arthrobacter globiformis* by chloroplast transformation method. This gene catalyzes the conversion of choline into GB. The transformed plants accumulated GB up to 0.3 and 1.21 μmol/g FW in their leaves and reproductive organs and were tolerant to cold stress. Similarly, rice was also transformed by Akami et al. (2008). They introduced two enzymes *wft2* and *wft1* which synthesizes wheat fructan; as a result transformed rice plants accumulated 16.2 mg/g FW of fructans. The transformed rice having *wtf2* accumulated a higher concentration of oligosaccharides and polysaccharides and thus were more tolerant to chilling stress.

3.6.1.3 Antifreeze Proteins

These proteins are present in almost all overwintering plants and play their role by inhibiting their growth and recrystallizing the ice present in intercellular spaces. At low temperature, plants conglomerate some of the proteins in their leaves or some cells called antifreeze proteins (AFPs). On the basis of mass, these AFPs vary from a range of 16–35 kD (Hon et al. 1994). The antifreeze proteins have the capability to inhibit the ice growth and get adsorbed onto the plant surface by the effect called Kelvin Effect. When antifreeze protein MpAFP149 was introduced into tobacco with the help of strong promoter CaMV 35 S resulting in the huge distribution of this protein as seen in transmission electron microscopy, thereby providing tobacco plant more tolerance to cold conditions (Wang et al. 2008).

3.6.1.4 Changes in Gene Expression

Plants respond to stress by showing a modification in their expression profile of genes. For making plants tolerant to stress conditions, many modifications can be done in their gene profile with the aid of genetic engineering technology. The alteration in gene profile can lead to upstream or downstream regulation of some of the genes or proteins involved in different mechanisms of the plant. Genetic engineering of plants for tolerance to cold stresses could be achieved by regulated expression of cold-induced transcription factors, which in turn would regulate expression of a large number of relevant downstream genes. Overexpression of ethylene response factor, TERF2 enhances transgenic rice tolerance to cold without affecting growth or agronomic traits. Physiological assays revealed that TERF2 could not only increase accumulation of osmotic substances and chlorophyll but also reduce reactive oxygen species (ROS) and malondialdehyde (MDA) content and decrease electrolyte leakage in rice under cold stress (Tian et al. 2011). Overexpression of CaPF1, which codes for an ERF transcription factor and is inducible by cold, subsequently activates expression of cold-responsive genes and confers tolerance against freezing temperature in Arabidopsis (Yi et al. 2004). The gene encoding a 1099-bp carrot antifreeze protein (AFP) was cloned into binary vector pCAMBIA2300 with the CaMV 35S promoter and used to transform tobacco NC82. PCR and Southern blot results verified integration of this gene into the genome of tobacco and reverse transcription-PCR verified that this gene had been expressed in transgenic tobacco. Experiments confirmed that transgenic tobacco plants displayed greater stamina than wild-type plants when subjected to cold treatment (Fan et al. 2002).

3.7 Flooding

Flooding also known as waterlogging is a virtually risky natural calamity that occurs because of the extreme levels of water on the ground or field. There are two kinds of flooding which prevail in the field: (1) one in which some part of shoot and roots are under the level of water and (2) one in which complete plant is under the level of water that is totally drowned in water. Flooding brings many changes in the plant like stunted growth, falling of leaves, loss of chlorophyll content, reduction in fixation of nitrogen, and sometimes death of the plant. To prevent plants from flooding, their genetic profile can be altered. Plants must have a mechanism to withstand more water level conditions. They must have a large number of aerenchyma cells, higher soluble sugar content, and antioxidant defense pathway for tolerating any stress related to flooding. Ethylene, a phytohormone plays a vital role in a plant system, when the plant is deprived of oxygen. It has been seen that ethylene helps in the formation of aerenchyma cells and in accelerating the fermentation and glycolytic pathways occurring in the plants.

3.7.1 Production of Flooding Tolerant Transgenic Plants

Tomato plants were made transgenic by inserting the ACC (1-aminocyclopropane-1-carboxylic acid) deaminase gene from bacteria with the help of a strong promoter 35S CaMV by Grichko and Glick (2001); they found that these transformed tomato plants had more tolerance power for flooding as compared to normal plants. Similarly, Cabello et al. (2016) targeted *Arabidopsis* with transcription factor HaHB11 from sunflower by fusing with GUS reporter gene and the targeted plants were transformed and showed tolerance to flooding. Chiang et al. (2015) introgressed rice plants with cDNA of ascorbate peroxidase originating from eggplant by *Agrobacterium*-mediated method of transformation. The rice transgenic plants have higher expression profile of ascorbate peroxidase in flooding stress.

3.8 Commercialized Varieties for Abiotic Stress Tolerance

S. No.	Crop name	Developed by	Trade name	GM trait	Gene introduced	Source	Product of introduced gene	Function of introduced gene
1	Maize	Monsanto Company and BASF	Genuity® DroghtGard™	Drought Tolerance	*CspB*	*B. subtalis*	Cold shock protein B	Normal cellular activity is maintained by keeping RNA stability and translation
2	Soybean	Verdeca	Verdeca HB4 soybean	Drought Tolerance	*Hahb-4*	*H. annus*	Transcription factor Hahb-4	Hahb-4 binds to dehydration regulating region
3	Sugarcane (Event Code: NXI-1 T)	Perseco	NA	Drought Tolerance	*EcBetA*	*E. coli*	Choline dehydrogenase	Glycine betaine production
4	Sugarcane (Event Code: NXI-4 T)	Perseco	NA	Drought Tolerance	*RmBetA*	*R. meliloti*	Choline dehydrogenase	Glycine betaine production
5	Sugarcane (Event Code: NXI-6 T)	Perseco	NA	Drought Tolerance	*RmBetA*	*R. meliloti*	Choline dehydrogenase	Glycine betaine production

Source: www.isaaa.org.in

3.9 Conclusion and Future Perspectives

Abiotic stress comprises mainly drought, salinity, temperature, flooding and light stress, etc. and is the major cause of yield losses worldwide and has become a major threat to agriculture. Therefore, the major aim of plant biotechnologists is to mitigate severe environmental stresses to reduce the demand for increasing food for the growing population. Due to multigenic nature, there is limited knowledge about the stress-associated metabolism, which is a major gap in improving stress tolerance in plants. To know about the key transcription factors involved in stress mitigation and their use in engineering plants for enhanced stress tolerance is a crucial area of research. Subtractive profiling of stress-related metabolites with another stress tolerant crop metabolism will play a key role in molecular breeding for stress tolerance. Differential expression analysis studies carried out with the help of transcription profiling and next generation sequencing will have a great potential in developing stress tolerant plants. The data generated from various genomic studies such as transcriptomics, proteomics, and metabolomics will help scientists to develop plants with enhanced traits to mitigate the problems caused by various stresses in near future. The response of crop plants to various biotic and abiotic stress can be better understood with the help of system biology approaches supplemented with "omics"-based data. Implementation of data generation of the next generation technologies to develop plants with better survival rate, stress tolerance, higher yield, and increased biomass will be a challenge in the near future.

References

Abebe T, Guenzi AC, Martin B, Cushman JC (2003) Tolerance of mannitol-accumulating transgenic wheat to water stress and salinity. Plant Physiol 131:1748–1755

Akami A, Sato KY, Yoshida M (2008) Genetic engineering of rice capable of synthesizing fructans and enhancing chilling tolerance. J Exp Bot 59(4):793–802

Apse M, Aharon P, Snedden GS, Blumwald E (1999) Salt tolerance conferred by overexpression of a vacuolar Na^+/H^+ antiport in *Arabidopsis*. Science 285:1256–1258

Bhauso TD, Radhakrishnan T, Kumar A et al (2014) Overexpression of bacterial *mtlD* gene in peanut improves drought tolerance through accumulation of mannitol. Sci World J1:1–10

Boston RS, Viitanen PV, Vierling E (1996) Molecular chaperones and protein folding in plants. Plant Mol Biol 32:191–222

Cabello JV, Giacomelli JI, Piattoni CV et al (2016) The sunflower transcription factor HaHB11 improves yield, biomass and tolerance to flooding in transgenic *Arabidopsis* plants. J Biotechnol 222:73–83

Chang C, Huang P, Lin H, Lu C (2007) Transactivation of protein expression by rice HSP101 in planta and using Hsp101 as a selection marker for transformation. Plant Cell Physiol 48:1098–1107

Chen S, Vaghchhipawala Z, Li W, Asard H, Dickman MB (2004) Tomato phospholipid hydroperoxide glutathione peroxidase inhibits cell death induced by Bax and oxidative stresses in yeast and plants. Plant Physiol 135:1630–1641

Chiang CM, Chen LFO, Shih SW, Lin KH (2015) Expression of eggplant ascorbate peroxidase increases the tolerance of transgenic rice plants to flooding stress. J Plant Biochem Biotechnol 3:257–267

Commercial GM trait: Abiotic stress tolerance (2015) www.isaaa.org.in
Craig W, Lenzi P, Scotti N et al (2008) Transplastomic tobacco plants expressing a fatty acid desaturase gene exhibit altered fatty acid profiles and improved cold tolerance. Transgenic Res 17:769–782
Fan Y, Liu B, Wang H et al (2002) Cloning of an antifreeze protein gene from carrot and its influence on cold tolerance in transgenic tobacco plants. Plant Cell Rep 21:296–301
Gill SS, Tuteja N (2010) Polyamines and abiotic stress tolerance in plants. Plant Signal Behav 5(1):26–33
Grichko VP, Glick BR (2001) Flooding tolerance of transgenic tomato plants expressing the bacterial enzyme ACC deaminase controlled by the 35*S*, *rolD* or PRB-1*b* promoter. Plant Physiol Biochem 39:19−25
Hayashi A, Alia Mustardy L, Deshnium P et al (1997) Transformation of *Arabidopsis thaliana* with the *codA* gene for choline oxidase: accumulation of glycine betaine and enhanced tolerance to salt and cold stress. Plant J 12:133–142
Holmstrom KO, Welin B, Mandal A, Kristiansdottir I et al (1994) Production of *Escherichia coli* glycine betaine dehydrogenase an enzyme required for the synthesis of osmoprotectant glycine betaine, in transgenic plant. Plant J 6:749−758
Holmstrom KO, Mantyla E, Welin B, Mandal A et al (1996) Drought tolerance in tobacco. Nature 37:683–684
Hon WC, Griffith M, Chong P, Yang DSC (1994) Extraction and isolation of antifreeze proteins from winter rye (*Secale cereale* L.) leaves. Plant Physiol 104:971–980
Katiyar-Agarwal S, Agarwal M, Grover A (2003) Heat-tolerant basmati rice engineered by overexpression of hsp101. Plant Mol Biol 51:677–686
Kishor KPB, Hong Z, Miao GH et al (1995) Overexpression of D1-pyrroline-5-carboxylate synthetase increases proline production and confers osmotolerance in transgenic plants. Plant Physiol 108:1387–1394
Ke Q, Wang Z, Ji CY, Jeong JC et al (2016) Transgenic poplar expressing *codA* exhibits enhanced growth and abiotic stress tolerance. Plant Physiol Biochem 10:75-–84
Moon SJ, Han SY, Kim DY, Yoon IS et al (2015) Ectopic expression of a hot pepper bZIP-like transcription factor in potato enhances drought tolerance without decreasing tuber yield. Plant Mol Biol 89:421–431
Park EJ, Jeknic Z, Sakamoto A et al (2004) Genetic engineering of glycine betaine synthesis in tomato protects seeds, plants, and flowers from chilling damage. Plant J 40:474–487
Patel M, Lewis SM, Zhang W, Templeton K et al (2015) Overexpression of ubiquitin-like LpHUB1 gene confers drought tolerance in perennial ryegrass. Plant Biotechnol J 13:689–699
Pilon-Smits EAH, Terry N, Sears T, Van Dun K (1999) Enhanced drought resistance in fructan-producing sugar beet. Plant Physiol Biochem 37:313–317
Prashanth SR, Sadhasivam V, Parida A (2008) Over expression of cytosolic copper/zinc superoxide dismutase from a mangrove plant *Avicennia marina* in indica rice var *Pusa Basmati*-1 confers abiotic stress tolerance. Transgenic Res 17:281–291
Qi Y, Wang H, Zou Y, Liu C, Liu Y et al (2011) Overexpression of mitochondrial heat shock protein 70 suppresses programmed cell death in rice. FEBS Lett 585:231–239
Qin H, Gu Q, Kuppu S, Sun L et al (2013) Expression of the *Arabidopsis* vacuolar H^+-pyrophosphatase gene *AVP1* in peanut to improve drought and salt tolerance. Plant Biotechnol Rep 7:345–355
Sakamoto A, Murata N (1998) Metabolic engineering of rice leading to biosynthesis of glycine betaine and tolerance to salt and cold. Plant Mol Biol 38:1011-–1019
Sakuma Y, Maruyama K, Quin F, Osakabe Y et al (2006) Dual function of an *Arabidopsis* transcription factor DREB2A in water-stress-responsive and heat stress responsive gene expression. Proc Natl Acad Sci U S A 103:18822-–18827
Surekha C, Kumari KN, Aruna LV, Suneetha G et al (2014) Expression of the *Vigna aconitifolia* P5CSF129A gene in transgenic pigeon pea enhances proline accumulation and salt tolerance. Plant Cell Tissue Organ Cult 116:27–36

Tang L, Kwon SY, Kim SH, Kim JS et al (2006) Enhanced tolerance of transgenic potato plants expressing both superoxide dismutase and ascorbate peroxidase in chloroplasts against oxidative stress and high temperature. Plant Cell Rep 25:1380–1386

Tian Y, Zhang H, Pan X, Chen X et al (2011) Overexpression of ethylene response factor TERF2 confers cold tolerance in rice seedlings. Transgenic Res 20:857–866

Vendruscoloa ECG, Schusterb I, Pileggic M et al (2007) Stress-induced synthesis of proline confers tolerance to water deficit in transgenic wheat. J Plant Physiol 164:1367–1376

Wang Y, Ying Y, Chen J, Wang X (2004) Transgenic *Arabidopsis* overexpressing Mn-SOD enhanced salt-tolerance. Plant Sci 167:671–677

Wang Y, Qiu L, Dai C, Wang J et al (2008) Expression of insect (*Microdera puntipennis dzungarica*) antifreeze protein MpAFP149 confers the cold tolerance to transgenic tobacco. Plant Cell Rep 27:1349–1358

Wang GP, Zhang XY, Li F et al (2010) Over accumulation of glycine betaine enhances tolerance to drought and heat stress in wheat leaves in the protection of photosynthesis. Photosynthetica 48:117–126

Wang X, Han H, Yan J et al (2015) A new AP2/ERF transcription factor from the oil plant *Jatropha curcas* confers sand drought tolerance to transgenic tobacco. Appl Biochem Biotechnol 176:582–597

Wei T, Deng K, Gao Y et al (2016) *Arabidopsis* DREB1B in transgenic *Salvia miltiorrhiza* increased tolerance to drought stress without stunting growth. Plant Physiol Biochem 104:17–28

Wu X, Shiroto Y, Kishitani S, Ito Y, Toriyama K (2009) Enhanced heat and drought tolerance in transgenic rice seedlings overexpressing OsWRKY11 under the control of HSP101 promoter. Plant Cell Rep 28:21–30

Xue ZY, Zhi DY, Xue GP, Zhang H et al (2004) Enhanced salt tolerance of transgenic wheat (*Triticum aestivum* L.) expressing a vacuolar Na^+/H^+ antiporter gene with improved grain yields in saline soils in the field and a reduced level of leaf Na^+. Plant Sci 167:849–859

Yamada M, Morishita H, Urano K, Shiozaki N et al (2005) Effects of free proline accumulation in petunias under drought stress. J Exp Bot 56(417):1975–1981

Yang X, Liang Z, Lu C (2005) Genetic engineering of the biosynthesis of glycine betaine enhances photosynthesis against high temperature stress in transgenic tobacco plants. Plant Physiol 138:2299–2309

Yi SY, Kim JH, Joung YH, Lee S et al (2004) The pepper transcription factor CaPF1 confers pathogen and freezing tolerance in *Arabidopsis*. Plant Physiol 136:2862–2874

Zhu B, Su J, Chang M, Verma DPS et al (1998) Overexpression of delta1-pyrroline-5-carboxylate synthase gene and analysis of tolerance to water and salt stress in transgenic rice. Plant Sci 199:41–48

Developing Climate Smart Aerobic Rice Varieties for Addressing the Problems of Water Scarcity and Global Warming

4

Nitika Sandhu, Virender Singh, Manvesh Kumar Sihag, Sunita Jain, and Rajinder Kumar Jain

Abstract

Growing population, rising demand for food, looming water crisis, climate change and global warming, reduced nutrient availability, higher cost of irrigation, and poor availability of labor have threatened the puddled transplanted system of rice (PTR). Rice needs two to three times more water than other cereals. Due to water scarcity, farmers may not be in a state to use the same amount of water for cultivation of rice and there is an urgent need to find out long-term effective and reliable methods to grow rice more efficiently. Use of machines, new technologies, coping strategies, genes for a range of agro-ecologies varying in edaphic and water regimes, and novel genomic techniques will allow us to move toward resource (labor, water, and energy) efficient and climate smart agriculture, especially in the case of rice. The identification of suitable traits and use of correlated genetic regions for seedling establishment and root traits that can potentially improve the nutrient uptake and grain yield under water-deficient conditions and pyramiding of these regions in a breeding program may lead to higher yield and adaptability of rainfed rice under aerobic conditions. Understanding of molecular mechanisms and successful exploitation of major

N. Sandhu
Division of Plant Breeding, International Rice Research Institute, DAPO, Box 7777, Metro Manila, Philippines

V. Singh
Indian Institute of Wheat and Barley Research, Karnal 132 001, Haryana, India

M.K. Sihag
Mansinhbhai Institute of Dairy and Food Technology (MIDFT), Mehsana 384 002, Gujarat, India

S. Jain • R.K. Jain (✉)
Department of Molecular Biology, Biotechnology and Bioinformatics, CCS Haryana Agricultural University, Hisar 125 004, Haryana, India
e-mail: rajinder.jain@gmail.com

S.K. Gahlawat et al. (eds.), *Plant Biotechnology: Recent Advancements and Developments*, DOI 10.1007/978-981-10-4732-9_4

effect genetic regions in various genetic backgrounds under variable environment and subsequent selection of lines with desired traits may result in development of novel water-efficient aerobic rice varieties. This chapter reviews the research including the development and use of innovative technologies and identification of useful traits and genetic regions associated with aerobic adaptation to achieve resource efficient cultivation of rice and to combat the effects of climate shift, water scarcity, and global warming.

4.1 Introduction

Rice, the major user of world's fresh water resources, accounts for about 24–30% withdrawal of the total fresh water and consumption of 34–43% of the total irrigation water. Water requirement for rice is two to three times more than that of other cereals such as wheat. While comparable evapo-transpirational rate of water loss has been reported in rice and wheat, the more water requirements of rice are due to puddling, leaching, and percolation losses coupled with traditional method of continuous flooding. Rice cultivation in puddled transplanted situation is also highly labor and energy intensive, and these resources are becoming increasingly scarce and expensive, making rice production less profitable. Apart from this, rapid economic growth and booming industrial sector in Asia have increased the demand of labor in nonagricultural sectors. These situations demand development of rice cultivation technologies which can support rice yield under decreasing availability of labor and irrigation water. Incidences of different kinds of drought and flood stress at different growth stages, long-standing aerobic phases, and soil infertility is likely to increase more and more in the near future. Progressing unpredictable growing environments as a result of alteration in climatic conditions demand adaptable crops that can maintain high yield across variable environment.

Amount, pattern, and duration of rainfall opt for the transplanting or direct-seeding establishment method in rice. Consequently, identification of suitable seedling stage establishment traits, traits that enhance nutrient uptake and phenotypic plasticity for root, traits accountable for yield stability and rice crop adaptation to the variable range of conditions, as well as genetic regions associated with these traits and understanding of biological and molecular mechanism of identified genetic regions may facilitate stability of rice genotypes to aerobic conditions and may have a strong effect on plant growth correlated with yield stability. This needs an improved understanding of factors and traits and proper management decisions that affect rice adaptability under variable conditions.

4.2 Aerobic Rice: A Step Toward Sustainability

Aerobic rice is a realistic substitute to conventional puddle transplanted rice (PTR) with great potential to save water and energy, reduce labor requirement, mitigate greenhouse gas (GHG) emissions and adapt to climatic risks, mechanization, high productivity in a well-managed crop (Kumar and Ladha 2011) that leads to high economic return and profitability. Aerobic rice practice involves direct cultivation of rice in dry, zero tillage fields with even slope under unpuddled/aerobic conditions. Aerobic rice has long, deep, thick, and robust root system; upright leaves; small to medium plant height; improved water use efficiency (WUE) with high biomass, harvest index (HI), and adaptability under aerobic conditions; low input requirement, crop manageability, yield competitiveness, profitability, early establishment which enhances productivity; and resistance to diseases and pests with weed competitiveness (Yadav et al. 1997; Ling et al. 2002). In fact, aerobic rice had lower total water use (~51%) and higher water productivity (32–88%) than continuous flooded rice (Bouman et al. 2005), significantly less labor requirement (Wang et al. 2002) in addition to less greenhouse gas emission from rice field (Mandal et al. 2010). Deeper root system facilitates plants to maintain better water potential and high water uptake from the deep soil layers, and direct to nonsignificant yield decline under aerobic system of cultivation (Mumbani and Lal 1983). Savings of up to 60% of labor and 12–35% of irrigation water in situations ranging from dry direct seeded flooded to aerobic under direct seeded aerobic rice comparable to transplanted rice have been observed. In addition, the same rice variety under aerobic conditions matures 10–12 days earlier than those under transplanted sown condition, thus providing better opportunities to plant the second crop after rice.

From 1980s to 2002, the area of upland aerobic rice in the North China Plain (NCP) spread out rapidly to about 80,000 ha (Wang et al. 2002). In northern China and Brazil, the development of temperate aerobic upland rice started in the mid-1980s. The grain yield potential of 6–7 t ha^{-1} has been reported in direct seeded rice varieties (Wang et al. 2002).

4.2.1 Aerobic vs. Drought Tolerance in Upland Rice

Aerobic rice can be rainfed, supplementary, or fully irrigated depending on the availability of water resources. In aerobic rice, the total amount of water supply probably needed is 600–800 mm over the season when the groundwater tables below the root zone. The drought-resistant upland varieties are well fitted with less than 400 mm water and deep groundwater tables. The "transition zones" between drought tolerant upland and aerobic rice and between aerobic and lowland rice production systems are quite site-soil type specific. Aerobic rice varieties were developed from crossing involving drought-resistant upland varieties with the high-yielding lowland varieties.

4.2.2 Problems Associated with Aerobic Rice

The varieties developed for transplanted situations have shown yield decline of 8–28.5% under aerobic conditions in India and Pakistan. The yield decline under aerobic conditions results from pre-emergence death of seedlings, delayed or non-uniform emergence, uneven plant population in the field, higher weed problem early in crop growth stage, and lower nutrient availability to rice under aerobic conditions with the existing rice root system and the prevalent nutrient management system that is more biased to transplanted flooded rice. There is a great need for rice seedlings with uniform and fast emergence from deep soil layer under suboptimal conditions. A genotype with long mesocotyls emerges through the soil efficiently and faster than a genotype with short mesocotyls. Reduced nutrient uptake especially immobilization of P in acidic soils, increased N leaching, and volatilization under aerobic cultivation conditions compared to continuous flooded conditions has been the key factor for yield reduction under aerobic cultivation than in flooded rice cultivation systems (Kumar and Ladha 2011). Additionally, deficiencies of micronutrients such as Mn, Fe, and Zn under soil pH condition >6.0 are common problems associated with the aerobic system of cultivation. The further gaps in aerobic rice cultivation include nonavailability of suitable adapted aerobic rice variety, site-specific nutrient and water management practices, nematode sickness, less knowledge, optimum sowing window, and effective weed management system.

4.2.3 Root Architecture—Designing Root System

Variations/adaptations in the rice root system comprise various changes in architecture, morphology, anatomy, physiology, and roots-shoots biomass distribution. The soil type, temperature, mineral–nutrient–water status, regulatory system of plant, and the above ground resources like radiation are major players in root architecture. This highlights the necessity of improving and better understanding the rice root system and how the root traits' QTLs (Quantitative trait loci) interact to affect the acquisition of nutrient and water uptake under variable range of environment. Designing better rice root system by including more number of nodal roots, more lateral root, maximum root length density (RLD), lateral root length, and branching with more root hairs similar to wheat root system is needed (Fig. 4.1).

4.2.4 Seedling Establishment, Root Traits, and Grain Yield: Correlation Studies

A better knowledge of the correlation among seedling traits, specifically, early vegetative vigor (EVV), nodal root number (NR), root hair length (RHL) and density (RHD), and grain yield may offer an innovative research platform to study the interaction of genes responsible for adaptation to aerobic condition. Sandhu et al. (2011) reported the superiority of the root system of aerobic rice genotypes than lowland

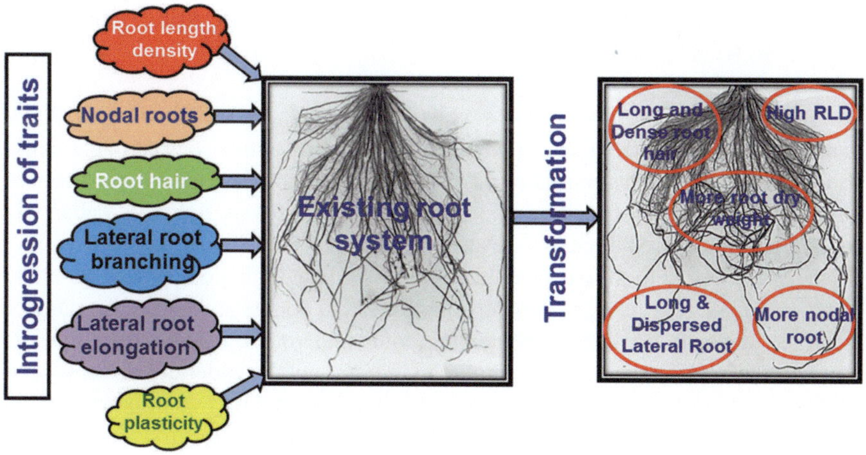

Fig. 4.1 Designing better and adaptable root system by introgressing suitable root traits

indica rice genotypes and a positive correlation of grain yield was found with root length and biomass signifying that root traits might be the main role players for grain yield improvement under water-saving aerobic conditions. The grain yield was found to be significantly and positively correlated with root length (RL), root number (RN), root volume (RV), and root biomass (RB), suggesting that these traits might be the key factor for improving the grain yield under aerobic cultivation conditions (Sandhu et al. 2013) possibly through enhanced water-nutrient uptake. Correlation of root hair density and lateral root growth with grain yield (Sandhu et al. 2015) showed that an intense root system will enhance water-nutrient uptake which ultimately enhances the grain yield potential under direct seeded conditions. Sandhu et al. (2015) reported a positive correlation between relative growth rate (RGR) and grain yield at seedling stage under direct seeded conditions. Architectural plasticity for root dry weight (RDW), % lateral roots (LR), and root length density (RLD) have been identified as the main root plasticity traits that make the rice root system stable and efficient for better uptake of different soil nutrient resources at different soil depth under cycles of aerobic to anaerobic situation to contribute to grain yield stability under variable growing conditions (Sandhu et al. 2016). Genetic regions associated with plasticity in root dry weight and % LR (lateral roots) were identified near a major-effect grain yield QTL under drought $qDTY_{1.1}$, conferring plastic responses to reproductive stage drought stress with better root and shoot growth (Vikram et al. 2015), in addition to an increase in the fraction of deep roots in the *OryzaSNP* panel under reproductive stage drought stress (Wade et al. 2015).

4.2.5 Genetic Analysis of Traits Promoting Aerobic Adaptation in Rice

Several QTLs for root traits (Zhang et al. 2001; Liu et al. 2008; Courtois et al. 2009; Sandhu et al. 2013, 2015) and grain yield under reproductive stage drought stress (Bernier et al. 2007; Venuprasad et al. 2009; Sandhu et al. 2014) have been reported; identification of new consistent effect QTLs/genetic regions for traits enhancing adaptation to dry direct seeded cultivation conditions, new donors, and new coping strategies may be helpful in developing climate resilient high yielding rice varieties.

Over the last few years, researchers have reported donors and QTLs for many of the traits (like early vigor, nodal root, total root length, root dry weight, root volume, root hair density, length, % lateral root, and nutrient uptake under dry direct seeded situation and grain yield under direct seeded situation) that may be important in the development of climate smart rice varieties with flexibility to be grown under variable situations ranging from dry and wet direct seeded to transplanted situations. On the other hand, recent advances in molecular technology have provided a unique opportunity to breeders to combine these traits/QTLs/genes efficiently to develop new varieties.

Efforts to find a viable solution to the decreased sustainability of rice yield and decreased adaptability of rice plants have shown some promising results in recent days. Sandhu et al. (2012) studied polymorphism in a panel including aerobic, Basmati, *indica*, and *japonica* rice varieties using 50 SSR (Simple sequence repeat) markers. Greater divergence was recorded between aerobic and Basmati rice varieties as compared to aerobic and *indica* rice varieties. Two SSR markers (RM302 and RM212) amplified specific alleles in all studied aerobic rice genotypes, which were absent in nonaerobic rice genotypes. Sandhu et al. (2013) reported two genetic regions on chromosome 2 and 8 (*qGY2.1*, R^2 value of 22.8% and *qGY8.1*, R^2 value of 34.0%) and one genetic loci on chromosome 2 (*qGY2.2*, R^2 value of 43.2%) for grain yield under aerobic cultivation conditions in aerobic (MASARB25) × Basmati (Pusa Basmati 1460) and HKR47 (*indica*) × MAS26 (aerobic), mapping populations, respectively. QTL for RL (root length) within the *qGY2.1* region (MacMillan et al. 2006) and for RT (root thickness) (Champoux et al. 1995) and GY (grain yield) under reproductive stage drought stress (Bernier et al. 2007) adjacent to *qGY2.2* have been reported. Two QTLs, *qRL8.2* (both mapping populations) and *qRL8.1* (MASARB25 × Pusa Basmati 1460), for root length have been reported by Sandhu et al. (2013). Qu et al. (2008) reported genetic region for fresh root weight (FRW) and root number (RN) coinciding with *qRL8.2*. Under aerobic conditions, a 25.1-cM segment on chromosome 6 locating between RM589 and RM314 affecting different root (root volume, root thickness, and fresh root weight) and shoot (fresh shoot weight and dry shoot weight) traits in aerobic (MASARB25) × Basmati (Pusa Basmati 1460) mapping population was reported in Sandhu et al. (2013) study. In the same region of chromosome 6, a study of 20 hydrological environments conducted by Venuprasad et al. (2009) reported *qDTY6.1*, a large-effect grain yield QTL in five mapping populations in three genetic backgrounds over a period of five seasons under aerobic environments using bulk-segregant analysis approach. QTL, *qPL12.1* for panicle length (RM28048) identified in study at CCSHAU (Chaudhary Charan Singh Haryana Agricultural University) in HKR47 (*indica*) × MAS26 (aerobic) mapping population showed an effect under aerobic system of cultivation. A major consistent effect genetic region *DTY12.1* (linked to RM28048 marker) related to grain yield under reproductive stage drought stress had shown effect over a series of environments in eastern India and the Philippines (Bernier et al. 2007, 2009). Bernier et al. (2007) showed the effectiveness of the genetic region (*DTY12.1*) not only in the Philippines, where it was primarily identified, but also in northern and eastern India; this validates the low QTL × environment interaction, one of the key requisites for the use of a genetic region in marker-assisted selection. Table 4.1 summarizes the identified genetic loci at CCS HAU.

Our study at CCS Haryana Agricultural University, Hisar, India, has led to the identification of several lines with good grain yield under aerobic cultivation conditions, high root length and dry root biomass, and grain length:breadth ratio, and aroma. The identified breeding lines will serve as fresh material for the selection of stable aerobic adapted Basmati rice varieties. Interesting and significant results were obtained from the experiments conducted at CCS HAU on introgression of aerobic traits in some popular mega low-land *indica* and Basmati rice varieties being widely cultivated in the northern-western part of India. Successful introgression of aerobic adaptation traits into this widely cultivated *indica* and Basmati rice varieties can play an important role toward national and international food security.

Unpredictable rain fall, heavy rain fall just after seeding, very low rain fall, weeds, less nutrient uptake, and lodging are major hurdles in aerobic rice cultivation practices. Root plasticity led to the better adaptation of the plant under cycles of flooded/aerobic conditions during crop growth stages. Donors for anaerobic germination (germination under stagnated water condition), early uniform emergence, high vegetative vigor, higher number of lateral roots (LR) and root hair density (RHD) for increased nutrient availability and lodging resistance have been identified. Research for trait development to develop better rice varieties for dry direct seeded situation has led to the identification of two major QTLs (*AG1, AG2*) (Angaji et al. 2010; Kretzschmar et al. 2015) for anaerobic germination, two QTLs for early uniform emergence (*qEMM1.1, qEMM11.1*), one QTL for lodging resistance (*qLDG3.1, qLDG4.1*) (Dixit et al. 2015), QTLs for early vigor (*qEVV9.1*), three QTLs for higher root hair density and better nutrient uptake (*qRHD1.1, qRHD5.1, qRHD8.1*), one QTL for higher number of nodal roots and nutrient uptake (*qNR4.1*), and three QTLs for grain yield under DSR (*qGY1.1, qGY6.1, qGY10.1*) (Sandhu et al. 2015). QTLs for higher nutrient uptake (Fe, Al, and P) and nutrient concentration (Fe, N, Al, and P) and, dense root hair were identified co-locating in the same region (Sandhu et al. 2015). The grouping and genomic co-location of root traits enhancing nutrient uptake, relative growth rate, and root:shoot ratio with grain yield specify the role of these traits in grain yield improvement under aerobic direct seeded conditions (Sandhu et al. 2015).

However, incorporation of selection criteria for these traits into conventional and marker-assisted breeding programs has been complicated due to insufficiency of

Table 4.1 Summary of identified genetic regions in aerobic × lowland indica and aerobic basmati derived mapping population

QTLs	Chr location	Flanking markers (Position cM)	MASARB25 × PB1460	HKR47 × MAS26	Remarks/References
qGY2.2	2	RM475-RM526 (92.5–136.3 cM)	–	+	Bernier et al. (2007), Venuprasad et al. (2009), Ghimire et al. (2012) (grain yield under reproductive stage drought stress)
qPL2.1	2	RM485-RM7033 (0.0–7.9 cM)	+	–	Bernier et al. (2007), Venuprasad et al. (2009) (grain yield under reproductive stage drought stress and root volume and weight)
qGY2.1		RM7562-RM279 (8.6–13.8 cM)			
qRV2.1		RM485-RM7033 (0.0–7.9 cM)			
qFRW2.1		RM485-RM7033 (0.0–7.9 cM)			
qDSW2.1		RM485-RM7033 (0.0–7.9 cM)			
qPL12.1	12	RM28048 (47.4 cM)	–	+	Bernier et al. (2007), Dixit et al. (2012) (grain yield under reproductive stage drought stress)
qRT1.1	1	RM488-RM237 (101.4–112.9 cM)	–	+	Bernier et al. (2007), Venuprasad et al. (2009), Ghimire et al. (2012) (grain yield under reproductive stage drought stress), Li et al. (2005) (root thickness)
qRL9.1	9	RM524-RM257 (13.2–66.1 cM)	–	+	Qu et al. 2008 (root length)
qRL8.2	8	RM547-RM331 (58.1–65.0 cM), RM310-RM547 (57.0–58.1 cM)	+	+	Qu et al. (2008) (root length)
qFRW2.2 *qDRW2.1*	2	RM174-RM300 (47.5–66.0 cM)	+	–	Venuprasad et al. (2009) (grain yield under reproductive stage drought stress)
qPH6.1	6	RM6836-RM527 (54.1–61.2 cM)	+	–	Lin et al. (1996) (plant height)

DSW dry shoot weight, *DRW* dry root weight, *FRW* fresh root weight, *GY* grain yield under aerobic condition, *PL* panicle length, *PH* plant height, *RL* root length, *RT* root thickness, +: Presence of QTL, –: Absence of QTL efficient and reliable screening protocols and techniques and the time-consuming and laborious nature of traits of interest with the existing facilities and techniques (Garrity et al. 1986). Root systems are difficult to uproot and measure unless studied in an artificial environment, and then they may develop and behave reasonably differently than in a field environment (Wojciechowski et al. 2009). Moreover, root architecture depends on several diverse and simultaneous processes in the roots (Lynch 1995) in addition to competence of resources (light, mineral nutrients, and water) for shoot activities (Husain et al. 1990). Innovation of high throughput screening system to access, screen, and measure under field condition can be a solution for addressing this problem.

The recent success with trait development, marker development for traits related to aerobic adaptation, and ability to carry marker-assisted breeding for large number of QTLs/genes with tightly linked markers will provide new opportunities to develop better rice varieties for aerobic situation using the MAS (Marker-assisted Selection) approach. Combining QTLs' information for traits related to adaptation to aerobic cultivation conditions, identification and validation of genomic coordinates for overlapped/related traits, differential expression of genes concerned in metabolic processes, signal transductions, and response of identified genes to dry direct seeded cultivation conditions will provide useful information to develop stable direct seeded rice varieties with high grain yield and adaptability for the farmers who depend on rice for food security. The challenges further on are the efficient use of identified QTLs/genetic regions/genes and their combinations in marker-assisted

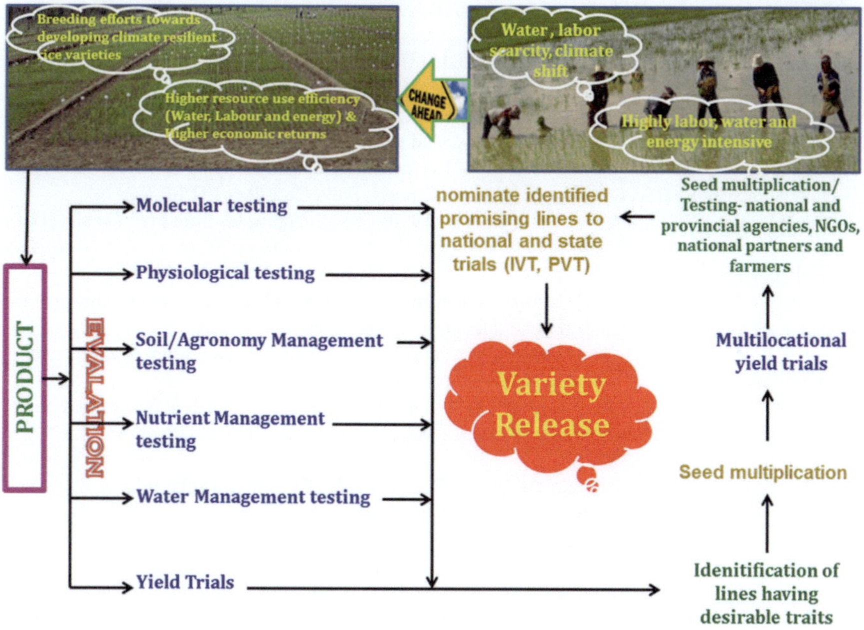

Fig. 4.2 Steps involved in developing climate smart rice varieties

breeding, fine mapping of QTLs to make possible specific introgression avoiding undesirable linkages, identification of useful candidate genes, their efficacy in other genetic backgrounds and variable environment, and understanding the molecular and physiological mechanisms linked with these major consistent effect genetic regions/QTLs/genes across variable environments. Development of a clear strategy by keeping all genetic and environmental factors, risks, management practices, dissemination method, knowledge, and adoption strategy (Fig. 4.2) in consideration may be a good step toward developing climate smart rice varieties.

References

Angaji S, Septiningsih E, Mackill D, Ismail A (2010) QTLs associated with tolerance to flooding during germination in rice (*Oryza sativa*). Euphytica 172:159–168

Bernier J, Kumar A, Venuprasad R, Spaner D, Atlin GN (2007) A large-effect QTL for grain yield under reproductive-stage drought stress in upland rice. Crop Sci 47:507–516

Bernier J, Kumar A, Spaner D, Verulkar S et al (2009) Characterization of the effect of rice drought tolerance qtl12.1 over a range of environments in Philippines and Eastern India. Euphytica 166:207–217

Bouman BAM, Peng S, Castaneda AR, Visperas RM (2005) Yield and water use of irrigated aerobic rice systems. Agric Water Manag 74:87–105

Champoux MC, Wang G, Sarkarung S et al (1995) Locating genes associated root morphology and drought avoidance in rice via linkage to molecular markers. Theor Appl Genet 90:969–981

Courtois B, Ahmadi N, Khowaja F et al (2009) Rice root genetic architecture: meta-analysis from a drought QTL database. Rice 2:115–128

Dixit S, Swamy BPM, Vikram P et al (2012) Fine mapping of QTL for rice grain yield under drought reveals sub-QTL conferring a response to variable drought severities. Theor Appl Genet 125:155–169

Dixit S, Grondin A, Lee CR, Henry A et al (2015) Understanding rice adaptation to varying agro-ecosystems: trait interactions and quantitative trait loci. BMC Genet 16:86

Garrity DP, Oldeman LR, Morris RA (1986) Rainfed lowland rice ecosystems: characterization and distribution. In: Progress in rainfed lowland rice. IRRI, Manila, pp 3–23

Ghimire KH, Quiatchon LA, Vikram P et al (2012) Identification and mapping of a QTL (*qDTY1.1*) with a consistent effect on grain yield under drought. Field Crop Res 131:88–96

Husain MM, Reid JB, Othman H, Gallagher JN (1990) Growth and water use of faba beans (*Vicia faba*) in a sub-humid climate I: root and shoot adaptations to drought stress. Field Crop Res 23:1–17

Kretzschmar T, Pelayo MAF, Trijatmiko KR et al (2015) A trehalose-6-phosphate phosphatase enhances anaerobic germination tolerance in rice. Nat Plants 1(9):15124

Kumar V, Ladha JK (2011) Direct seeding of rice: recent developments and future research needs. Advances in agronomy, vol 111. Newark, pp 297–413

Li ZC, Mu P, Li CP, Zhang HL et al (2005) QTL mapping of root traits in a doubled haploid population from a cross between upland and lowland japonica rice in three environments. Theor Appl Genet 110:1244–1252

Lin HX, Qian HR, Zhuang JY et al (1996) RFLP mapping of QTL for yield and related characters in rice (*Oryza sativa* L.) Theor Appl Genet 92:920–927

Ling ZM, Li ZC, Yu R, Mu P (2002) Agronomic root characters of upland rice and paddy rice (*Oryza sativa* L.) J Chin Agric Univ 7:7–11

Liu L, Mu P, Li X, Qu Y et al (2008) Localization of QTL for basal root thickness in japonica rice and effect of marker-assisted selection for a major QTL. Euphytica 164:729–737. http://dxdoiorg/101007/s10681-008-9695-4

Lynch J (1995) Root architecture and plant productivity. Plant Physiol 109:7–13
MacMillan K, Emrich K, Piepho HP et al (2006) Assessing the importance of genotype x environment interaction for root traits in rice using a mapping population II: conventional QTL analysis. Theor Appl Genet 113:953–964
Mandal DK, Mandal C, Raja P, Goswami SN (2010) Identification of suitable areas for aerobic rice cultivation in the humid tropics of eastern India. Curr Sci 99:227–231
Mumbani B, Lal R (1983) Response of upland rice varieties to drought stress. Plant Soil 73(1):73–94. http://dxdoiorg/101007/BF02197758
Qu Y, Ping M, Hongliang Z, Charles Y et al (2008) Mapping QTL of root morphological traits at different growth stages in rice. Genetica 133:187–200
Sandhu N, Jain S, Kumar A, Battan KR, Jain RK (2011) Aerobic rice genotypes displayed greater adaptation to water-limited cultivation and tolerance to polyethyleneglycol-6000 induced stress. Physiol Mol Biol Plant 18(1):33–43
Sandhu N, Jain S, Chowdhury VK, Jain RK (2012) Microsatellite diversity among aerobic and low land *indica* rice genotypes with differential water requirements. Indian J Genet 72(4):463–467
Sandhu N, Jain S, Kumar A, Mehla BS, Jain RK (2013) Genetic variation, linkage mapping of QTL and correlation studies for yield, root, and agronomic traits for aerobic adaptation. BMC Genet 14:104–119
Sandhu N, Torres RO, Sta Cruz MT et al (2014) Identification and mapping of stable QTL with main and epistasis effect on rice grain yield under upland drought stress. BMC Genet 15:63
Sandhu N, Torres RN, Cruz MTS et al (2015) Traits and QTL for development of dry direct-seeded rainfed rice varieties. J Exp Bot 66(1):225–244
Sandhu N, Raman KA, Torres RO et al (2016) Rice root architectural plasticity traits and genetic regions for adaptation to variable cultivation and stress conditions. Plant Physiol 171:2562–2576
Venuprasad R, Bool ME, Dalid CO et al (2009) Genetic loci responding to two cycles of divergent selection for grain yield under drought stress in a rice breeding population. Euphytica 167:261–269
Vikram P, Swamy M, Dixit S, Singh R et al (2015) Drought susceptibility of modern rice varieties: an effect of linkage of drought tolerance with undesirable traits. Sci Rep 5:14799
Wade L, Bartolome V, Mauleon R et al (2015) Environmental response and genomic regions correlated with rice root growth and yield under drought in the *OryzaSNP* panel across multiple study systems. PLoS One 10:e0124127
Wang H, Bouman BAM, Zhao D et al (2002) Aerobic rice in northern China: opportunities and challenges. In: Bouman BAM, Hengsdijk H, Hardy B et al (eds) Water-wise rice production. International Rice Research Institute, Los Baños, pp 143–154
Wojciechowski T, Gooding MJ, Ramsay L, Gregory PJ (2009) The effects of dwarfing geneson seedling root growth of wheat. J Exp Bot 60:2565–2573
Yadav R, Courtois B, Huang N, McLaren G (1997) Mapping genes controlling rootmorphology and root distribution in a doubled-haploid population of rice. Theor Appl Genet 94:619–632. http://dxdoiorg/101007/s001220050459
Zhang J, Zheng HG, Aarti A, Pantuwan G et al (2001) Locating genomic regions associated with components of drought resistance in rice: comparative mapping within and across species. Theor Appl Genet 103:19–29

Biotechnological Aspects for Enhancement of Mineral Bioavailability from Cereals and Legumes

5

Prince Chawla, Latika Bhandari, Sanju B. Dhull, Pardeep Kumar Sadh, Surinder Paul Sandhu, Ravinder Kaushik, and Navnidhi

Abstract

India is bound to be the global leader in terms of production and consumption of cereals and legumes and food products based on cereals and legumes. Cereals and legumes are a major source of edible proteins, carbohydrates and total dietary fiber and therefore considered to be highly valuable in human and animal nutrition. Micronutrient deficiency is a serious risk to the health and efficiency of billions of people of both developed and developing countries. Minerals (iron, zinc and calcium) are important microelements that play various fundamental roles in the human body. Besides this, cereals and legumes also contribute to the consumption of various minerals, including calcium, iron and zinc, in daily intake of food. Nutritional quality of cereals and legumes is directly correlated with the presence of antinutritional factors, and therefore, several methods have been introduced to combat these problems in foods to improve the nutritional

P. Chawla (✉) • R. Kaushik
School of Bioengineering and Food Technology, Shoolini University,
Solan 173 229, Himachal Pradesh, India
e-mail: princefoodtech@gmail.com

L. Bhandari
National Dairy Research Institute, Karnal 132 001, Haryana, India

S.B. Dhull
Department of Food Science and Technology, Chaudhary Devi Lal University,
Sirsa 125 055, Haryana, India

P.K. Sadh • S.P. Sandhu
Department of Biotechnology, Chaudhary Devi Lal University, Sirsa 125 055, Haryana, India

Navnidhi
Department of Bio and Nanotechnology, Guru Jambheshwar University,
Hisar 125 001, Haryana, India

S.K. Gahlawat et al. (eds.), *Plant Biotechnology: Recent Advancements and Developments*, DOI 10.1007/978-981-10-4732-9_5

value of both cereals and legumes. Also, numerous biotechnological tools and various pretreatments have been included to improve the bioavailability of minerals. Therefore, this chapter is concerned with the mineral bioavailability from different cereals and legumes consumed in India and worldwide. Effects of various treatments of cereals and legumes on mineral bioavailability and bioaccessibility have also been discussed.

5.1 Introduction

India is predominantly vegetarian in dietary habits and heavily depends upon vegetative sources to meet its daily nutrient requirement; therefore, India is bound to be the global leader in terms of production and consumption of cereals and legumes and food products based on cereals and legumes (Singh et al. 2015). Cereals and legumes are a major source of edible proteins, carbohydrates and total dietary fiber and therefore considered to be highly valuable in human and animal nutrition (Singh et al. 2016; Torres et al. 2016). They play a significant role in the agriculture and in the diet of most of the people worldwide and are the major source of food (Onwurafor et al. 2014; Kaur et al. 2010). Micronutrient deficiency is a major risk to the health and efficiency of billions of people worldwide (Powell et al. 2014). Cereals and legumes contribute to the ingestion of various minerals, including calcium, iron and zinc (Viadel et al. 2006).

Nutritive quality of cereal legumes is always characterized by the presence of antinutritional factors such as phytates oxalates and complex polysaccharides, i.e. starch, fibers and lignins (Afify et al. 2011). These antinutritional components combine with minerals and form complexes which affect their bioavailability (Ghavidel and Prakash 2007). Iron, zinc and calcium as important microelements play various fundamental roles in the human body. These are essential transition metals for virtually all life and vital for cellular survival (Lane et al. 2015; Afify et al. 2011). Iron-deficiency anemia occurs due to insufficient intake of iron by cells (Powell et al. 2014). Iron-deficiency anemia is a major threat for developing and developed countries; it is categorized by the absence of stored form of iron, low concentrations of serum hemoglobin, increased count of platelets, and reduced saturated form of transferrin with obvious increase in total iron-binding capacity (Nakao et al. 2015). Zinc deficiency is characterized by deprived growth, miserable immune function, increased liability and severity of infections and neurobehavioral deformities (Hemalatha et al. 2007a). Free form of inorganic salts participates in generation of reactive oxygen species in the presence of oxidants. Calcium is a vital micronutrient which is essential for significant biological functions such as nerve conduction, muscle contraction, mitosis, blood coagulation and structural support of the skeleton (Miller et al. 2001). Low ingestion of foods containing calcium, reduced bioavailability of dietary calcium due to the presence of antinutritional factors in food, inability of calcium metabolization in body are the major factors for calcium deficiency, which is characterized by numerous chronic ailments and disorders such as osteoporosis, osteomalacia, hypertension, colon cancer and obesity (Berner et al. 1990; Tunick 1987; Zemel and Miller 2004). Humans and other living creatures require micronutrients in minor quantities to orchestrate all physiological functions of their body, as they cannot produce these by themselves (Lieberman and Bruning 1990). Therefore, this chapter deals with mineral bioavailability and bioaccessibility from cereal grains and legumes consumed in India as staple food. Studies of bioavailability of minerals from cereals and legumes have been discussed.

5.2 Current Scenario of Cereals and Legumes in India

5.2.1 Cereals

India is one of the largest producers of cereals in the world, and cereal grains are the major staple food in every region of India. Several authors have reported the importance of cereals in Indian diet by describing nutritional value and production data of the cereals. From the civilization of human beings cereal grains have played a significant role in the survival of human population. Major cereals, including rice, wheat, maize, sorghum and millets, are the principal contributors of human diets all around the world. More than half of the world's population fulfill their caloric intake requirements by consuming cereals and cereal-based food products. India produced about 228.4 million tons of different cereal grains in year 2015–2016, including major as well as minor cereal crops. After green revolution, India is now self-reliant in food grains production (Ahmad et al. 2015). In addition, Swaminathan and Bhavani (2013) reported about the scenario of grain production in India and the increase in the production of grains in India due to the green revolution. According to them, production rate of grains (i.e. rice, wheat and other cereals and pulses) between 1950–1951 and 2006–2007 increased significantly, with an annual rate of 2.5% with increasing growth rate of population. Also, they stated that rapid increase in population caused decline in the per capita availability of cereals and pulses. In 1990–1991, consumption of grains was 468 g per capita per day, while it was recorded as 412 g per capita per day in 2005–2006. This data clearly demonstrates a 13% decline in the gram per capita per day during respective period. Availability of food grains also declined by 4.5% between the two periods 1991–2000 and 2001–2005. According to the report of, India is the second largest wheat producer in the world, with Punjab, Haryana, Western Uttar Pradesh, Rajasthan and Madhya Pradesh being its leading producers. In 2011–2012, the production of major cereals like rice, wheat, maize and bajra stood at 104.32 MT, 93.90 MT, 21.57 MT and 10.06 MT, respectively. Also, India's export of cereals was in excess of Rs 200,000 million during the same period.

5.2.2 Legumes

India contributes the major share of total production of pulses in the world, with chickpea, pigeon pea, mung bean and urd bean as principal legume crops. Owing to the fact that India is foremost in vegetarian dietary habits, production of legumes in India is done in almost every region. In this perspective, Singh et al. (2015) reported that India is the largest producer and consumer of pulses in the world. In India 29% area is used for the production of pulses which contributes to a total of 19% of the world's production. As discussed earlier its 2.9 % for Tamilnadu, every region of India contributes to the production of pulses and legumes, and among these, major zones that produce pulses are in the states of Madhya Pradesh, Maharashtra, Rajasthan, Uttar Pradesh, Karnataka, Andhra Pradesh, Chhattisgarh, Bihar and Tamil Nadu, which contribute 20.3, 13.8, 16.4, 9.5, 9.3, 7.9, 3.8, 2.6 and 2.9% respectively. In this context India produced 19.5 MT pulses, which included chickpea, pigeon pea, mung bean and urd bean, and they contributed 39, 21, 11, and 10% respectively of the total production of pulses in the world during 2013–2014. Uttar Pradesh contributes significantly to the pulses production, and its share to the national pulses security is 21.8% with 3.196 MT. Other two leading states are Madhya Pradesh and Rajasthan with 19.5% and 13.6% production share of India. Only eight states contribute 90% of total pulses production. As per these estimates, the deficit of pulses will be to the tune of 24.9 MT by 2020.

Although India is the largest producer of the pulses, to fulfill the demand of the population, India is importing 2.5–3.5 MT pulses every year. Now the major apprehension for India is the increasing trend to import the pulses from other countries, which is directly affecting the economy of India to a large extent. With this situation the requirement of the pulses will be increased tremendously, and by 2050, it would be 26.50 MT. To overcome this problem, several strategies need to be introduced, and to fulfill the domestic requirements, additional production of pulses would be the most vital objective of Indian dogma and socioeconomic policy. Therefore, one of the major strategies to increase the productivity is effective pest management control with major concerning factors such as inhibition of abiotic stresses, unforeseen climatic variations, introduction of biotechnological tools for the development of insect- and pest-resistant species and enhancement of vital and valuable micronutrients in the soil. The foremost breach between domestic demand and supply of pulses is mainly highlighted and lack in production of pulses has been accredited to number of factors. Also, other monetary factors such as only few assured markets for procurement of pulses, futile policy of Indian government for the procurement of pulses from farmers, lack of least sustenance price of pulses and trade liberalization make pulses cultivation less remunerative in comparison with other crops and agricultural produces (Narayan and Kumar 2015).

5.3 Mineral Deficiency and Bioavailability

Minerals (iron, zinc and calcium) are important microelements that play various fundamental roles in the human body (Chawla et al. 2016). Micronutrient deficiency is a serious risk to the health and efficiency of billions of people of both developed and developing countries, even though it is largely preventable (Diosady and Rutkowski 2007). In several locations of India as well as in the rest of the world, micronutrient malnutrition is very severe. India is predominantly vegetarian in dietary habits and heavily depends upon vegetative sources to meet its daily nutrient requirements (Singh et al. 2015). Therefore low intake and less bioavailability of micronutrients are the major reasons for micronutrient malnutrition, which leads to many life-threatening diseases. Women and children require more micronutrients for reproduction and growth, respectively (Dostal et al. 2012).

5.3.1 Iron Deficiency and Bioavailability

National productivity of a country is always measured or determined by the nutritional status of the population, which directly influences the eminence and efficiency of the people. Iron-deficiency anemia is a major risk for the billions of people of every age, and it is in the top target disease for cure and prevention for World Health Organization (Pereira et al. 2015). In developing countries, every second pregnant woman and about 40% of preschool children are estimated to be anemic (WHO 2008). Strategies to combat mineral deficiency include supplementation with medicinal form as well as fortification of foods with micronutrients (Hurrell 1997). Food fortification is considered as the most cost-effective, sustainable and optimal strategy in the battle to combat these micronutrient deficiencies (Gupta et al. 2015). Being a member of the transition metal series, soluble mineral salts of iron, zinc and calcium are highly reactive and potentially affect the taste, color and physicochemical stability of fortified foods (lipid oxidation). Also, the problem of their precipitation during gastrointestinal transit and absorption in duodenum results into lower bioavailability of added minerals. Efficacy of types of iron compounds that are used for supplementation or fortification is strongly influenced by mineral absorption enhancers and inhibitors in the diet (Carrier et al. 2002; Sazawal et al. 2006; Werner et al. 2011; Dostal et al. 2012). In addition, iron-deficiency anemia is a major cause of maternal deaths in all over the world. Almost 50% of maternal deaths are directly related to iron-deficiency anemia, which is a serious concern across the world. In India, National Family Health Survey (NFHS 2005–2006) revealed that iron-deficiency anemia occurs in 70% of children aged 6–59 months and 55.3% of married women. It was found that breastfeeding mothers (63.2%) were more prone to iron-deficiency anemia and 58.7% of pregnant women were also found severely anemic (Anand et al. 2015). Furthermore, National Nutrition Monitoring Bureau also studied the status of hemoglobin of diverse physiological groups and they found that the level of hemoglobin was very low, compared to that recommended by World Health Organization (WHO), and anemia was also diagnosed to a large extent. According to their report, lactating mothers (78%), pregnant women (75%), adolescent girls (about 70%) and preschool children (67%) were more susceptible to iron-deficiency anemia. Therefore, it can be concluded that Indian population is living with iron-deficiency anemia, imperiling child growth, which directly influences the eminence and efficiency of the people and in turn affects the development and economic productivity of the country. In India, the major reason for iron-deficiency anemia is insufficient ingestion of dietary iron and the resultant low bioavailability of iron from foodstuffs. In this context, Tripathi and Patel (2013) revealed the Indian status of iron-deficiency anemia. According to their observation, preschool children (60–70%), pregnant women (70–80%) and lactating women (about 50%) suffered from iron-deficiency anemia and the major reason for iron-deficiency anemia is the ingestion of plant-based diets at a large extent with reduced bioavailability of iron.

5.3.2 Calcium Deficiency and Bioavailability

Calcium is a vital microelement of mineralized tissues and is essential for the normal growth and development of bones for people of all age groups. Intake of calcium is significant for the health of human bones and to minimize bone loss among the elderly (Matkovic et al. 2005). The major component of human bones and teeth is calcium (99%); serum, extravascular fluid, muscles, and other tissues of body contain the remaining 1% calcium for their functioning (Baker et al. 1999). Therefore, intake of sufficient amounts of calcium always increases the metabolization of bones and also significantly improves overall vigor of bones (Anderson 2001). Moreover, calcium is also significant for various vital functions of human body such as nerve conduction, muscle contraction and relaxation, mitosis, blood coagulation and structural sustenance of the skeleton system (Miller et al. 2001). Therefore, calcium is one of the major minerals required to human body and must be provided regularly in the diet (Allen et al. 2006). With this perspective, according to a survey conducted by the number of hip fractures across the world due to calcium deficiency is 1.66 million, and it is well anticipated that the number will increase four times in 2050 with increasing number of elderly people across the world. Low intake of foods rich in calcium and the inability to absorb and metabolize calcium, or kidney, intestinal and liver diseases are the major causes of deficiency in children and adolescents because of increased demand for minerals during development. Furthermore, deficiency of dietary calcium and low intake of calcium lead to various bone-related diseases such as osteoporosis, osteomalacia, hypertension, colon cancer and obesity (Tunick 1987; Berner et al. 1990; Zemel and Miller 2004). In India, the number of people who are not able to take sufficient amount of dietary calcium which does not meet with RDA value of calcium is increasing day by day, which results in bone-related diseases. In recent years, Indian government and people are more concerned about calcium deficiency and are also aware of the importance of calcium for healthier bones and teeth. Therefore, calcium-fortified food products are available at large extent in Indian markets. The mean accretion of calcium is 80 mg/day during first year of life, rising slightly to 100 mg/day during the second year of life, and it remains consistent until 4–5 years and rises much more rapidly during puberty, when it can reach levels as high as 300–400 mg/d (Abrams 2010). Calcium accumulation in bone during childhood is about 150–200 mg/day and may be as high as 400–500 mg/day during pubertal growth spurt (Matkovic 2005). There is general agreement that extra dietary calcium is required for optimum skeletal growth at this time, with recommended intakes at puberty around 1200 mg or more.

5.3.3 Zinc Deficiency and Bioavailability

Zinc is one of the important transition metals which is essentially required for the growth of vital cells, tissue differentiation, maintenance of bone health, integrity of skin, proper functioning of immune system and sexual maturation. Also, it is a vital component of various enzymes and plays a significant role in various enzymatic reactions. Zinc has a rapid discrepancy which is mainly differentiated on the bases of human immune system and digestive tract (Gropper and Smith 2012). Zinc deficiency is a serious threat and it is more predominant across the world (Etcheverry et al. 2004). In developing countries, zinc deficiency and its related diseases are a major and serious health issue to the population of every age group, and every year, more people are diagnosed with zinc deficiency and its related diseases. Moreover, pregnant women are more prone to zinc deficiency, and infants born to zinc-deficienct women had neural tube defects with increased rate of infant and child mortality. Respiratory tract disease like pneumonia (Barnett et al. 2010; Dey et al. 2010), diarrhea (Luabeya et al. 2007) and high incidence of child stunting (Umeta et al. 2000) are mainly attributed to zinc deficiency. Bioavailability of minerals is always influenced by the antinutritional factors present in the foods. Therefore, bioavailability of zinc is always affected by the food composition and antinutritional factors present in dietary foods (Chawla et al. 2016). Research on zinc bioavailability suggested that introduction of animal proteins always improves intestinal absorption of zinc during simulated gastrointestinal digestion, which results in increased zinc bioavailability (Sandstrom and Lonnerdal 1989). International Zinc Nutrition Consultative Group (IZiNCG) published a report on zinc bioavailability from plant-based diets which have phytate contents. If dietary zinc intake is equivalent to the recommended daily allowance from which only 27% of intestinal zinc occur in adult male and for adult women are 35% respectively which is directly influenced by the phytate content and molar ratio of zinc. Increased phytate to molar ratio of zinc always resulted in reduced bioavailability of zinc and absorption rates (Holt and Brown 2004).

In India, the status of zinc deficiency is very serious and is more predominant in children, adolescent girls and pregnant women. Recent and advanced researches on zinc deficiency revealed that children from low socioeconomic groups are more prone to zinc deficiency and its related diseases. In India, five states like Orissa (51.3%), followed by Uttar Pradesh (48.1%), Gujarat (44.2%), Madhya Pradesh

(38.9%), and Karnataka (36.2%) are facing severe zinc deficiency, and it was also found that 45% population of the adolescent girls are facing low concentration of zinc in plasma. In Haryana, almost 64.6% pregnant women are diagnosed with severe zinc deficiency due to low ingestion of dietary zinc and reduced bioavailability. Moreover, nonpregnant women are also principally influenced by the zinc deficiency, and 41% of this population is diagnosed with significant zinc deficiency with low serum zinc concentrations. Therefore, the present status of zinc in populations of different age groups in India significantly requires more government and scientific policies. Also, extensive surveys are essentially required to identify the factors triggering zinc deficiency to a large extent. More studies are required to estimate and diagnose zinc deficiency among different age groups and to contrive approaches to combat this nutritional problem in India (Akhtar 2013).

5.4 Mineral Bioavailability from Cereals and Legumes

Cereals and legumes are principal sources of edible proteins, dietary fibers and various transition metals for vegetarian population. Despite these essential factors, cereals and legumes contain large amount of polyphenols and flavonoids. These components are known as antinutritional components which are the principal reason for reduced bioavailability of vital minerals from cereals and legumes. Various scientific approaches revealed that these polyphenols and flavonoids have a tendency to bind with minerals like iron, zinc and calcium and reduce their solubility and absorption in gastrointestinal tract, which results in reduced bioavailability of minerals. Also, reduced bioavailability in turn decreases the nutritional value of cereals- and legumes-based food products like infant formulas during industrial production (Sandberg 2002). Numerous studies revealed bioavailability and bioaccessibility of minerals from different legumes; however, there are only few reports on mineral bioavailability from cereals and legumes which are mainly consumed in India. In a study of bioavailability of iron and zinc from 25 raw and cooked food samples of legumes, cereal, meat and milk products of Fiji, Singh et al. (2016) revealed reduced bioavailability of minerals with lots of variations in bioavailability of minerals. Highest variation of iron content in raw food samples was observed and it ranged from 2.19 ± 0.04 to 0.93 ± 0.03 mg/100 g in legumes. The raw black eye bean, cheese and fish showed high zinc content up to 8.85 ± 0.01, 12.93 ± 0.26 and 172.03 ± 5.09 mg/100 g, respectively. Pulses and cereals showed high level of ionizable iron. They stated that zinc showed reduced bioavailability in cereals as compared to pulses and bioavailability in yellow split and Bengal gram was observed to be 4.02% and 17.40%, respectively. Moreover, in cheese samples, 17.40% zinc bioavailability was recorded, whereas cooked rice showed significantly increased bioavailability in comparison with white bread and milk, respectively. Lowest mineral bioavailability, i.e. 0.84%, was observed in fish. Hemalatha et al. (2007a) studied iron and zinc contents and their bioaccessibility in cereals and pulses consumed in

India. They selected cereals rice (*Oryza sativa*), finger millet (*Eleusine coracana*), sorghum (*Sorghum vulgare*), wheat (*Triticum aestivum*) and maize (*Zea mays*); and pulses chickpea (*Cicer arietinum*) – whole and decorticated, green gram (*Phaseolus aureus*) – whole and decorticated, decorticated black gram (*Phaseolus mungo*), decorticated red gram (*Cajanus cajan*), cowpea (*Vigna catjang*) and French bean (dry; *Phaseolus vulgaris*). They also reported variation in the mineral content of different selected cereals and legumes. Moreover, dialyzability of minerals was also observed, and they found low bioavailability of cereals in comparison with legumes. A significant negative correlation between antinutritional factors and zinc bioavailability was observed in pulses. In the same year, Perales et al. (2007) reported the availability of iron from milk-based formulas and fruit juices containing milk and cereals. Solubility and dialyzability of iron were estimated using the combination of simulated gastrointestinal method and Caco-2 cells. Cells were used to estimate cellular retention, transport and uptake. Erdman (1981) reported bioavailability of trace minerals from cereals and legumes. He reported that foods of plant origin always show less or reduced bioavailability of trace minerals in comparison with animal origin. Furthermore, endogenous and exogenous factors of cereals and legumes are responsible for reduction of mineral absorption from cereal and legumes.

In addition, Wang et al. (2009) investigated the effect of cooking and dehulling on nutrients and antinutritional factors of several varieties of lentils (*Lens culinaris*) and observed significant difference in nutritional value and mineral content, respectively. Also they reported the influence of cooking and dehulling on compositional factors and minerals of legumes. They suggested that the influence of antinutritional factors was high, which resulted in decreased nutritional value. Hemalatha et al. (2007b) evaluated the influence of exogenous iron and calcium equivalent to their supplemental levels on the bioaccessibility of zinc from food grains and observed significant reduction in bioavailability of respective mineral content. Also, negative stimulus of exogenous iron and calcium was observed almost comparable for both raw and cooked grains. However, there was no negative effect on the bioavailability of respective minerals in adequate level of exogenous iron and calcium. Lestienne et al. (2005) also assessed different processing treatments of cereals and legumes and the effect of these treatments was observed on different minerals of respective cereals and legumes. In their research, they observed negative effects of processing treatments on mineral content of respective cereals and legumes. Chawla et al. (2016) also observed the impact of solid- state fermentations on mineral content and bioavailability of iron, zinc and calcium of black-eyed pea. Significant improvement in bioavailability due to solid-state fermentation was assessed and the bioavailability was evaluated by the combination of simulated gastrointestinal method and Caco-2 cells.

5.5 Biotechnological Tools for Enhancement of Mineral Bioavailability

In recent years, due to modernization in the field of agriculture for increasing productivity and sustainability of crops, different biotechnological processes, tools and methods are used. These biotechnological processes introduced a new aspect to such modernism, offering efficient and cost-effective means to produce a various array of novel and tools.

A variety of technical, regulatory and indeed perception challenges inherent in the perceived and real challenges of complex modifications are involved in developing plants with improved traits. Both traditional and modern plant breeding–based techniques are used to produce plants with desired quality traits. Molecular and genomic technologies are improved continually, contributing to the acceleration of product development (McGloughlin 2008). The newly recognized harvest plus consortium, that is made up of different review groups on international agricultural research (the CGIAR) centers and other national and international organizations, is focusing on combining traditional plant-breeding methods with biotechnological techniques to provide micronutrient-enriched staple food crops (rice, wheat, maize, bean, sweet potato and cassava) for distribution to farmers in developing countries (Welch 2005). Several researchers have introduced various biotechnological aspects to improve the mineral bioavailability from and nutritional values of food grains and pulses. To overcome iron deficiency, Goto et al. (2000) and Lucca et al. (2001) introduced the gene for ferritin under the control of a constitutive endosperm-specific promoter which resulted in the production of grains that contained three times more iron than normal rice. Further, an attempt had been carried out to modify iron transport within the plant to enhance the iron content in the grain (Lucca et al. 2001). Drakakaki et al. (2005) demonstrated that endosperm-specific co-expression of recombinant soybean ferritin and *Aspergillus* phytase in maize significantly improved the bioavailability of iron. Goto et al. (2000) also found enhanced iron content in lettuce. In addition, Welch (2005) also revealed that genetic modification of cereals and legumes significantly increased the bioavailability of micronutrients from their edible portions as they are eaten; this method holds much promise as a sustainable solution to decrease malnutrition due to micronutrients.

Lonnerdal (2003) disclosed that biotechnology is often employed in many ways to increase the element content of cereals and legumes. This might be achieved by insertion of genes that code for trace element–binding proteins, overexpression of storage proteins already present and/or increased expression of proteins that are responsible for trace element uptake in plants. However, even higher levels of expression might not significantly increase the iron and zinc contents unless several atoms of trace elements are bound per protein molecule. Another prospect is to introduce a macromolecule that specifically enhances trace element absorption even within the presence of natural inhibitors, thus improving bioavailability. Genetically modifying plants to reduce the amount of inhibitors of trace element absorption like phytate is another approach. Increasing the expression of antioxidants that enhance trace element absorption like ascorbic acid is also a possibility, though this has received restricted attention thus far. Iron absorption could also be increased by higher ascorbic or citric acid content; however, it needed overexpression of enzymes that are involved within the synthetic pathways. Finally, a combination of all of these approaches complemented with conventional breeding techniques might prove productive. For biofortification of minerals, various researchers targeted polished rice because it is the most consumed staple food in the world and rice proteins have low allergenicity. Some scientists enhanced the iron accumulation in seeds of rice by the expression of ferritin genes Soyfer H1 and Soyfer H2 under the control of endosperm-specific promoters (Goto et al. 1999; Qu et al. 2005; Paul et al. 2012); they also enhanced the iron level by the overexpression of NAS gene in rice plants (Masuda et al. 2008; Ishimaru et al. 2010; Johnson et al. 2011). Suzuki et al. (2008) also increased iron uptake and translocation in rice plants by the ISD3 genes.

Several researchers have contributed to enhance the mineral level in staple crops and succeeded in this by using biotechnological tools. Still in India, food safety and cost-effectiveness are the major points to be solved.

5.6 Conclusion

Micronutrients are important microelements which play various fundamental roles in the human body and are equally essential for other living beings in small quantities to orchestrate the physiological functions of their body. Cereal grains and legumes are rich source of minerals, but the presence of antinutritional factors and mineral-binding components reduce the bioavailability of their minerals. Several biotechnological techniques and processing treatments have been used tremendously to improve the mineral bioavailability of cereals and legumes. The present status of minerals in populations of different age groups in India essentially required more government and scientific policies and approaches. Also, extensive surveys are essential to identify the factors triggering mineral deficiencies. More studies are required to estimate and diagnose mineral deficiencies among different age groups and to contrive approaches to combat this nutritional problem in India.

References

Abrams SA (2010) Calcium absorption in infants and small children: methods of determination and recent findings. Nutrition 2(4):474–480

Afify AEMMR, El-Beltagi HS, El A, Salam SM, Omran AA (2011) Bioavailability of iron, zinc, phytate and phytase activity during soaking and germination of white sorghum varities. PLoS One 6(10):e25512

Ahmad IM, Samuel E, Makama SA et al (2015) Trend of area, production and productivity of major cereals: India and Nigeria scenario. Res J Agric For Sci 3(2):10–15

Akhtar S (2013) Zinc status in south Asian populations-an update. J Health Popul Nutr 31(2):139–149

Allen L, Benoist BD, Dary O, Hurrell R (2006) Guidelines on food fortification with micronutrients. World Health Organization and Food and Agriculture Organization of the United Nations (WHO/FAO)

Anand T, Rahi M, Sharma P, Ingle GK (2015) Issues in prevention of iron deficiency anemia in India. Nutrition 30(7–8):764–770

Anderson JJ (2001) Calcium requirements during adolescence to maximize bone health. J Am Coll Nutr 20(2):186–191

Baker SS, Cochran WJ, Flores CA et al (1999) American academy of pediatrics, committee on nutrition, calcium requirements of infants, children, and adolescents. Pediatrics 104:1152–1157

Barnett BJ, Hamer D, Meydani NS (2010) Low zinc status: a new risk factor for pneumonia in the elderly. Nutr Rev 68(1):30–37

Berner LA, McBean LD, Lofgren PA (1990) Calcium and chronic disease prevention: challenges to the food industry. Food Technol 44(3):50–70

Carrier J, Aghdassi E, Cullen J, Allard JP (2002) Iron supplementation increases disease activity and vitamin E ameliorates the effect in rats with dextran sulfate sodium-induced colitis. J Nutr 132(10):3146–3150

Chawla P, Bhandari L, Sadh PK, Kaushik R (2016) Impact of solid state fermentation (*Aspergillus oryzae*) on functional properties and mineral bioavailability of black eyed pea (*Vigna unguiculata*) seed flour. Cereal Chem. doi.1094/CCHEM-05-16-0128-R

Dey CA, Shahidullah M, Mannan AM et al (2010) Maternal and neonatal serum zinc level and its relationship with neural tube defects. J Health Popul Nutr 28(4):343–350

Diosady LL, Rutkowski K (2007) Vitamin a stability in triple fortified salt. Food Res Int 40(1):147–152

Dostal A, Chassard C, Hilty FM et al (2012) Iron depletion and repletion with ferrous sulfate or electrolytic iron modifies the composition and metabolic activity of the gut microbiota in rats. J Nutr 142(2):271–277

Drakakaki G, Marcell S, Glahn RP, Lund EK et al (2005) Endosperm-specific co-expression of recombinant soybean ferritin and *Aspergillus* phytase in maize results in significant increases in the levels of bioavailable iron. Plant Mol Bio 59:869–880

Erdman JW (1981) Bioavailability of trace minerals from cereals and legumes. Cereal Chem 58(1):21–26

Etcheverry P, Miller DD, Glahn PR (2004) A low-molecular-weight factor in human milk whey promotes iron uptake by caco-2 cells. Human Nutr Meta 134(1):93–98

Ghavidel RH, Prakash J (2007) The impact of germination and dehulling on nutrients, antinutrients, *in vitro* and calcium bioavailability and *in vitro* starch and protein digestibility of some legume seeds. LWT Food Sci Technol 40:1292–1299

Goto F, Yoshihara T, Shigemoto N, Toki S, Takaiwa F (1999) Iron fortification of rice seed by the soybean ferritin gene. Nat Biotechnol 17:282–286

Goto F, Yoshihara T, Saiki H (2000) Iron accumulation and enhanced growth in transgenic lettuce plants expressing the iron-binding protein ferritin. Theor Appl Genet 100:658–664

Gropper SS, Smith JL (2012) Advanced nutrition and human metabolism, 6th edn. Wadsworth Cengage Learning, Belmont

Gupta C, Chawla P, Arora S, Tomar SK, Singh AK (2015) Iron microencapsulation with blend of gum arabic, maltodextrin and modified starch using modified solvent evaporation method–milk fortification. Food Hydrocoll 43:622–628

Hemalatha S, Platel K, Srinivasan K (2007a) Zinc and iron contents and their bioaccessibility in cereals and pulses consumed in India. Food Chem 102:1328–1336

Hemalatha S, Platel K, Srinivasan K (2007b) Influence of heat processing on the bioaccessibility of zinc and iron from cereals and pulses consumed in India. J Trace Elem 21:1–7

Holt C, Brown KH (2004) International zinc nutrition consultative group (IZiNCG). Technical document. Assessment of the risk of zinc deficiency in populations and options for its control. Food Nutr Bull 25(2):94–203

Hurrell RF (1997) Preventing iron deficiency through food fortification. Nutr Rev 55(6):210–222

Ishimaru Y, Masuda H, Bashir K et al (2010) Rice metal-nicotianamine transporter, OsYSL2, is required for the long-distance transport of iron and manganese. Plant J 62:379–390

Johnson AAT, Kyriacou B, Callahan DL, Carruthers L et al (2011) Constitutive overexpression of the OsNAS gene family reveals single gene strategies for effective iron- and zinc-biofortification of rice endosperm. PLoS One 6:e24476
Kaur M, Sandhu KS, Lim ST (2010) Microstructure, physicochemical properties and in vitro digestibility of starches from different Indian lentil (*Lens culinaris*) cultivers. Carbohydr Polym 79:349–355
Lane DJR, Merlot AM, Huang MLH et al (2015) Cellular iron uptake, trafficking and metabolism: key molecules and mechanisms and their roles in disease. Biochim Biophys Acta 1853:1130–1144
Lestienne I, Icard-Verni CE, Mouquet C et al (2005) Effects of soaking whole cereal and legume seeds on iron, zinc and phytate contents. Food Chem 89:421–425
Lieberman S, Bruning N (1990) The real vitamins and minerals book. In: The definitive guide to designing your personal supplement program, 4th edn. Avery publishers (USA, New York
Lonnerdal BO (2003) Trace element undernutrition: biology to interventions. J Nutr 133:1490–1493
Luabeya KK, Mpontshane N, Mackay M et al (2007) Zinc or multiple micronutrient supplementation to reduce diarrhea and respiratory disease in South African children: a randomized controlled trial. PLoS One 2(6):e541
Lucca P, Hurrel R, Potrykus I (2001) Approaches to improving the bioavailability and level of iron in rice seeds. Theor Appl Genet 102:392–397
Masuda H, Suzuki M, Morikawa KC et al (2008) Increase in iron and zinc concentrations in rice grains via the introduction of barley genes involved in phytosiderophore synthesis. Rice 1:100–108
Matkovic V, Goel PK, Badenhop-Stevens NE et al (2005) Calcium supplementation and bone mineral density in females from childhood to young adulthood: a randomized controlled trial. Am J Clin Nutr 81(1):175–188
McGloughlin MN (2008) Nutritionally improved agricultural crops. Plant Physiol 147:939–953
Miller GD, Jarvis JK, MeBean ID (2001) The importance of meeting of calcium needs with foods. J Am Coll Nutr 20(2):168–185
Nakao M, Yamamoto H, Nakahasi O, Ikeda S et al (2015) Dietary phosphate supplementation delays the onset of iron deficiency anemia and affects iron status in rat. Nutr Res 35:1016–1024
Narayan P, Kumar S (2015) Constraints of growth in area production and productivity of pulses in India: an analytical approach to major pulses. Indian J Agric Res 49(2):114–124
Onwurafor EU, Onweluzo JC, Ezeoke AM (2014) Effect of fermentation methods on chemical and microbial properties of mung bean (*Vigna radiata*) flour. Niger Food J 32(1):89–96
Paul S, Ali N, Gayen D et al (2012) Molecular breeding of Osfer2 gene to increase iron nutrition in rice grain. GM Crops Food 3:310–316
Perales S, Barbera R, Lagarda MJ, Farre R (2007) Availability of iron from milk-based formulas and fruit juices containing milk and cereals estimated by *in vitro* methods (solubility, dialysability) and uptake and transport by Caco-2 cells. Food Chem 102:1296–1303
Pereira DI, Lederer B, Powell JJ (2015) A balanced salt solution that prevents agglomeration of nano iron oxo-hydroxides in serum-free cellular assays. Mat Res Exp. doi:10.1088/2053-1591/2/1/015403
Powell JJ, Bruggraber S, Faria N, Poots LK et al (2014) A nano disperse ferritin-core mimetic that efficiency corrects anemia without luminal iron redox activity. Nanomed Nanotech BioMed 10:1529–1538
Qu LQ, Yoshihara T, Ooyama A, Goto F, Takaiwa F (2005) Iron accumulation does not parallel the high expression level of ferritin in transgenic rice seeds. Planta 222:225–233
Sandberg AS (2002) Bioavailability of minerals in legumes. Br J Nutr 88(3):281–285
Sandstrom B, Lonnerdal B (1989) Promoters and antagonists of zinc absorption. ILSI Hum Nutr Rev 1:57–78
Sazawal S, Black RE, Ramsan M et al (2006) Effects of routine prophylactic supplementation with iron and folic acid on admission to hospital and mortality in preschool children in a high

malaria transmission setting: community-based, randomized placebo-controlled trial. Lancet 367(9505):133–143
Singh AK, Singh SS, Prakash V et al (2015) Pulses production in India: present status, bottleneck and way forward. J Agric Search 2(2):75–83
Singh P, Prasad S, Aalbersberg W (2016) Bioavailability of Fe and Zn in selected legumes, cereals, meat and milk products conjumed in Fiji. Food Chem 207:125–131
Suzuki M, Morikawa KC, Nakanishi H et al (2008) Transgenic rice lines that include barley genes have increased tolerance to low iron availability in a calcareous paddy soil. J Soil Sci Plant Nutr 54:77–85
Swaminathan MS, Bhavani RV (2013) Food production & availability – essential prerequisites for sustainable food security. Indian J Med Res 138(3):383–391
Torres J, Rutherfurd SM, Munoz LS et al (2016) The impact of heating and soaking on the *in vitro* enzymatic hydrolysis of protein varies in different species of tropical legumes. Food Chem 194:377–382
Tripathi B, Platel K (2013) Feasibility in fortification of sorghum (*Sorghum bicolor* L. *Moench*) and pearl millet (*Pennisetum glaucum*) flour with iron. LWT Food Sci Technol 50(1):220–225
Tunick MH (1987) Calcium in dairy products. J Dairy Sci 70(11):2429–2438
Umeta M, West CE, Haidar J et al (2000) Zinc supplementation and stunted infants in Ethiopia: a randomised controlled trial. Lancet 355(9220):2021–2026
Viadel B, Barbera R, Farre R (2006) Effect of cooking and legume species upon calcium, iron and zinc uptake by Caco-2 cells. J Trace Elem Med Biol 20:115–120
Wang N, Hatcher DW, Toews R, Gawalko EJ (2009) Influence of cooking and dehulling on nutritional composition of several varieties of lentils (*Lens culinaris*). LWT Food Sci Technol 42:842–848
Welch MR (2005) Biotechnology, biofortification and global health. Food Nutr Bull 26:4
Werner T, Wagner SJ, Martinez I, Walter J et al (2011) Depletion of luminal iron alters the gut microbiota and prevents Crohn's disease like ileitis. Gut 60(3):325–333
WHO (2008) The World health report on: reducing risks, promoting healthy life. WHO, Geneva
Zemel MB, Miller SL (2004) Dietary calcium and dairy modulation of adiposity and obesity risk. Nutr Rev 62(4):125–131

Recent Biotechnological Approaches to Study Taxonomy of Legume Nodule Forming Rhizobia

6

Pooja Suneja, Joginder Singh Duhan, Namita Bhutani, and Surjit Singh Dudeja

Abstract

Nodule forming bacteria commonly referred as rhizobia establish N_2-fixing symbiosis with legumes. Legumes and their rhizobial symbionts are of prime importance due to their ability to improve the nitrogen fertility of soils. To discriminate rhizobia into species and subspecies, identification and classification of bacteria and particularly rhizobia using polyphasic approaches is becoming the most accepted technology. Phenotypic and biochemical analysis includes: Gram staining, cell morphology and motility, oxidase and catalase activity, Biolog tests, NaCl tolerance, antibiotic, pH and temperature profiling, fatty acid composition, nodulation and nitrogen fixation studies. Genetic analysis includes DNA–DNA hybridization, G + C contents, and PCR analysis using large number of genes including housekeeping genes. Sequence analysis of 16SrRNA, 16-23S rRNA ITS region, other housekeeping genes, and complete genome sequencing and comparison with existing database using various bioinformative techniques is frequently being used. Further multilocus sequence analysis and matrix-assisted laser desorption ionization are widely used to study taxonomy of legume nodule forming rhizobia. In the early 1970s only 1 rhizobial genera (*Rhizobium*) with 6–7 species was recognized. But now the use of these multiple approaches has resulted in the identification of 16 α and β rhizobial genera. These genera belong to α and β proteobacteria – *Rhizobium*, *Shinella*, *Ensifer (Sinorhizobium)*,

P. Suneja • N. Bhutani
Department of Microbiology, Maharishi Dayanand University, Rohtak 124 001, Haryana, India

J.S. Duhan
Department of Biotechnology, Chaudhary Devi Lal University, Sirsa 125 005, Haryana, India

S.S. Dudeja (✉)
Department of Bio and Nanotechnology, Guru Jambheshwar University of Science and Technology, Hisar 125 001, Haryana, India
e-mail: ssdudeja@gmail.com

S.K. Gahlawat et al. (eds.), *Plant Biotechnology: Recent Advancements and Developments*, DOI 10.1007/978-981-10-4732-9_6

Aminobacter, *Neorhizobium*, *Allorhizobium*, *Mesorhizobium*, *Phyllobacterium*, *Bradyrhizobium*, *Methyllobacterium*, *Microvirga*, *Azorhizobium*, *Devosia*, *Ochrobacterum*, *Burkholderia*, and *Cuprivadus* belonging to seven families. About 176 species belonging to these rhizobial genera nodulating different legumes have been reported from different parts of the World. Most of the Indian legumes including pulse crops are nodulated by more than one species belonging to one or even more than one genera, though reports from India are very limited.

6.1 Introduction

Fabaceae, also known as *Leguminosae* consists of about 751 genera and about 20,000 species of trees, shrubs, vines, and herbs distributed over a wide range of ecological conditions of the world (Lewis et al. 2005; Fornasero et al. 2014). This includes important pulses, forage legumes, and medicinal plants. Legumes establishes nitrogen-fixing symbiosis with rhizobia, this plant family and their bacterial symbionts are the focus of intensive investigation to improve the nitrogen fertility of soils and model systems for studying plant microbe interactions at different levels of OMICS. In spite of this only about 20% of the legumes have been explored for nodulation and their ability to fix atmospheric nitrogen. The situation in India is still worse as only 1200 species of legumes have been explored. Therefore, for sustainable agricultural practices, concerted efforts are needed to explore new rhizobial-legume systems.

Symbiotic nitrogen fixation is established by series of large number of interactive steps between rhizobia and legumes. Better understanding of this association is very important for enhancing productivity in agriculture, as well as for exploitation of plant organogenesis, molecular processes controlling microbe recognition, and pathogen defense mechanism (Sheokand et al. 2012; Dudeja 2016). Legumes encode and synchronize all functions necessary for nodule development (Dudeja et al. 2012a). Commonly all these nitrogen-fixing bacteria belonging to different genera that form nodules on legume plants are called rhizobia. Majority of these bacteria belong to proteobacteria with α and β subclass to distinguish phylogenetic lineages of nodule-symbiotic proteobacteria (Dudeja and Narula 2008). Taxonomy of prokaryotes including α and β proteobacteria is becoming more important due to the exploration of more and more legumes. Phylogenetic studies are important to determine the exact taxonomic position of organisms in relations to other groups (EnTao et al. 2000; Lapage et al. 1992). Repositories of 16S rRNA genes and other housekeeping genes are increasing day by day indicating that 16S rRNA gene sequence is mainly being used in prokaryote taxonomy and phylogeny. However, 16S rRNA gene of α and β proteobacteria is highly conserved but it is difficult to delineate different species (Azevedo et al. 2015), so other complimentary approaches are also used.

6.2 Approaches Being Used to Address the Taxonomy of Rhizobia

Polyphasic taxonomy is the best taxonomic consensus to address the taxonomy of rhizobia, and in this phenotypic, genotypic, and phylogenetic data and information are combined. Strains sharing a high degree of similarity can be discriminated using this approach (Vandamme et al. 1996). The phenotypic data involves biochemical and physiological traits. Genotypic charaterization includes the information obtained from DNA as well as RNA: GC contents, DNA hybridization; restriction-fragment-length polymorphism; gene sequencing; and PCR DNA finger printing (Stackebrandt et al. 2002). Characterization and screening of nodule forming bacteria; using repetitive DNA sequences like BOX, ERIC sequences, 16S-23S rRNA intergenic spacer is also being used. However, limited database is available for comparison of results from different studies.

To further improve the identification process, analysis of conserved housekeeping genes is being used to more precisely detect the diversity. Housekeeping genes include the symbiotic genes which are nodulation genes (*nod* genes) and nitrogen fixation genes (*nif and fix*, *x* genes), which are usually located within symbiotic plasmids or genomic regions called symbiosis islands. The *nod* genes are classified as *nod*A, *nod*B, *nod*C, *nod*I, and *nod*J genes and are involved in the biosynthesis of *nod* factors (van Rhijn and Vanderleyden 1995; Suominen et al. 2001). The common *nod* genes form an operon in most α rhizobial strains and are often represented as *nod*ABCIJ. Due to the limited number of rhizobial genomes in databases, other genes like *atp*D, *dna*K, *gap*, *gln*A, *gln*II, *glt*A, *gyr*B, *pnp*, *rec*A, *rpo*B, and *thr*C of rhizobia have also been selected largely in the belief that they were housekeeping genes (Aoki et al. 2013). No comparisons between the phylogenies of these genes and phylogenies derived from whole genome data were made before using them. Although different subsets of these possible housekeeping genes are carefully chosen for further analysis in each genus, incongruent phylogenetic signals among these loci have been reported.

Another approach being used is MLSA (multilocus sequence analysis) methodology. Identification of prokaryotic organisms and the delineation of species for the taxonomy of microorganisms is the main focus of this analysis (Azevedo et al. 2015). In case of highly related species, where analysis of the 16S rRNA genes shows low resolution, there use of MLSA can properly discriminate the species into subspecies(Werner et al. 2015).

Mass spectrometry is another approach to identify and characterize bacteria (Jia et al. 2015). Matrix-assisted laser desorption ionization mass spectra were highly comparable with the 16S rDNA phylogenetic tree. Further attempts are being made to develop mass spectral database for further discrimination of all rhizobial species for accurate identification and classification.

6.3 Current Status of Rhizobial Classification

Rhizobia belongs to seven families of classes α and β proteobacteria, commonly referred as α and β rhizobia. Different rhizobial genera under α Rhizobia families are (1) *Rhizobiaceae* – *Rhizobium*, *Shinella*, *Ensifer (Sinorhizobium)*, *Allorhizobium*, *Aminobacter*, *Neorhizobium*; (2) *Phyllobacteriaceae* – *Mesorhizobium*, *Phyllobacterium*; (3) *Bradyrhizobiaceae* – *Bradyrhizobium*; (4) *Methyllobacteriaceae* – *Methyllobacterium*, *Microvirga*; (5) *Hypomicrobiaceae* – *Azorhizobium*, *Devosia*; (6) *Brucellaceae* – *Ochrobactrum* (Dudeja et al. 2012b; Weir 2012; ICSP 2013), while β rhizobia are from the family *Burkholderiaceae* – *Burkholderia*, *Cuprivadus* (Fig. 6.1).

6.3.1 Alpha Proteobacteria or α Rhizobia

6.3.1.1 Rhizobiaceae

All the rhizobia are not nodule forming; few species have been reported to be non-nodulating also, but in this review specifically nodulating species are being included.

6.3.1.1.1 *Rhizobium*

The genus *Rhizobium* was given by Frank (1889). All nodulating bacteria, till 1980, were classified under *Rhizobium* genus with different species but afterwards the concept of new genera based on growth characteristic and phylogenetic analyses was given. A large number of hosts nodulated by this genus are detailed in Table 6.1.

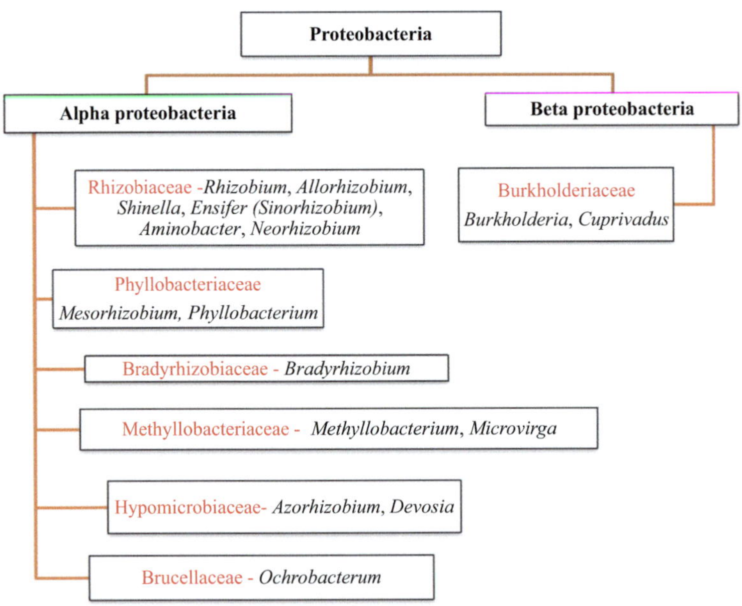

Fig. 6.1 Classification of rhizobia forming nodules in legumes

Table 6.1 Alpharhizobia belonging to Rhizobiaceae

S.No.	Rhizobial genera and spp.	Nodulating host	References
Rhizobium			
1.	*Rhizobium acidisoli*	*Phaseolus vulgaris*	Roman-Ponce et al. (2016)
2.	*Rhizobium alamii*	*Medicago ruthenica*	Berge et al. (2009)
3.	*Rhizobium alkalisoli*	*Caragana intermedia*	Lu et al. (2009a)
4.	*Rhizobium anhuiense*	*Vicia faba, Pisum sativum*	Zhang et al. (2015)
5.	*Rhizobium azibense*	*Phaseolus vulgaris*	Mnasri et al. (2014)
6.	*Rhizobium calliandrae*	*Caragana grandiflora*	Rincon-Rosales et al. (2013)
7	*Rhizobium cauense*	*Kummerowia stipulacea*	Liu et al. (2012)
8.	*Rhizobium ecuadorense*	*Phaseolus vulgaris*	Ribeiro et al. (2015)
9.	*Rhizobium endophyticum*	*Phaseolus vulgaris*	Lopez-Lopez et al. (2010)
10.	*Rhizobium etli*	*Phaseolus vulgaris* L.	Segovia et al. (1993)
11.	*Rhizobium fabae*	*Vicia faba*	Tian et al. (2008)
12.	*Rhizobium freirei*	*Phaseolus vulgaris*	Dall'Agnol et al. (2013)
13.	*Rhizobium galegae*	*Galega orientalis, G. officinalis*	Lindstrom (1989)
14.	*Rhizobium gallicum*	*Phaseolus vulgaris*	Amarger et al. (1997)
15.	*Rhizobium giardinii*	*Phaseolus vulgaris*	Amarger et al. (1997)
16.	*Rhizobium grahamii*	*Dalea leporine*	Lopez-Lopez et al. (2012)
17.	*Rhizobium hainanense*	Tropical legumes	Chen et al. (1997)
18.	*Rhizobium herbae*	Various legumes	Ren et al. (2011b)
19.	*Rhizobium huautlense*	*Sesbania herbacea*	Wang et al. (1998)
20.	*Rhizobium indigoferae*	*Kummerowia stipulacea*	Wei et al. (2002)
21.	*Rhizobium jaguaris*	*Calliandra grandiflora*	Rincon-Rosales et al. (2013)
22.	*Rhizobium laguerreae*	*Vicia faba*	Saidi et al. (2014)
23.	*Rhizobium leguminosarum*	*Cicer canariense*	Martinez-Hidalgo et al. (2015b)
24.	*Rhizobium leucaenae*	*Leucaena* spp.	Ribeiro et al. (2012)
25.	*Rhizobium loessense*	*Astragalus, Lespedeza* spp.	Wei et al. (2003)
26.	*Rhizobium lusitanum*	*Phaseolus vulgaris*	Valverde et al. (2006)
27.	*Rhizobium mayense*	*Calliandra grandiflora*	Rincon-Rosales et al. (2013)
28.	*Rhizobium mesoamericanum*	*Phaseolus vulgaris*	Lopez-Lopez et al. (2012)
29.	*Rhizobium mesosinicum*	*Albizia, Kummerowia, Dalbergia* spp.	Lin et al. (2009)
30.	*Rhizobium miluonense*	*Lespedeza* spp.	Gu et al. (2008)
31.	*Rhizobium mongolense*	*Medicago ruthenica* (L.)	van Berkum et al. (1998)

(continued)

Table 6.1 (continued)

S.No.	Rhizobial genera and spp.	Nodulating host	References
32.	*Rhizobium multihospitium*	Multiple leguminous hosts	Han et al. (2008)
33.	*Rhizobium paranaense*	*Phaseolus vulgaris*	Dall'agnol et al. (2014)
34.	*Rhizobium phaseoli*	*Phaseolus vulgaris*	Ramirez-Bahena et al. (2008)
35.	*Rhizobium pisi*	*Phaseolus vulgaris*	Ramirez-Bahena et al. (2008)
36.	*Rhizobium pongamiae*	*Pongamia pinnata*	Kesari et al. (2013)
37.	*Rhizobium puerariae*	*Pueraria candollei*	Boonsnongcheep et al. (2016)
38.	*Rhizobium qilianshanense*	*Oxytropis ochrocephala*	Xu et al. (2013)
39.	*Rhizobium sophorae*	*Sophora flavescens*	Jiao et al. (2015)
40	*Rhizobium sophoriradicis*	*Sophora flavescens*	Jiao et al. (2015)
41.	*Rhizobium sphaerophysae*	*Sphaero physasal sula*	Xu et al. (2011)
42.	*Rhizobium* spp.	Mungbean, urdbean, pigeonpea	Jagiriti (2005)
43.	*Rhizobium sullae*	*Hedysarum coronarium* L.	Squartini et al. (2002)
44.	*Rhizobium taibaishanense*	*Kummerowia striata*	Yao et al. (2012)
45.	*Rhizobium tibeticum*	*Oxytropis glabra*	Zhang et al. (2011)
46.	*Rhizobium tubonense*	*Neptunia natans*	Young et al. (2001)
47.	*Rhizobium undicola*	*Phaseolus vulgaris, Mimosa pudica, Indigo feraspicata*	Wang et al. (2011)
48.	*Rhizobium vallis*	*Phaseolus vulgaris, Mimosa pudica, Indigo feraspicata*	Wang et al. (2011)
49.	*Rhizobium vignae*	Multiple legume species	Ren et al. (2011a)
50.	*Rhizobium yanglingense*	Wild legumes	Tan et al. (2001)
Shinella			
1.	*Shinella kummerowiae*	*Kummerowia stipulacea*	Lin et al. (2008)
Ensifer			
1.	*Ensifer abri*	*Abrus precatorius*	Ogasawara et al. (2003)
2.	*Ensifer arbores*	*Acacia, Senegal, Prosopis chilensis*	Nick et al. (1999)
3.	*Ensifer fredii*	*Glycine max*	Scholla and Elkan (1984)
4.	*Ensifer garamanticus*	*Argyrolobium uniflorum, Lotus creticus, Medicago sativa, Lotus arabicus*	Merabet et al. (2010)
5.	*Ensifer glycinis*	*Astragalus mongholicus, Glycine max*	Yan et al. (2016)
6.	*Ensifer indiaense*	*Sesbania rostrata*	Ogasawara et al. (2003)
7.	*Ensifer kostiensis*	*Acacia senegal, Prosopis chilensis*	Nick et al. (1999)
8.	*Ensifer kummerowiae*	*Kummerowia* spp.	Wei et al. (2002)
9.	*Ensifer medicae*	*Medicago* spp.	Rome et al. (1996)

(continued)

Table 6.1 (continued)

S.No.	Rhizobial genera and spp.	Nodulating host	References
10.	*Ensifer meliloti*	*Medicago sativa*	de Lajudie et al. (1994)
11.	*Ensifer mexicanus*	*Acacia angustissima*	Lloret et al. (2007)
12.	*Ensifer numidicus*	*Argyrolobium uniflorum, Lotus creticus, Medicago sativa, Lotus arabicus*	Merabet et al. (2010)
14.	*Ensifer psoraleae*	*Psoralea corylifolia*	Wang et al. (2013a)
15.	*Ensifer saheli*	*Sesbania, Acacia* spp.	de Lajudie et al. (1994)
16.	*Ensifer sesbaniae*	*Sesbania cannabina*	Wang et al. (2013a)
17.	*Ensifer sojae*	*Glycine max*	Li et al. (2011)
18.	*Ensifer terangae*	*Sesbania, Acacia* spp.	de Lajudie et al. (1994)
19.	*Sinorhizobium americanum*	*Acacia* species	Toledo et al. (2003)
20.	*Sinorhizobium morelense*	*Leucaena leucocephala*	Wang et al. (2002)
Aminobacter			
1.	*Aminobacter anthyllidis*	*Anthyllis vulneraria*	Maynaud et al. (2012)
Neorhizobium			
1.	*Neorhizobium galegae*	*Galega, Astragalus, Caragana, Lotus, Medicago, Sesbania, Vigna*	Osterman et al. (2014) Mousavi et al. (2014)
Allorhizobium			
1.	*Allorhizobium undicola*	*Neptunia natans*	de Lajudie et al. (1998a)

Currently this genus consists of 52 rhizobial species. Different important Indian legumes being nodulated by this genera are *Phaseolus vulgaris*, *Vicia faba*, *Cicer canariense*, *Medicago* spp., *Trigonella*, *Pisum sativum*, *Sesbania*, mungbean, urdbean, and pigeonpea.

6.3.1.1.2 *Shinella*

The genus Shinella has only one species *Shinella kummerowiae* isolated from *Kummerowia stipulacea* and found to nodulate this legume (Table 6.1). This genus belongs to family *Rhizobiaceae* of α proteobacteria.

6.3.1.1.3 *Ensifer*

The *Sinorhizobium* genus was described by Chen et al. in 1988, whereas *Ensifer* genus was described earlier and both belong to a single taxon. This controversy was resolved and *Ensifer* was declared correct, so in the present review *Ensifer* is being used. This genus belongs to the family *Rhizobiaceae* of α proteobacteria. At present there are 20 species of *Ensifer* which are isolated from different hosts and have been verified to nodulate these hosts as detailed in the Table 6.1. The important Indian legume crops being nodulated by different species of this genus include *Glycine max*, *Medicago sativa*, *Sesbania rostrata*, and even other species of *Medicago*, *Sesbania*, and *Vigna.* Tree legumes commonly grown under Indian conditions like

Acacia, *Senegal*, *Prosopis*, *Leucaena*, and even Lotus are being nodulated by this genus. Interesting single host like *Acacia* or soybean is being nodulated by different species of this genus in different parts of the world.

6.3.1.1.4 *Aminobacter*

This genus belongs to the family *Rhizobiaceae* of α proteobacteria. This genus comprises single species *Aminobacter anthyllidis* which is known to be isolated and nodulates *Anthyllis vulneraria* (Maynaud et al. 2012).

6.3.1.1.5 *Neorhizobium*

Phylogenomic analyses conducted by Ormeno-Orrillo et al. (2015) showed the presence of two major groups within the family Rhizobiaceae. These two groups are of *Rhizobium/Agrobacterium* and *Shinella/Ensifer*. In the *Rhizobium/Agrobacterium* group, four distinct genera including newly proposed genus *Neorhizobium* was included. So in this new genus there is only one species *Neorhizobium galegae* which nodulates *Galega*, *Astragalus*, *Caragana*, *Lotus*, *Medicago*, *Sesbania*, and *Vigna* (Table 6.1).

6.3.1.1.6 *Allorhizobium*

Allorhizobium (meaning other *Rhizobium*) was proposed by (de Lajudie et al. 1998a). Only one nitrogen-fixing species *Allorhizobium undicola* was isolated from *Neptunia natans*, a waterlogged stem nodulating tropical legume (Table 6.1).

6.3.1.2 Phyllobacteriaceae

6.3.1.2.1 *Mesorhizobium*

Jarvis et al. (1997) described the genus *Mesorhizobium*. *Mesorhizobium* species have been reported in Africa, Asia, Australia, Europe, South and North America, and even in the Arctic (Poinsot et al. 2001). This genus belongs to the family *Phyllobacteriaceae* of α proteobacteria. To date 36 species of *Mesorhizobium* are known (Table 6.2). These different species nodulate various tree legumes including most important Indian pulses crop chickpea (Dudeja and Nidhi 2013).

6.3.1.2.2 *Phyllobacterium*

Mesorhizobium and another root nodulating genus *Phyllobacterium* belong to *Phyllobacteriaceae* family of α proteobacteria in the category of α rhizobia. This genus has three species *Phyllobacterium trifolii*, *P. ifriqiyense*, and *P. leguminum*; these are known to nodulate *Trifolium*, *Lupinus*, *Lathyrus*, and *Astragalus* (Table 6.2).

6.3.1.3 Bradyrhizobiaceae

6.3.1.3.1 *Bradyrhizobium*

Jordan (1982) has described the genus *Bradyrhizobium* and this genus is placed in the third family *Bradyrhizobiaceae* of α proteobacteria with 31 species nodulating diverse legumes (Table 6.3). *Bradyrhizobium* strains are also photosynthetic, which have

Table 6.2 Alpharhizobia belonging to *Phyllobacteriaceae*

S. No.	Rhizobial genera and spp.	Nodulating host	References
Mesorhizobium			
1.	*Mesorhizobium abyssinicae*	Woody legumes	Degefu et al. (2013)
2.	*Mesorhizobium acacia*	*Acacia melanoxylon*	Zhu et al. (2015)
3.	*Mesorhizobium albiziae*	*Albizia kalkora*	Wang et al. (2007)
4.	*Mesorhizobium alhagi*	*Alhagi sparsifolia*	Chen et al. (2010)
5.	*Mesorhizobium amorphae*	*Amorpha fruticosa*	Wang et al. (1999)
6.	*Mesorhizobium australicum*	*Biserrula pelecinus* L.	Nandasena et al. (2009)
7.	*Mesorhizobium calcicola*	*Sophora* spp.	De Meyer et al. (2016)
8.	*Mesorhizobium camelthorni*	*Alhagi sparsifolia*	Chen et al. (2011)
9.	*Mesorhizobium cantuariense*	*Sophora microphylla*	De Meyer et al. (2015)
10.	*Mesorhizobium caraganae*	*Caragana* spp.	Guan et al. (2008)
11.	*Mesorhizobium chacoense*	*Prosopis alba*	Velazquez et al. (2001)
12.	*Mesorhizobium ciceri*	*Cicer arietinum* L.	Jarvis et al. (1997), NBAIM (2011)
13.	*Mesorhizobium erdmanii*	*Lotus corniculatus*	Martinez-Hidalgo et al. (2015b)
14.	*Mesorhizobium hawassense*	Woody legumes	Degefu et al. (2013)
15.	*Mesorhizobium huakuii*	*Astragalus sinicus*	Jarvis et al. (1997)
16.	*Mesorhizobium jarvisii*	*Lotus corniculatus*	Martinez-Hidalgo et al. (2015a, b)
17.	*Mesorhizobium kowhai*	*Cicer arietinum*	Zhang et al. (2012a)
18.	*Mesorhizobium loti*	*Lotus corniculatus*	Jarvis et al. (1997), Martinez-Hidalgo et al. (2015a)
19.	*Mesorhizobium mediterraneum*	*Cicer arietinum*	Jarvis et al. (1997), Dudeja and Singh (2008), Suneja et al. (2016)
20.	*Mesorhizobium metallidurans*	*Anthyllis vulneraria*	Vidal et al. (2009)
21.	*Mesorhizobium muleiense*	*Cicer arietinum*	Zhang et al. (2012a)
22.	*Mesorhizobium newzealandense*	*Sophora* spp.	De Meyer et al. (2016)
23.	*Mesorhizobium opportunistum*	*Biserrula pelecinus* L.	Nandasena et al. (2009)
24.	*Mesorhizobium plurifarium*	*Acacia* spp.	de Lajudie et al. (1998b)
25.	*Mesorhizobium qingshengii*	*Astragalus sinicus*	Zheng et al. (2013)
26.	*Mesorhizobium robiniae*	*Robinia pseudoacacia*	Zhou et al. (2010)

(continued)

Table 6.2 (continued)

S. No.	Rhizobial genera and spp.	Nodulating host	References
27.	*Mesorhizobium sangaii*	*Astragalus luteolus*	Zhou et al. (2013)
28.	*Mesorhizobium septentrionale*	*Astragalus adsurgens*	Gao et al. (2004)
29.	*Mesorhizobium shangrilense*	*Caragana bicolour*, *C. erinacea*	Lu et al. (2009b)
30.	*Mesorhizobium shonense*	Woody legumes	Degefu et al. (2013)
31.	*Mesorhizobium silamurunense*	*Astragalus* spp.	Zhao et al. (2012)
32.	*Mesorhizobium sophorae*	*Sophora* spp.	De Meyer et al. (2016)
33.	*Mesorhizobium tamadayense*	*Anagyris latifolia*, *Lotus berthelotii*	Ramirez-Bahena et al. (2012)
34.	*Mesorhizobium temperatum*	*Astragalus adsurgens*	Gao et al. (2004)
35.	*Mesorhizobium waimense*	*Sophora* spp.	De Meyer et al. (2016)
36.	*Mesorhizobium waitakense*	*Sophora* spp.	De Meyer et al. (2016)
Phyllobacterium			
37.	*Phyllobacterium ifriqiyense*	*Lathyrus*, *Astragalus*	Mantelin et al. (2006)
38.	*Phyllobacterium leguminum*	*Astragalus*	Mantelin et al. (2006)
39.	*Phyllobacterium trifolii*	*Trifolium*, *Lupinus*	Valverde et al. (2005)

been isolated from stem and root nodules of *Aeschynomene*. World over widely cultivated legumes like soyabean, peanut, mungbean, and urdbean nodulating rhizobia have been identified. At present 6–7 different species of *Bradyrhizobium* are known to nodulate these cultivated legumes apart from other genera like *Ensifer* or *Rhizobium*.

6.3.1.4 Methyllobacteriaceae

6.3.1.4.1 *Methylobacterium*

So far there is only one mehthylotrophic rhizobial species *Methylobacterium nodulans* and this genus is placed in the fourth family *Methylobacteriaceae* of α proteobacteria. This rhizobial strain has been isolated from root nodules of *Lotononis bainesii* (Sy et al. 2001a, b, 2005; Jourand et al. 2004, 2005; Menna et al. 2006).

6.3.1.4.2 *Microvirga*

Methylobacterium genus and another root nodulating genus *Microvirga* belong to *Methylobacteriaceae* family of α proteobacteria. This methylotrophic genus has three species: *Microvirga lupine*, *M. lotononidis* and *M. zambiensis.* This genus forms root nodules of *Listia angolensis* and *Lupinus texensis* (Table 6.3).

Table 6.3 Alpharhizobia belonging to *Bradyrhizobiaceae*, *Methyllobacteriaceae*, *Hypomicrobiaceae*, and *Brucellaceae*

S. No.	Rhizobial genera and spp.	Nodulating host	References
Bradyrhizobium			
1.	*Bradyrhizobium arachidis*	*Arachis hypogaea*	Wang et al. (2013b)
2.	*Bradyrhizobium canariense*	*Genistoid legumes*	Vinuesa et al. (2005)
3.	*Bradyrhizobium cytisi*	*Cytisus villosus*	Chahboune et al. (2011)
4.	*Bradyrhizobium daqingense*	*Glycine max*	Wang et al. (2013c)
5.	*Bradyrhizobium elkanii*	*Glycine* spp.	Kuykendall et al. (1992)
6.	*Bradyrhizobium embrapense*	*Desmodium heterocarpon*	Delamuta et al. (2015)
7.	*Bradyrhizobium erythrophlei*	*Erythrophleum fordii*	Yao et al. (2015)
8.	*Bradyrhizobium ferriligni*	*Erythrophleum fordii*	Yao et al. (2015)
9.	*Bradyrhizobium ganzhouense*	*Acacia melanoxylon*	Lu et al. (2014)
10.	*Bradyrhizobium guangdongense*	*Arachis hypogaea*	Li et al. (2015)
11.	*Bradyrhizobium guangxiense*	*Arachis hypogaea*	Li et al. (2015)
12.	*Bradyrhizobium huanghuaihaiense*	*Glycine max*	Zhang et al. (2012b)
13.	*Bradyrhizobium icense*	*Phaseolus lunatus*	Duran et al. (2014)
14.	*Bradyrhizobium iriomotense*	*Entadakoshunensis*	Islam et al. (2008)
15.	*Bradyrhizobium japonicum/diazoefficiens*	*Glycine* spp.	Jordan (1982), Delamuta et al. (2013)
16.	*Bradyrhizobium jicamae*	*Pachyrhizus erosus*	Ramirez-Bahena et al. (2009)
17.	*Bradyrhizobium kavangense*	*Vigna unguiculata, Vigna subterranean*	Gronemeyer et al. (2015b)
18.	*Bradyrhizobium lablabi*	*Lablab purpureus, Arachis hypogaea*	Chang et al. (2011)
19.	*Bradyrhizobium liaoningense*	*Glycine soja, Glycine max*	Xu et al. (1995)
20.	*Bradyrhizobium manausense*	*Centrolobium paraense, Vigna unguiculata*	Zilli et al. (2014), Siva et al. (2014)
21.	*Bradyrhizobium neotropicale*	*Centrolobium paraense*	Zilli et al. (2014)
22.	*Bradyrhizobium ottawaense*	*Glycine max*	Yu et al. (2014)
23.	*Bradyrhizobium oligotrophicum*	*Aeschynomene indica*	Okubo et al. (2013)
24.	*Bradyrhizobium pachyrhizi*	*Pachyrhizus erosus*	Ramirez-Bahena et al. (2009)
25.	*Bradyrhizobium paxllaeri*	*Phaseolus lunatus*	Duran et al. (2014)
26.	*Bradyrhizobium retamae*	*Cytisusvillosus*	Chahboune et al. (2012)
27.	*Bradyrhizobium rifense*	*Retama sphaerocarpa, Retama monosperma*	Guerrouj et al. (2013)

(continued)

Table 6.3 (continued)

S. No.	Rhizobial genera and spp.	Nodulating host	References
28.	*Bradyrhizobium subterraneum*	*Vigna unguiculata, Vigna subterranean*	Gronemeyer et al. (2015a)
29.	*Bradyrhizobium tropiciagri*	*Neonotonia wightii*	Delamuta et al. (2015)
30.	*Bradyrhizobium vignae*	*Vigna unguiculata, Vigna subterranean, Arachis hypogaea*	Gronemeyer et al. (2016) and Bejarano et al. (2014)
31.	*Bradyrhizobium viridifuturi*	*Centrosema pubescens*	Ramirez-Bahena et al. (2013) and Helene et al. (2015)
32	*Bradyrhizobium yuanmingense*	*Lespedeza* spp.	Yao et al. 2002
Methylobacterium			
1.	*Methylobacterium nodulans*	*Crotalaria pedocarpa, Lotononis bainesii*	Sy et al. (2001a, b, 2005), Jourand et al. (2004) and Ardley et al. (2013)
Microvirga			
1.	*Microvirga lupine*	*Listia angolensis, Lupinus texensis*	Ardley et al. (2012)
2.	*Microvirga lotononidis*	*Listia angolensis, Lupinus texensis*	Ardley et al. (2012)
3.	*Microvirga zambiensis*	*Listia angolensis, Lupinus texensis*	Ardley et al. (2012)
Azorhizobium			
1.	*Azorhizobium doebereineraee*	*Sesbania virgata*	Moreira et al. (2006)
2.	*Azorhizobium caulinodans*	*Sesbania rostrata*	Menna et al. (2006) and Dreyfus et al. (1988)
Devosia			
1.	*Devosia neptuniae*	*Devosia neptuniae*	Rivas et al. (2002) and Rivas et al. (2003)
Ochrobactrum			
1.	*Ochrobactrum cytisi*	*Cytisus scoparius*	Zurdo-Pineiro et al. (2007)
2.	*Ochrobactrum lupine*	*Lupinus hon*	Trujillo et al. (2005)

6.3.1.5 Hypomicrobiaceae

6.3.1.5.1 *Azorhizobium*

Dreyfus et al. (1988) described the *Azorhizobium* genus. This genus currently contains species *A. caulinodans* and *A. doebereinerae*. This genus forms photosynthetic stem nodules on *Sesbania* species like *S. sesban*, *S. aegyptica*, *S. virgata*, and *S. rostrata* (Table 6.3). This genus along with other root-nodule forming bacteria – *Devosia* – are placed in the family *Hyphomicrobiaceae* of α proteobacteria.

6.3.1.5.2 *Devosia*

Only single rhizobial species of *Devosia neptuniae* has been reported. *Neptunia natans* is an aquatic legume and develops root-nodule symbiosis on floating stems (Rivas et al. 2002, 2003). This strain from India belongs to *Devosia* species and was named *Devosia neptuniae* sp. *nov*.

6.3.1.6 Brucellaceae

6.3.1.6.1 *Ochrobactrum*

The *Ochrobactrum* genus presently contains two rhizobial species *Ochrobactrum cystisi* and *O. lupini*. This genus along with other root-nodule forming bacteria is placed in the family *Brucellaceae* of α proteobacteria. These rhizobia were isolated from nodules of *Cytisus scoparius* and *Lupinus honoratus* (Trujillo et al. 2005; Zurdo-Pineiro et al. 2007).

6.3.2 Beta Proteobacteria or β Rhizobia

6.3.2.1 Burkholderiaceae

6.3.2.1.1 *Burkholderia*

The *Burkholderia* genus and *Cupriavidus* genus belong to β proteobacteria; β rhizobia are placed in *Burkholderiaceae* family. *Burkholderia* genus contains 14 species and others two as *Burkholderia* sp. which has been reported to nodulate many legumes (Table 6.4). Vandamme et al. 2002 described two novel species: *Bu. tuberum* isolated from the legume *Aspalathus carnosa* and *Bu. phymatum* from *Machaerium lunatum*. Genus *Burkholderia* was reported to be a predominant root-nodule symbiont for *Mimosa* and *Dalbergia* tree legumes (Barrett and Parker 2005; Chen et al. 2005; Menna et al. 2006; Rasolomampianina et al. 2005; Bournand et al. 2013).

6.3.2.1.2 *Cupriavidus*

Cupriavidus genus currently contains three rhizobial species, *C. taiwanensis*, *C. eutopha*, *C. necator*, and one as *Cupriavidus* sp. These were isolated from different species of *Mimosa, Lupinus*, *Macroptilium atropurpureum*, *Phaseolus vulgaris*, and *Vigna unguiculata* (Table 6.4). *Cupriavidus* genus has also been reported from North and South India and was isolated from the nodules of *Mimosa pudica* (Verma et al. 2004).

6.4 Conclusion

To discriminate rhizobia in species and subspecies, identification and classification using polyphasic methodology for identification has led to the identification of world over only about 20–25% legumes being nodulated by different rhizobia. Interestingly under different ecological niches and environmental conditions

Table 6.4 Beta rhizobia belonging to *Burkholderiaceae*

S. No	Rhizobial genera and spp.	Nodulating host	References
Burkholderia			
1.	*Burkholderia aspalathi*	*Aspalathus abietina* Thunb	Mavengere et al. (2014)
2.	*Burkholderia caballeronis*	Mimosoideae subfamily	Gyaneshwar et al. (2011)
3.	*Burkholderia caribensis*	Mimosoideae subfamily	Gyaneshwar et al. (2011)
4.	*Burkholderia diazotrophica*	Mimosoideae subfamily	Gyaneshwar et al. (2011)
5.	*Burkholderia dilworthii*	*Lebeckia ambigua*	De Meyer et al. (2014)
6.	*Burkholderia dipogonis*	*Dipogon lignosus*	Sheu et al. (2015)
7.	*Burkholderia mimosarum*	*Mimosa pudica*	Liu et al. (2011) and Chen et al. (2006)
8.	*Burkholderia nodosa*	*Mimosa pudica*	Chen et al. (2007)
9.	*Burkholderia phymatum*	*Mimosa pudica*	Liu et al. (2011) and Vandamme et al. (2002)
10.	*Burkholderia rhynchosiae*	*Rhynchosia ferulifolia*	Sprent and James (2008) and De Meyer et al. (2013b)
11.	*Burkholderia* sp.	*Lebeckia ambigua*	Howieson et al. (2013)
12.	*Burkholderia* sp.	*Parapiptadenia rigida*	Beukes et al. (2013)
13.	*Burkholderia sabiae*	*Mimosa caesalpiniifolia*	Chen et al. (2008)
14.	*Burkholderia sprentiae*	*Lebeckia ambigua*	De Meyer et al. (2013a)
15.	*Burkholderia symbiotica*	*Mimosa cordistipula, M. misera*	Sheu et al. (2012)
16.	*Burkholderia tuberum*	*Mimosa pudica*	Liu et al. (2011) and Vandamme et al. (2002)
Cupriavidus			
1.	*Cupriavidus taiwanensis*	*Mimosa pudica, M. diplotricha*	Chen et al. (2003)
2.	*Cupriavidus eutopha*	*Mimosa*	
3.	*Cupriavidus* sp.	*Mimosa*	Barrett and Parker (2006)
4.	*Cupriavidus necator*	*Mimosa caesalpiniaefolia, L. leucocephala, Macroptilium atropurpureum, Phaseolus vulgaris, Vigna unguiculata*	da Silva et al. (2012)

different rhizobial genera are predominant. Same legume under different conditions may be nodulated by different rhizobial genera and same species may be nodulating different legumes. Still lots of efforts are to be made to identify and classify the enormous wealth of rhizobia and their infecting legume. What are different parameters/characters which make it suitable for its predominance and to nodulate a legume? What is common between different genera and different species able to nodulate one host and vice versa different legumes being nodulated by same genera and same species?

References

Amarger N, Macheret V, Laguerre G (1997) *Rhizobium gallicum* sp. nov and *Rhizobium giardinii* sp nov, from *Phaseolus vulgaris* nodules. Int J Syst Bacteriol 47:996–1006

Aoki S, Ito M, Iwasaki W (2013) From β- to α-Proteobacteria: the origin and evolution of rhizobial nodulation genes nodI. Mol Biol Evol 30:2494–2508

Ardley JK, Parker MA, De Meyer SE et al (2012) *Microvirga lupini* sp. nov, *Microvirga lotononidis* sp. nov and *Microvirga zambiensis* sp nov are alphaproteobacterial root-nodule bacteria that specifically nodulate and fix nitrogen with geographically and taxonomically separate legume hosts. Int J Syst Evol Microbiol 62:2579–2588

Ardley JK, Reeve WH, O'Hara GW, Yates RJ, Dilworth MJ, Howieson JG (2013) Nodule morphology, symbiotic specificity and association with unusual rhizobia are distinguishing features of the genus *Listia* within the southern African crotalarioid clade *Lotononis* sl. Ann Bot 112:1–15

Azevedo LB, Van Zelm R, Leuven R et al (2015) Combined ecological risks of nitrogen and phosphorus in European freshwaters. Environ Pollut 200:85–92

Barrett CF, Parker MA (2005) Prevalence of *Burkholderia* sp. nodule symbionts on four mimosoid legumes from Barro Colorado Island, Panama. Syst Appl Microbiol 28:57–65

Barrett CF, Parker MA (2006) Coexistence of *Burkholderia*, *Cupriavidus* and *Rhizobium* sp nodule bacteria on two *Mimosa* sp in Costa Rica. Appl Environ Microbiol 72:1198–1206

Bejarano A, Ramirez-Bahena MH, Velazquez E, Peix A (2014) *Vigna unguiculata* is nodulated in Spain by endosymbionts of Genisteae legumes and by a new symbiovar (vignae) of the genus *Bradyrhizobium*. Syst Appl Microbiol 37:533–540

Berge O, Lodhi A, Brandelet G, Santaella C et al (2009) *Rhizobium alamii*sp. nov, an exopolysaccharide producing species isolated from legume and non-legume rhizospheres. Int J Syst Evol Microbiol 59:367–372

Beukes CW, Venter SN, Law IJ, Phalane FL, Steenkamp ET (2013) South African papilionoid legumes are nodulated by diverse *Burkholderia* with unique nodulation and nitrogen-fixation Loci. PLoS One 8:e68406

Boonsnongcheep P, Prathanturarug S, Takahashi Y, Matsumoto A (2016) *Rhizobium puerariae* sp nov, an endophytic bacterium from the root nodules of medicinal plant *Pueraria candollei* var candollei. Int J Syst Evol Microbiol 66:1236–1241

Bournaud C, de Faria SM, dos Santos JMF et al (2013) *Burkholderia* species are the most common and preferred nodulating symbionts of the piptadenia group (tribe mimosaceae). PLoS One 8(5):e63478

Chahboune R, Carro L, Peix A et al (2011) *Bradyrhizobium cytisi* sp. nov, isolated from effective nodules of *Cytisus villosus*. Int J Syst Evol Microbiol 61:2922–2927

Chahboune R, Carro L, Peix A et al (2012) *Bradyrhizobium rifense* sp. nov isolated from effective nodules of *Cytisus villosus* grown in the Moroccan Rif. Syst Appl Microbiol 35:302–305

Chang YL, Wang JY, Wang ET, Liu HC, Sui XH, Chen WX (2011) *Bradyrhizobium lablabi* sp. nov, isolated from effective nodules of *Lablab purpureus* and *Arachis hypogaea*. Int J Syst Evol Microbiol 61:2496–2502

Chen WX, Yan GH, Li JL (1988) Numerical taxonomic study of fast-growing soybean rhizobia and a proposal that *Rhizobium fredii* be assigned to *Sinorhizobium* gen. nov. Int J Syst Bacteriol 38:392–397

Chen WX, Tan ZY, Gao JL, Li Y, Wang ET (1997) *Rhizobium hainanense* sp. nov, isolated from tropical legumes. Int J Syst Bacteriol 47:870–873

Chen WM, Moulin L, Bontemps C et al (2003) Legume symbiotic nitrogen fixation by beta proteobacteria is widespread in nature. J Bacteriol 185:7266–7272

Chen WM, de Faria S, Straliotto MR et al (2005a) Proof that *Burkholderia* strains form effective symbioses with legumes: a study of novel *Mimosa* nodulating strains from South America. Appl Environ Microbiol 71:7461–7471

Chen WM, James EK, Chou JH, Sheu SY, Yang SZ et al (2005b) Betarhizobia from Mimosa pigra, a newly discovered invasive plant in Taiwan. Newphytol 168:661–675

Chen WM, James EK, Coenye T, Chou JH et al (2006) *Burkholderia mimosarum* sp. nov, isolated from root nodules of *Mimosa* sp. from Taiwan and South America. Int J Syst Evol Microbiol 56:1847–1851

Chen WM, de Faria SM, James EK et al (2007) *Burkholderia nodosa* sp. nov isolated from root nodules of the woody Brazilian legumes *Mimosa bimucronata* and *Mimosa scabrella*. Int J Syst Evol Microbiol 57:1055–1059

Chen WM, de Faria SM, Chou JH, James EK et al (2008) *Burkholderia sabiae* sp. nov isolated from root nodules of *Mimosa caesalpiniifolia*. Int J Syst Evol Microbiol 58:2174–2179

Chen WM, Zhu WF, Bontemps C, Young JPW, Wei GH (2010) *Mesorhizobium alhagi* sp nov, isolated from wild *Alhagi sparsifolia* in north-western China. Int J Syst Evol Microbiol 60:958–962

Chen WM, Zhu WF, Bontemps C, Young JPW, Wei GH (2011) *Mesorhizobium camelthorni* sp. nov, isolated from *Alhagi sparsifolia*. Int J Syst Evol Microbiol 61:574–579

da Silva K, Florentino LA, da Silva KB et al (2012) *Cupriavidus necator* isolates are able to fix nitrogen in symbiosis with different legume species. Syst Appl Microbiol 35:175–182

Dall'agnol RF, Ribeiro RA, Ormeno-Orrillo E et al (2013) *Rhizobium freirei* sp. nov, a symbiont of *Phaseolus vulgaris* that is very effective at fixing nitrogen. Int J Syst Evol Microbiol 63:4167–4173

Dall'agnol RF, Ribeiro RA, Delamuta JRMO et al (2014) *Rhizobium paranaense* sp. nov, an effective N2-fixing symbiont of common bean (*Phaseolus vulgaris* L.) with broad geographical distribution in Brazil. Int J Syst Evol Microbiol 64:3222–3229

de Lajudie P, Willems A, Pot B, Dewettinck D et al (1994) Polyphasic taxonomy of rhizobia: emendation of the genus *Sinorhizobium* and description of *Sinorhizobium meliloti* comb. nov, *Sinorhizobium saheli* sp. nov, and *Sinorhizobium teranga* sp. nov. Int J Syst Bacteriol 44:715–733

de Lajudie P, Laurent-Fulele E, Willems A et al (1998a) *Allorhizobium undicola* gen. nov. sp. nov. nitrogen-fixing bacteria that efficiently nodulate *Neptunia natans* in Senegal. Int J Syst Bacteriol 48(4):1277–1290

de Lajudie P, Willems A, Nick G, Moreira F et al (1998b) Characterization of tropical tree rhizobia and description of *Mesorhizobium plurifarium* sp. nov. Int J Syst Bacteriol 48:369–382

De Meyer SE, Cnockaert M, Ardley JK et al (2013a) *Burkholderia rhynchosiae* sp. nov, isolated from *Rhynchosia ferulifolia* root nodules. Int J Syst Evol Microbiol 63:3944–3949

De Meyer SE, Cnockaert M, Ardley JK et al (2013b) *Burkholderia sprentiae* sp. nov isolated from *Lebeckia ambigua* root nodules. Int J Syst Evol Microbiol 63:3950–3957

De Meyer SE, Cnockaert M, Ardley JK et al (2014) *Burkholderia dilworthii* sp. nov isolated from *Lebeckia ambigua* root nodules. Int J Syst Evol Microbiol 64:1090–1095

De Meyer SE, Tan HW, Heenan PB et al (2015) *Mesorhizobium waimense* sp. nov, isolated from *Sophora longicarinata* root nodules and *Mesorhizobium cantuariense* sp. nov, isolated from *Sophora microphylla* root nodules. Int J Syst Evol Microbiol 65:3419–3426

De Meyer SE, Tan HW, Heenan PB, Andrews M, Willems A (2016) *Mesorhizobium calcicola* sp. nov, *Mesorhizobium waitakense* sp. nov, *Mesorhizobium sophorae* sp. nov, *Mesorhizobium newzealandense* sp. nov and *Mesorhizobium kowhaii* sp nov, isolated from *Sophora* root nodules. Int J Syst Evol Microbiol 66:786–795

Degefu T, Wolde-Meskel E, Liu B, Cleenwerck I et al (2013) *Mesorhizobium shonense* sp nov, *Mesorhizobium hawassense* sp nov and *Mesorhizobium abyssinicae* sp nov, isolated from root nodules of different agroforestry legume trees. Int J Syst Evol Microbiol 63:1746–1753

Delamuta JR, Ribeiro RA, Ormeno-Orrillo E, Melo IS et al (2013) Polyphasic evidence supporting the reclassification of *Bradyrhizobium japonicum* group Ia strains as *Bradyrhizobium diazoefficiens* sp nov. Int J Syst Evol Microbiol 63:3342–3345

Delamuta JR, Ribeiro RA, Ormeno-Orrillo E et al (2015) *Bradyrhizobium tropiciagri* sp. nov and *Bradyrhizobium embrapense* sp. nov, nitrogen-fixing symbionts of tropical forage legumes. Int J Syst Evol Microbiol 12:4424–4433

Dreyfus B, Garcia JL, Gillis M (1988) Characterization of *Azorhizobium caulinodans* gen nov sp. nov, a stem-nodulating nitrogen-fixing bacterium isolated from *Sesbania rostrata*. Int J Syst Bacteriol 38:89–98

Dudeja SS (2016) Beneficial effects and molecular diversity of endophytic bacteria in legume and non legumes. In: Singh DP et al (eds) Microbial inoculants in sustainable agricultural productivity, pp 245–256

Dudeja SS, Narula N (2008) Molecular diversity of root nodule forming bacteria (review). In: Khachatourians GG, Arora DK, Rajendran TP, Srivastava AK (eds) Agriculturally important microorganisms, vol II. Academic World International, Bhopal, pp 1–24

Dudeja SS, Nidhi (2013) Molecular diversity of rhizobial and non rhizobial bacteria from nodules of cool season legumes. In: Salar RK, Gahlawat SK, Siwach P, Duhan JS (eds) Biotechnology: prospects and applications. Springer, New Delhi, pp 113–125

Dudeja SS, Singh PC (2008) High and low nodulation in relation to molecular diversity of Chickpea Mesorhizobia in Indian soils. Arch Agron Soil Sci 54:109–120

Dudeja SS, Sheokand S, Kumari S (2012a) Legume root nodule development and functioning under tropics and subtropics: perspectives and challenges. Legum Res 35:85–103

Dudeja SS, Giri R, Saini R, Suneja-Madan P, Kothe E (2012b) Interaction of endophytic microbes with legumes. J Basic Microbiol 52:248–260

Duran D, Rey L, Mayo J, Zuniga-Davila D et al (2014) *Bradyrhizobium paxllaeri* sp nov and *Bradyrhizobium icense* sp. nov, nitrogen-fixing rhizobial symbionts of Lima bean (*Phaseolus lunatus* L.) in Peru. Int J Syst Evol Microbiol 64:2072–2078

EnTao W, Minez-Romero E, Triplett E et al (2000) Phylogeny of root-and stem nodule bacteria associated with legumes. Prokaryotic nitrogen fixation: a model system for the analysis of a biological process, pp 177–186

Fornasero LV, Del Papa MF, Lopez JL, Albicoro FJ, Zabala JM et al (2014) Phenotypic, molecular and symbiotic characterization of the Rhizobial Symbionts of *Desmanthus paspalaceus* (Lindm) Burkart that grow in the Province of Santa Fe, Argentina. PLoS One 9(8):e104636

Frank B (1889) Ueber die Pilzsymbiose der Leguminosen. Ber Dtsch Bot Ges 7:332–346

Gao JL, Turner SL, Kan FL, Wang ET et al (2004) *Mesorhizobium septentrionale* sp. nov and *Mesorhizobium temperatum* sp. nov, isolated from *Astragalus adsurgens* growing in the northern regions of China. Int J Syst Evol Microbiol 54:2003–2012

Gronemeyer JL, Hurek T, Bünger W, Reinhold-Hurek B (2016) *Bradyrhizobium vignae* sp. nov, a nitrogen-fixing symbiont isolated from effective nodules of *Vigna* and *Arachis*. Int J Syst Evol Microbiol 66:62–69

Gronemeyer JL, Chimwamurombe P, Reinhold-Hurek B (2015a) *Bradyrhizobium subterraneum* sp. nov, a symbiotic nitrogen-fixing bacterium from root nodules of groundnuts. Int J Syst Evol Microbiol 65:3241–3247

Gronemeyer JL, Hurek T, Reinhold-Hurek B (2015b) *Bradyrhizobium kavangense* sp nov, a symbiotic nitrogen-fixing bacterium from root nodules of traditional Namibian pulses. Int J Syst Evol Microbiol 65:4886–4894

Gu CT, Wang ET, Tian CF, Han TX et al (2008) *Rhizobium miluonense* sp. nov, a symbiotic bacterium isolated from *Lespedeza* root nodules. Int J Syst Evol Microbiol 58:1364–1368

Guan SH, Chen WF, Wang ET, Lu YL et al (2008) *Mesorhizobium caraganae* sp. nov, a novel rhizobial species nodulated with *Caragana* sp. in China. Int J Syst Evol Microbiol 58:2646–2653

Guerrouj K, Ruiz-Diez B, Chahboune R, Ramirez-Bahena MH et al (2013) Definition of a novel symbiovar (sv retamae) within *Bradyrhizobium retamae* sp. nov nodulating *Retama sphaerocarpa* and *Retama monosperma*. Syst Appl Microbiol 36:218–223

Gyaneshwar P, Hirsch AM, Moulin L, Chen WM et al (2011) Legume-nodulating betaproteobacteria: diversity, host range, and future prospects. Mol Plant-Microbe Interact 24:1276–1288

Han TX, Wang ET, Wu LJ, Chen WF et al (2008) *Rhizobium multihospitium* sp nov, isolated from multiple legume species native of Xinjiang, China. Int J Syst Evol Microbiol 58:1693–1699

Helene LC, Delamuta JR, Ribeiro RA et al (2015) *Bradyrhizobium viridifuturi* sp nov, encompassing nitrogen-fixing symbionts of legumes used for green manure and environmental services. Int J Syst Evol Microbiol 12:4441–4448

Howieson JG, De Meyer SE, Vivas-Marfisi A et al (2013) Novel *Burkholderia* bacteria isolated from *Lebeckia ambigua* a perennial suffrutescent legume of the fynbos. Soil Biol Biochem 60:55–64

ICSP Subcommittee on the taxonomy of *Rhizobium* and *Agrobacterium* – diversity, phylogeny and systematics (2013) Rhizobial taxonomy up-to-date Submitted by vinuesa. 20:23

Islam MS, Kawasaki H, Muramatsu Y, Nakagawa Y, Seki T (2008) *Bradyrhizobium iriomotense* sp nov, isolated from a tumor-like root of the legume *Entada koshunensis* from Iriomote Island in Japan. Biosci Biotechnol Biochem 72:1416–1429

Jagiriti (2005) Characterization of chickpea, pigeonpea and mungbean rhizobia by RFLP of PCR amplified products. In: Ph.D. thesis submitted to CCS Haryana Agricultural University, Hisar, India

Jarvis BDW, Van Berkum P, Chen WX et al (1997) Transfer of *Rhizobium loti*, *Rhizobium huakuii*, *Rhizobium ciceri*, *Rhizobium mediterraneum*, and *Rhizobium tianshanense* to *Mesorhizobium* gen nov. Int J Syst Bacteriol 47:895–898

Jia RZ, Zhang RJ, Wei Q, Chen WF et al (2015) Identification and classification of rhizobia by matrix-assisted laser desorption/ionization time-of-flight mass spectrometry. J Proteomics Bioinform 8:98–107

Jiao YS, Yan H, Ji ZJ, Liu YH et al (2015) *Rhizobium sophorae* sp nov and *Rhizobium sophoriradicis* sp nov, nitrogen-fixing rhizobial symbionts of the medicinal legume *Sophora flavescens*. Int J Syst Evol Microbiol 65:497–503

Jordan DC (1982) Transfer of *Rhizobium japonicum* Buchanan 1980 to *Bradyrhizobium* gen nov, a genus of slow-growing, root nodule bacteria from leguminous plants. Int J Syst Bacteriol 32:136–139

Jourand P, Giraud E, Bena G, Sy A et al (2004) *Methylobacterium nodulans* sp. nov, for a group of aerobic, facultatively methylotrophic, legume root-nodule forming and nitrogen-fixing bacteria. Int J Syst Evol Microbiol 54:2269–2273

Jourand P, Renier A, Rapior S, Miana de Faria S, Prin Y, Galiana A, Giraud E, Dreyfus B (2005) Role of methylotrophy during symbiosis between Methylobacterium nodulans and Crotalaria podocarpa. Mol Plant-Microbe Interact 18(10):1061–1068

Kesari V, Ramesh AM, Rangan L (2013) *Rhizobium pongamiae* sp. nov, from root nodules of *Pongamia pinnata*. Biomed Res Int 2013:9

Kuykendall LD, Saxena B, Devine TE, Udell SE (1992) Genetic diversity in *Bradyrhizobium japonicum* Jordan 1982 and a proposal for *Bradyrhizobium elkanii* sp. nov. Can J Microbiol 38:501–505

Lapage SP, Sneath PHA, Lessel EF et al (1992) International code of nomenclature of bacteria: bacteriological code, 1990 revision. ASM Press, Washington, DC

Lewis GP, Schrire B, Mackinder B, Lock M (2005) Introduction. In: Lewis GP, Schrire B, Mackinder B, Lock M (eds) Legumes of the world. Royal Botanic Gardens, Kew, pp 1–20

Li QQ, Wang ET, Chang YL, Zhang YZ et al (2011) *Ensifer sojae* sp. nov, isolated from root nodules of *Glycine max* grown in saline-alkaline soils. Int J Syst Evol Microbiol 61:1981–1988

Li YH, Wang R, Zhang XX, Young JP et al (2015) *Bradyrhizobium guangdongense* sp. nov and *Bradyrhizobium guangxiense* sp. nov, isolated from effective nodules of peanut. Int J Syst Evol Microbiol 12:4655–4661

Lin DX, Wang ET, Tang H, Han TX et al (2008) *Shinella kummerowiae* sp. nov, a symbiotic bacterium isolated from root nodules of the herbal legume *Kummerowia stipulacea*. Int J Syst Evol Microbiol 58:1409–1413

Lin DX, Chen WF, Wang FQ, Hu D et al (2009) *Rhizobium mesosinicum* sp. nov, isolated from root nodules of three different legumes. Int J Syst Evol Microbiol 59:1919–1923

Lindstrom K (1989) *Rhizobium galegae*, a new species of legume root nodule bacteria. Int J Syst Bacteriol 39:365–367

Liu XY, Wu W, Wang ET, Zhang B, Macdermott J, Chen WX (2011) Phylogenetic relationships and diversity of β-rhizobia associated with *Mimosa* species grown in Sishuangbanna, China. Int J Syst Evol Microbiol 61:334–342

Liu TY, Li Y Jr, Liu XX, Sui XH et al (2012) *Rhizobium cauense* sp. nov, isolated from root nodules of the herbaceous legume *Kummerowia stipulacea* grown in campus lawn soil. Syst Appl Microbiol 35:415–420

Lloret L, Ormeno-Orrillo E, Rinco NR, Martinez-Romero J et al (2007) *Ensifer mexicanus* sp. nov, a new species nodulating *Acacia angustissima* (mill) Kuntze in Mexico. Syst Appl Microbiol 30:280–290

Lopez-Lopez A, Rogel MA, Ormeno-Orrillo E et al (2010) *Phaseolus vulgaris* seed-borne endophytic community with novel bacterial species such as *Rhizobium endophyticum* sp. nov. Syst Appl Microbiol 33:322–327

Lopez-Lopez A, Rogel-Hernandez MA, Barois I et al (2012) *Rhizobium grahamii* sp. nov, from nodules of *Dalealeporina*, *Leucaena leucocephala* and *Clitoria ternatea* and *Rhizobium mesoamericanum* sp. nov, from nodules of *Phaseolus vulgaris*, siratro, cowpea and *Mimosa pudica*. Int J Syst Evol Microbiol 62:2264–2271

Lu YL, Chen WF, Han LL, Wang ET, Chen WX (2009a) *Rhizobium alkalisoli* sp. nov, isolated from *Caragana intermedia* growing in saline-alkaline soils in the north of China. Int J Syst Evol Microbiol 59:3006–3011

Lu YL, Chen WF, Wang ET, Han LL et al (2009b) *Mesorhizobium shangrilense* sp. nov, isolated from root nodules of *Caragana* species. Int J Syst Evol Microbiol 59:3012–3018

Lu JK, Dou YJ, Zhu YJ, Wang SK, Sui XH, Kang LH (2014) *Bradyrhizobium ganzhouense* sp. nov, an effective symbiotic bacterium isolated from *Acacia melanoxylon* R Br nodules. Int J Syst Evol Microbiol 64:1900–1905

Mantelin S, Fischer-Le Saux M, Zakhia F, Bena G et al (2006) Emended description of the genus *Phyllobacterium* and description of four novel species associated with plant roots: *Phyllobacterium bourgognense* sp. nov, *Phyllobacterium ifriqiyense* sp. nov, *Phyllobacterium leguminum* sp. nov and *Phyllobacterium brassicacearum* sp. nov. Int J Syst Evol Microbiol 56:827–839

Martinez-Hidalgo P, Ramirez-Bahena MH, Flores-Felix JD et al (2015a) Revision of the taxonomic status of type strains of *Mesorhizobium loti* and reclassification of strain USDA 3471^T as the type strain of *Mesorhizobium erdmanii* sp. nov and ATCC 33669^T as the type strain of *Mesorhizobium jarvisii* sp nov. Int J Syst Evol Microbiol 65:1703–1708

Martinez-Hidalgo P, Flores-Felix JD, Menendez E, Rivas R et al (2015b) *Cicer canariense*, an endemic legume to the Canary Islands, is nodulated in mainland Spain by fast-growing strains from symbiovar trifolii phylogenetically related to *Rhizobium leguminosarum*. Syst Appl Microbiol 38:346–350

Mavengere NR, Ellis AG, Le Roux JJ (2014) *Burkholderia aspalathi* sp. nov, isolated from root nodules of the South African legume *Aspalathus abietina* Thunb. Int J Syst Evol Microbiol 64:1906–1912

Maynaud G, Willems A, Soussou S, Vidal C et al (2012) Molecular and phenotypic characterization of strains nodulating *Anthyllis vulneraria* in mine tailings, and proposal of *Aminobacter anthyllidis* sp. nov, the first definition of *Aminobacter* as legume-nodulating bacteria. Syst Appl Microbiol 35:65–72

Menna P, Hungria M, Barcellos FG, Bangel EV, Hess PN, Martinez-Romero E (2006) Molecular phylogeny based on the 16S rRNA gene of elite rhizobial strains used in Brazilian commercial inoculants. Syst Appl Microbiol 29:315–332

Merabet C, Martens M, Mahdhi M, Zakhia F, Sy A et al (2010) Multilocus sequence analysis of root nodule isolates from *Lotus arabicus* (Senegal), *Lotus creticus*, *Argyrolobium uniflorum* and *Medicago sativa* (Tunisia) and description of *Ensifer numidicus* sp. nov and *Ensifer garamanticus* sp nov. Int J Syst Evol Microbiol 60:664–674

Mnasri B, Liu TY, Saidi S, Chen WF, Chen WX, Zhang XX, Mhamdi R (2014) *Rhizobium azibense* sp nov, a nitrogen fixing bacterium isolated from root-nodules of *Phaseolus vulgaris*. Int J Syst Evol Microbiol 64:1501–1506

Moreira FMS, Cruz L, Faria SM, Marsh T et al (2006) *Azorhizobium doebereinerae* sp nov, microsymbiont of *Sesbania virgata* (Caz) Pers. Syst Appl Microbiol 29:197–206

Mousavi SA, Osterman J, Wahlberg N, Nesme X, Lavire C et al (2014) Phylogeny of the *Rhizobium-Allorhizobium-Agrobacterium* clade supports the delineation of *Neorhizobium* gen nov. Syst Appl Microbiol 37:208–215

Nandasena KG, O'hara GW, Tiwari RP, Willems A et al (2009) *Mesorhizobium australicum* sp. nov and *Mesorhizobium opportunistum* sp. nov, isolated from *Biserru lapelecinus* L. in Australia. Int J Syst Evol Microbiol 59:2140–2147

NBAIM Annual Report (2011–2012) http://nbaim.org.in/more.aspx?code=5&dtype=0

Nick G, De Lajudie P, Eardly BD, Suomalainen S et al (1999) *Sinorhizobium arboris* sp. nov and *Sinorhizobium kostiense* sp. nov, isolated from leguminous trees in Sudan and Kenya. Int J Syst Bacteriol 49:1359–1368

Ogasawara M, Suzuki T, Mutoh I, Annapurna K, Arora NK et al (2003) *Sinorhizobium indiaense* sp. nov and *Sinorhizobium abri* sp. nov, isolated from tropical legumes, *Sesbania rostrata* and *Abrus precatorius* respectively. Symbiosis 34:53–68

Okubo T, Fukushima S, Itakura M, Oshima K et al (2013) Genome analysis suggests that the soil oligotrophic bacterium *Agromonas oligotrophica* (*Bradyrhizobium oligotrophicum*) is a nitrogen-fixing symbiont of *Aeschynomene indica*. Appl Environ Microbiol 79:2542–2551

Ormeno-Orrillo E, Servin-Garciduenas LE, Rogel MA, Gonzalez V et al (2015) Taxonomy of rhizobia and agrobacteria from the Rhizobiaceae family in light of genomics. Syst Appl Microbiol 38(4):287–291

Osterman J, Marsh J, Laine PK, Zeng Z, Alatalo E, Sullivan JT et al (2014) Genome sequencing of two *Neorhizobium galegae* strains reveals a *noeT* gene responsible for the unusual acetylation of the nodulation factors. BMC Genomics 15:500

Poinsot V, Bélanger E, Laberge S, Yang GP et al (2001) Unusual methyl-branched α, ß-unsaturated acyl chain substitutions in the *Nod* factors of an Arctic *Rhizobium*, *Mesorhizobium* sp. strain N33 (*Oxytropicarctobia*). J Bacteriol 183:3721–3728

Ramirez-Bahena MH, Garcia-Fraile P, Peix A et al (2008) Revision of the taxonomic status of the species *Rhizobium leguminosarum* (Frank 1879) Frank 1889^{AL}, *Rhizobium phaseoli* 1926^{AL} and *Rhizobium trifolii* Dangeard 1926^{AL}. *R. trifolii* is a later synonym of *R. leguminosarum*. Reclassification of the strain *R. leguminosarum* DSM 30132 (=NCIMB 11478) as *Rhizobium pisi* sp. nov. Int J Syst Evol Microbiol 58:2484–2490

Ramirez-Bahena MH, Peix A, Rivas R, Camacho M et al (2009) *Bradyrhizobium pachyrhizi* sp. nov and *Bradyrhizobium jicamae* sp nov, isolated from effective nodules of *Pachyrhizus erosus*. Int J Syst Evol Microbiol 59:1929–1934

Ramirez-Bahena MH, Hernandez M, Peix A et al (2012) *Mesorhizobial* strains nodulating *Anagyris latifolia* and *Lotus berthelotii* in Tamadaya ravine (Tenerife, Canary Islands) are two symbiovars of the same species, *Mesorhizobium tamadayense* sp. nov. Syst Appl Microbiol 35:334–341

Ramirez-Bahena MH, Chahboune R, Velazquez E et al (2013) *Centrosema* is a promiscuous legume nodulated by several new putative species and symbiovars of *Bradyrhizobium* in various American countries. Syst Appl Microbiol 36:392–400

Rasolomampianina R, Bailly X, Fetiarison R et al (2005) Nitrogen-fixing nodules from rose wood legume trees (*Dalbergia* spp.) endemic to Madagascar host seven different genera belonging to α- and β-Proteobacteria. Mol Ecol 14:4135–4146

Ren DW, Chen WF, Sui XH, Wang ET, Chen WX (2011a) *Rhizobium vignae* sp. nov, a symbiotic bacterium isolated from multiple legume species. Int J Syst Evol Microbiol 61:580–586

Ren DW, Wang ET, Chen WF, Sui XH et al (2011b) *Rhizobium herbae* sp. nov and *Rhizobium giardinii*-related bacteria, minor microsymbionts of various wild legumes in China. Int J Syst Evol Microbiol 61:1912–1920

Ribeiro RA, Rogel MA, Lopez-Lopez A, Ormeno-Orrillo E et al (2012) Reclassification of *Rhizobium tropici* type A strains as *Rhizobium leucaenae* sp nov. Int J Syst Evol Microbiol 62:1179–1184

Ribeiro RA, Martins TB, Ormeno-Orrillo E, Marcon Delamuta JR et al (2015) *Rhizobium ecuadorense* sp nov, an indigenous N_2-fixing symbiont of the Ecuadorian common bean (*Phaseolus vulgaris* L) genetic pool. Int J Syst Evol Microbiol 65:3162–3169

Rincon-Rosales R, Villalobos-Escobedo JM, Rogel MA et al (2013) *Rhizobium calliandrae* sp. nov, *Rhizobium mayense* sp. nov and *Rhizobium jaguaris* sp. nov, rhizobial species nodulating the medicinal legume *Calliandra grandiflora*. Int J Syst Evol Microbiol 63:3423–3429

Rivas R, Velazquez E, Willems A, Vizcaino N et al (2002) A new species of *Devosia* that forms a unique nitrogen-fixing root nodule symbiosis with the aquatic legume *Neptunia natans* (L.f.) Druce. Appl Environ Microbiol 68:5217–5222

Rivas R, Willems A, Subba-Rao NS, Vizcanino N et al (2003) Description of *Devosia neptuniae* sp nov, that nodulates and fixes nitrogen in symbiosis with *Neptunia natans*, an aquatic legume from India. Syst Appl Microbiol 26:47–53

Roman-Ponce B, Jing Zhang Y, Soledad Vasquez-Murrieta M et al (2016) *Rhizobium acidisoli* sp. nov, isolated from root nodules of *Phaseolus vulgaris* in acid soils. Int J Syst Evol Microbiol 66:398–406

Rome S, Fernandez MP, Brunel B, Normand P et al (1996) *Sinorhizobium medicae* sp. nov, isolated from annual *Medicago* spp. Int J Syst Bacteriol 46:972–980

Saidi S, Ramirez-Bahena MH, Santillana N, Zuniga D et al (2014) *Rhizobium laguerreae* sp. nov nodulates *Vicia faba* on several continents. Int J Syst Evol Microbiol 64:242–247

Scholla MH, Elkan GH (1984) *Rhizobium fredii* sp. nov, a fast-growing species that effectively nodulates soybeans. Int J Syst Bacteriol 34:484–486

Segovia L, Young JPW, Martinez-Romero E (1993) Reclassification of American *Rhizobium leguminosarum* biovar Phaseoli type I strains as *Rhizobium Etli* sp. nov. Int J Syst Bacteriol 43:374–377

Sheokand S, Dudeja SS, Kumari S (2012) Nitrogen fixation in tropical environments – adaptive response and benefits. Res Crops 13:743–753

Sheu SY, Chou JH, Bontemps C, Elliott GN et al (2012) *Burkholderia symbiotica* sp nov isolated from root nodules of *Mimosa* spp native to north-east Brazil. Int J Syst Evol Microbiol 62:2272–2278

Sheu SY, Chen MH, Liu WY et al (2015) *Burkholderia dipogonis* sp. nov, isolated from root nodules of *Dipogon lignosus* in New Zealand and Western Australia. Int J Syst Evol Microbiol 65:4716–4723

Sprent JI, James EK (2008) Legume–rhizobial symbiosis: an anorexic model? New Phytol 179:3–5

Squartini A, Struffi P, Doring H et al (2002) *Rhizobium sullae* sp. Nov (formerly *'Rhizobium hedysari'*), the root-nodule microsymbiont of *Hedysarum coronarium* L. Int J Syst Evol Microbiol 52:1267–1127

Stackebrandt E, Frederiksen W, Garrity GM, Grimont PAD et al (2002) Report of the ad hoc committee for the re-evaluation of the species definition in bacteriology. Int J Syst Evol Microbiol 52:1043–1047

Suneja P, Piplani S, Dahiya P, Dudeja SS (2016) Molecular characterization of rhizobia from revertants of non-nodulating cultivar and normal cultivar of chickpea. J Agric Sci Technol 18:763–773

Suominen L, Roos C, Lortet G, Paulin L, Lindstrom K (2001) Identification and structure of the *Rhizobium galegae* common nodulation genes: evidence for horizontal gene transfer. Mol Biol Evol 18:907–916

Sy A, Giraud E, Samba R, de Lajudie P, Gillis M, Dreyfus B (2001a) Certaines legumineuses du genre *Crotalaria* sont specifiquement nodulees par une nouvelle espece de *Methylobacterium*. Can J Microbiol 47:503–508

Sy A, Giraud E, Jourand P, Garcia N, Willems A et al (2001b) Methylotrophic *Methylobacterium* bacteria nodulate and fix nitrogen in symbiosis with legumes. J Bacteriol 183:214–220

Sy A, Timmers ACJ, Knief C, Vorholt JA (2005) Methylotrophic metabolism is advantageous for *Methylobacterium extorquens* during colonization of *Medicago truncatula* under competitive conditions. Appl Environ Microbiol 71:7245–7252

Tan ZY, Kan FL, Peng GX, Wang ET et al (2001) *Rhizobium yanglingense* sp. nov, isolated from arid and semi-arid regions in China. Int J Syst Evol Microbiol 51:909–914

Tian CF, Wang ET, Wu LJ, Han TX et al (2008) *Rhizobium fabae* sp. nov, a bacterium that nodulates *Vicia faba*. Int J Syst Evol Microbiol 58:2871–2875

Toledo I, Lloret L, Martinez-Romero E (2003) *Sinorhizobium americanus* sp. nov, a new sinorhizobium species nodulating native *Acacia* spp. in Mexico. Syst Appl Microbiol 26:54–64

Trujillo ME, Willems A, Abril A, Planchuelo AM et al (2005) Nodulation of *Lupinus albus* by strains of *Ochrobactrum lupini* sp nov. Appl Environ Microbiol 71:1318–1327

Valverde A, Velazquez E, Fernandez-Santos F, Vizcaino N et al (2005) *Phyllobacterium trifolii* sp nov, nodulating *Trifolium* and *Lupinus* in Spanish soils. Int J Syst Evol Microbiol 55:1985–1989

Valverde A, Igual JM, Peix A, Cervantes E, Velazquez E (2006) *Rhizobium lusitanum* sp nov, a bacterium that nodulates *Phaseolus vulgaris*. Int J Syst Evol Microbiol 56:2631–2637

Van Berkum P, Beyene D, Bao G, Campbell TA, Eardly BD (1998) *Rhizobium mongolense* sp. nov is one of three rhizobial genotypes identified which nodulate and form nitrogen-fixing symbioses with *Medicago ruthenica* (L) Ledebour. Int J Syst Bacteriol 48:13–22

van Rhijn P, Vanderleyden J (1995) The rhizobium–plant symbiosis. Microbiol Rev 59:124–142

Vandamme P, Pot B, Gillis M, De Vos P, Kersters K, Swings J (1996) Polyphasic taxonomy, a consensus approach to bacterial systematics. Microbiol Rev 60(2):407–438

Vandamme P, Goris J, Chen WM, De Vos P, Willems A (2002) *Burkholderia tuberum* sp. nov and *Burkholderia phymatum* sp. nov, nodulate the roots of tropical legumes. Syst Appl Microbiol 25:507–512

Velazquez E, Igual JM, Willems A, Fernandez MP et al (2001) *Mesorhizobium chacoense* sp. nov, a novel species that nodulates *Prosopis alba* in the Chaco Arido region (Argentina). Int J Syst Evol Microbiol 51:1011–1021

Verma SC, Chowdhury SP, Tripathi AK (2004) Phylogeny based on 16S rDNA and nifH sequences of Ralstonia taiwanensis strains isolated from nitrogen-fixing nodules of Mimosa pudica, in India. Can J Microbiol 50(5):313–322

Vidal C, Chantreuil C, Berge O, Maure L et al (2009) *Mesorhizobium metallidurans* sp. nov, a metal-resistant symbiont of *Anthyllis vulneraria* growing on metallicolous soil in Languedoc, France. Int J Syst Evol Microbiol 59:850–855

Vinuesa P, Leon-Barrios M, Silva C, Willems A et al (2005) *Bradyrhizobium canariense* sp. nov, an acid-tolerant endosymbiont that nodulates endemic genistoid legumes (Papilionoideae: Genisteae) from the Canary Islands, along with *Bradyrhizobium japonicum* bv *genistearum, Bradyrhizobium* genospecies alpha and *Bradyrhizobium* genospecies beta. Int J Syst Evol Microbiol 55:569–575

Wang ET, Van Berkum P, Beyene D, Sui XH, Dorado O et al (1998) *Rhizobium huautlense* sp. nov, a symbiont of *Sesbania herbacea* that has a close phylogenetic relationship with *Rhizobium galegae*. Int J Syst Bacteriol 48:687–699

Wang ET, Van Berkum P, Sui XH, Beyene D, Chen WX, Martinez-Romero E (1999) Diversity of rhizobia associated with *Amorpha fruticosa* isolated from Chinese soils and description of *Mesorhizobium amorphae* sp. nov. Int J Syst Bacteriol 49:51–65

Wang ET, Tan ZY, Willems A, Fernandez-Lopez M et al (2002) *Sinorhizobium morelense* sp. nov, a *Leucaena leucocephala*-associated bacterium that is highly resistant to multiple antibiotics. Int J Syst Evol Microbiol 52:1687–1693

Wang FQ, Wang ET, Liu J, Chen Q et al (2007) *Mesorhizobium albiziae* sp. nov, a novel bacterium that nodulates *Albizia kalkora* in a subtropical region of China. Int J Syst Evol Microbiol 57:1192–1199

Wang F, Wang ET, Wu LJ, Sui XH, Li Y Jr, Chen WX (2011) Rhizobium vallis sp. nov, isolated from nodules of three leguminous species. Int J Syst Evol Microbiol 61:2582–2588

Wang YC, Wang F, Hou BC, Wang ET, Chen WF et al (2013a) Proposal of *Ensifer psoraleae* sp. nov, *Ensifer sesbaniae* sp. nov, *Ensifer morelense* comb nov and *Ensifer americanum* comb nov. Syst Appl Microbiol 36:467–473

Wang R, Chang YL, Zheng WT, Zhang D et al (2013b) *Bradyrhizobium arachidis* sp. nov, isolated from effective nodules of *Arachis hypogaea* grown in China. Syst Appl Microbiol 36:101–105

Wang JY, Wang R, Zhang YM, Liu HC, Chen WF et al (2013c) *Bradyrhizobium daqingense* sp. nov, isolated from soybean nodules. Int J Syst Evol Microbiol 63:616–624

Wei GH, Wang ET, Tan ZY, Zhu ME, Chen WX (2002) *Rhizobium indigoferae* sp. nov and *Sinorhizobium kummerowiae* sp. nov, respectively isolated from *Indigofera* spp. and *Kummerowi astipulacea*. Int J Syst Evol Microbiol 52:2231–2239

Wei GH, Tan ZY, Zhu ME, Wang ET et al (2003) Characterization of rhizobia isolated from legume species within the genera *Astragalus* and *Lespedeza* grown in the Loess Plateau of China and description of *Rhizobium loessense* sp nov. Int J Syst Evol Microbiol 53:1575–1583

Weir BS (2012) The current taxonomy of rhizobia. New Zealand rhizobia website. http://www.rhizobia.co.nz/taxonomy/rhizobia.html

Werner GD, Cornwell WK, Cornelissen JH, Kiers ET (2015) Evolutionary signals of symbiotic persistence in the legume-rhizobia mutualism. Proc Natl Acad Sci U S A 112(33):10262–10269

Xu LM, Ge C, Cui Z, Li J, Fan H (1995) *Bradyrhizobium liaoningense* sp. nov, isolated from the root nodules of soybeans. Int J Syst Bacteriol 45:706–711

Xu L, Shi JF, Zhao P, Chen WM, Qin W, Tang M, Wei GH (2011) *Rhizobium sphaerophysae* sp. nov, a novel species isolated from root nodules of *Sphaerophysa salsula* in China. Antonie Van Leeuwenhoek 99:845–854

Xu L, Zhang Y, Deng ZS, Zhao L, Wei XL, Wei GH (2013) *Rhizobium qilianshanense* sp. nov, a novel species isolated from root nodule of *Oxytropis ochrocephala* Bunge in China. Antonie Van Leeuwenhoek 103:559–565

Yan H, Yan J, Sui XH, Wang ET et al (2016) *Ensifer glycinis* sp. nov., a rhizobial species associated with species of the genus *Glycine* Int J Syst Evol Microbiol 66(8):2910–2916

Yao ZY, Kan FL, Wang ET, Wei GH, Chen WX (2002) Characterization of rhizobia that nodulate legume species of the genus *Lespedeza* and description of *Bradyrhizobium yuanmingense* sp. nov. Int J Syst Evol Microbiol 52:2219–2230

Yao LJ, Shen YY, Zhan JP, Xu W, Cui GL, Wei GH (2012) *Rhizobium taibaishanense* sp. nov, isolated from a root nodule of *Kummerowia striata*. Int J Syst Evol Microbiol 62:335–341

Yao Y, Sui XH, Zhang XX, Wang ET, Chen WX (2015) *Bradyrhizobium erythrophlei* sp. nov and *Bradyrhizobium ferriligni* sp. nov, isolated from effective nodules of *Erythrophleum fordii*. Int J Syst Evol Microbiol 65:1831–1837

Young JM, Kuykendall LD, Martinez-Romero E, Kerr A, Sawada HA (2001) Revision of *Rhizobium* Frank 1889, with an emended description of the genus, and the inclusion of all species of *Agrobacterium* Conn 1942 and *Allorhizobium undicola* de Lajudie et al. 1998 as new combinations: *Rhizobium radiobacter*, *R. rhizogenes*, *R. rubi*, *R. undicola* and *R. vitis*. Int J Syst Evol Microbiol 51:89–103

Yu X, Cloutier S, Tambong JT, Bromfield ES (2014) *Bradyrhizobium ottawaense* sp. nov, a symbiotic nitrogen fixing bacterium from root nodules of soybeans in Canada. Int J Syst Evol Microbiol 64:3202–3207

Zhang RJ, Hou BC, Wang ET, Li Y Jr, Zhang XX, Chen WX (2011) *Rhizobium tubonense* sp. nov, isolated from root nodules of *Oxytropis glabra*. Int J Syst Evol Microbiol 61:512–517

Zhang JJ, Liu TY, Chen WF, Wang ET et al (2012a) *Mesorhizobium muleiense* sp. nov, nodulating with *Cicer arietinum* L. Int J Syst Evol Microbiol 62:2737–2742

Zhang YM, Li Y Jr, Chen WF, Wang ET et al (2012b) *Bradyrhizobium huanghuaihaiense* sp. nov, an effective symbiotic bacterium isolated from soybean (*Glycine max* L.) nodules. Int J Syst Evol Microbiol 62:1951–1957

Zhang YJ, Zheng WT, Everall I, Young JP et al (2015) *Rhizobium anhuiense* sp. nov, isolated from effective nodules of *Vicia faba* and *Pisum sativum*. Int J Syst Evol Microbiol 65:2960–2967

Zhao CT, Wang ET, Zhang YM, Chen WF et al (2012) *Mesorhizobium silamurunense* sp. nov, isolated from root nodules of *Astragalus species*. Int J Syst Evol Microbiol 62:2180–2186

Zheng WT, Li Y, Wang R, Sui XH, Zhang XX et al (2013) *Mesorhizobium qingshengii* sp. nov, isolated from effective nodules of *Astragalus sinicus*. Int J Syst Evol Microbiol 63:2002–2007

Zhou PF, Chen WM, Wei GH (2010) *Mesorhizobium robiniae* sp. nov, isolated from root nodules of *Robinia pseudoacacia*. Int J Syst Evol Microbiol 60:2552–2556

Zhou S, Li Q, Jiang H, Lindstrom K, Zhang X (2013) *Mesorhizobium sangaii* sp. nov, isolated from the root nodules of *Astragalus luteolus* and *Astragalus ernestii*. Int J Syst Evol Microbiol 63:2794–2799

Zhu YJ, Kun J, Chen YL, Wang SK, Sui XH, Kang LH (2015) *Mesorhizobium acaciae* sp. nov, isolated from root nodules of *Acacia melanoxylon* R Br. Int J Syst Evol Microbiol 65:3558–3563

Zilli JE, Barauna AC, Da Silva K, De Meyer SE et al (2014) *Bradyrhizobium neotropicale* sp. nov, isolated from effective nodules of *Centrolobium paraense*. Int J Syst Evol Microbiol 64:3950–3957

Zurdo-Pineiro JL, Rivas R, Trujillo ME, Vizcaino N et al (2007) *Ochrobactrum cytisi* sp. nov, isolated from nodules of *Cytisus scoparius* in Spain. Int J Syst Evol Microbiol 57:784–788

Biotechnological Applications of *Trichoderma* Species for Environmental and Food Security

7

Mohd Kashif Kidwai and Manju Nehra

Abstract

This chapter reviews diverse biotechnological applications of *Trichoderma* species as a biofungicide for plant disease control, biofertilizer for plant growth promotory effects resulting in a high yield and productivity ensuring food security, along with environmental security by reducing the use of hazardous agrochemicals, production of industrially important chemicals, and having a potential for bioremediation for environmental cleanup activities. Use of chemical pesticide is one of the best strategies applied in intensive conventional agriculture for the control of pest and achieving high yields. However, pesticides are widely known for their adverse impacts on human health, and environmental challenges related to pesticides are a broad spectrum of toxicity affecting the biodiversity, persistence issues affecting food chain, residues in food chain, biomagnification, and acute and chronic effects on humans. Among beneficial microorganisms, *Trichoderma* species have attracted the attention because of their multipronged beneficial actions for plants in agriculture for their biocontrol potential, for plant growth promotion, for biostimulation of defense compounds in various crops, as well as for the production of important enzymes and drugs in industrial sector. *Trichoderma* species as a successful biocontrol agent employ complex mechanisms, i.e., competition, mycoparasitism, antibiosis, and induced systemic resistance against various phytopathogenic organisms. Several *Trichoderma*-based formulations are commercially available across the globe. Apart from the successful use of *Trichoderma* species as biocontrol agents (BCA) in agriculture, it also has ample potential for various industrial

M.K. Kidwai (✉)
Department of Energy and Environmental Sciences, Chaudhary Devi Lal University, Sirsa 125 055, Haryana, India
e-mail: kashif357313@yahoo.co.in

M. Nehra
Department of Food Science and Technology, Chaudhary Devi Lal University, Sirsa 125 055, Haryana, India

S.K. Gahlawat et al. (eds.), *Plant Biotechnology: Recent Advancements and Developments*, DOI 10.1007/978-981-10-4732-9_7

applications such as in paper and pulp industry, pharmaceutical industry, aromatic industry, and food and beverage industry. Intensive research is needed to develop *Trichoderma*-based formulations having prolonged shelf life, which is a constraint in the current scenario. Diverse biotechnological applications of *Trichoderma* species ensure food as well as environmental security direly needed for supporting life on this planet.

7.1 Introduction

Agriculture is a human activity carried out primarily to produce food, fodder, etc. (Morris and Burgess 2012). The origin of agriculture dates back to 10,000 years (Martin and Saurborn 2013). The population of earth is expected to cross 9.5 billion people till the end of the twenty-first century, and such a rise in population will ultimately increase the demand of all agricultural commodities globally (Abhilash et al. 2016). Agriculture in India involves majority of the human resource and is a source of livelihood for a large population living in villages; agriculture contributes 17% of GDP (Arjun 2013) and employment for about 58% of Indian population (Pandey and Singh 2012). After the onset of green revolution, conventional agriculture enabled the nation for producing enough food, thereby ensuring food security for the expanding Indian population. Green revolution conquered hunger and starvation of the nation and also provided the capacity to produce surplus yields for economic gains. Successful green revolution resulted in increased agricultural production, which is a landmark, but several environmental issues and challenges are surfaced regarding the toxic effect of synthetic agrochemicals due to an extensive application of fertilizers and pesticides. Some of the commercially successful pesticides are in the list of persistent organic pollutants reported to pose serious health hazards to humans and other biotic components of the ecosystem. Several pesticides are reported to cause acute as well as chronic adverse health effects on humans (Michel et al. 2006; Odukkathil and Vasudevan 2013).

Use of pesticide is one of the best strategies in conventional agriculture for attaining high yields with maximum economic gain. Global consumption of pesticides is approximately 2 million tons/year, of which two thirds is used by European nations and the USA and the remaining one third is used by the rest of the world (Syed et al. 2014). India is among the top ten pesticide-consuming countries. Though the average consumption of pesticides in India is comparatively less than developed nations, the issue of pesticide residue in India is very big; tropical nations such as India experience high crop losses due to the high pest population because of high temperature and humidity, which suits the pest population to increase and is one of the basic reasons of intensive application of pesticides (Verma et al. 2014). Nearly 80,000 tons of pesticide is annually consumed in Indian agriculture for the cultivation of various crops like cotton, rice, and fruits (Kalia and Gosal 2011). Pesticides like endosulfan, phorate, Mancozeb, methyl parathion, monocrotophos, carbendazim, sulfur, butachlor, isoproturon, cypermethrin, chlorpyrifos, and dichlorvos are widely consumed in India (Verma et al. 2014). Approximately

700 pesticides, including insecticides, herbicides, and fungicides, act on 95 biochemical targets in pest insects, weeds, and destructive fungi, and are classified according to their mode of action. Pesticides play a significant role in achieving food security as they are applied against different disease-causing organisms such as fungi, bacteria, nematodes, mites, and weeds on the entire crops worldwide. Some of the common environmental issues associated with pesticides are as follows: (1) a broad spectrum of toxicity adversely affecting the biodiversity (Niewiadomska 2004; Qingxiang et al. 2009), (2) pesticide residues affecting food chain (Khoo et al. 2012; Verma et al. 2014), (3) biomagnification, and (4) acute and chronic effects on humans (Levillian et al. 2012; Odukkathil and Vasudevan 2013); among pesticides, organochlorine pesticides, popularly known for their adverse impacts on human health and environment, are globally banned (Levillian et al. 2012). Pesticides persist in soil for a very long duration and pose negative impacts on soil microbial flora affecting soil health and productivity (Kalia and Ghosal 2011).

Fungal diseases are one of the major constraints for productivity in agriculture. Fungal-borne diseases cause reduction in the yield of approximately 20% globally (Petit et al. 2012). Indian agriculture suffers losses of approximately 22–25% in crops due to fungal diseases each year. The economic damage due to fungal diseases is estimated to be approximately Rs. 50,000 crores annually (Srivastava et al. 2016). Among pesticides, fungicide is a specific chemical applied to control fungal pest population causing fungal-borne disease. On the basis of the mechanism of action, they are further classified into two broad categories: (1) contact fungicides, e.g., sulfur-based fungicides, and (2) systemic fungicides, e.g., benzionidazoles. Several studies reported induction of stress in plants due to fungicide-causing alteration in various metabolic functions such as carbon metabolism, nitrogen metabolism, carbohydrate metabolism, and photosynthesis (Tiyagi et al. 2004; Petit et al. 2012). Rad and Jagannath (2011) observed reduction in the speed of seed germination, root and shoot elongation, and dry weight due to the use of the herbicide imazethapyr. Chemical fungicides like carbendazim and thiram are reported to pose adverse effect on soil microbial activity, i.e., nitrogenase activity and microbial diversity (Niewiadomska 2004; Qingxiang et al. 2009). Keeping in view of the alarming situation, agricultural scientists recommend to practice agriculture with sustainability, i.e., sustainable agriculture to counter the environmental hazards and health risks associated with conventional agriculture, which involves high ecological cost, pollution of soil, water, air, etc. (Tikhonovich and Provorov 2011).

7.2 Sustainable Agriculture

Sustainable agriculture is an ecofriendly approach as it ensures food security as well as environmental security at the same time. Sustainable agriculture covers various dimensions such as social, economic, technological, and ecological. Organic farming, integrated pest management (IPM), and integrated nutrient management (INM) are ecofriendly concepts to counter the environmental challenges posed due to conventional farming. Organic farming is considered to be the key solution for various ecological issues raised due to chemical inputs in conventional agriculture. The objectives of organic farming are to produce high-quality food with minimal environmental impacts, enhancement of biogeochemical cycles, maintenance of long-term soil fertility, safety of water resources, sustenance of biological and genetic diversity, and minimization of soil, air, and water pollution (Liefeld and Fuhrer 2010; Pandey and Singh 2012), but it is practically not possible to completely switch over to organic farming. In such situations, concepts like integrated pest management (IPM) and integrated nutrient management (INM) should be adopted for effective disease management for high yields (Arbelizi and Fuentes 2007). Organic farming, integrated pest management (IPM), and integrated nutrient management (INM) involve the use of biopesticides and biofertilizers which are derived from plants (Khater 2012), animals, and microorganisms (Copping and Menn 2000).

7.3 Application of Biotechnology in Agriculture

Biotechnology is the use of biological system, living organism, or their products for specific use (Soetan 2011). Biotechnology is providing novel solutions for various environmental problems. Biotechnological strategies play a significant role in the substitution of agrochemicals in agriculture and bioconversion of agricultural waste into value-added by-products. Among all types of living organisms, microorganisms play a significant role due to their multipotential abilities. Microorganisms improve nutritional requirement in crops and induce the ability in plants to resist both biotic and abiotic stress (Andrews et al. 2012). Several microbes are established as beneficial microorganisms, such as in the case of bacteria, e.g., *Bacillus* spp., *Pseudomonas* spp., *Rhizobium* species, and *Serratia* spp. Many fungal agents proved to be important microbes applied in agriculture as biofungicides and biofertilizers, e.g., *Aspergillus* spp., *Gliocladium* spp., *Metarhizium* spp., *Beauveria* spp., *Trichoderma* spp., and *Clonostachys* spp. Viruses also contribute, i.e., NPV (nuclear polyhedrosis virus) and CPV (cytoplasmic polyhedrosis virus) (Trikhonovich and Provorov 2011; Al-Naemi et al. 2016). Approximately 60 bacterial species, 60 fungal species, and 29 virus/phage species are being used in biopesticide products worldwide (Andrews et al. 2012). Natamycin, blasticidin, kasugamycin, mildiomycin, streptomycin, milbemectin, polynactin, bilanafos, etc., are commercially available microbial biopesticides (Copping and Menn 2000).

7.4 Biological Control Agents

Biological control agents (BCA) are defined as beneficial microorganisms, such as specialized fungi, bacteria, actinomycetes, used to control different disease-causing plant pathogens. Different mechanisms are employed by biocontrol agents (BCA) to control various pathogenic microorganisms (Sharma et al. 2012). Use of fungal biocontrol agents is considered to be an ecofriendly strategy for displacement of hazardous chemical fungicides. Biocontrol agents offer an ecologically safe alternative to chemical control of phytopathogens causing various diseases in a wide variety of crops ultimately reducing the yield (Kidwai 2007; Al-taweil et al. 2009).

7.4.1 *Trichoderma* Species

Among fungal biocontrol agents (BCA), genus *Trichoderma* has attracted most attention in agriculture sector because of its multipronged actions beneficial for plants, having biocontrol potential against various diseases, biostimulation of defense compounds, plant growth promotory effects in various crops, resulting in high productivity, as well as for the production of industrially important enzymes, secondary metabolites, drugs, etc., in the industrial sector (Papavizas 1982; Baker 1989; Vinale et al. 2006; Kidwai 2007; Mishra and Nautiyal 2009; Pandya et al. 2011; Ratnakumari et al. 2011; Singh et al. 2012; Linke et al. 2015). Diverse features of *Trichoderma* species, such as rhizosphere competence; ability to produce enzymes and secondary metabolites; plant growth promotion; utilization of different substrates; compatible with agrochemicals such as phorate, chlorpyrifos, carbendazim, and topsin (Roca et al. 1996; Jebakumar et al. 2000; Lane and Bowen 2005; Kidwai et al. 2006, 2007; Khan and Shahzad 2007; Khan et al. 2013); tolerance to adverse climatic conditions; survival in a wide pH range; long-term survival in fields; and ecologically safe, make it a commercially successful BCA for different agroecosystems (Naher et al. 2012; Saxena et al. 2014). Due to these features, genus *Trichoderma* is one of the most studied and widely applied fungal biocontrol agents (BCA) (Kidwai 2007; Ommati and Zaker 2012; Singh et al. 2013; Kumar et al. 2014). *Trichoderma* species are widely known biofungicides for soil-borne, seed-borne, and foliar pathogens but also have the potential to be developed and commercialized as bionematicides, bioherbicides, and bioinsecticides.

7.4.2 Systematics

Genus *Trichoderma* is a filamentous fungus belonging to Ascomycetes. It is cosmopolitan and has been reported to be found in diverse ecosystems on this planet such as forests, grasslands, agricultural fields, and lake water (Samuels 2006; Kidwai 2007; Bourguignon 2008; Oros et al. 2011; Pandya et al. 2011; Motlagh and Samimi 2013; Jones et al. 2016). *Trichoderma* is a beneficial symbiotic fungus which acts as an avirulent strong opportunistic symbiont in the rhizosphere of many plants such as in soybean (John et al. 2010). Some species of *Trichoderma* are also reported as endophytes (Samuels 2006; Hermosa et al. 2012; Rinu et al. 2013). *Trichoderma* was first proposed as a genus by Persoon in 1794, and its disease-controlling ability was reported in the early 1930s (Pandya et al. 2011; Ommati and Zaker 2012). Later on Rifai classified genus *Trichoderma* (Samuels 2006) in the year 1991; Bisset revised the classification on the basis of morphological characteristics, and four sections were identified as *Trichoderma*, *Pachybasium*, *Longibrachiatum*, and *Hypocreanum*. As per classification, genus *Trichoderma* is an imperfect filamentous fungus with teleomorphs belonging to Hypocreaceae family, Hypocreales order of Ascomycota division (Samuels 2006; Kidwai 2007; Gajera et al. 2013). The sexual (teleomorph) stage of genus *Trichoderma* is known as *Hypocrea*, which is an Ascomycete of order Hypocreales. Hypocrea was first described by Elia Fries in 1825 (Motlagh and Samimi 2013). *Trichoderma* teleomorphs possess morphological characteristics of *Trichoderma* anamorphs and can sexually reproduce to form ascospores. Links of *Trichoderma viride* with *Hypocrea rufa* were earlier reported in 1865. Majorly *Trichoderma* is reported to produce not only asexual spores but also chlamydospores and conidia from pustules (Samuels 2006). More than 200 *Trichoderma* species are reported from all over the globe (Skoneczny et al. 2015), and the phylogenetic tree is still growing. On the basis of appearance, *Trichoderma* species appears to be cottony, floccose colonies of different colors (white, yellow, green). Due to the color of conidia, conidial pigmentation varies among species. In India, *Trichoderma* was first isolated by Thakur and Norris in 1928 from Madras (Pandya et al. 2011; Bourguignon 2008). Various molecular techniques have been employed for phylogenetic analysis, which include random amplified polymorphic DNA (RAPD) analysis, restriction fragment length polymorphism (RFLP) analysis, amplified fragment length polymorphism (AFLP) analysis, and sequence-characterized amplified region (SCAR) analysis. With the development of new molecular tools like TRICHOBlast and TRICHOkey, new species are identified and characterized on the basis of rDNA-ITS1 analysis, UP-PCR, fungal oligonucleotide barcode, etc. (Hermosa et al. 2000; Cumagun et al. 2000; Druzhinina et al. 2006; Kim et al. 2013; Błaszczyk et al. 2014; Shahid et al. 2014; Skoneczny et al. 2015). Majority of *Trichoderma*-based formulations commercially available in India are developed from *Trichoderma viride*.

Biotechnological applications of genus *Trichoderma* as a biocontrol agent are not new as the antagonistic nature of genus *Trichoderma* was demonstrated more than 70 years ago (Elad et al. 1981; Samuels 2006; Kidwai 2007; Sharma et al. 2009; Pandya et al. 2011). *Trichoderma* species are well studied as a biological control agent (Gil et al. 2009; Nazanza et al. 2012) and have been found to play a significant role in plant disease management. Several researchers reported the disease control ability of *Trichoderma* species to parasitize and control several plant pathogenic fungi and invasive pathogens, such as *Sclerotinia sclerotiorum, Sclerotinia americana, Sclerotium rolfsii, Sclerotium cepivorum, Rhizoctonia solani kuhn, Rhizoctonia bataticola, Fusarium oxysporum, Rhizopus oryzae, Fusarium solani, Fusarium culmorum, Fusarium graminearum, Fusarium udum, Phytophthora infestans, Phytophthora capsici, Phytophthora parasitica, Pythium aphanidermatum, Pythium ultimum, Pythium indicum, Pythium arrhenomanes, Pythium aphanidermatum, Phoma lingam, Botrytis cinerea, Gaeumannomyces graminis* var. *tritici, Alternaria alternata, Armillaria mellea, Macrophomina phaseolina, Moniliophthora roreri, Corticium sasakii, Colletotrichum capsici, Colletotrichum acutatum, Crinipellis perniciosa, Cochliobolus sativus, Calonectria pauciramosa, Cytospora chrysosperma, Cylindrocladium floridanum, Drechslera sorokiniana, Gibberella xylarioides, Ceratocystis radicicola, Ganoderma* species, and *Botrytis fabae* (Weindling 1934; Ahmad and Baker 1987; Chet 1987; Sivan and Chet 1989; Krishamoorthy and Baskaran 1990; Velikanov et al. 1994; Adejumo et al. 1999;

Kidwai 2007; Verma et al. 2007; John et al. 2010; Mulaw et al. 2010; Inch et al. 2011; Meki et al. 2011; Mishra et al. 2011; Oros et al. 2011; Kredics et al. 2012; Mukherjee et al. 2012; Naher et al. 2012; Vitale et al. 2012; Motlagh and Samimi 2013; Gajera et al. 2013; Yang et al. 2013; Singh et al. 2013; Crozier et al. 2015; Degenkolb et al. 2015; Al-Naemi et al. 2016; Jones et al. 2016; Mbazia et al. 2016; Martínez et al. 2016; Srivastava et al. 2016).

Some of the successful *Trichoderma* species reported for biocontrol potential are *Trichoderma harzianum*, *Trichoderma viride*, *Trichoderma koningii*, *Trichoderma longibrachiatum*, *Trichoderma virens*, *Trichoderma pseudokoningii*, *Trichoderma atroviride*, *Trichoderma aureoviride*, *Trichoderma asperellum*, *Trichoderma reesei*, *Trichoderma lignorum*, *Trichoderma hamatum*, *Trichoderma polysporum*, *Trichoderma tomentosum*, *Trichoderma rossicum*, *Trichoderma gamsii*, *Trichoderma koningiopsis*, *Trichoderma roseum*, *Trichoderma paucisporum*, *Trichoderma theobromicola*, *Trichoderma martiale*, *Trichoderma arundinaceum*, *Trichoderma guizhouense*, *Trichoderma saturnisporum*, *Trichoderma spirale*, *Trichoderma aggressivum*, *Trichoderma pleurotum*, *Trichoderma pleuroticola*, *Trichoderma brevicompactum*, *Trichoderma crassum*, *Trichoderma cerinum*, and *Trichoderma ovalisporum* (Roy 1977; Papavizas and Lewis 1983; Ahmad and Baker 1987; Sivan and Chet 1987, 1989; Trutmann and Keane 1990; Mukhopadhyay 1994; Aziz and El-foulyKhalaf 1997; Khan and Gupta 1998; Rocco and Perez 2001; Prasada et al. 2002; Kredics et al. 2004, 2012; Mclean et al. 2005; Rabeendran et al. 2005; Verma and Dohroo 2005; Samuels 2006; Jash and Pan 2007; Kidwai 2007; Rojo et al. 2007; Sharma et al. 2009; Mukherjee et al. 2012; Vitale et al. 2012; Gajera et al. 2013; Yang et al. 2013; Saxena et al. 2014; Strakowska et al. 2014; Degen-Kolb et al. 2015; Al-Naemi et al. 2016; Islam et al. 2016; Marin-Guirao et al. 2016; Martínez et al. 2016; Moosa et al. 2016; Sain and Pandey 2016: Srivastava et al. 2016), which are discussed in Table 7.1.

Trichoderma harzianum ATCC 20476 was the first fungus registered in the USA for control of plant diseases; more than 50 registered *Trichoderma*-based commercial products are available all over the globe (Sharma et al. 2012). Some of the popular *Trichoderma*-based commercial formulations are Ecofit, Funginil, Trichogourd, Biocon, Bioderma, Ecoderma, Tricho-X, T-22 Root Shield™, T-39 Tridodex 20SP™, Trifender WP, and Trichovab, which are applied against a variety of phytopathogens causing diseases on different crops (Kumar et al. 2014; Dawidziuk et al. 2016).

7.4.3 Mechanism of *Trichoderma* Species for Biocontrol

For effective and successful biocontrol activity, *Trichoderma* spp. employ diverse synergistic mechanisms for attacking phytopathogenic organisms. These mechanisms are (1) competition for space and nutrients, (2) mycoparasitism, (3) antibiosis, and (4) induced systemic resistance (Howell 2003; Benitez et al. 2004; Harman et al. 2004; Harman 2006; Vinale et al. 2006; Kidwai 2007; El-Hassan et al. 2009; Hasanloo et al. 2010; Sharma et al. 2012; Singh et al. 2012; Bernal-Vicente et al. 2015; Al-Naemi et al. 2016; Martínez et al. 2016).

Table 7.1 *Trichoderma* species used as a biocontrol agent against different plant pathogens

S. No.	*Trichoderma*	Plant	Pathogen	Reference(s)
1	*Trichoderma harzianum*	Tomato	*F. oxysporum*	Ozbay and Newmann (2004)
			S. rolfsii	Varshney and Chaube (2001)
			B. cinerea	Cardoza et al. (2014)
		Potato	*R. solani*	Wilson et al. (2008)
		Cocoa	*Crinipellis perniciosa*	De Marco et al. (2000)
		Carnation	*C. pauciramosa*	Vitale et al. (2012)
		Mung bean	*R*. solani	Dubey et al. (2011)
		Date palm	*C. radicicola*	Al-Naemi et al. (2016)
		Cacao	*Moniliophthora roreri*	Crozier et al. (2015)
		Gladiolus	*F. oxysporum* f. sp. *gladioli*	Nosir (2016)
		Chickpea	*R. bataticola*	Dubey et al. (2011)
			F. oxysporum f. sp. *ciceri*	Mukhopadhyay and Kaur (1990), Jayalakshmi et al. (2009) and Meki et al. (2011)
			S. rolfsii	Mukhopadhyay and Kaur (1990)
			R. solani	
		Sunflower	*R. solani*	Singh et al. (2013)
		Pigeon Pea	*F. udum*	Niranjana et al. (2009)
		Onion	*S. cepivorum*	Lane and Bowen (2005)
		Wheat	*Gibberella zeae*	Inch et al. (2011)
			S. rolfsii	Islam et al. (2007)
		Safflower	*F. oxysporum* f. sp. *carthami*	Govinndappa et al. (2011)
		Bean	*F. oxysporum* f. sp. *phaseoli*	Sallam et al. (2008)
		Cucumber	*F. oxysporum* f. sp. *cucumerinum*	Chen et al. (2012)
		Maize	*F. verticillioides*	Ferrigo et al. (2014)
		Soybean	*R. solani, S. rolfsii*	Menedez and Godeas (1998)
			M. phaseolina	Mishra et al. (2011)
		Lettuce	*S. sclerotiorum*	Melo (1991)

(continued)

Table 7.1 (continued)

S. No.	*Trichoderma*	Plant	Pathogen	Reference(s)
		Lentil	*S. rolfsii*	Iqbal et al. (1995)
		Ginger	*P. aphanidermatum*	Dohroo (2001)
		Pea	*S. sclerotiorum*, *R. solani*	Velikanov et al. (1994)
			F. oxysporum, *F. solani*	
		Apple	*B. berengeriana* f. sp. *piricola*	Kexiang et al. (2002)
		Cowpea	*P. phaseoli*	Adejumo et al. (1999)
		Coffee	*G. xylarioides*	Mulaw et al. (2010)
		Egg plant	*M. phaseolina*	Khan and Gupta (1998)
			R. solani	Lewis and Larkin (1997)
		Black pepper	*P. capsici*	Jebakumar et al. (2000)
		Sugar beet	*S. rolfsii*	Upadhayay and Mukhopadhyay (1986)
		Mung bean	*R. solani*	Dubey et al. (2011)
2	*Trichoderma viride*	Ground nut	*S. rolfsii*	Parmar et al. (2015)
		Gladiolus	*F. oxysporum* f. sp. *gladioli*	Nosir (2016)
		Chickpea	*R. bataticola*	Dubey et al. (2011)
		Soybean	*P. arrhenomanes*	John et al. (2010)
			F. oxysporum f. sp. *adzuki*	John et al. (2010)
		Bean	*F. oxysporum* f. sp. *phaseoli*	Sallam et al. (2008)
			P. ultimum	Papavizas and Lewis (1983)
			S. rolfsii	Papavizas and Lewis (1983)
		Groundnut	*S. rolfsii*	Manjula et al. (2004)
		Rice	*R. solani*	Mathivanan et al. (2005)
			C. sasakii	Roy (1977)

(continued)

Table 7.1 (continued)

S. No.	*Trichoderma*	Plant	Pathogen	Reference(s)
		Eggplant	*M. phaseolina*	Khan and Gupta (1998)
			R. solani	Mathivanan et al. (2000)
			S. rolfsii	Lewis and Larkin (1997)
		Okra	*R. solani*	Mathivanan et al. (2000)
		Sunflower	*S. rolfsii*	
		Lentil	*S. rolfsii*	Iqbal et al. (1995)
		Sugarcane	*C. falcatum*	Reddy and Krishnamma Narayana (2009)
3	*Trichoderma hamatum*	Cucumber	*Phytophthora* sp.	Khan et al. (2004)
		Cabbage	*S. sclerotiorum*	Rabeendran et al. (2005)
		Egg plant	*M. phaseolina*	Khan and Gupta (1998)
			R. solani	Lewis and Larkin (1997)
4	*Trichoderma atroviride*	Tea	*P. theae*	Anita and Ponmurugan (2011)
5	*Trichoderma koningii*	Bean	*F. oxysporum* f. sp. *phaseoli*	Otadoh et al. (2011)
		Cowpea	*P. phaseoli*	Adejumo et al. (1999)
6	*Trichoderma virens*	Mung bean	*R. solani*	Dubey et al. (2011)
	Trichoderma virens	Chickpea	*R. bataticola*	Dubey et al. (2011)
7	*Trichoderma reesei*	Bean	*F. oxysporum* f. sp. *phaseoli*	Otadoh et al. (2011)
8	*Trichoderma asperellum*	Cucumber	*R. solani*	Trillas et al. (2006)
		Common bean	*F. solani*	Azarmi et al. (2011)
9	*Trichoderma spirale*	Bean	*F. oxysporum* f. sp. *phaseoli*	Sallam et al. (2008)
10	*Trichoderma saturnisporum*	Carnation	*F. oxysporum* f. sp. *dianthi*	Sharma and Shanmugam (2012)
10	*Trichoderma longibrachiatum*	Ground nut	*A. flavus*	Anjaiah et al. (2006)
		Potato	*F. solani*	Ommati and Zaker (2012)
11	*Trichoderma asperellum*	Oilseed rape	*Phoma lingam*	Dawidzuik et al. (2016)
			Botrytis cinerea	
12	*Trichoderma guizhouense*	Banana	*Fusarium oxysporum* f. sp. *cubense* 4	Zhang et al. (2016)
13	*Trichoderma lignorum*	Bean	*R. solani*	Aziz and El-foulyKhalaf (1997)

7.4.3.1 Competition

Competition with pathogen for space, substrates, and nutrients is one of the widely used mechanisms by *Trichoderma* species (Gajera et al. 2013; Vos et al. 2015; Al-Naemi et al. 2016). Due to its effective competitive ability, many fungal phytopathogens encounter starvation, resulting in enhanced biocontrol mechanism against them. *Trichoderma* is reported to have the ability to colonize on a wide variety of substrates (Kidwai 2007; Kumar et al. 2014; El- Shishtawy et al. 2015; Srivastava et al. 2016). Some species of *Trichoderma* are reported to produce high-quality siderophores of low molecular weight, which chelate the iron and discourage the growth of other pathogenic fungi (Heidrum et al. 1991; Qi and Zhao 2012; Gajera et al. 2013). *Trichoderma* species such as *Trichoderma harzianum, Trichoderma virens*, and *Trichoderma viride* are reported for high phosphate-solubilizing potential due to the production of acids such as citric acid, gluconic acid, and fumaric acid, which are needed for decreasing soil pH for phosphate solubilization (Akter et al. 2013; Khan et al. 2010; Akladious and Abbas 2014), and indole acetic acid (IAA)- and hydrogen cyanide (HCN)-producing potential (Rinu et al. 2013; Akladious and Abbas 2014). *Trichoderma* species are also reported to solubilize nutrients needed in plant metabolic functions such as iron (Fe), zinc (Zn), manganese (Mn), and magnesium (Mg) by oxidative dissolution, chelation, and reduction processes (Altomare et al. 1999; Khan et al. 2010; Akladious and Abbas 2014). The quality of being an effective competitor helps *Trichoderma* in employing other biological control mechanisms against a variety of plant pathogens.

7.4.3.2 Mycoparasitism

Mycoparasitic potential of *Trichoderma* species is the most explored biocontrol mechanism (Sivan and Chet 1989; Inbar et al. 1996; Kredics et al. 2001; Joshi et al. 2010; Kumar and Sharma 2011; Gajera et al. 2013; Yang et al. 2013; Vos et al. 2015). *Trichoderma* species possess the ability to remotely sense and parasitize various fungal pathogens, leading to death, which has been the reason behind the success for their development as commercial biofungicides. In this mechanism, *Trichoderma* directly attacks the other pathogenic fungus by completely parasitizing it. Mycoparasitism involves stages such as recognition of pathogen by use of lectins, coiling, penetration, and digestion of pathogen by using cell-wall-degrading lytic enzymes (Weindling 1932). *Trichoderma* species are grouped in necrotrophic parasites (Kidwai 2007). A series of cell-wall-degrading enzymes (CWDE) produced by *Trichoderma* species are cellulase (exo-β-1,4-glucanases, endo-β-1,4-glucanases, β-glucosidases), chitinase, pectinase, xylanase, glucanase, lipase, amylase, protease, etc. (Soglio et al. 1998; Seidl et al. 2005; Shafique et al. 2009; Yang et al. 2013; Blaszczyk et al. 2014; Cardoza et al. 2014; Strakowska et al. 2014; Ramada et al. 2016), which help to identify and attack the pathogen. *Trichoderma* species are reported to produce extracellular enzymes with retained enzymatic activity even in heavy-metal-polluted agricultural fields (Verma et al. 2007). Due to their effective mycoparasitic ability, *Trichoderma* species are also reported for their bionematicidal potential as they control the plant parasitic nematodes (Siddiqui et al. 2001; Suarez et al. 2004; Goswami et al. 2008; Bokhari 2009; Khan and Haque 2011; Mukherjee et al. 2012), bioinsecticidal potential as they are reported to control spruce budworm insect larvae (Verma et al. 2007), and also explored for bioherbicidal potential for control of weeds (Hutchinson 1999; Javaid and Ali 2011).

7.4.3.3 Antibiosis

Antibiosis is another important biocontrol mechanism used by *Trichoderma* species (Vinale et al. 2006, 2016; Nemcovic et al. 2008; El-Hassan et al. 2009; Vos et al. 2015; Mutawila et al. 2016) in which low-molecular-weight diffusible compounds, i.e., antibiotics, are secreted, resulting in the inhibition of the growth of fungal and bacterial phytopathogens. In this process, indirect antagonism takes place. *Trichoderma* species are prolific producers of different secondary metabolites (Cardoza et al. 2014; Malmierca et al. 2015; Zeilinger et al. 2016). Over 180 secondary metabolites produced by genus *Trichoderma* have been characterized (Blaszczyk et al. 2014; Vos et al. 2015; Dawidziuk et al. 2016). *Trichoderma* is a popularly known producer of both volatile and nonvolatile fungicidal metabolites, and six major groups have been defined, i.e., polyketides, pyrones, terpenoids, isonitriles, diketopiperazines, and peptaibols (Bourguignon 2008; Inch et al. 2011). Metabolite like ketoaldehyde named Tandyukisin is recently derived from *Trichoderma harzianum* (Yamada et al. 2014). Dual plate assay is popularly used for the demonstration of the antagonistic activity of *Trichoderma* species due to antibiotics. Various *Trichoderma* species are reported to produce a series of antibiotics (Table 7.2) responsible for biocontrol activities against several phytopathogens.

7.4.3.4 Induced Resistance

Trichoderma spp. have the ability to induce resistance in plants, i.e., induce systemic resistance (ISR) and systemic acquired resistance (SAR) by eliciting and promoting plant defense responses that restrict disease development in several economically important agricultural crops reported by several researchers (Kredics et al. 2001; Khan et al. 2004; Soresh et al. 2005; Hoitink et al. 2006; Djonovic et al. 2007; Govindappa et al. 2011; Hermosa et al. 2012; Qi and Zhao 2012; Cai et al. 2013; Singh et al. 2013; Bernal-Vicente et al. 2015; Levy et al. 2015; Shukla et al. 2015; Srivastava et al. 2016). *Trichoderma* spp. induce antioxidative metabolism in plants (Zheng and Shetty 2000), resulting in induction in total phenolic content (TPC), phenylalanine ammonia-lyase (PAL), polyphenol oxidase (PPO), peroxidase (PO), and cinnamyl alcohol dehydrogenase (CAD) activities are reported to play significant roles in many defense processes (Singh et al. 2013; Shukla et al. 2015; Vos et al. 2015), including direct antimicrobial action, lignin formation, phytoalexin production, induction in pathogenesis-related (PR) protein (Yedidia et al. 1999; Yedidia et al. 2000), generation of hypersensitive responses, reactive oxygen species (ROS)-scavenging enzymes, chloroplastic and cytosolic enzymes, toxic phenolic antioxidant compounds, reprogramming of oxidant and antioxidant metabolites, oxidation of phenolic compounds resulting in the formation of quinine, which are considered to be more toxic to microbial community, and hydrolysis of fungal cell wall components, i.e., chitin and beta-1, 3-glucan by PR-2 and PR-3 proteins (Singh et al. 2013).

Table 7.2 Antibiotics produced by different *Trichoderma* species

S. No.	*Trichoderma* species	Secondary metabolite	Reference(s)
1	*Trichoderma harzianum*	Trichoharzianin A1	Hajji et al. (1987)
		Trichoharzianin B1	Bourguignon (2008)
		Trichorzin HA, MA	Blaszczyk et al. (2014)
		Trichorzin PA	Degenkolb et al. (2015)
		Trichokinin	Corley et al. (1994)
		Trichothene	Degenkolb et al. (2015)
		Trichokindins	
		Trichotoxins A-50	
		6-Pentyl-α-pyrone	Leobardo et al. (2006)
		Anthraquinone	Vinale et al. (2006)
		Azaphilone	Vos et al. (2015)
		Butenolide	
		Trichosetin	Gajera et al. (2013)
		Harzionolide	Cutler and Jacyno (1991)
		Harzianopyridone	Bourguignon (2008)
			Degenkolb et al. (2015)
			Vos et al. (2015)
		Isonitrin A	Srivastava et al. (2016)
		Trichodiene	Malmierca et al. (2015)
		6-Pentyl-α-pyrone	Mutawila et al. (2016)
		Hypomurocins B	Degenkolb et al. (2015)
		Trichosimmonsins	
2	*Trichoderma viride*	6-Pentyl-α-pyrone	
		Dermadin	Srivastava et al. (2016)
		Viridofungin A	Harris et al. (1993)
		Trichotoxin A	Blaszczyk et al. (2014)
		Trichorovin, Trichocellins	Blaszczyk et al. (2014)
3	*Trichoderma virens*	Gliotoxin	Blaszczyk et al. (2014)
		Glioviren	Bourguignon (2008)
		Gliocladic	Scharf et al. (2016)
4	*Trichoderma longibrachiatum*	Longibrachins	Blaszczyk et al. (2014)
5	*Trichoderma koningii*	Trichokonin	Blaszczyk et al. (2014)
		6-Pentyl-α-pyrone	
6	*Trichoderma atroviride*	Neoatroviridin A–D	Blaszczyk et al. (2014)
		Atroviridin A–C	
		Hypomurocin B	Degenkolb et al. (2015)
		6-Pentyl-2H-pyran-2-one	Garnica-vergara et al. (2016)
7	*Trichoderma stromaticum*	Trichostromaticins	Degenkolb et al. (2015)
8	*Trichoderma asperellum*	Trichotoxin	Degenkolb et al. (2015)
9	*Trichoderma arundinaceum*	Harzianum A	Malmieraca et al. (2015)

(continued)

7.5 *Trichoderma* Species as Biofertilizer

Trichoderma species are also reported as successful commercial biofertilizers. Plant growth promotion due to the application of *Trichoderma* species is reported by many researchers on various crops such as wheat, cucumber, carnation, maize, and rice (Windham et al. 1986; Christopher 2000; Gasso and Prosperi 2001; Kredics et al. 2001; Lo and Lin 2002; Bjorkmann 2004; Harman et al. 2004; Khan et al. 2005; Adam et al. 2007; Mishra and Nautiyal 2009; Hasanloo et al. 2010; John et al. 2010; Viterbo et al. 2010; Nazanza et al. 2012; Singh et al. 2012; Qi and Zhao 2012; Akter et al. 2013; Rinu et al. 2013; Topolovec-pintaric et al. 2013; Triveni et al. 2013; Akladious and Abbas 2014; Doni et al. 2014; Shukla et al. 2015; Martínez et al. 2016; Marin-Guirao et al. 2016; Vinale et al. 2016). Some *Trichoderma* species are reported to have significant influence on germination, seedling vigor, and plant growth parameters like plant vigor, root length, shoot length, fresh weight, dry weight, leaf area, phytohormone content, and nucleic acid content (Mckenzie and Starman 1995; Zheng and Shetty 2000; Yedidia et al. 2001; Hermosa et al. 2012; Akladious and Abbas 2014). *Trichoderma* species are reported to influence the physiological process in rice plants by increasing photosynthetic rate, stomatal conductance, transpiration, etc., resulting in overall plant growth promotion (Doni et al. 2014). Shortening of crop cycle by reduction in plant growth time is correlated with enhanced productivity by the application of *Trichoderma* species as biofertilizer (Akladious and Abbas 2014). Hasanloo et al. (2010) reported that *Trichoderma* spp. produce various organic acids, such as fumaric, gluconic, and citric acids, that decrease the soil pH and allow the process of phosphate solubilization and bioavailability of micronutrients and mineral cations, including iron, zinc, manganese, and magnesium, considered to be useful for different metabolic processes in plants. Xylanase secreted by several *Trichoderma* species was reported to induce ethylene production and plant defense responses. Several proteins are involved in the induction of terpenoid phytoalexin biosynthesis, and peroxidase activity in cotton was found to be produced by strains of *Trichoderma virens* (Harman et al. 2004). Several studies revealed that root colonization by various *Trichoderma* strains resulted in the induction of defense-related plant enzymes, including various peroxidases, chitinases, β-1,3-glucanases, and the lipoxygenase-pathway hydroperoxide lyase. It is clear that *Trichoderma* spp. have developed a symbiotic rather than a parasitic relationship with plants. Root elongation and lateral root formation are also supported by *Trichoderma* species, which sustain high elongation rate and high phytohormonal production, e.g., IAA, GA_3, indole-3-ethanol (Tweddell et al. 2007; Akladious and Abbas 2014). Plant growth-promoting activity due to *Trichoderma asperellum* was observed in cucumber seedlings, with increased biomass and chlorophyll pigments even when the plant was experiencing salt-induced stress (Qi and Zhao 2012). *Trichoderma viride* along with other microorganisms used as biofilm resulted in higher biochemical activity under in vitro conditions correlated for plant growth promotion (Triveni et al. 2013). *Trichoderma* species are reported to enhance starch and protein contents in crops, thereby enhancing the nutritional status of crops (Akladious and Abbas 2014).

7.6 Mass Production of *Trichoderma* Species

For the commercial development of biofertilizer, biofungicide, and other industrially important products, large quantities of *Trichoderma* species are needed. *Trichoderma* species can be produced by both liquid-state fermentation (LSF) or submerged fermentation (SF) and solid-state fermentation (SSF) system for conidial mass, mycelial biomass, chlamydospores, industrially important enzymes, value-added product such as citric acid, etc.

7.6.1 Liquid-State Fermentation

Trichoderma species are grown in liquid medium. Molasses, V-8 juice, molasses-soy medium, jaggery-soy medium, slaughterhouse wastewater, cheese industry wastewater, pulp-and-paper industry wastewater, etc., are popularly used in liquid-state fermentation (Verma et al. 2007; Al-taweil et al. 2009). This process is considered to be less time-consuming by following aeration, agitation, temperature control, etc. Biomass of *Trichoderma* spp. may be achieved in 4 days (Srivastava et al. 2016).

7.6.2 Solid-State Fermentation

Solid-state fermentation is one of the most popular methods for mass multiplication of *Trichoderma* species. *Trichoderma* species are reported to grow and mass-multiply on various substrates, i.e., agricultural waste such as wheat bran, wheat flour, wheat husk, wheat straw, rice husk, pea seed cover, arhar seed cover, sorghum grain, rice straw, tea leaf waste, coffee husk, coconut coir pith, neem cake, bajra grain, mustard cake, sugarcane bagasse, sawdust, orange peel, maize spent cob, linseed cake, sesame cake, pongamia cake, cowdung, groundnut hull, chickpea seed cover, vegetable waste, corn flour, cranberry pomace, press mud, banana waste, palm leaves, date seeds, farmyard manure, and chicken manure (Chad 1999; Hutchinson 1999; Prakash et al. 1999; Kidwai 2007; Singh et al. 2007; Al-taweil et al. 2009; Niranjana et al. 2009; Iqbal et al. 2011; Onilude et al. 2013; Kumar et al. 2014, 2016; Pandey et al. 2015; El- Shishtawy et al. 2015; Srivastava et al. 2016). Solid substrate fermentation is considered to be better than liquid substrate fermentation due to better yield, low risk of contamination, toleration of downstream processing, and cost-effectiveness (Pandey et al. 2015; El-Shishtawy et al. 2015). By solid-state fermentation, the cellulosic substrates like agri waste are converted into monomeric glucose by the process of hydrolysis. Superior strains of *Trichoderma* are identified, characterized, and formulated.

However, novel methodology like simple scrapping technique of *Trichoderma harzianum* NBRI-1055 was established for the production of concentrated, stable, effective, and cost-efficient formulation of *Trichoderma harzianum* for seed application (Singh and Nautiyal 2012), where neither solid substrate fermentation nor liquid-state fermentation is required for the development of *Trichoderma*-based formulations.

Abiotic factors such as temperature, nutrients, light, and pH influence the growth and production of different *Trichoderma* species. Among *Trichoderma atroviride*, *Trichoderma hamatum*, and *Trichoderma pleuroticola* used as biocontrol agents, temperature is one of the most important limiting factors in their production and bioactivity. Temperature range may vary from species to species; for example, *Trichoderma polysporum* can grow in very low temperature (0° C) and *Trichoderma koningii* can grow in high temperature (40 °C). The optimum temperature for most *Trichoderma* spp. is in the 25–30 °C range (Onilude et al. 2013; Daryaei et al. 2016).

Suitable carriers, stickers, adjuvants, and emulsions are incorporated in *Trichoderma*-based formulations for better on-field performance; some of the popular commercial carriers are talc, carboxymethyl cellulose, potassium sorbate, sodium metabisulfite vermiculite, Pesta granules, alginate prill, pressmud, vegetable oil, mineral oil, etc., which increase the shelf life for on-field success of the commercial formulations (Verma et al. 2007; Kumar et al. 2014). On-field performance of *Trichoderma*-based formulations mainly depends upon the delivery methods. Some of the popular delivery methods for the application of *Trichoderma*-based formulations are (1) seed treatment, (2) seed biopriming, (3) soil application, (4) seeding dip, and (5) aerial spraying for foliar pathogens (Kumar et al. 2014; Srivastava et al. 2016).

7.7 Molecular Approaches

According to Mukherjee et al. (2012), genome sequences of five species of *Trichoderma* are available, i.e., *Trichoderma reesei*, *Trichoderma atroviride*, *Trichoderma virens*, *Trichoderma harzianum*, and *Trichoderma asperellum*. Difference in genome size among species is reported; for example, *Trichoderma reesei* has the smallest genome size of 34.1 Mb, 9129 gene models and *Trichoderma harzianum* has the largest genome size of 40.98 Mb, 14,095 gene models. Several molecular approaches involving genetic manipulations are employed using genetic recombination techniques for the development of superior *Trichoderma* strains, such as protoplast fusion, *Agrobacterium*-mediated transformation, biolistic transformation, insertional mutagenesis, transposon mutagenesis, mutation, and sexual crossing approach such as meiosis and postmeiotic mitosis for diverse biotechnological applications (Yang and Liu 2005; Schuster and Schmoll 2010; Gusakov 2011; Kumar et al. 2014; Chuang et al. 2015; Hasan et al. 2014). Genetically transformed *Trichoderma virens* strain is found to have enhanced competitive ability than the wild-type strain essentially needed for effective biocontrol activity (Weaver and Kenerley 2008).

7.8 Biotechnological Applications of *Trichoderma* Species in Industries

Apart from the use of *Trichoderma* species as biocontrol agents (BCA) in agriculture, they also have ample potential for various industrial applications as discussed in Table 7.3. *Trichoderma* species are popularly known for the biodegradation of plant polymers like cellulose (Cardoza et al. 2014). In genus *Trichoderma*,

Table 7.3 Some of the industrial applications of *Trichoderma* species

Industrial application of *Trichoderma* species	Reference(s)
Bioremediation of heavy metals, dyes, hydrocarbons, polycyclic aromatic hydrocarbon, pentachlorophenol, benzo[a]pyrene, phenol, cyanide, nitrates	Altomare et al. (1999), Kredics et al. (2001), Ezzi and Lynch (2002), Rigot and Matsumura (2002), Ezzi and Lynch (2005), Lopez and Vazquez (2003), Verdin et al. (2005), Anand et al. (2006), Adams et al. (2007), Zhou et al. (2007), Dasilva et al. (2008), Mishra and Nautiyal (2009), Yazdani et al. (2010), Prasad et al. (2013), Blaszczyk et al. (2014), Rybczynska and Kornillowicz-Kowalska (2015), Siddiquee et al. (2015), Zafra and Cortes-Espinosa (2015) and Chew and Ting (2016)
Used in wine industry for the enhancement of flavor	Blaszczyk et al. (2014)
Used in animal husbandry for the enhancement of the nutritive value of animal feed	Galante et al. (1998b), Blaszczyk et al. (2014) and Bayitse et al. (2015)
Production of industrially important biochemicals, i.e., lipases, pectinases, cellulases, and secondary metabolites from different *Trichoderma* species	Reczey et al. (1996), Coradi et al. (2013), Irshad et al. (2014), Pandey et al. (2015) and Kumar et al. (2016)
Production of bioethanol from crop waste with the help of β-glucosidases, cellulase, cellobiohydrolases and endo-β-1,4-glucanases by *Trichoderma reesei*	Shafique et al. (2009), Gusakov (2011), Iqbal et al. (2011), Kawai et al. (2013), Rodriguez-Gomez and Hobley (2013), Abo-State et al. (2014), Blaszczyk et al. (2014), Pandey et al. (2015), Bischof et al. (2016) and Santos et al. (2016)
Bioconversion of lignocellulosic waste with enzymatic activities into valuable products by *Trichoderma reesei*	Damisa et al. (2012), Bayitse et al. (2015), Kumar et al. (2016) and Shida et al. (2016)
Use in paper and pulp industry	Bayitse et al. (2015)
Use in aromatic industry due to the production of coconut-like aroma	Blaszczyk et al. (2014) and Fadel et al. (2015)
Production of enriched protein products	Bayitse et al. (2015)
Production of citric acid by *Trichoderma viride*	Kumar et al. (2016)
Production of metallic nanoparticles by *Trichoderma harzianum*	Singh and Raja (2011)
Used in pharmaceutical industry for the production of camptothecin (CPT), an anticancer drug by *Trichoderma atroviride* LY357, anticancer potential compound 1 6-methylheptadecanoic acid methyl ester from marine *Trichoderma* spp.	You et al. (2010), Qureshi et al. (2011), Pu et al. (2013), Scharf et al. (2016) and Sarvankumar et al. (2015)

(continued)

Table 7.3 (continued)

Industrial application of *Trichoderma* species	Reference(s)
Used in pharmaceutical industry as a promising biocatalyst and has the potential for industrial production of (R)-[3,5-bis (trifluoromethyl) phenyl] ethanol ((R)-BTPE) by *Trichoderma asperellum* ZJPH0810	Li et al. (2013)
Production of high-quality washing powder	Blaszczyk et al. (2014)
In textile industry for biostoning, biofinishing, softening, and conditioning of textile	Galante et al. (1998a) and Blaszczyk et al. (2014)

Trichoderma reesei is widely explored and considered as a biotechnological workhorse due to the proficient production of cellulases (Reczey et al. 1996; Schuster et al. 2012) and hemicellulases used in various industries such as food, medicine, aromatic, and other industries (Gusakov 2011).

7.9 Conclusion

The development of viable alternatives to regulate the excessive use of agrochemicals is a big challenge. Biotechnological applications of genus *Trichoderma* as biopesticide and biofertilizer are promising alternatives to discourage, reduce, substitute, and eliminate the use of agrochemicals for agronomic and environmental benefits. In the future, *Trichoderma*-based commercial formulations should be improved, standardized, and developed with enhanced shelf life, consistent field performance in extreme climatic conditions, and high compatibility with various agrochemicals to meet the objectives of integrated pest management (IPM), integrated nutrient management (INM), and integrated disease management (IDM). In the current era of climate change, more efficient *Trichoderma*-based formulations are needed, which may be developed by employing molecular techniques, harnessing of genomic information for better understanding of various biochemical mechanisms, i.e., intensive metabolomic and genomic studies may elucidate the role of genes responsible for various mechanisms, exploring the role of novel compounds, gene encoding for the development of transgenic plants resistant for various diseases. The development of superior competitive strains by the application of modern techniques like the development of green fluorescent protein fusants with enhanced expression of enzymatic activities will ultimately help to develop effective *Trichoderma*-based biocontrol formulations. Bioprospection of various enzymes and secondary metabolites produced by *Trichoderma* species should be explored for novel agricultural and industrial applications especially in food and pharmaceutical industries for the exploration of new aromatic flavors and drugs. Proper use of biocontrol agents compatible with chemical treatments along with modern agrotechnical methods may reduce the use of harmful pesticides ultimately reducing the agrochemical induced pollution in the environment. *Trichoderma* species successfully play an active role in many environmental cleanup processes by bioremediation due to their enzymatic potential: even heavy-metal-polluted soils may be reclaimed by the use of *Trichoderma* species; heavy-metal-resistant *Trichoderma* mutants may be developed as commercial formulations for metal-contaminated soils. Many reports discussed the potential of *Trichoderma* species in the production of second-generation biofuels utilizing diverse agrowaste; this is a silver lining for the entire human population depending on nonrenewable energy resources. The commercial production of *Trichoderma*-based formulations should be restricted to well-trained and technically sound personnel, and farmers should also be made aware to minimize the risks associated with fungal infection and allergies. Molecular identification based on the sequencing of Internal transcribed spacer (ITS) is not enough; it requires at least a second marker gene such as tef1, ef1728, or rpb2, and reliable screening techniques are needed to identify *Trichoderma* species reported for causing human pathogenicity for health safety aspects. In the larger interest of farmers and technicians, *Trichoderma* species, e.g., *Trichoderma aggressivum*, *Trichoderma pleurotum*, and *Trichoderma pleuroticola*, are reported to cause diseases like green mold in mushroom which result in economic losses to the farmers. Many value-added biological and chemical products may be developed from biotechnological conversion of different agroindustrial waste, which is a cost-effective and environmentally sustainable approach needed on this planet to earn money from waste as well as reduction in pollution. Industrial collaborations, training programs, and sharing of technologies should be encouraged, which will produce better employment opportunities for young professionals ensuring social security. Registration of commercial formulations of *Trichoderma* as biofertilizers and biopesticides is still a herculean task, which should be made easy and less time-consuming. *Trichoderma* is one of the fungi explored for its diverse biotechnological applications in agricultural as well as industrial sectors. Awareness for the adoption of sustainable strategies like the use of biofertilizers and biopesticides should be encouraged among farming community as environmental pollution is more devastating and lethal than hunger and starvation because of the fact that pollution affects both present and future generations. The use of *Trichoderma* species ensures food security as well as environmental security at the same time, which is direly needed for the existence of life on this planet.

References

Abhilash PC, Tripathi V, Edrisi SA et al (2016) Sustainability of crop production from polluted lands. Energy Ecol Environ. doi:10.1007/s40974-016-0007-x

Abo-state MA, Ragab AME, El-gendy N, Farahat LA, Madian HR (2014) Bioethanol production from rice straw enzymatically saccharified by fungal isolates *Trichoderma viride* F-94 and *Aspergillus* F-98. Soft 3:19–29. doi:10.4236/soft.2014.32003

Adam P, De-leij FAAM, Lynch JM (2007) *T. harzianum* Rifai 1295-22 mediates growth promotion of crack willow (*Salix fragilis*) sapling in both clean and metal contaminated soil. Microb Ecol 54:306–313. doi:10.1007/s00248-006-9203-0

Adejumo TO, Ikotun T, Florini D (1999) Biological control of *Protomycopsis phaseoli* the causal agent of leaf smut of Cowpea. J Phytopathol 147:371–375

Ahmad JS, Baker R (1987) Rhizosphere competence of *T. harzianum*. Phytopathology 77:182–189

Akladious AS, Abbas SA (2014) Application of *Trichoderma harzianum* T22 as a biofertilizer potential in maize growth. J Plant Nutr 37:30–49

Akter Z, Weinmann M, Neumann G, Romheld V (2013) An *in vitro* screening methods to study the activity potential of biofertilizers based on *Trichoderma* and *Bacillus* species. J Plant Nutr 36:1439–1452

Al-Naemi FA, Ahmed TA, Nishad R, Radwan O (2016) Antagonistic effects of *Trichoderma harzianum* isolates against *Ceratocystis radicicola* pioneering a biocontrol strategy against black scorch disease in date palm trees. J Phytopathol 164:464–475. doi:10.1111/jph.12472

Al-taweil HI, Osman MB, Aidil AH, Yussof WMW (2009) Optimizing of *Trichoderma viride* cultivation in submerged state fermentation. Am J Appl Sci 6(7):1277–1281

Altomare C, Norvell WA, Bjorkman T, Harman GE (1999) Solubulization of phosphates and micronutrients by the plant growth promoting and biocontrol fungus *Trichoderma harzianum* rifai 1295-22. Appl Environ Microbiol 65:2926–2933

Anand P, Isar J, Saran S, Saxena RK (2006) Bioaccumulation of copper by *T. viride*. Bioresour Technol 97(8):1018–1025

Andrews M, Cripps MG, Edwards GR (2012) The potential of beneficial microorganisms in agricultural systems. Ann Appl Biol 160:1–5. doi:10.1111/j.1744-7348.2011.00519.x

Anita S, Ponmurugan P (2011) *In vitro* evaluation of *Trichoderma atroviride* against *Phomopsis theae* a causal agent of collar canker disease in tea plants. Int J Agric Res 6(8):620–631. doi:10.3923/ijar.2011.620.631

Anjaiah V, Thakur RP, Koedam N (2006) Evaluation of bacteria and *Trichoderma* for biocontrol of preharvest seed infection by *Aspergillus flavus* in groundnut. Biocontrol Sci Tech 16(4):431–436. doi:10.1080/09583150500532337

Arbeli V, Fuentes Cilia L (2007) Accelerated biodegradation of pesticides: an overview of the phenomenon, its basis and possible solutions; and a discussion on the tropical dimension. Crop Prot 26:1733–1746. doi:10.1016/j.cropro.2007.03.009

Arjun KM (2013) Indian agriculture, status, importance and role in Indian economy. Int J Agric Food Sci Technol 4(4):343–346

Azarmi R, Hajieghrari B, Giglou A (2011) Effect of *Trichoderma* isolates on tomato seedling growth response and nutrient uptake. Afr J Biotechnol 10(31):5850–5855

Aziz NH, El-foulyKhalaf LA (1997) Influence of bean seedling root exudates on the rhizosphere colonization by *T. lignorum* for the control of *R. solani*. Bot Bull Acad Sin 38:33–37

Baker R (1989) Improved *Trichoderma* species for promoting crop productivity. Trends Biotechnol 7:34–38

Bayitse R, Hou X, Laryea G, Bjerre AB (2015) Protein enrichment of cassava residue using *Trichoderma pseudokoningii* (ATCC 26801). AMB Express 5:80. doi:10.1186/s13568-015-0166-8

Benitez T, Rincon AM, Limon MC, Codon AC (2004) Biocontrol mechanism of *Trichoderma* strains. Int Microbiol 7:249–260

Bernal-Vicente A, Pascual JA, Tittarelli F, Hernández JA, Diaz-Vivancos P (2015) *Trichoderma harzianum* T-78 supplementation of compost stimulates the antioxidant defence system in melon plants. J Sci Food Agric 95:2208–2214. doi:10.1002/jsfa.6936

Bischof RH, Ramoni J, Seiboth B (2016) Cellulase and beyond: the first 70 years of the enzyme producer *Trichoderma reesii*. Microb Cell Factories 15:106. doi:10.1186/s12934-016-0507-6

Bjorkman T (2004) Effect of *Trichoderma* colonization of auxin mediated regulation of root elongation. Plant Growth Regul 43:89–92

Błaszczyk L, Siwulski M, Sobieralski K, Lisiecka J, Jędryczka M (2014) *Trichoderma* spp. – application and prospects for use in organic farming and industry. J Plant Protect Res 54(4):309–317

Bokhari FM (2009) Efficacy of some *Trichoderma* species in the control of *Rotylen chulusreniformis* and *Meloidogyne javanica*. Arch Phytopathol Plant Protect 42(4):361–369

Bourguignon E (2008) Ecology and diversity of indigenous *Trichoderma* species in vegetable cropping Systems. Ph.D. Thesis, National Centre for Advanced Bio-Protection Technologies, Lincoln University, Canterbury, New Zealand

Cai F, Guanghui Y, Ping W, Zhong W, Lin F, Qirong S, Wei C (2013) Harzianolide, a novel plant growth regulator and systemic resistance elicitor from *Trichoderma harzianum*. Plant Physiol Biochem 73:106–113

Cardoza RE, Malmierca MG, Gutierrez S (2014) Overexpression of erg1 gene in *Trichoderma harzianum* CECT 2413: effect on the induction of tomato defence-related genes. J Appl Microbiol 117:812–823. doi:10.1111/jam.12574

Chad MH (1999) T. virens inoculated composted chicken manure for biological weed control. Biol Control 16:217–222

Chen L, Huang X, Zhang F et al (2012) Application of *Trichoderma harzianum* SQR-T037 bio-organic fertilizer significantly controls *Fusarium* wilt and affects the microbial communities of continuously cropped soil of cucumber. J Sci Food Agric 92:2465–2470

Chet I (1987) *Trichoderma*: application, mode of action and potential as biocontrol agent of soil borne plant pathogenic fungi. In: Chet I (ed) Innovative approaches to plant disease control. Wiley, New York, pp 137–160

Chew SY, Ting ASY (2016) Biosorption behaviour of alginate-immobilized *Trichoderma asperellum*, a common microfungus in single and multi-metal systems. Sep Sci Technol 51(5):743–748. doi:10.1080/01496395.2015.1130059

Christopher MSM (2000) Increased plant growth induced by *Trichoderma* apps (Ecoderma) in paddy. Pestology 24:40–41

Chuang YC, Li WC, Chen C, Hsu PW, Tung S et al (2015) *Trichoderma reesei* meiosis generates segmentally aneuploid progeny with higher xylanase producing capability. Biotechnol Biofuels 8:30. doi:10.1186/s13068-015-0202-6

Copping LG, Menn JJ (2000) Biopesticides: a review of their action, applications and efficacy. Pest Manag Sci 56:651–676

Coradi GV, Visitaçao VL, DeLima EA et al (2013) Comparing submerged and solid-state fermentation of agro-industrial residues for the production and characterization of lipase by *Trichoderma harzianum*. Ann Microbiol 63:533–540. doi:10.1007/s13213-012-0500-1

Corley DG, Miller-Wideman M, Durley RC (1994) Isolation and structure of harzianum a: a new trichothecene from *Trichoderma harzianum*. J Nat Prod 57(3):422–425

Crozier J, Arroyo C, Morales H, Melnick RL, Strem MD et al (2015) The influence of formulation on *Trichoderma* biological activity and frosty pod rot management in *Theobroma cacao*. Plant Pathol 64:1385–1395

Cumagun CJR, Hockenhull J, Lubeck M (2000) Characterization of *Trichoderma* isolates from Philippines rice fields by UP-PCR and rDNA –ITS1 analysis: identification of UP-PCR markers. J Phytopathol 148:109–115

Cutler HG, Jacyno GM (1991) Biological activity of (–) *Harziano pyridone* isolated from *T. Harzianum*. Agric Biol Chem 55:2629–2631

Damisa D, Sule EI, Moneme S (2012) Cellulase production from waste paper using *Trichoderma* species isolated from rhizospheric soil. Afr J Biotechnol 11(97):16342–16346. doi:10.5897/AJB12. 2555

Daryaei A, Jones EE, Ghazalibiglar H, Glare TR, Falloon RE (2016) Effects of temperature, light and incubation period on production, germination and bioactivity of *Trichoderma atroviride*. J Appl Microbiol 120:999–1009. doi:10.1111/jam.13076

DaSilva M, Passarini MRZ, Bonugli RC, Settle LD (2008) Cnidarian derived filamentous fungi from Brazil: isolation, characterization and RBBR decolourisation screening. Environ Technol 29:1331–1339

Dawidziuk A, Popiel D, Kaczmarek J, Strakowska J, Jedryczka M (2016) Optimal *Trichoderma* strains for control of stem canker of brassicas: molecular basis of biocontrol properties and azole resistance. Biol Control. doi:10.1007/s10526-016-9743-2

De Marco JL, Lima LH, Sousa MV, Felix CR (2000) A *Trichoderma harzianum* chitinase destroys the cell wall of the phytopathogen *Crinipellis perniciosa*, the causal agent of witches' broom disease of cocoa. World J Microbiol Biotechnol 16:383–386

Degenkolb T, Nielsen KF, Dieckmann R, Branco-Rocha F et al (2015) Peptaibol, secondary-metabolite, and hydrophobin pattern of commercial biocontrol agents formulated with species of the *Trichoderma harzianum* complex. Chem Biodivers 12:662–668

Djonovic S, Vargas WA, Kolomiets MV, Horndeski M, Weist A, Kenerley CM (2007) Aproteinaceous elicitor Sm1 from the beneficial fungus *T. virens* is required for systemic resistance in maize. Plant Physiol 145:875–889

Dohroo NP (2001) Study of population dynamics of naturally occurring *T. harzianum* Rifai and its antagonistic potential against rhizome rot of ginger. Indian J Plant Pathol 19:39–43

Doni F, Isahak A, Zain CRCM, Yusoff WMW (2014) Physiological and growth response of rice plants (*Oryza sativa* L.) to *Trichoderma* spp. inoculants. AMB Express 4:45

Druzhinina IS, Kopchinskiy AG, Kubicek CP (2006) The first 100 *Trichoderma* species characterized by molecular data. Mycoscience 47:55–64

Dubey SC, Bhavani R, Singh B (2011) Integration of soil application and seed treatment formulations of *Trichoderma* species for management of wet root rot of mungbean caused by *Rhizoctoniasolani*. Pest Manag Sci 67:1163–1168

Elad Y, Chet I, Henis Y (1981) A selective medium for improving quantitative isolation of *Trichoderma*spp from soil. Phytoparasitica 9(1):59–67

El-Hasan A, Walker F, Schone J, Buchenauer H (2009) Detection of viridofungin A and other antifungal metabolites secreted by *Trichoderma harzianum* active against different plant pathogens. Eur J Plant Pathol 124:457–470

El-Shishtawy RM, Mohamed SA, Asiri AM et al (2015) Saccharification and hydrolytic enzyme production of alkali pre-treated wheat bran by *Trichoderma virens* under solid state fermentation. BMC Biotechnol 15:37. doi:10.1186/s12896-015-0158-4

Ezzi MI, Lynch JM (2002) Cyanide catabolizing enzymes in *Trichoderma* spp. Enzym Microb Technol 31:1042–1047

Ezzi MI, Lynch JM (2005) Biodegradation of cyanide by *Trichoderma* spp. and *Fusarium* spp. Enzym Microb Technol 36:849–854

Fadel HHM, Mahmoud MG, Asker MMS, Lofty SN (2015) Characterization and evaluation of coconut aroma produced by *Trichoderma viride* EMCC-107 in solid state fermentation on sugarcane bagasse. Electron J Biotechnol 18(1):5–9

Ferigo D, Raiola A, Piccolo E, Scopel C, Causin R (2014) *Trichoderma harzianum* T22 induces in maize systemic resistance against *Fusarium verticillioides*. J Plant Pathol 96(1):133–142

Gajera H, Domadiya R, Patel S, Kapopara M, Golakiya B (2013) Molecular mechanism of *Trichoderma* as bio-control agents against phytopathogen system – a review. Curr Res Microbiol Biotechnol 1(4):133–142

Galante Y, De Conti A, Monteverdi (1998a) Application of *Trichoderma* enzymes in the textile industry. In: Harman G, Kubicek C (eds) *Trichoderma* and *Gliocladium*, enzymes, biological control and commercial applications, vol 2, pp 311–326

Galante Y, De Conti A, Monteverdi (1998b) Application of *Trichoderma* enzymes in the food and feed industry. In: Harman G, Kubicek C (eds) *Trichoderma* and *Gliocladium*, enzymes, biological control and commercial applications, vol 2, pp 327–342

Garnica-vergara A, Barrera-ortiz S, Munoz-Parra E et al (2016) The volatile 6-pentyl-2H-pyran-2-one from *Trichoderma atroviride* regulates *Arabidopsis thaliana* root morphogenesis via auxin signaling and ethylene insensitive functioning. New Phytol 209:1496–1512

Gasso A, Prosperi A (2001) Growth promotion with *Trichoderma* spp. formulation in four crops during early stages. Argentenian Plant Pathol 21:56

Gil SV, Pastor S, March GJ (2009) Quantitative isolation of biocontrol agents *Trichoderma* spp., *Gliocladium* spp. and actinomycetes from soil with culture media. Microbiol Res 164:196–205

Goswami J, Pandey RK, Tewari JP, Goswami BK (2008) Management of root knot nematode on tomato through application of fungal antagonists, *Acremonium strictum* and *Trichoderma harzianum*. J Environ Sci Health B 43:237–240

Govindappa M, Rai VR, Lokesh S (2011) *In vitro* and *in vivo* responses of different treating agents against wilt disease of safflower. J Cereals Oilseeds 2(1):16–25

Gusakov AV (2011) Alternatives to *Trichoderma reesei* in biofuel production. Trends Biotechnol 29(9):419–425

Hajji EM, Rebuffat S, Lecommandeur D, Bodo B (1987) Isolation and determination of trichorzianines an antifungal peptide from *T.harzianum*. Int J Pept Protein Res 29:207–215

Harman GE (2006) Overview of mechanism and uses of *Trichoderma* spp. Phytopathology 96:190–194

Harman GE, Howell CR, Viterbo A, Chet I, Lorito M (2004) *Trichoderma* species- opportunistic, avirulent plant symbionts. Nat Rev Microbiol 2:43–56

Harris GH, Turner ET, Meinz MS et al (1993) Isolation and structure elucidation of virido fungins A, B and C. Tetrahedron Lett 34(33):5235–5238

Hasan MM, Ismail IA, Sorour AA (2014) Phylogeny and antagonistic activity of some protoplast fusants in *Trichoderma* and Hypocrea. Int J Appl Sci Biotechnol 2(2):146–151

Hasanloo T, Kowsari M, Naraghi MS, Bagheri O (2010) Study of different *Trichoderma* strains on growth characteristics and silymarin accumulation of milk thistle plant. J Plant Interact 5(1):45–49

Heidrum A, Josaine Kinn Karlerik B, Olov S (1991) Production of siderophores by strains of genus *Trichoderma*. Biometals 4:176–180

Hermosa MR, Grondona I, Iturriaga EA (2000) Molecular charachterization and identification of biocontrol isolates of *Trichoderma* spp. Appl Environ Microbiol 166:1890–1898

Hermosa R, Viterbo A, Chet I, Monte E (2012) Plant beneficial effects of *Trichoderma* and of its genes. Microbiology 158:17–25

Hoitink HAJ, Madden LV, Dorrance AE (2006) Systemic resistance induced by *Trichoderma* spp. interaction between the host, the pathogen, the biocontrol agent and the soil organic matter quality. Phytopathology 96(2):186–189

Howell CR (2003) Mechanism employed by *Trichoderma* spp. in the biological control of plant diseases: the history and evolution of current concepts. Plant Dis 87(1):4–10

Hutchinson CM (1999) *Trichoderma virens*-inoculated composted chicken manure for biological weed control. Biol Control 16:217–222

Inbar J, Menedez A, Chet I (1996) Hyphal interaction between *T. harzianum* and *S. sclerotiorum* and its role in biological control. Soil Biol Biochem 28(6):757–763

Inch S, Walker DJ, Gilbert J et al (2011) The development of a model to predict the potential efficacy of *Trichoderma harzianum* isolates on perithecial production of *Gibberella zeae* based on secondary metabolite production. Can J Plant Pathol 33(3):337–346

Iqbal SM, Bakhsh A, Hussain S, Malik BA (1995) Microbiological antagonism against *S. rolfsii* the cause of collar rot of Lentil. Lens News Lett 22(1):48–49

Iqbal HMN, Ahmed I, Zia MA, Irfan M (2011) Purification and characterization of the kinetic parameters of cellulase produced from wheat straw by *Trichoderma viride* under SSF and its detergent compatibility. Adv Biosci Biotechnol 2:149–156

Irshad M, Anwar Z, Mahmood Z, Aqil T, Mehmmod S, Nawaz H (2014) Bio-processing of agro-industrial waste orange peel for induced production of pectinase by *Trichoderma viride*; its purification and characterization. Turk J Biochem 39(1):9–18

Islam MN, Shamsuddula AM, Ahmed HP (2007) Comparative effectiveness of *Trichoderma* colonized organic waste in controlling foot and root rot (*Sclerotium rolfsii*) disease of wheat. Afr Crop Sci Conf Proc 8:2079–2082

Islam MM, Hossain DM, Rahman MME, Suzuki K et al (2016) Native *Trichoderma* strains isolated from Bangladesh with broad spectrum antifungal action against fungal phytopathogens. Arch Phytopathol Plant Protect 49(1–4):75–93

Jash S, Pan S (2007) Variability in antagonistic activity and root colonizing behaviour of *Trichoderma* isolates. J Trop Agric 45(1–2):29–35

Javaid A, Ali S (2011) Herbicidal activity of culture filtrates of *Trichoderma* spp. against two problematic weeds of wheat. Nat Prod Res 25(7):730–740

Jayalakshmi SK, Raju S, Usha R et al (2009) *Trichoderma harzianum* L1 as a potential source for lytic enzymes and elicitor of defense responses in chickpea (C*icer arietinum*) against wilt disease caused by *Fusarium oxysporum* f. sp. *ciceri*. Aust J Crop Sci 3(1):44–52

Jebakumar RS, Anandraj M, Sharma YR (2000) Compatability of phorate and chloropyriphos with *T. harzianum* (rifai.) applied for integrated disease management in black pepper (*Piper nigrum* L.) J Spices Aromatic Crops 9(2):111–115

John RP, Tyagi RD, Prévost D, Brar Satinder K et al (2010) Mycoparasitic *Trichoderma viride* as a biocontrol agent against *Fusarium oxysporum* f. sp. *adzuki* and *Pythium arrhenomanes* and as a growth promoter of soybean. Crop Prot 29:1452–1459

Jones EE, Bienkowski DA, Stewart A (2016) The importance of water potential range tolerance as a limiting factor on *Trichoderma* spp. bio control of *Sclerotinia sclerotiorum*. Ann Appl Biol 168:41–51

Joshi BB, Bhatt RP, Bahukhandi D (2010) Antagonistic and plant growth activity of *Trichoderma* isolates of Western Himalayas. J Environ Biol 31(6):921–928

Kalia A, Gosal SK (2011) Effect of pesticide application on soil microorganisms. Arch Agron Soil Sci 57(6):569–596

Kawai T, Nakazawa HI, da N Okada H, Ogasawara W et al (2013) A comprehensive analysis of the effects of the main component enzymes of cellulase derived from *Trichoderma reesei* on biomass saccharification. J Ind Microbiol Biotechnol 40:805–810

Kexiang G, Xiaoguang L, Yonghong L, Tiantro Z, Shuliang W (2002) Potential of *T. harzianum* and *T. atroviride* to control *B. berengeriana* f. sp. *piricola* the cause of apple ring rot. J Phytopathol 150:271–276

Khan MR, Gupta J (1998) Antagonistic efficacy of *Trichoderma* species against *M. phaseolina* on eggplant. J Plant Disease Protect 105(4):387–393

Khan MR, Haque Z (2011) Soil application of *Pseudomonas flourescens* and *Trichoderma harzianum* reduces root-knot nematode *Meloidogyne incognita* on tobacco. Phytopathol Mediterr 50:257–266

Khan MO, Shahzad S (2007) Screening of *Trichoderma* species for tolerance to fungicides. Pak J Bot 39:945–951

Khan J, Ooka J, Miller S, Madden A, Hoitink HA (2004) Systemic resistance induced by *T. hamatum* 382 in cucumber against *Phytophthora* crown rot and leaf blight. Plant Dis 88:280–286

Khan AA, Sinha AP, Rathi YPS (2005) Plant growth promoting activity of *Trichoderma harzianum* in rice seed germination and seedling vigour. Indian J Agric Res 39(4):256–262

Khan MS, Zaidi A, Ahemad M et al (2010) Plant growth promotion by phosphate solubilizing fungi – current perspective. Arch Agron Soil Sci 56(1):73–98

Khan MO, Shah N, Shahzad S, Khan MN (2013) Induce tolerance in *Trichoderma* species against fungicide. PJLS 1:35–42

Khater HF (2012) Prospects of botanical biopesticides in insect pest management. J Appl Pharm Sci 2(05):244–259

Khoo GM, Clausen MR, Pedersen HL, Larsen E (2012) Bioactivity and chemical composition of black currant (*Ribes nigrum*) cultivars with and without pesticide treatment. Food Chem 132:1214–1220

Kidwai MK (2007) Selection, characterization and application of *Trichoderma* for control of soil borne plant pathogens. Ph.D. Thesis, University of Lucknow, Lucknow

Kidwai MK, Vikas SS, Singh HB (2006) Compatibility of *T. harzianum* to selected fungicides. J Ecofriendly Agric 1(2):156–161

Kidwai MK, Vikas SS, Srivastava SC, Singh HB (2007) Compatibility of *T. harzianum* to selected insecticides. J Ecofriendly Agric 2(2):170–174

Kim CS, Shirouzu T, Nakagiri A, Sotome K, Maekawa N (2013) *Trichoderma eijii* and *T. pseudolacteum*, two new species from Japan. Mycol Prog 12:739–753

Kredics L, Antal Z, Manczinger L, Nagy E (2001) Breeding of mycoparasitic *Trichoderma* strains for heavy metal resistance. Lett Appl Microbiol 33:112–116

Kredics L, Manczinger L, Antal Z, Penzes Z et al (2004) *In vitro* water activity and pH dependence of mycelial growth and extracellular enzyme activities of *Trichoderma* strains with biocontrol potential. J Appl Microbiol 96:491–498

Kredics L, Láday M, Körmöczi P, Manczinger L et al (2012) Genetic and biochemical diversity among *Trichoderma* isolates in soil samples from winter wheat fields of the great hungarian plain. Acta Biologica Szegediensis 56(2):141–149

Krishnamoorthy AS, Baskaran R (1990) Biological control of damping off disease of tomato caused by *Pythium indicum*. J Biol Control 4:52–54

Kumar M, Sharma P (2011) Molecular and morphological characters: an appurtenance for antagonism in *Trichoderma* spp. Afr J Biotechnol 10(22):4532–4543

Kumar S, Thakur M, Rani A (2014) *Trichoderma*: mass production, formulation, quality control delivery and its scope in commercialization in India for the management of plant diseases. Afr J Agric Res 9(53):3838–3852

Kumar A, Gautam A, Dutt D (2016) Biotechnological transformation of lignocellulosic biomass into industrial products: an overview. Adv Biosci Biotechnol 7:149–168

Lane SD, Bowen NJ (2005) Revisiting the use of Iprodione and *Trichoderma* in the integrated management of onion white rot. Arch Phytopathol Plant Protect 38(2):133–138

Leifeld J, Fuhrer J (2010) Organic farming and soil carbon sequestration: what do we really know about the benefits. Ambio 39:585–599

Leobardo S, Carreon VJ, Antonio R, Estrada M, Enrique G (2006) Shake flask to stirred fermenters: scale up of an extractive fermentation process for 6 pentyl-α-pyrone production by *Trichoderma harzianum* using volumetric power input. Process Biochem 41:1347–1352

Levillian J, Cattan P, Colin F, Voltz M, Cabidoche (2012) Analysis of environmental and farming factors of soil contamination by POP chlordecone in a banana production area of French west indies. Agric Ecosyst Environ 159:123–132

Levy NO, Harell YM, Haile ZM et al (2015) Induced resistance to foliar diseases by soil solarization and *Trichoderma harzianum*. Plant Pathol 64:365–374

Lewis JA, Larkin RP (1997) Extruded granular formulation with biomass of biocontrol *Gliocladium virens* and *Trichoderma* spp. to reduce damping-off of eggplant caused by *Rhizoctonia solani* and saprophytic growth of the pathogen in soil-less mix. Biocontrol Sci Tech 7:49–60

Li J, Wang P, He J, Huang J, Tang J (2013) Efficient biocatalytic synthesis of (R)-[3, 5-bis (trifluoromethyl) phenyl] ethanol by a newly isolated *Trichoderma asperellum* ZJPH0810 using dual cosubstrate: ethanol and glycerol. Appl Microbiol Biotechnol 97:6685–6692

Linke R, Thallinger GG, Haarmann T, Eidner J et al (2015) Restoration of female fertility in *Trichoderma reesei* QM6a provides the basis for inbreeding in this industrial cellulose producing fungus. Biotechnol Biofuels 8:155

Lo CT, Lin CY (2002) Screening strains of *Trichoderma* species for plant growth enhancement in Taiwan. Plant Pathol Bull 11:215–220

Lopez EE, Vazquez C (2003) Tolerance and uptake of heavy metals by *Trichoderma atroviride* isolated from sludge. Chemosphere 50:137–143

Malmierca MG, Mccormick SP, Cardoza RE et al (2015) Production of trichodiene by *Trichoderma harzianum* alters the perception of this biocontrol strain by plants and antagonized fungi. Environ Microbiol 17(8):2628–2646

Manjula K, Kishore GK, Girish AG, Singh SD (2004) Combined application of *Pseudomonas fluorescens* and *Trichoderma viride* has an improved biocontrol activity against stem rot in groundnut. Plant Pathol J 20(1):75–80

Marin-Guirao JI, Rodrıguez-Romera P, Rodrıguez Lupion B et al (2016) Effect of *Trichoderma* on horticultural seedlings growth promotion depending on inoculum and substrate type. J Appl Microbiol 1:1–8

Martin K, Saurborn J (2013) Origin and development of agriculture. Agroecology 2:9–48

Martínez FD, Santos M, Carretero F, Marín F (2016) *Trichoderma saturnisporum,* a new biological control agent. J Sci Food Agric 96:1934–1944

Mathivanan N, Srinivasan K, Chelliah S (2000) Biological control of soil-borne diseases of cotton, eggplant, okra and sunflower with *Trichoderma viride.* J Plant Dis Protect 107:235–244

Mathivanan N, Prabhavathy VR, Vijayanandraj VR (2005) Application of talc formulations of *Pseudomonas fluorescens* migula and *Trichoderma viride* Pers. ex S.F. Gray decrease the sheath blight disease and enhance the plant growth and yield in rice. J Phytopathol 153:697–701

Mbazia A, Youssef NOB, Kharrat M (2016) Tunisian isolates of *Trichoderma* spp. and *Bacillus subtilis* can control *Botrytis fabae* on faba bean. Biocontrol Sci Tech 26(7):915927

Mckenzie AJ, Starman T (1995) Enhanced root and shoot growth of chrysanthemum cutting propogated with *T. harzianum.* Hortic Sci 30(3):496–498

Mclean KL, Saminathan JC, Hunt F, Ridgway HJ, Stewart A (2005) Effect of formulation on the rhizosphere competence and biocontrol ability of *T. atroviride* C52. Plant Pathol 54:212–218

Meki S, Ahmed S, Sakhuja PK (2011) Control of chickpea wilt (*Fusarium oxysporum* f. sp. *ciceris)* using *Trichoderma* spp. in Ethiopia. Arch Phytopathol Plant Protect 44(5):432–440

Melo IS (1991) Antagonism of *Trichoderma* species against *S. sclerotiorum* and *S. minor in vitro.* Rev Microbiol 22(2):147–150

Menedez AB, Godeas A (1998) Biological control of *S. sclerotiorum* attacking soybean plants degradation of cell wall by *T.harzianum* (BAFC-742). Mycopathologia 142:153–160

Michel M, Gnusowski B, Buszewski B (2006) Comparison of various extraction techniques to determine fungicide residue in wheat grain. J Liq Chromatogr Relat Technol 29:247–261

Mishra A, Nautiyal CS (2009) Functional diversity of the microbial community in the rhizosphere of chickpea grown in diesel fuel-spiked soil amended with *Trichoderma reesei* using sole-carbon-source utilization profiles. World J Microbiol Biotechnol 25:1175–1180

Mishra BK, Mishra RK, Mishra RC, Tiwari AK et al (2011) Biocontrol efficacy of *Trichoderma viride* isolates against fungal plant pathogens causing disease in *Vigna radiate* L. Arch Appl Sci Res 3(2):361–369

Moosa A, Sahi ST, Haq IU, Farzand A, Khan SA, Javaid K (2016) Antagonistic potential of *Trichoderma* isolates and manures against fusarium wilt of tomato. Int J Vegetable Sci:1–12. doi:10.1080/19315260.2016.1232329

Morris J, Burgess PJ (2012) Modern agriculture and implication of land use and management. Environ Impacts Modern Agric 34:1–34

Motlagh MRS, Samimi Z (2013) Evaluation of *Trichoderma* spp. as biocontrol agents in some of plant pathogens. Ann Biol Res 4(3):173–179

Mukherjee M, Mukherjee PK, Horwitz BA et al (2012) *Trichoderma*–plant–pathogen interactions: advances in genetics of biological control. Indian J Microbiol 52(4):522–529

Mukhopadhaya AN (1994) Biocontrol of soilborne fungal pathogens current status, future prospects and potential limitation. Ind Phytopathol 47(1):1–8

Mukhopadhaya AN, Kaur NP (1990) Biological control of soil borne funal plant pathogens current status, future prospects and potential limitations. Indian Phytopath 42(2):119–126

Mulaw TB, Kubicek CP, Druzhinina IS (2010) The rhizosphere of *Coffea arabica* in its native highland forests of Ethiopia provides a niche for a distinguished diversity of *Trichoderma*. Diversity 2:527–549

Mutawila C, Vinale F, Haleen F, Lorito M, Mostert L (2016) Isolation, production and *in vitro* effect of major secondary metabolite produced by *Trichoderma* species used for control of grapevine trunk diseases. Plant Pathol 65:104–113

Naher L, Yusuf UK, Siddiquee S, Ferdous J, Aminur Rahman M (2012) Effect of media on growth and antagonistic activity of selected *Trichoderma* strains against *Ganoderma*. Afr J Microbiol Res 6(48):7449–7453

Nemcovik M, Jakubikova L, Viden I, Farkas V (2008) Induction of conidiation by endogenous volatile compounds in *Trichoderma* spp. FEMS Microbiol Lett 284:231–236

Niewiadomska A (2004) Effect of carbendazim, imazeptapir, and thiram on nitrogenase activity, number of microorganism in soil and the yield of red clover (*Trifolium pretense* L.) Pol J Environ Stud 13(4):403–410

Niranjana SR, Lalitha S, Hariprasad P (2009) Mass multiplication and formulations of biocontrol agents for use against *Fusarium* wilt of pigeonpea through seed treatment. Int J Pest Manag 55(4):317–324

Nosir WS (2016) *Trichoderma harzianum* as growth promoter and biocontrol agent against *Fusarium oxysporum* f. sp. *tuberosi*. Adv Crop Sci Technol 4:217

Nzanza B, Marais D, Soundy P (2012) Response of tomato (*Solanum lycopersicum* L.) to nursery inoculation with *Trichoderma harzianum* and arbuscularmycorrhizal fungi under field conditions. Acta Agric Scandinavica Sect B Soil Plant Sci 62:209–215

Odukkathil G, Vasudevan N (2013) Toxicity and bioremediation of pesticides in agricultural soil. Rev Environ Sci Biotechnol 12:421–444

Ommati F, Zaker M (2012) *In vitro* and greenhouse evaluations of *Trichoderma* isolates for biological control of potato wilt disease (*Fusarium solani*). Arch Phytopathol Plant Protect 45(14):1715–1723

Onilude AA, Tayo BCA, Odeniyi AO et al (2013) Comparative mycelia and spore yield by *Trichoderma viride* in batch and fed-batch cultures. Ann Microbiol 63:547–553

Oros G, Naar Z, Cserhati T (2011) Growth response of *Trichoderma* species to organic solvents. Mol Inf 30:276–285

Otadoh JA, Okoth SA, Ochanda J, Kahindi J (2011) Assessment of *Trichoderma* isolates for virulence efficacy on *Fusarium oxysporum* f. sp. *phaseoli*. Tropical Subtropical Agroecosyst 13:99–107

Ozbay N, Newman SE (2004) Biological control with *Trichoderma* spp. with emphasis on *T. harzianum*. Pak J Biol Sci 7(4):478–484

Pandey J, Singh A (2012) Opportunities and constraints in organic farming: an Indian perspective. J Sci Res 56:47–72

Pandey S, Srivastava M, Shahid M, Kumar V, Singh A, Trivedi S, Srivastava YK (2015) *Trichoderma* species cellulase produced by solid state fermentation. J Data Mining Genomic Proteonomic 6:2. dx/doi.org/10.4172/2153-0602.1000170

Pandya JR, Sabalpara AN, Chawda SK (2011) *Trichoderma*: a particular weapon for biological control of phytopathogens. J Agric Technol 7(5):1187–1191

Papavizas GC (1982) Survival of *Trichoderma harzianum* in soil and pea and bean rhizospheres. Ecol Epidemiol 72(1):121–125

Papavizas GC, Lewis JA (1983) Physiological and biological characterization of stable mutants of *T. viride* resistant to MBC fungicides. Phytopathology 73:407–411

Parmar HJ, Hasan MM, Bodar NP et al (2015) *In vitro* antagonism between phytopathogenic fungi *Sclerotium rolfsii* and *Trichoderma* strains. Int J Appl Sci Biotechnol 3(1):16–19

Petit AN, Fontane F, Vasta P, Clement C, Gaveau NV (2012) Fungicide impact on photosynthesis in crop plants. Photosynth Res 44(3):224–230

Prakash MG, Gopal KV, Anandraja M, Sarma YR (1999) Evaluation of substrates for mass multiplication of fungal biocontrol agents *T. harzianum* and *T. virens*. J Spices Aromatic Crops 8(2):207–210

Prasad ASA, Varatharaju G, Anushri C, Dhivyasree S (2013) Biosorption of lead by *Pleurotus florida* and *Trichoderma viride*. British Biotechnol J 3(1):66–78

Prasada RD, Rangeshwarana R, Hegdeb SV, Anuroop CP (2002) Effect of soil and seed application of *T. harzianum* on pigeon pea wilt caused by *F. indicum* under field conditions. Crop Prot 26:782–787

Pu X, Qu X, Chen F, Bao J, Zhang G, Luo Y (2013) Camptothecin-producing endophytic fungus *Trichoderma atroviride* LY357: isolation, identification, and fermentation conditions optimization for camptothecin production. Appl Microbiol Biotechnol 97:9365–9375

Qi W, Zhao L (2012) Study of the siderophore producing *T. asperellum* Q1 on cucumber growth promotion under salt stress. J Basic Microbiol 52:1–10

Qingxiang Y, Jing Z, Kanfang Z, Hao Z (2009) Influence of oxytetracycline on the structure and activity of microbial community in wheat rhizospheric soil. J Environ Sci 21:954–959

Qureshi SA, Hira SV, Ara J, Haque SE (2011) Cytotoxic potential of fungi associated with rhizosphere and rhizoplane of wild and cultivated plants. Pak J Bot 43(6):3025–3028

Rabeendran N, Moot DJ, Jones EE, Steward A (2005) Evaluation of fungal isolates for the control of *S. sclerotiorum* using cabbage pot bioassays. N Z Plant Protect 58:251–255

Rad MH, Jagannath S (2011) Effect of herbicide imazethapyr on chickpea seed germination. Arch Phytopathol Plant Protect 44(3):224–230

Ramada MHS, Steindorff AS, Block C Jr, Ulhoa CJ (2016) Secretome analysis of the myco parasitic fungus *Trichoderma harzianum* ALL42 cultivated in different media supplemented with *Fusarium solani* cell wall or glucose. Proteomics 16:477–490

Ratnakumari RY, Nagamani A, Bhramaramba S et al (2011) Nonvolatile and volatile metabolites of antagonistic *Trichoderma* against collar rot pathogen of *Mentha arvensis*. Adv Nat Appl Sci 5(1):55–57

Reczey K, Szengyel Z, Eklund R, Zachhi G (1996) Cellulase production by *T. reesei*. Bioresour Technol 57:25–30

Reddy K, Krishnamma Narayana P (2009) Efficacy of *T. viride* against *Colletotrichum falcatum* in sugarcane. Indian J Plant Protect 37:111–115

Rigot J, Matsumura F (2002) Assessment of the rhizosphere competency and pentachlorophenol metabolizing ability of a pesticide degrading strain of *Trichoderma harzianum* introduced into the root zone of corn seedlings. J Environ Sci Health 37(3):201–210

Rinu K, Sati P, Pandey A (2013) *Trichoderma gamsii* (NFCCI 2177): a newly isolated endophytic, psychrotolerant, plant growth promoting and antagonistic fungal strain. J Basic Microbiol 1:1–10

Roca MF, Cristani C, Vanacii G (1996) Senstivity of *Trichoderma* isolates and selected mutants to DMI fungicides. Crop Prot 15(7):615–620

Rocco A, Perez LM (2001) *In vitro* biocontrol activity of *T. harzianum* on *A. alternata* in the presence of growth regulators. Electron J Biotechnol 4(2):68–73

Rodriguez-Gomez D, Hobley TJ (2013) Is an organic nitrogen source needed for cellulase production by *Trichodermareesei* Rut-C30. World J Microbiol Biotechnol 29:2157–2165

Rojo FG, Reynoso Maria M, Marcela F, Chulze SN, Torres AM (2007) Biological control by *Trichoderma* species of *F. solani* causing peanut brown root rot under field condition. Crop Prot 26:549–555

Roy AK (1977) Parasitic activity of *T. viride* on the sheath blight fungus of rice *Corticum sasaki*. J Plant Disease Protect 84(11):675–683

Rybczynska K, Kornillowicz-kowalska (2015) Evaluation of dye compounds decolourization capacity of selected *H. haematococca* and *T. harzianum* strains by principal component analysis (PCA). Water Air Soil Pollut:226–228. doi:10.1007/s11270-015-2473-8

Sain SK, Pandey AK (2016) Biological spectrum of *Trichoderma harzianum* Rifai isolates to control fungal diseases of tomato (*Solanum lycopersicon* L.) Arch Phytopathol Plant Protect. doi: 10.1080/03235408.2016.1242393

Sallam NMA, Abo-Elyousr, Hasan MAE (2008) Evaluation of *Trichoderma* species as biocontrol agents of damping off and wilt disease of *Phaseolus vulgaris* L. and efficacy of suggested formula. Egypt J Phytopathol 36(1–2):81–93

Samuels G (2006) *Trichoderma*: systematics, the sexual stage and ecology. Phytopathology 96(2):195–206

Santos CA, Zanphorlin LM, Crucello A, Tonoli CCC et al (2016) Crystal structure and biochemical characterization of the recombinant ThBgl, a GH1 β-glucosidase overexpressed in *Trichoderma harzianum* under biomass degradation conditions. Biotechnol Biofuels 9:71. doi:10.1186/s13068-016-0487-0

Sarwankumar K, Vivek R, Boopathy NS, Yaqian L, Kathiresan K, Chen J (2015) Anticancer potential of bioactive 1 6 methylheptadecanoic acid methyl ester derived from marine *Trichoderma*. J Appl Biomed 13(3):199–212

Saxena A, Raghuwanshi R, Singh HB (2014) *Trichoderma* species mediated differential tolerance against biotic stress of phytopathogens in *Cicer arietinum* L. J Basic Microbiol 55:195–206

Scharf DH, Brakhage AA, Mukherjee PK (2016) Gliotoxin- bane or boon. Environ Microbiol 18(4):1096–1109

Schuster A, Schmoll M (2010) Biology and biotechnology of *Trichoderma*. Appl Microbiol Biotechnol 87:787–799

Schuster A, Bruno KS, Collett JR, Baker SE et al (2012) A versatile toolkit for high throughput functional genomics with *Trichoderma reesei*. Biotechnol Biofuels 5(1):1

Seidl V, Huemer B, Seiboth B, Kubicek CP (2005) A complete survey of *Trichoderma* chitinases reveals three distinct subgroups of family 18 chitinases. FEBS J 272:5923–5939

Shafique S, Bajwa R, Shafique S (2009) Cellulase biosynthesis by selected *Trichoderma* species. Pak J Bot 41(2):907–916

Shahid M, Singh A, Srivastava M, Srivastava DK (2014) Molecular characterization of *Trichoderma viride* isolated from rhizospheric soils of Uttar Pradesh based on rDNA markers and analysis of their PCR-ISSR profiles. J Mol Biomark Diagn 5:169. doi:10.4172/2155-9929.1000169

Sharma V, Shanmugam V (2012) Purification and characterization of an extracellular 24 kDa chitobiosidase from the mycoparasitic fungus *Trichoderma saturnisporum*. J Basic Microbiol 52:324–331

Sharma K, Mishra AK, Mishra RS (2009) Morphological, biochemical and molecular characterization of *Trichoderma harzianum* isolates for their efficacy as biocontrol agents. J Phytopathol 157:51–56

Sharma R, Joshi A, Dhaker RC (2012) A brief review on mechanism of *Trichoderma* fungus use as biological control agent. Int J Innovations BioSci 2(4):200–210

Shida Y, Furukawa T, Ogasawara W (2016) Deciphering the molecular mechanisms behind cellulase production in *Trichoderma reesei*, the hyper-cellulolytic filamentous fungus. Biosci Biotechnol Biochem 80(9):1712–1729

Shukla N, Awasthi RP, Rawat L, Kumar J (2015) Seed biopriming with drought tolerant isolates of *T. harzianum* promote growth and drought tolerance in *Triticum aestivum*. Ann Appl Biol 166:171–182

Siddiquee S, Rovina K, Azad SA, Naher L et al (2015) Heavy metal contaminants removal from wastewater using the potential filamentous fungi biomass: a review. J Microbial Biochem Technol 7(6):384–393

Siddiqui IA, Zareen A, Zaki MJ, Shaukat SS (2001) Use of *Trichoderma* species in the control of *Meloidogyne javanica* root knot nematode in okra and mungbean. Pak J Biol Sci 4(7):846–848

Singh PC, Nautiyal CS (2012) A novel method to prepare concentrated conidial biomass formulation of *Trichoderma harzianum* for seed application. J Appl Microbiol 113:1442–1450

Singh P, Raja RB (2011) Biological synthesis and characterization of silver nanoparticles using the fungus *Trichoderma harzianum*. Asian J Exp Biol Sci 2(4):600–605
Singh A, Srivastava S, Singh HB (2007) Effect of substrate on growth and shelf life of *Trichoderma harzianum* and its use in biocontrol of diseases. Bioresour Technol 98:470–473
Singh HB, Singh BN, Singh SP, Sarma BK (2012) Exploring different avenues of *Trichoderma* as a potent biofungicidal and plant growth promoting candidate—an overview. Rev Plant Pathol 5:315–426
Singh BN, Singh A, Singh BR, Singh HB (2013) *Trichoderma harzianum* elicits induced resistance in sunflower challenged by *Rhizoctonia solani*. J Appl Microbiol:1–13. doi:10.1111/jam.12387
Sivan A, Chet I (1987) Biological control of *Fusarium* crown rot of tomato by *T. harzianum* under field conditions. Plant Dis 71:589–592
Sivan A, Chet I (1989) Degradation of fungal cell walls by lytic enzymes of *Trichoderma harzianum*. J Gen Microbiol 135:675–682
Skoneczny D, Oskiera M, Szczech M, Bartoszewski G (2015) Genetic diversity of *Trichoderma atroviride* strains collected in Poland and identification of loci useful in detection of within-species diversity. Folia Microbiol 60:297–307
Soetan KO (2011) The role of biotechnology towards attainment of a sustainable and safe global agriculture and environment. Biotechnol Mol Biol Rev 6(5):109–117
Soglio FK, Bertagnolli BL, Sinclair JB, Yu GY, Eastburn DM (1998) Production of chitinolytic enzymes and endoglucanase in the soybean rhizosphere in presence of *T. harzianum* and *R. solani*. Biol Control 12:111–117
Soresh M, Yedidia I, Chet I (2005) Involvement of jasmonic acid/ethylene signaling pathways in systemic resistance induced in cucumber by *T. asperellum* T-203. Phytopathology 95:76–84
Srivastava M, Kumar V, Shahid M, Pandey S, Singh A (2016) *Trichoderma*- a potential and effective bio fungicide and alternative source against notable phytopathogens: a review. Afr J Agric Res 11(5):310–316
Strakowska J, Błaszczyk L, Chełkowski J (2014) The significance of cellulolytic enzymes produced by *Trichoderma* in opportunistic lifestyle of this fungus. J Basic Microbiol 54(1):12–13
Suarez B, Rey M, Castillo P, Monte E, Llobell A (2004) Isolation and characterization of PRA1, a trypsin like protease from the biocontrol agent *Trichoderma harzianum* CECT 2413 displaying nematicidal activity. Appl Microbiol Biotechnol 65(1):46–55
Syed JH, Alamdar A, Mohammad A, Ahad K, Shabir Z et al (2014) Pesticide residues in fruits and vegetables from Pakistan: a review of the occurrence and associated human health risks. Environ Sci Pollut Res 21:13367–13393
Tiyagi SA, Ajaz S, Azam MF (2004) Effect of some pesticides on plant growth, root nodulation and chlorophyll content of chickpea. Arch Agron Soil Sci 50:529–533
Topolovec-pintaric S, Zutic E, Dermic E (2013) Enhanced growth of cabbage and red beet by *Trichoderma viride*. Acta Agric Slovenica 101(1):87–92
Trikhonovich IA, Provorov NA (2011) Microbiology is the basis of sustainable agriculture: an opinion. Ann Appl Biol 159:155–168. doi:10.1111/j.1744-7348.2011.00489.x
Trillas M, Isabel CE, Aviles M et al (2006) Compost from agriculture waste and *T. asperellum* strain T-34 suppress *R. solani* in cucumber seedlings. Biol Control 39:32–38
Triveni S, Prasanna R, Shukla L, Saxena AK (2013) Evaluating the biochemical traits of novel *Trichoderma*- based biofilms for use as plant growth-promoting inoculants. Ann Microbiol 63:1147–1156
Trutmann P, Keane PJ (1990) *T. koningii* as a biocontrol agent for *S. sclerotiorum* in southern Australia. Soil Biol Biochem 22:43–50
Tweddell RJ, Valerie G, Hani A (2007) Growth stimulation and fruit yield of greenhouse tomato plants by inoculation with *P. putida* and *T. atroviride*: possible role of indole acetic acid (IAA). Soilbiol Biochem 39:968–977

Upadhyay JP, Mukhopadhyay AN (1986) Biological control of *S. rolfsii* by *T. harzianum* in sugar-beet. Trop Pest Manag 32:215–220

Varshney S, Chaube HS (2001) Mycorrhizal, rhizobacterial and fungal antagonists: interactions in microbes and plants. In: Sinha A (ed). Campus Books International, New Delhi, pp 226–238

Velikanov LL, Cukhonosenko E, Yu-Nikolaeva SI, Zevelishko IA (1994) Comparasion of hyper-parasitic and antibiotic activity of genus *Trichoderma* Pers. Fr. And *Gliocladium virens* towards the pathogens causing root rot of pea. Mikologia I Fitopatol 28(6):52–56

Verdin A, Loune A, Sanaroua H et al (2005) Polyclyclic aromatic hydrocarbon storage by *F. solani* in intracellular lipid vesiciles. Environ Pollut 133:283–291

Verma S, Dohroo NP (2005) Biocontrol of *Fusarium* wilt of autumn pea caused by *F. oxysporum* f. sp. *pisi*. Indian J Plant Pathol 23(1&2):58–62

Verma M, Brar SK, Tyagi RD, Surampalli RY, Valero JR (2007) Industrial wastewaters and dewatered sludge: rich nutrient source for production and formulation of biocontrol agent, *Trichoderma viride*. World J Microbiol Biotechnol 23:1695–1703

Verma JP, Jaiswal DK, Sagar R (2014) Pesticide relevance and their microbial degradation: a-state-of-art. Rev Environ Sci Biotechnol 13:429–466

Vinale F, Marra R, Scala F, Ghibsalberti EL et al (2006) Major secondry metabolites produced by two commercial *Trichoderma* strains active against different phytopathogen. Lett Appl Microbiol 43:143–148

Vinale F, Strakowska J, Mazzei P, Piccolo A, Marra R et al (2016) Cremenolide, a new antifungal, 10-member lactone from *Trichoderma cremeum* with plant growth promotion activity. Nat Prod Res 30(22):2575–2581

Vitale A, Cirvilleri G, Castello I et al (2012) Evaluation of *Trichoderma harzianum* strain T22 as biological control agent of *Calonectriapauci ramosa*. Biol Control 57:687–696

Viterbo A, Landau U, Kim S, Chernin L, Chet I (2010) Characterization of ACC deaminase from the biocontrol and plant growth-promoting agent *Trichoderma asperellum* T203. FEMS Microbiol Lett 305:42–48

Vos CMF, Cremer KD, Cammue PA, Coninck BD (2015) The tool box of *Trichoderma* spp. in the biocontrol of *Botrytis cineria* disease. Mol Plant Pathol 16(4):400–412

Weaver MA, Kenerley CM (2008) Competitiveness of a genetically engineered strain of *Trichoderma virens*. Mycopathologia 166(1):51–59

Weindling R (1932) *Trichoderma harzianum* as a parasite of other soil fungi. Phytopathology 22:837–845

Weindling R (1934) Studies on a lethal principle effective in parasitic action *T. lignorum* on *R. solani* and other soil fungi. Phytopathology 24:1153–1179

Wilson PS, Ahvenniemi PM, Lehtonen MJ et al (2008) Biological and chemical control and their combined use to control different stages of the *Rhizoctonia* disease complex on potato through the growing season. Ann Appl Biol 153:307–320

Windham MT, Elad Y, Baker R (1986) A mechanism for increased plant growth induced by *Trichoderma* species. Phytopathology 6:518–521

Yamada T, Mizutani Y, Umebayashi Y et al (2014) Tandyukisin, a novel ketoaldehyde decalin derivative, produced by marine sponge- derived *Trichoderma harzianum*. Tetrahedron Lett 55(3):662–664

Yang Q, Liu P (2005) Identification of genes with a biocontrol function in *T. harzianum* mycelium using the expressed sequence tag approach. Res Microbiol 156:416–423

Yang X, Cong H, Song J, Zhang J (2013) Heterologous expression of an aspartic protease gene from biocontrol fungus *Trichoderma asperellum* in *Pichia pastoris*. World J Microbiol Biotechnol 29:2087–2094

Yazdani M, Yap CK, Abdullah F (2010) Adsorption and absorption of cu in *Trichoderma atro-viride*. Pertanika J Trop Agric Sci 33(1):71–77

Yedidia I, Benhamou N, Chet I (1999) Induction of defense responses in cucumber plants (*Cucumis sativus* L.) by the biocontrol agent *T. harzianum*. Appl Environ Microbiol 65:1061–1070

Yedidia I, Benhamou N, Kapulnik Y, Chet I (2000) Induction and accumulation of PR protein activity during early stages of root colonization by the mycoparasitic *T. harzianum* strain T-20. Plant Physiol Biochem 38:863–873

Yedidia I, Srivastava AK, Kapulnik Y, Chet I (2001) Effect of *T. harzianum* on microelement concentrations and increased growth of cucumber plants. Plant Soil 235:235–242

You J, Dai H, Chen Z, Liu G, He Z et al (2010) Trichoderone, a novel cytotoxic cyclopentenone and cholesta-7,22-diene-3β,5α,6β-triol, with new activities from the marine-derived fungus *Trichoderma* sp. J Ind Microbiol Biotechnol 37:245–252

Zafra G, Cortes-Espinosa DV (2015) Biodegradation of polycyclic aromatic hydrocarbons by *Trichoderma* species: a mini review. Environ Sci Pollut Res 22(24):19426–19433

Zeilinger S, Gruber S, Mukherjee PK (2016) Secondary metabolism in *Trichoderma* chemistry meets genomics. Fungal Biol Revi 30(2):74–90

Zhang J, Akcapinar BG, Atanasova L et al (2016) The neutral metallopeptidase NMP1 of *Trichoderma guizhouense*is required for mycotrophy and self-defence. Environ Microbiol 18(2):580–597

Zheng Z, Shetty K (2000) Enhancement in pea (*Pisum sativum*) seedling vigour and associated phenolic content by extract of apple pomace fermented with *Trichoderma* spp. Process Biochem 36:79–84

Zhou X, Xu S, lexing L, Chen J (2007) Degradation of cyanide by *Tichoderma* mutants constructed by restriction enzyme mediated integration (REMI). Bioresour Technol 98(15):2958–2962

Fermentation of Cereals: A Tool to Enhance Bioactive Compounds

8

Kawaljit Singh Sandhu, Sneh Punia, and Maninder Kaur

Abstract

Popularity of antioxidant-rich food products are increasing day by day. Cereals being a major source of nutrients are still deficient in some basic nutritional components. So to improve their nutritional value, functional and sensory properties, interest has been shown to develop bioprocesses for the production of bioactive compounds and their applications in the field of food, chemical and pharmaceutical industries. Solid-state fermentation (SSF) has received greater attention as fermentation has the potential to release phenolic compounds from plant-based matrices. The potential application of fermentation process to convert the profile of phenolic compounds is mainly due to the release of bound phenolic compounds as a consequence of the degradation of the cell wall structure by microbial enzymes into unique metabolites through different bioconversion pathways.

8.1 Introduction

Cereals occupy 73% of the total world's harvested area with 60% of world's food production. They provide carbohydrates, proteins, energy, dietary fibre minerals and vitamins (Charalampopoulos et al. 2002) and significant levels of antioxidants (Baublis et al. 2000; Emmons et al. 1999). Cereals contain a wide range of chemical classes with antioxidant activity (Adom and Liu 2002). They are rich in phenolic

K.S. Sandhu (✉) • S. Punia
Department of Food Science and Technology, Chaudhary Devi Lal University, Sirsa 125 055, Haryana, India
e-mail: kawsandhu@rediffmail.com

M. Kaur
Department of Food Science and Technology, Guru Nanak Dev University, Amritsar 143 001, Punjab, India

S.K. Gahlawat et al. (eds.), *Plant Biotechnology: Recent Advancements and Developments*, DOI 10.1007/978-981-10-4732-9_8

acids and saponins, while phytoestrogens and flavonoids are present in small quantities (Senter et al. 1983). Most of the cereal phenolic compounds (PCs) are bound to cell wall polysaccharides. The amount of bound phenolic compounds in wheat, maize, rice and oats are 90, 87, 71 and 58%, respectively (Adom and Liu 2002). However, the nutritional quality and sensorial properties of cereals and their products are poor in comparison with dairy products because of the coarse nature, lower protein content, deficiency of some essential amino acids, the presence of antinutrients i.e., phytic acid, tannins and polyphenols (Chavan and Kadam 1989; Guenter 1997; Blandino et al. 2003). Due to limited protein content, low content of iron and zinc and swelling of starch upon cooking, nutritional value of cereals has been restricted (Nout 1993; Mouquet et al. 2008). The β-glucosidases and esterases produced by human microflora are unable to degrade the highly cross-linked water-insoluble dietary fibres (Bhanja Dey and Kuhad 2014). So it is advisable that prior to human consumption, insoluble cereal phenolics are converted into soluble form to improve health benefits of cereal phenolic compounds and their bioavailability (Moore et al. 2007; Vitaglione et al. 2008). Many methods have been engaged for improving the nutritional qualities of cereals (Blandino et al. 2003) and many processing technologies have been employed to improve the nutritional properties of cereals, although the best one is fermentation (Mattila-Sandholm 1998; Pugalenthi and Vadivel 2005). Cereals are regarded as most suitable substrates for the development of foods containing probiotic microorganisms and prebiotic properties (Charalampopoulos et al. 2009; Kedia et al. 2007).

Fermentation is known as the oldest form of biotechnology which dates back to about 5000 years ago when barley was converted to beer (Borgstrom 1968). It is a process when plants and animal tissues are under the action of microbes and enzymes to give desirable changes and to modify food quality. Worldwide fermented food contributes about one third of the diet (Campbell-Platt 1994). The quality proteins (Wang and Fields 1978; Chavan et al. 1988), nutritive value (Hamad and Fields 1979) and vitamin B (Chavan and Kadam 1989) of cereals may be improved by fermentation. Fermentation has multiple effects on the nutritional value of food, and it offers acceptability, improved nutritional value, food preservation, food safety, enhanced flavour and reduction in antinutritional compounds (Westby et al. 1997). Microbial fermentation leads to a decrease in the level of carbohydrates, poly- and oligosaccharides; synthesis of some amino acids and improvement in availability of B group vitamins (Nout and Ngoddy 1997), increasing the content of the essential amino acids lysine, methionine and tryptophan (Adams 1990). Fermentation processes have two systems: submerged fermentation (SmF) consists of liquid medium and solid-state fermentation (SSF) consists of growth of microbes on solid particles. SSF has received more interest as this process may lead to utilization of low cost agricultural and agro-industrial residues, low capital costs, higher production and better product characteristics. A variety of microorganisms (fungi, yeasts and bacteria) may be used in fermentation process; however, fungi and yeasts are the most commonly used microorganisms due to their ability to grow in low moisture content in the fermentation media (Martins et al. 2011). The potential of SSF has been evaluated for the improvement of total phenolics content and antioxidant potential of wheat (Sandhu et al. 2016), buckwheat, wheat germ, barley and rye (Dordevic et al. 2010), wheat bran (Moore et al. 2007), maize (Da-ker et al. 2008), rice (Bhanja et al. 2008), wheat grains (Bhanja et al. 2009) and oats (Cai et al. 2012a, b), employing several microorganisms. About 22-fold improvement of total phenolic content (TPC) was observed in *A. oryzae* fermented wheat grain (Bhanja et al. 2009). During fermentation of corn, riboflavin and niacin levels doubled but thiamine remains unchanged (Steinkraus 1994).

8.2 Fermentation by Microorganisms (Bacteria, Fungi and Yeasts)

A positive influence has been noticed by fermentation on the phenolic content and antioxidant activities of cereals; however, the degree of influence depends on what type (species) of microorganism is used. Microorganisms have mechanism to protect themselves by forming enzymatic antioxidants which scavenge reactive oxygen species (ROS) (Farr and Kogoma 1991; Kim et al. 2005). Initially, mostly fungi were used for fermentation process (as these microorganisms were considered to be very optimally active in very low water activity). Later, many bacterial species and yeasts were used to carry out such fermentations also (Singhania et al. 2009). A list of commonly used microorganism is described in Table 8.1.

8.2.1 Bacteria

The common fermenting bacteria are species of *Leuconostoc*, *Lactobacillus*, *Streptococcus*, *Pediococcus*, *Micrococcus* and *Bacillus* (Mensah 1997; Steinkraus 1998). In food fermentations, lactic acid bacteria are widely used (Jakobsen and Lei 2004; Ng et al. 2011). Species of *Lactobacillus* play a significant role in fermentation of cereals (Sanni 1993). *Lactobacillus plantarum* is the species which is frequently used to ferment food products (Rodriguez et al. 2009), as they hydrolyze complex polyphenols to simpler ones and biologically active compounds are formed (Duenas et al. 2005). Lactic acid bacteria have the ability to release isoflavone aglycone which acts as an antioxidant and includes *Lactobacillus* (L.) *acidophilus*, *L. plantarum*, *L. casei*, *L. fermentum*, and others for fermenting food products (Marazza et al. 2009). Rice bran has been used many times in bacterial fermentation (Tanaka et al. 2006; Gao et al. 2008; Li et al. 2012; Watanabe et al. 2013). Because of the lack of amylase in rice, it is fermented by *Lactobacillus bulgaricus* and *Streptococcus thermophillus*, which is essential for saccharification of rice starch (Lee et al. 1999).

Table 8.1 Studies of solid-substrate fermentation using different microorganisms

Microorganisms	Substrate support	Reference/s
Fungi		
Aspergillus oryzae	Wheat (WH-711) fraction	Duhan et al. (2016)
Aspergillus awamori	Wheat	Sandhu et al. (2016)
Aspergillus oryzae	Rice bran	Razak et al. (2015)
Rhizopus oryzae	Rice bran	Schmidt et al. (2014)
Rhizopus oryzae	Wheat	Bhanja Dey and Kuhad (2014)
Aspergillus oryzae var. effuses, *Aspergillus oryzae* and *Aspergillus niger*	Oats	Cai et al. (2012b)
Aspergillus niger	Wheat bran,	Orzua et al. (2009)
Aspergillus oryzae	Rice bran	Chutmanop et al. (2008)
Aspergillus sojae	Maize meal	Ustok et al. (2007)
Bjerkandera adusta, Phlebia rufa	Wheat straw	Dinis et al. (2009)
Phanerochaete chrysosporium	Rice straw	Yu et al. (2009)
Yeasts		
Saccharomyces cerevisiae	Corn	Zhao and Xia (2009)
Baker yeast AF37X	Sorghum	Yu et al. (2008)
Saccharomyces cerevisiae	Wheat bran	Moore et al. (2007)
Bacteria		
Lactobacillus johnsonii, Lactobacillus reuteri, Lactobacillus acidophilus	Whole grain barley, Oat groat	Hole et al. (2012)
Lactobacillus rhamnosus	Rye, Buckwheat	Dordevic et al. (2010)
Nocardia lactamdurans	Wheat bran, rice	Kagliwal et al. (2009)
Bacillus sphaericus	Wheat bran	El-Bendary et al. (2008)
Bacillus subtilis	Wheat bran	Gupta et al. (2008)

8.2.2 Fungi

Filamentous fungi are commonly used microorganisms as they have great potential to produce bioactive compounds by SSF (Topakas et al. 2003; Aguilar et al. 2008; Martins et al. 2011). They have also received great attention due to production of thermostable enzymes of high commercial value (Christakopoulos et al. 1990; Martins et al. 2002) such as glycoside hydrolase, cellulose or xylan degrading enzymes and esterase, which are capable of softening the kernel structure, breaking down cereal cell walls, and releasing bound compounds (Cai et al. 2012b). Furthermore, by the catalytic action of β-glucosidase, the liberation of lipophilic aglycones from isoflavone glucosides during fermentation by fungi increases antioxidant activity in extracts (Lin et al. 2006). The enzymes produced by fungus have potential to hydrolyze the glucosidic bonds of phenolic compounds and hydrolysis of phenolic glucosides is a promising way to increase the amount of bound to free polyphenols to enhance the bioactivity of various foods (Georgetti et al. 2009). The fungi genera *Aspergillus*, *Paecilomyces*, *Cladosporium*, *Fusarium*, *Penicillium* and *Trichothecium* are the most frequently found in certain products (Steinkraus 1998). For SSF, food-grade filamentous fungi (*Aspergillus* sp. and *Rhizopus* sp.) are used for the enhancement of phenolic and anthocyanin compounds (Karaffa et al. 2001; Lee et al. 2008). These fungi also produce highly digestible proteins without any toxic substance being generated (Razak et al. 2015). *Aspergillus oryzae* has been used in the production of protease through the fermentation of rice bran (Chutmanop et al. 2008).

8.2.3 Yeasts

Generally fermentation by yeast is commonly used for the preparation of alcoholic beverages but now the potential of yeast to enhance the bioactive property of cereals and whole grains have also been evaluated. Since yeast can grow on low moisture content it can be used for SSF process using substrates like rice bran, wheat bran, etc. (Prabhu et al. 2014). Fermentation of cereals with yeast increases the protein content (Svanberg and Lorri 1997). The common fermenting yeasts are species of *Saccharomyces*, which results in alcoholic fermentation (Steinkraus 1998). For SSF potentially known organisms is *Saccharomyces cerevisiae* and regarded as a safe organism when use in food products (Grange et al. 1996). It has ability to produce enzymes, i.e., carboxylesterases, β-glucosidases and esterases (Hernandez et al. 2003; Coghe et al. 2004), which increases antioxidant activity.

8.3 Production of Enzymes

During growth of microorganisms due to the action of cellulolytic, pectinolytic and ligninolytic enzymes, the profile of phenolic compounds is changed as described in flow diagram 1, 2 and 3, and an overview of enzymes which are involved in release of phenolic components by softening the kernel structure, by breaking down of cell wall matrix and produced by microorganism are given in Table 8.2. Among these, β-glucosidase has been widely reported as an enzyme responsible for catalyzing the hydrolysis of glycosidic linkages. These enzymes have capability of breaking down cell wall's chemical components. As a consequence, the free as well as bound phenolic compounds are released more efficiently (Huynh et al. 2014). Similar results were observed when filamentous fungi in SSF of cereal sources, such as oat (Cai et al. 2012b) and rice bran (Schmidt et al. 2014), caused an increase in the content of phenolic acids. The changes in phenolic profile were observed as a result of microbial fermentation of plant-based matrices. The fermentation with filamentous fungi, lactic acid bacteria, and yeast may be considered as a potential process to increase the release of phenolic compounds contributing to the production of food products with an added value (Figs. 8.1, 8.2, and 8.3).

Table 8.2 Enzyme system produced by commonly used microorganism to degrade the cell wall matrix

Microorganisms	Enzymes	References
Fungi		
Aspergillus awamori nakazawa	Xylanase, α-L-arabinofuranosidase, feruloyl esterase	Bhanja et al. (2009) and Georgetti et al. (2009)
Aspergillus niger	Cellulase, esterase, β-glucosidase, xylanase	Hur et al. (2014), Bhanja et al. (2009) and Georgetti et al. (2009)
Aspergillus oryzae	Cellulase, β-glucosidase, xylanase, pectinase	Wu et al. (2011)
Rhizopus oligosporus	β-glucosidase, β-glucuronidase, xylanase	McCue et al. (2003), Vattem et al. (2004) and Correia et al. (2004)
Rhizopus oryzae	β-glucosidase, tannase, pectinase	Yadav et al. (2013)
Bacteria		
Lactobacillus lactis	Esterase, decarboxylase	Hur et al. (2014)
Lactobacillus plantarum	β-Glucosidase, decarboxylase	Hur et al. (2014)
Lactobacillus rhamnosus	Cellulase, esterase, β-glucosidase	Hur et al. (2014) and Marazza et al. (2009)
Bacillus cereus	Cellulase, tannase	Hur et al. (2014)
Bacillus subtilis	Cellulase, β-glucanase	Hur et al. (2014)
Bacillus thuringiensis	Cellulase, tannase	Hur et al. (2014)
Yeast		
Saccharomyces cerevisiae	β-Glucosidase, feruoylesterase	Hur et al. (2014) and Coghe et al. (2004)

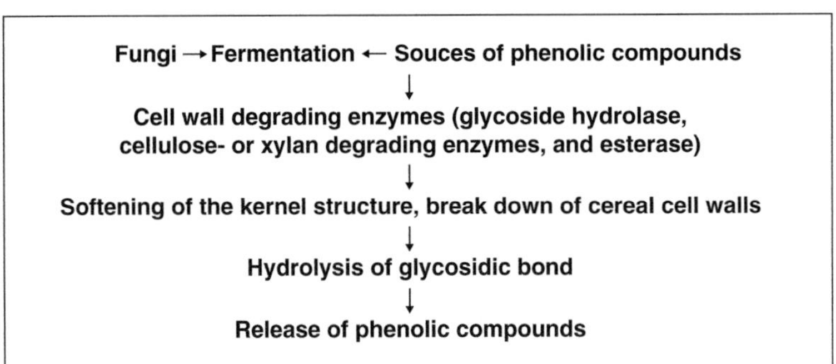

Fig. 8.1 Flow diagram of effect of fermentation by fungi on the antioxidative activity (Huynh et al. 2014)

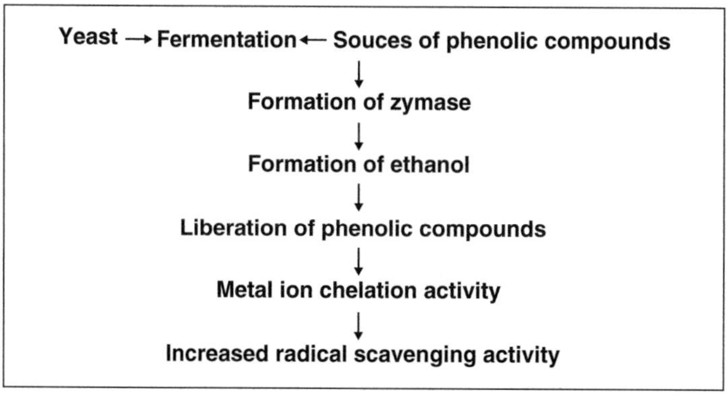

Fig. 8.2 Flow diagram of effect of fermentation by yeast on the antioxidative activity (Hur et al. 2014)

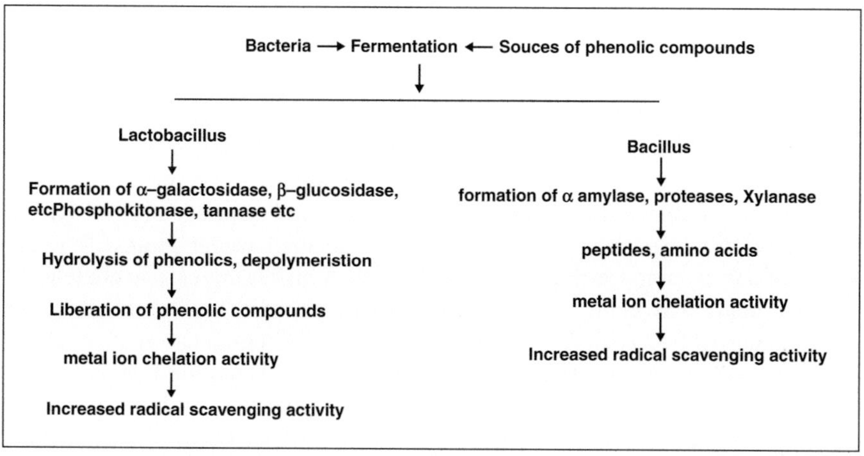

Fig. 8.3 Flow diagram of effect of fermentation by bacteria on the antioxidative activity (Hur et al. 2014)

8.4 Antioxidant Activity of Phytochemicals by Fermentation

Phenolic compounds considered as reducing agents, hydrogen donors and singlet oxygen quenchers (Miller and Rice-Evans 1997) are present as free phenolic compounds (in the vacuoles of plant cells) and bound phenolic compounds (combined with cellulose, hemicellulose, lignin, pectin and protein) through covalent bonds (Pinelo et al. 2006; Cerda et al. 2013; Yadav et al. 2013). Free phenolic compounds may be easily extracted through conventional techniques, while to enhance the release of bound phenolics, hydrolysis processes have been used and fermentation is considered as one of the best processes to obtain bioactive compounds with a high quality and a high activity (Martins et al. 2011).

Fermentation has a positive influence on phytochemicals of cereals, but the degree of influence depended on microorganism species (Kariluoto et al. 2006). Fermentation increases the amount of phenolic compounds (Katina et al. 2007) due to which glycemic index lowers (Solomon and Blannin 2007). Several studies reported that during fermentation, phenolic profile of plant sources is influenced. Phenolic component's cross-linking between polysaccharides and lignin are responsible for cell wall structures of cereals (Faulds and Williamson 1999) and by fermentation process, structural break down of cereal cell walls may occur, leading to the increased extraction of bioactive compounds (Heinio et al. 2003). Zheng and Shetty (2000) reported that improvement in phenolic compounds is due to action of enzymes such as β-glucosidase, α-amylase and lactase, whereas Ju et al. (2009) reported that fermenting microbes cause degradation of phenolic compounds. The potential of SSF has been evaluated for the improvement of total phenolics content and antioxidant potential of wheat bran (Moore et al. 2007), maize (Daker et al. 2008), buckwheat, wheat germ, barley and rye (Dordevic et al. 2010), rice (Bhanja et al. 2008) wheatgrains (Bhanja et al. 2009) and oat (Cai et al. 2012a, b) employing several microorganisms.

Bioactivity and chemical composition of rice were improved by SSF with *Cordyceps sinensis* (Zhang et al. 2008). Bhanja et al. (2009) used *Aspergillus oryzae* and *Aspergillus awamori* in SSF of wheat grains and reported that these fungi are very effective for the enhancement of phenolic content and antioxidant properties of wheat grains and considered fermented wheat grains as richer source of antioxidant healthier food supplement. A higher increase in contents of phenolic acids, flavonoids and aglycone isoflavone with free hydroxyl groups achieved during SSF with *Trichoderma harzianum* for fermentation of soyabean has been reported (Singh et al. 2010). Razak et al. (2015) reported phenolic acid i.e., caffeic, protocatechuic, vanillic and 4-hydroxybenzoic acids in fermented rice bran when compared with unfermented rice bran. Cai et al. (2011) fermented oats by three different fungi i.e., *Aspergillus oryzae* var. *effuses*, *Aspergillus oryzae* and *Aspergillus niger* at 25 °C for 3 days and reported that there was improvement in phenolic acid profile and *Aspergillus oryzae* increased the content of caffeic acid and ferulic acid in oats (*Avena sativa* L.) up to about 2.7–3 folds and 5.59 folds, respectively, in comparison with non-fermented oats. Schmidt et al. (2014) investigated the increased effect of fermentation by *Rhizopus oryzae* on the profile of phenolic acids of rice bran. Prabhu et al. (2014) observed an increase in phenolic, flavonoid content and antioxidant properties, whereas decrease in antinutritional factor when fermentation of rice bran was done with *Saccharomyces cerevisiae.*

Fermentation of rice bran by yeast resulted in increase in flavonoid content which may be due to the increase in acidic value during fermentation that is liberating bound flavonoid components and making it more bioavailable (Prabhu et al. 2014). In wheat bran, ferulic acid is the most abundant phenolic compound and structural component in cell walls. Since ferulic acid is covalently bound to the cell wall structures, its bioavailability is quite low (Singh et al. 2015). Bioavailability of ferulic acid can be increased by processing of cereals with fermentation (Mateo et al. 2009b; Napolitano et al. 2009). Mateo et al. (2009a) bioprocessed wheat bran with yeast fermentation and observed an increase in bioaccessibility of phenolic compounds (Mateo et al. 2009a). Dordevic et al. (2010) reported an increase in antioxidant activity from 45.0% to 50.4% in rye, 36.6% to 42.9% in barley and 31.0% to 35.9% in wheat after fermentations with *L. rhamnosus*.

8.5 Conclusion

The nutritional and sensorial properties of cereals may be enhanced by fermentation which leads to their improved nutritional value and better digestibility. Through fermentation, due to action of cell wall degrading enzymes, and bioconversion of flavonoids into their metabolites, phenolic compounds are increased. Thus, microbial fermentation could be considered as a potential technology for releasing phenolic compounds from natural resources, as well as for producing new bioactive compounds. Although some fermentation processes are applied in the production of phenolic compounds, the yield of bioconversion is variable, depending on the fermentation parameters used (e.g., microorganisms, medium, temperature and pH) as well as due to differences in the plant matrix itself. Therefore, further research into optimal process is required.

8.6 Future Prospects

Solid substrate fermentation is a technology of great potential to increase the content of bioactive phenolic compounds in food products. Due to the increased consumer desire for healthy and nutritionally rich food, SSF has great potential to expand in a near future. Fermentation processes contain variety of microflora but potential of these microorganisms towards the metabolism of secondary plant compounds are unknown and still there is a lack of knowledge on the conversion potential, identification and quantification of phenolic compounds. So, more studies in this area are needed to evaluate the microbial pathways of bioactivity determinations, flavonoid conversion and identification of the metabolites. Furthermore, in depth insight in the reasons for the conversion of bioactive compounds by microorganisms is needed. Also, there is still a lack of knowledge on the metabolic pathways as well as the relationship between specific metabolites and their corresponding bioactivity, bioaccessibility and bioavailability, which thus demands more research in this field. Therefore, more studies dealing with this research gap can lead to the discovery of a huge potential of new bioactive natural compounds.

References

Adams MR (1990) Topical aspects of fermented foods. Trends Food Sci Technol 8:140–144

Adom KK, Liu RH (2002) Antioxidant activity of grains. J Agric Food Chem 50:6182–6187

Aguilar CN, Aguilera-Carbo A, Robledo A et al (2008) Production of antioxidants nutraceuticals by solid-state cultures of pomegranate (*Punica granatum*) peel and creosote bush (*Larrea tridentata*) leaves. Food Technol Biotechnol 46:218–222

Baublis A, Decker EA, Clydesdale FM (2000) Antioxidant effects of aqueous extracts from wheat-based ready-to-eat breakfast cereals. Food Chem 68:1–6

Bhanja Dey T, Kuhad RC (2014) Upgrading the antioxidant potential of cereals by their fungal fermentation under solid-state cultivation conditions. Lett Appl Microbiol 59:493–499

Bhanja T, Rout S, Banerjee R et al (2008) Studies on the performance of a new bioreactor for improving antioxidant potential of rice. LWT Food Sci Technol 41:459–1465

Bhanja T, Kumari A, Banerjee R (2009) Enrichment of phenolics and free radical scavenging property of wheat koji prepared with two filamentous fungi. Bioresour Technol 100:2861–2866

Blandino A, Al-Aseeri ME, Pandiella SS et al (2003) Cereal-based fermented foods and beverages. Food Res Int 36:527–543

Borgstrom G (1968) Principals of food science-ll, food microbiology and biochemistry. Macmillan, New York

Cai S, Gao F, Zhang X et al (2012a) Evaluation of c-aminobutyric acid, phytate and antioxidant activity of tempeh-like fermented oats (*Avena sativa* L.) prepared with different filamentous fungi. J Food Sci Technol 51(10):2544–2551

Cai S, Wang O, Wu W et al (2012b) Comparative study of the effects of solid-state fermentation with three filamentous fungi on the total phenolics content (TPC), flavonoids, and antioxidant activities of subfractions from oats (*Avena sativa* L.) J Agric Food Chem 60:507–513

Cai S, Wang O, Wu W et al (2011) Comparative study of the effects of solid-state fermentation with three filamentous fungi on the total phenolics content (TPC), flavonoids, and antioxidant activities of subfractions from oats (*Avena sativa* L.) J Agric Food Chem 60:507–513

Campbell-Platt G (1994) Fermented foods- a world perspective. Food Res Int 27:253–257

Cerda A, Martinez ME, Soto C et al (2013) The enhancement of antioxidant compounds extracted from *Thymus vulgaris* using enzymes and the effect of extracting solvent. Food Chem 139:138–143

Charalampopoulos D, Wang R, Pandiella SS et al (2002) Application of cereals and cereal components in functional foods. Int J Food Microbiol 79:131–141

Charalampopoulos D, Vazquez JA, Pandiella SS (2009) Modelling and validation of *Lactobacillus plantarum* fermentations in cereal-based media with different sugar concentrations and buffering capacities. Biochem Eng J 44:96–105

Chavan JK, Kadam SS (1989) Nutritional improvement of cereals by fermentation. CRC Crit Rev Food Sci Nutrition 28:349–400

Chavan UD, Chavan JK, Kadam SS (1988) Effect of fermentation on soluble proteins and *in vitro* protein digestibility of sorghum, green gram and sorghum green gram blends. J Food Sci 53:1574

Christakopoulos P, Macris BJ, Kekos D (1990) Exceptionally thermostable α- galactosidases and β-galactosidases from *Aspergillus niger* separated in one step. Process Biochem Int 25:210–212

Chutmanop J, Chuichulcherm S, Chisti Y et al (2008) Protease production by *Aspergillus oryzae* in solid-state fermentation using agroindustrial substrates. J Chem Technol Biotechnol 83:1012–1018

Coghe S, Benoot K, Delvaux F et al (2004) Ferulic acid release and 4-vinylguaiacol formation during brewing and fermentation: indications for feruloyl esterase activity in *Saccharomyces cerevisiae*. J Agric Food Chem 52:602–608

Correia RTP, McCue P, Magalhaes MMA et al (2004) Phenolic antioxidant enrichment of soy flour-supplemented guava waste by *Rhizopus oligosporus*-mediated solid-state bioprocessing. J Food Biochem 28:404–418

Daker M, Abdullah N, Vikineswary S et al (2008) Antioxidant from maize and maize fermented by *Marasmiellus* sp. as stabiliser of lipid- rich foods. Food Chem 107:1092–1098

Dey TB, Kuhad RC (2014) Enhanced production and extraction of phenolic compounds from wheat by solid-state fermentation with *Rhizopus oryzae* RCK2012. Biotechnol Rep 4:120–127

Dinis MJ, Bezerra RMF, Nunes F et al (2009) Modification of wheat straw lignin by solid state fermentation with white-rot fungi. Bioresour Technol 100:4829–4835

Dordevic TM, Siler-Marinkovic SS, Dimitrijevic-Brankovi SI (2010) Effect of fermentation on antioxidant properties of some cereals and pseudo cereals. Food Chem 119:957–963

Duenas M, Fernández D, Hernández T et al (2005) Bioactive phenolic compounds of cowpeas (*Vigna sinensis* L). Modifications by fermentation with natural microflora and with *Lactobacillus plantarum* ATCC 14917. J Sci Food Agric 85:297–304

Duhan JS, Mehta K, Sadh PK, Saharan P, Surekha (2016) Bio-enrichment of phenolics and free radicals scavenging activity of wheat (WH-711) fractions by solid state fermentation with *Aspergillus oryzae*. Afr J Biochem Res 10(2):12–19. doi:10.5897/AJBR2015.0854

El-Bendary MA, Monoram D, Foda MS (2008) Efficient mosquitocidal toxin production by *Bacillus sphaericus* using cheese whey permeate under both submerged and solid state fermentations. J Invertebr Pathol 98:46–53

Emmons CL, Peterson DM, Paul GL (1999) Antioxidant capacity of oat (*Avena sativa* L.) extracts. 2. *In vitro* antioxidant activity and contents of phenolic and tocol antioxidants. J Agric Food Chem 47:4894–4898

Farr SB, Kogoma T (1991) Oxidative stress responses in *Escherichia coli* and *Salmonella typhimurium*. Microbiol Rev 55:561–585

Faulds M, Williamson G (1999) The role of hydroxycinnamates in the plant cell wall. J Sci Food Agric 79:393–395

Gao M, Kaneko M, Hirata M et al (2008) Utilization of rice bran as nutrient source for fermentative lactic acid production. Bioresour Technol 99:3659–3664

Georgetti SR, Vicentini FT, Yokoyama CY et al (2009) Enhanced in vitro and in vivo antioxidant activity and mobilization of free phenolic compounds of soybean flour fermented with different beta-glucosidase-producing fungi. J Appl Microbiol 106:459–466

Grange DC, Pretorius IS, Zyl WH (1996) Expression of Trichoderma Reesi B-xylanase gene (XYN2) in *Saccharomyces cerevisiae*. Appl Environ Microbiol 62:1036–1044

Guenter W (1997) Phytases in cereals and hemicelluloses in canola (rape seed) meal and lupins. In: Marquardt RR, Han Z (eds) Enzymes in poultry and swine nutrition. International Development Research Center, Ottawa, pp 99–114

Gupta S, Kapoor M, Sharma KK et al (2008) Production and recovery of an alkaline exo-polygalacturonase from *Bacillus subtilis* RCK under solid-state fermentation using statistical approach. Bioresour Technol 99:937–945

Hamad AM, Fields ML (1979) Evaluation of the protein quality and available lysine of germinated and fermented cereal. J Food Sci 44:456–459

Heinio RL, Katina K, Wilhelmson A et al (2003) Relationship between sensory perception and flavour-active volatile compounds of germinated, sourdough fermented and native rye following the extrusion process. LWT Food Sci Technol 36:533–545

Hernandez LF, Espinosa JC, Fernandez-Gonzalez M et al (2003) Glucosidase activity in a *Saccharomyces cerevisiae* wine strain. Int J Food Microbiol 80:171–176

Hole AS, Rud I, Grimmer S et al (2012) Improved bioavailability of dietary phenolic acids in whole grain barley and oat groat following fermentation with probiotic *Lactobacillus acidophilus*, *Lactobacillus johnsonii*, and *Lactobacillus reuteri*. J Agric Food Chem 60:6369–6375

Hur SJ, Lee SY, Kim YC et al (2014) Effect of fermentation on the antioxidant activity in plant-based foods. Food Chem 160:346–356

Huynh NT, Van Camp J, Smagghe G et al (2014) Improved release and metabolism of flavonoids by steered fermentation processes: a review. Int J Mol Sci 15:19369–19388

Jakobsen M, Lei V (2004) Microbiological characterization and probiotic potential of koko and koko sour water, an African spontaneously fermented millet porridge and drink. J Appl Microbiol 96:384–397

Ju HK, Cho EJ, Jang MH et al (2009) Characterization of increased phenolic compounds from fermented Bokbunja (*Robus coreanus* Miq.) and related antioxidant activity. J Pharm Biomed Anal 49:820–827

Kagliwal LD, Survase SA, Singhal RS (2009) A novel medium for the production of cephamycin C by Nocardia lactamduransm using solid-state fermentation. Bioresour Technol 100:2600–2606

Karaffa L, Sandor E, Fekete E et al (2001) The biochemistry of citric acid accumulation by *Aspergillus niger*. Acta Microbiol Immunol Hung 48:429–440

Kariluoto S, Aittamaa M, Korhola M et al (2006) Effects of yeasts and bacteria on the levels of folates in rye sourdoughs. Int J Food Microbiol 106:137–143

Katina-Laitil A, Juvonen R, Liukkonen KH et al (2007) Bran fermentation as a means to enhance technological properties and bioactivity of rye. Food Microbiol 24:175–186

Kedia G, Wang R, Patel H et al (2007) Use of mixed cultures for the fermentation of cereal-based substrates with potential probiotic properties. Process Biochem 42:65–70

Kim HS, Chae HS, Jeong SG et al (2005) Antioxidant activity of some yogurt starter cultures. Asian-Australas J Anim Sci 18:255–258

Lee JH, Lee SK, Park KI et al (1999) Fermentation of rice using amylolytic *Bifidobacterium*. Int J Food Microbiol 50:155–161

Lee IH, Hung YH, Chou CC (2008) Solid-state fermentation with fungi to enhance the antioxidative activity, total phenolic and anthocyanin contents of black bean. Int J Food Microbiol 121:150–156

Li Z, Lu J, Yang Z et al (2012) Utilization of white rice bran for production of lactic acid. Biomass Bioenergy 39:53–58

Lin CH, Wei YT, Chou CC (2006) Enhanced antioxidative activity of soybean koji prepared with various filamentous fungi. Food Microbiol 23:628–633

Marazza JA, Garro MS, Savoy de Giori G (2009) Aglycone production by *Lactobacillus rhamnosus* CRL981 during soymilk fermentation. Food Microbiol 26:333–339

Martins ES, Silva D, Da Silva R et al (2002) Solid state production of thermostable pectinases from thermophilic *Thermoascus aurantiacus*. Process Biochem 37:949–954

Martins S, Mussatto SI, Martínez-Avila G et al (2011) Bioactive phenolic compounds: production and extraction by solid-state fermentation: a review. Biotechnol Adv 29:365–373

Mateo AN, Selinheimo E, Havenaar R et al (2009a) Bioprocessing of wheat bran improves *in vitro* bioaccessibility and colonic metabolism of phenolic compounds. J Agric Food Chem 57:6148–6155

Mateo AN, Berg R, Havenaar R et al (2009b) Bioavailability of ferulic acid is determined by its bioaccessibility. J Cereal Sci 49:296–300

Mattila-Sandholm T (1998) VTT on lactic acid bacteria. VTT Symposium 156:1–10

McCue P, Horii A, Shetty K (2003) Solid-state bioconversion of phenolic antioxidants from defatted soybean powders by *Rhizopus oligosporus*: role of carbohydrate-cleaving enzymes. J Food Biochem 27:501–514

Mensah P (1997) Fermentation-the key to food safety assurance in Africa. Food Control 8:271–278

Miller NJ, Rice-Evans CA (1997) Factors influencing the antioxidant activity determined by the ABT - radical cation assay. Free Radic Res 26:195–199

Moore J, Cheng Z, Hao J et al (2007) Effects of solid-state yeast treatment on the antioxidant properties and protein and fibre compositions of common hard wheat bran. J Agric Food Chem 55:10173–10182

Mouquet RC, Icard VC, Guyot JP et al (2008) Consumption pattern, biochemical composition and nutritional value of fermented pearl millet gruels in Burkina Faso. Int J Food Sci Nutr 59:716–729

Napolitano A, Costabile A, Martin-Pelaez S et al (2009) Potential prebiotic activity of oligosaccharides obtained by enzymatic conversion of durum wheat in soluble dietary fibre into soluble dietary fibre. Nutr Metab Cardiovasc Dis 19:283–290

Ng CC, Wang CY, Wang YP et al (2011) Lactic acid bacterial fermentation on the production of functional antioxidant herbal *Anoectochilus formosanus* Hayata. J Biosci Bioeng 111:289–293

Nout MJR (1993) Processed weaning foods for tropical climates. Int J Food Sci Nutr 43:213–221

Nout MJR, Ngoddy PO (1997) Technological aspects of preparing affordable fermented complementary foods. Food Control 8:279–287

Orzua MC, Mussatto SI, Contreras-Esquivel JC et al (2009) Exploitation of agro industrial wastes as immobilization carrier for solid-state fermentation. Ind Crop Prod 30:24–27

Pinelo M, Arnous A, Meyer AS (2006) Upgrading of grape skins: significance of plant cell-wall structural components and extraction techniques for phenol release. Trends Food Sci Technol 17:579–590

Prabhu AA, Mrudula CM, Rajesh J (2014) Effect of yeast fermentation on nutraceutical and antioxidant properties of rice bran. Int J Agric Food Sci 4:59–65

Pugalenthi M, Vadivel V (2005) Nutritional evaluation and the effect of processing methods on antinutritional factors of sword bean *(Canavalia gladiata*). J Food Sci Technol 42:510–516

Razak DLA, Rashid NYA, Jamaluddin A et al (2015) Cosmeceutical potentials and bioactive compounds of rice bran fermented with single and mix culture of *Aspergillus oryzae* and *Rhizopus oryzae*. J Saudi Soc Agric Sci 04:001. doi org/101016/jjssas2015

Rodríguez H, Curiel JA, Landete JM et al (2009) Food phenolics and lactic acid bacteria. Int J Food Microbiol 132:79–90

Sandhu KS, Punia S, Kaur M (2016) Effect of duration of solid state fermentation by *Aspergillus awamorinakazawa* on antioxidant properties of wheat cultivars. LWT Food Sci Technol 71:323–328

Sanni AI (1993) The need for process optimization of african fermented foods and beverages. Int J Food Microbiol 18:85–95

Schmidt CG, Gonçalves LM, Prietto L et al (2014) Antioxidant activity and enzyme inhibition of phenolic acids from fermented rice bran with fungus *Rhizopus oryzae*. Food Chem 146:371–377

Senter SD, Horvat RJ, Forbus WR (1983) Comparative GLCMS analysis of phenolic acids of selected tree nuts. J Food Sci 48:798–803

Singh HB, Singh BN, Singh SP et al (2010) Solid-state cultivation of *Trichoderma Harzianum* NBRI-1055 for modulating natural antioxidants in soybean seed matrix. Bioresour Technol 101:6444–6453

Singh AK, Rehal J, Kaur A et al (2015) Enhancement of attributes of cereals by germination and fermentation: a review. Crit Rev Food Sci Nutr 55:1575–1589

Singhania RR, Patel AK, Soccol CR et al (2009) Recent advances in solid-state fermentation. Biochem Eng J 44:13–18

Solomon TPJ, Blannin AK (2007) Effects of short-term cinnamon ingestion on in vivo glucose tolerance. Diabetes Obes Metab 9:895–901

Steinkraus KH (1994) Nutritional significance of fermented food. Food Res Int 27:259–267

Steinkraus KH (1998) Bio-enrichment: production of vitamins in fermented foods. In: Wood JB (ed) Microbiology of fermented foods. Blackie Academic and Professional, London, pp 603–619

Svanberg U, Lorri W (1997) Fermentation and nutrient availability. Food Control:319–327

Tanaka T, Hoshina M, Tanabe S et al (2006) Production of D-lactic acid from defatted rice bran by simultaneous saccharification and fermentation. Bioresour Technol 97:211–217

Topakas E, Kalogeris E, Kekos D et al (2003) Bioconversion of ferulic acid into vanillic acid by the thermophilic fungus *Sporotrichum thermophile*. LWT Food Sci Technol 36:561–565

Ustok FI, Tari C, Gogus N (2007) Solid-state production of polygalacturonase by *Aspergillus sojae* ATCC 20235. J Biotechnol 127:322–334
Vattem DA, Lin YT, Labbe RG et al (2004) Antimicrobial activity against select food-borne pathogens by phenolic antioxidants enriched in cranberry pomace by solid-state bioprocessing using the food grade fungus *Rhizopus oligosporus*. Process Biochem 39:1939–1946
Vitaglione P, Aurora N, Fogliano V (2008) Cereal dietary fibre: a natural functional ingredient to deliver phenolic compounds into the gut. Trends Food Sci Technol 19:451–463
Wang YD, Fields ML (1978) Feasibility of home fermentation to improve the amino acid balance of corn meal. J Food Sci 43:1104
Watanabe M, Makino M, Kaku N et al (2013) Fermentative L-(+)-lactic acid production from non-sterilized rice washing drainage containing rice bran by a newly isolated lactic acid bacteria without any additions of nutrients. J Biosci Bioeng 115:449–452
Westby A, Reilly A, Zoe B (1997) Review of the effect of fermentation on naturally occurring toxins. Food Control 8:329–339
Wu Z, Song L, Huang D (2011) Food grade fungal stress on germinating peanut seeds induced phytoalexins and enhanced polyphenolic antioxidants. J Agric Food Chem 59:5993–6003
Yadav G, Singh A, Bhattacharya P et al (2013) Comparative analysis of solid-state bioprocessing and enzymatic treatment of finger millet for mobilization of bound phenolics. Bioprocess Biosyst Eng 36:1563–1569
Yu J, Zhang X, Tan T (2008) Ethanol production by solid state fermentation of sweet sorghum using thermotolerant yeast strain. Fuel Process Technol 89:1056–1059
Yu M, Zeng G, Chen Y et al (2009) Influence of *Phanerochaete chrysosporium* on microbial communities and lignocellulose degradation during solid-state fermentation of rice straw. Process Biochem 44:17–22
Zhang Z, Lei Z, Lu Y et al (2008) Chemical composition and bioactivity changes in stale rice after fermentation with *Cordyceps sinensis*. J Biosci Bioeng 106:188–193
Zhao J, Xia L (2009) Simultaneous saccharification and fermentation of alkaline-pretreated corn stover to ethanol using recombinant yeast strain. Fuel Process Technol 99:1193–1197
Zheng Z, Shetty K (2000) Solid-state bioconversion of phenolics from cranberry pomace and role of lentinus edodes β-glucosidase. J Agric Food Chem 48:895–900

Molecular Structure, Biological Functions, and Metabolic Regulation of Flavonoids

9

Sudhanshu Dwivedi, Chanchal Malik, and Vinod Chhokar

Abstract

Flavonoids, constituting about 5–10% of the known secondary metabolites, are important for plant growth and development and thus can have a large impact on agricultural productivity. Plants are specialized in synthesizing and accumulating specific combinations of about 5000 flavonoids. The richness of flavonoids in plant sources cannot be underestimated and they are found closely associated with human daily dietary intake; hence, it is significant to understand the biochemistry of flavonoids. The flavonoid biosynthetic pathway is under tight developmental control, and multiple environmental factors like temperature lights stress and fertilizers affect the expression of flavonoid biosynthetic genes. Flavonoids exhibit a wide range of biological effects such as antibacterial, antiviral, antifungal, antiosteoporotic, antitumor, antithrombogenic, anti-inflammation, antiatherosclerotic, and antihepatotoxic activities. They also have a regulatory role on different hormones like estrogens, androgens, and thyroid hormones. Due to their diverse pharmacological activities, functional artificial polymeric flavonoids, flavonoid polymers, and amine-containing polymer-flavonoid conjugates have also been developed. As research advances in natural products, further achievements will certainly lead to a new age of flavonoids as either food supplements or pharmaceuticals.

S. Dwivedi • C. Malik • V. Chhokar (✉)
Department of Bio and Nanotechnology, Guru Jambheshwar University of Science and Technology, Hisar 125 001, Haryana, India
e-mail: vinodchhokar@yahoo.com

S.K. Gahlawat et al. (eds.), *Plant Biotechnology: Recent Advancements and Developments*, DOI 10.1007/978-981-10-4732-9_9

9.1 Introduction

Flavonoids are the most common group of plant polyphenols found naturally in all fruits, vegetables, and medicinal plants. These compounds functionally evolved in the plant as a defense mechanism against herbivores, microbes, viruses or competing plants, and signal compounds (Foti and Amorati 2016). They are thus important for the plant's survival and reproductive fitness. In recent years, plant-derived phytochemical compounds gained a lot of importance because of their potential health benefits arising from the antioxidant, anticancerous, and numerous therapeutic properties (Gurib-Fakim 2006). As a dietary component, flavonoids are thought to have the largest share among phytochemical compounds due to their availability and diversity in the plant kingdom (Yao et al. 2004).

Flavonoids in general are universally distributed in higher plants. They also occur in many lower plant groups, notably in mosses, ferns, etc., but there are no records so far from bacteria. Flavonoids have been extracted from leaves, roots, wood, bark, pollen, fruit skin, and seeds. They have been detected during the fermentation of tea (Dwyer and Peterson 2013) and wine (Friedman 2014) as well as the manufacturing of cocoa (Zhang et al. 2013).

9.2 Structure/Subclasses

The name "flavonoid" is derived from the Latin word *flavus* meaning yellow and thus takes into account the fact that most of these substances have a yellowish color in their pure form. The pigments that color most flowers, fruits, and seeds are flavonoids (Beecher 2003). These metabolites are produced as the result of the secondary metabolism of plants and are commonly found with sugars moieties (glycosides) and other conjugates, thus being water-soluble in nature. Occasionally, flavonoids also occur in plants as aglycones.

Flavonoids represent largest group of polyphenolic compounds and categorized into six major subclasses: flavones (apigenin, luteolin), flavonols (quercetin, myricetin), flavanones (naringenin, hesperidin), flavanols or catechins (epicatechin, gallocatechin), anthocyanidins (cyanidin, pelargonidin), and isoflavones (genistein, daidzein). Classification of flavonoids is based upon their chemical and structural complexity. Structurally, flavonoids have a common C6–C3–C6 structure that contains two aromatic rings (A and B rings) that are connected by a three-carbon bridge (C-ring), creating an oxygenated heterocycle (Fig. 9.1).

Subclasses of the flavonoids (Fig. 9.2), viz., flavones, flavonols, flavanones, anthocyanins, isoflavones, and flavan-3-ols are defined by differences in the oxidation state and functional groups of the C-ring as well as by the assembly of the B-ring to the C-ring. The structural complexities among flavonoid subclasses are due to their potential for substitution in the A- and B-rings with phenolic hydroxyls, O-sugars, methoxy groups, sulfates, and glucuronides. In a few cases, the six-membered heterocyclic ring C occurs in an isomeric open form or is replaced by a five-membered ring (Beecher 2003). The first flavonoid, identified as rutin, was

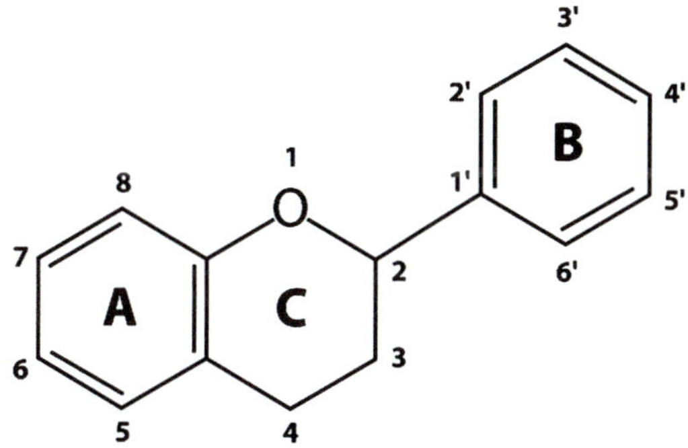

Fig. 9.1 Basic flavonoid structure

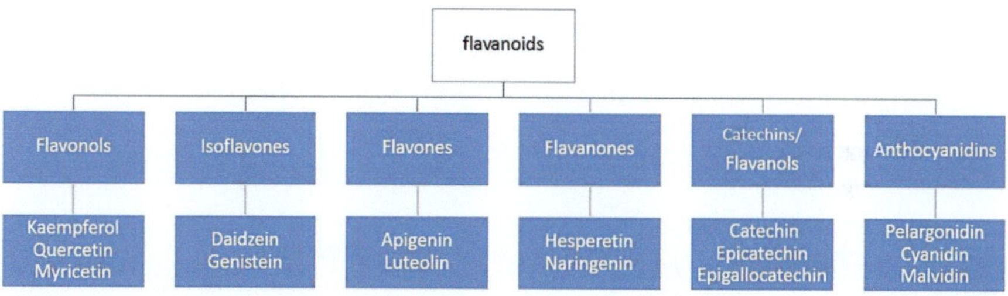

Fig. 9.2 Major subclasses of flavonoids

isolated from orange (genus: *Citrus*) in 1930; since then a number of flavonoids and their derivatives have been isolated from plants and investigated with respect to biochemical and biological activities.

9.2.1 Flavones

Flavones are one of the important subgroups of flavonoids. Flavones are widely present as glucosides in leaves, flowers, and fruits. Celery, parsley, red peppers, chamomile, mint, and *Ginkgo biloba* are among the major sources of flavones. Luteolin, apigenin, and tangeritin belongs to this subclass of flavonoids (Fig. 9.3). The peels of citrus fruits are rich in the polymethoxylated flavones tageretin, nobiletin, and sinensetin. Flavones exhibit various biological functions including antiallergic, antiviral, anti-inflammatory, antioxidation, anticancer, and antitumor effects (Lin et al. 2015).

9.2.2 Flavonone

Flavonone is another important class of flavonoids which is generally present in all the citrus fruits such as oranges, lemons, grapes, etc. Hesperitin, naringenin, and eriodictyol are examples of this class of flavonoids. Flavonones are associated with

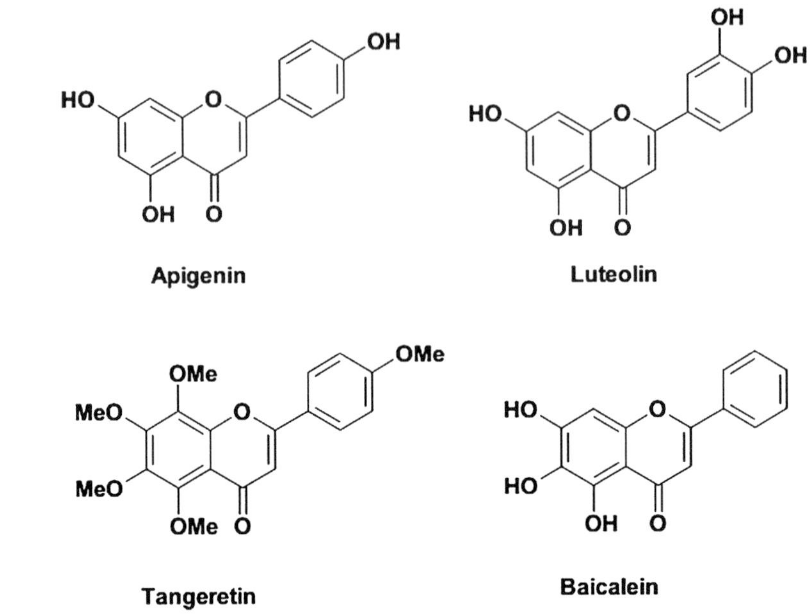

Fig. 9.3 Structures of some commonly studied flavones

a number of health benefits because of their free-radical-scavenging properties. Hesperidin and naringin are two glycosides of hesperetin and naringenin (Fig. 9.4) (Paredes et al. 2003). These compounds are accountable for the bitter taste of the juice and peel of citrus fruits. Citrus flavonoids exhibit pharmacological effects as antioxidant, anti-inflammatory, and cholesterol-lowering agents (Tripoli et al. 2007).

9.2.3 Flavonols

Flavonols are flavonoids containing a ketone group. They are the building blocks of proanthocyanins. Flavonols occur widely in a variety of fruits and vegetables. The most studied flavonols are kaempferol, quercetin, myricetin, and fisetin (Fig. 9.5). Onions, lettuce, tomato, apple, grape, and berries are good sources apart from fruits and vegetables. Beverages such as tea and red wine are also important sources of flavonols. Intake of flavonols is found to be associated with a wide range of health benefits which includes antioxidant potential and reduced risk of vascular disease (Pietta et al. 2003) (Table 9.1).

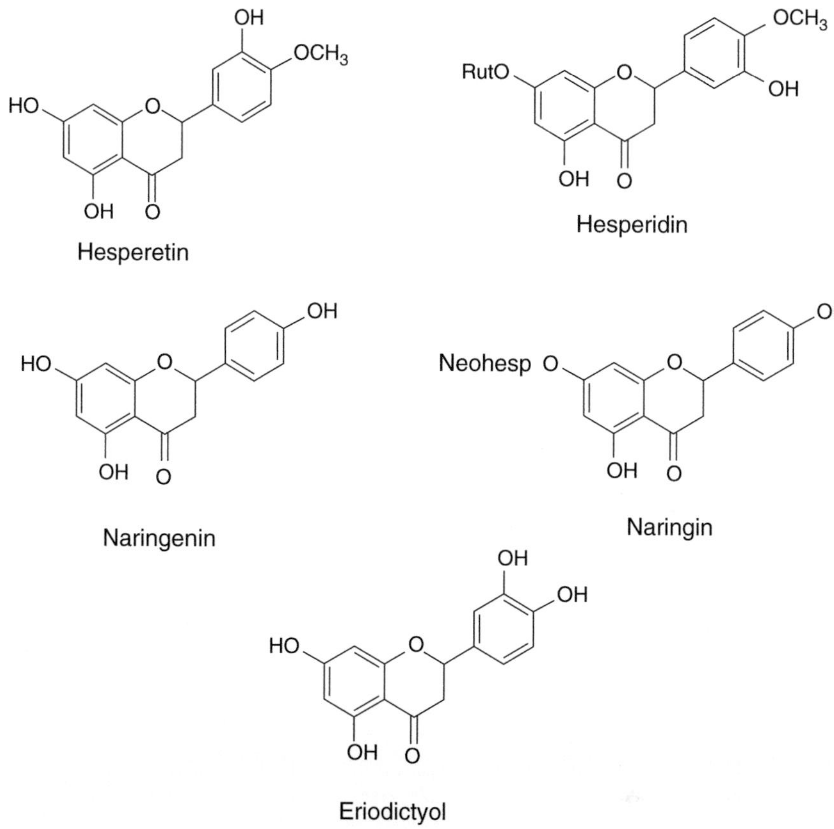

Fig. 9.4 Structures of some commonly studied flavanones

9.2.4 Anthocyanins

Anthocyanins are pigments responsible for colors in plants, flowers, and fruits. Cyanidin, delphinidin, malvidin, pelargonidin, and peonidin are the most commonly studied anthocyanins (Fig. 9.6). They occur predominantly in the outer cell layers of various fruits such as cranberry, black currant, red grape, merlot, raspberry, strawberry, blueberry, bilberry, and blackberry. Stability coupled with the health benefits of these compounds enable them to be used in the food industry in a variety of applications. Anthocyanins display wide range of biological activities including antioxidant, anti-inflammatory, antimicrobial, and anticarcinogenic activities (Weng and Yen 2012).

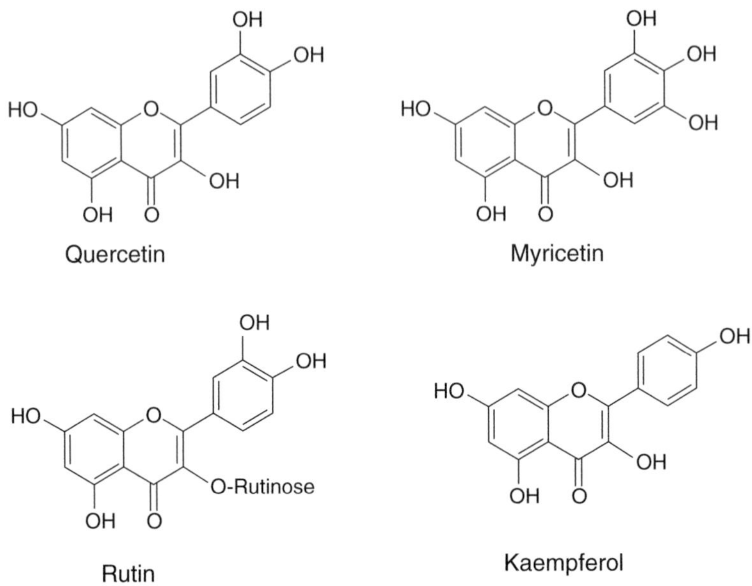

Fig. 9.5 Structures of some commonly studied flavonols

Table 9.1 Flavonoids commonly occuring as compounds in plant foods

Sr. No.	Subclass	Source	Compounds
1.	Flavones	Apple skins, berries, broccoli, celery, fruit peels, cranberries, grapes, lettuce, olives, onions, parsley, and *Ginkgo biloba*	Apigenin, chrysin, luteolin, sibelin, and (diosmin, flavoxate are synthetic)
2.	Flavanones	Citrus fruits and tomatoes	fisetin, hesperetin, narigin, neohesperidin, neoeriocitrin, naringenin, taxifolin
3.	Flavonols	Nearly ubiquitous in all plant sources	isorhamnetin, kaempferol, myricetin, quercetin, rutin, and gossypetin
4.	Anthocyanins	Berries (blueberries, Red grapes, Strawberries)	cyanidin, delphinidin, malvidin, peonidin, petunidin, pelargonidin
5.	Isoflavonids	*Trifolium pratense*, Soybean and soy foods, legumes	biochanin a, daidzein, formononentin, genistein, and glyciten

9.2.5 Isoflavonoids

Isoflavonoids are a large and a very distinctive subgroup of flavonoids. Isoflavonoids enjoy only a limited distribution in the plant kingdom and are predominantly found in soybeans and other leguminous plants. Some isoflavonoids have also been

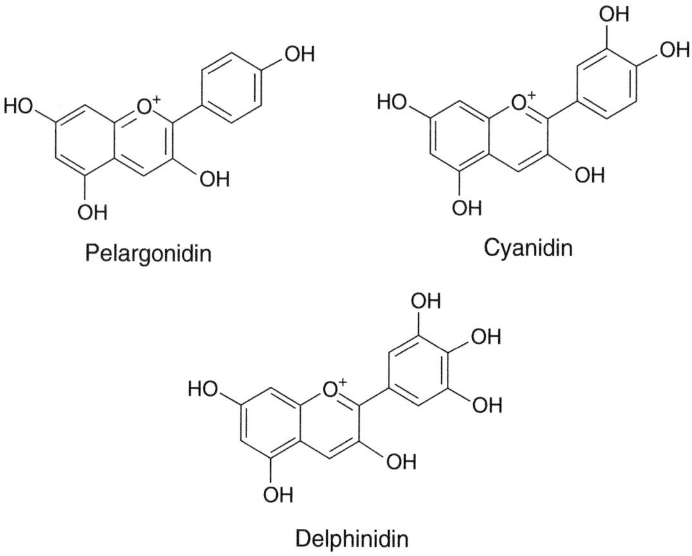

Fig. 9.6 Structures of some commonly studied anthocyanins

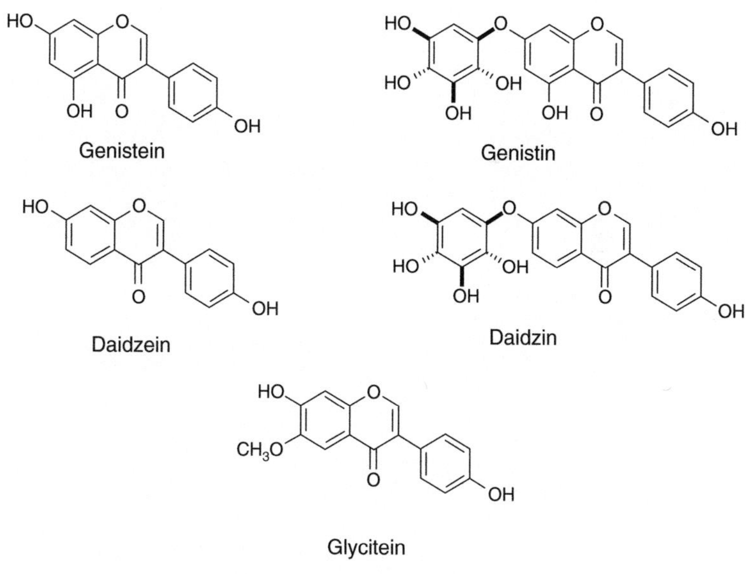

Fig. 9.7 Structures of some commonly studied isoflavonoids

reported to occur in microbial organisms. They are also found to play an important role as precursors for the development of phytoalexins during plant-microbe interactions. Isoflavonoids exhibit tremendous potential to fight a number of diseases. Isoflavones such as genistein and daidzein are commonly regarded to be phytoestrogens because of their estrogenic activity in certain animal models (Fig. 9.7) (Barnes 2010; Du et al. 2010).

9.3 Flavonoid Biosynthesis

Flavonoids are among the most prevalent class of metabolites in plant kingdom, and their distribution is almost ubiquitous. It is estimated that several hundred thousand secondary metabolites exist and approximately 20% of the carbon fixed by photosynthesis is believed to be channeled into the phenylpropanoid pathway, which encompasses transformation of phenylalanine into a variety of important secondary products, including such as flavonoids, lignins, phenolic acids, etc. Many efforts have been made to decipher the biosynthetic pathway of polyphenols and flavonoids using genetic and combinatorial approaches (Huang et al. 2008).

Flavonoids constitute a relatively diverse family of aromatic molecules that are derived from phenyalanine and malonyl-coenzyme A (phenylpropanoid pathway). Phenylalanine ammonia lyase (PAL) catalyzes the conversion of phenylalanine to cinnamate. The cinnamate 4-hydroxylase catalyzes the synthesis of p-hydroxycinnamate from cinnamate and 4-coumarate: CoA ligase converts p-coumarate to its coenzyme A ester, activating it for reaction with malonyl CoA (Kim et al. 2013).

The flavonoid biosynthetic pathway (Fig. 9.8) starts with the condensation of one molecule of 4-coumaroyl-CoA and three molecules of malonyl-CoA, yielding naringenin chalcone. This reaction is carried out by the enzyme chalcone synthase.

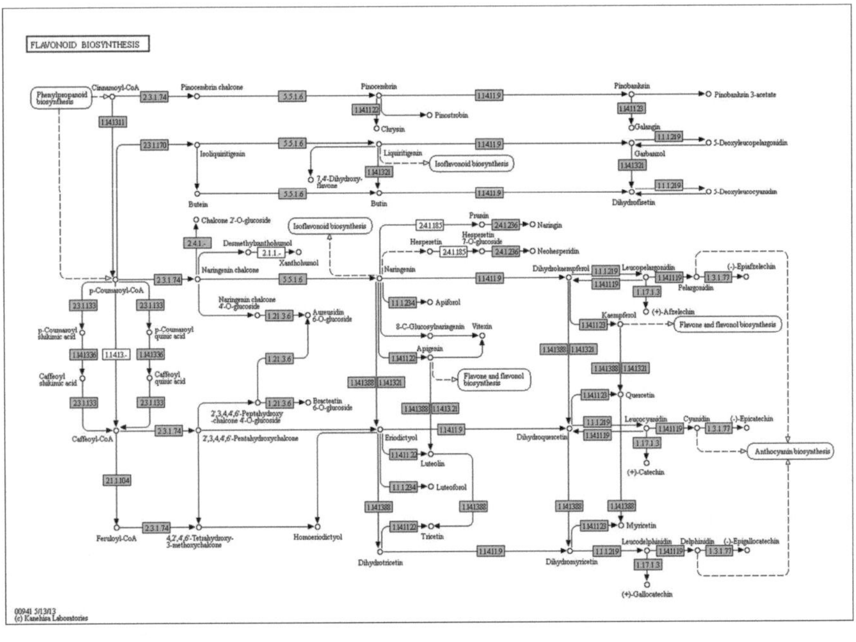

Fig. 9.8 Flavonoid biosynthesis pathway (Adapted from http://www.genome.jp/kegg-bin/show_pathway?org_name=ko&mapno=00941&mapscale=&show_description=show)

Chalcone is isomerized to a flavanone by the enzyme chalcone flavanone isomerase (CHI). From these central intermediates, the pathway diverges into several side branches, each resulting in a different class of flavonoids. Flavanone 3-hydroxylase (F3H) catalyzes the hydroxylation of (2S)-flavanones to dihydroflavonols (Ghasemzadeh et al. 2012). For the biosynthesis of anthocyanins, dihydroflavonol reductase catalyzes the reduction of dihydroflavonols to flavan-3, 4-diols, which are converted to anthocyanidins by anthocyanidin synthase. The formation of glucosides is catalyzed by the enzyme UDP-glucose flavonoid 3-o-glucosyl transferase, which stabilizes the anthocyanidins by 3-O-glucosylation. For the biosynthesis of flavonols, flavanone 3-hydroxylase converts the dihydroflavonol to dihydroquercetin which is then converted into flavonol quercetin via the activity of the enzyme flavonol synthase (Li et al. 2012). As the finishing step of flavonol biosynthesis, glucosyl transferases catalyzes the transfer of glucose and rhamnose from UDP-glucose, UDP-rhamnose for the synthesis of flavonol glycoside, i.e., rutin.

9.4 Regulation

The flavonoid biosynthetic pathway is under tight developmental control, and multiple environmental conditions, like light and hormones, affect the expression of flavonoid biosynthetic genes (Besseau et al. 2007). Indeed, a decrease of flavonoid biosynthesis has been observed when either endogenous (e.g., plant hormones), or exogenous factors (e.g., water and temperature stress, light, fertilizer, etc.) are limiting or excessive (Gupta et al. 2014). Accumulation of flavonoids increased in leaves of ginger grown in shade revealing that flavonoids are light dependent and their biosynthetic rate is related to light intensity (Karimi et al. 2013). It is important to note that flavonol biosynthesis is stimulated by light, and therefore they accumulate in the outer and aerial tissue of fruits. Interestingly, differences in concentration can exist among fruits on the same tree and even between different sides of a single piece of fruit, depending on exposure to sunlight.

Environmental factors, like temperature and light stress, and fertilizers that increase plant vigor are also reported to affect flavonoid content in grape berry (Braidot et al. 2008). Nitrogen and potassium at high levels caused the enhancement of vegetative growth, the delay of ripening, and a decrease of color in grape berry. Impact of temperature on proanthocyanidin (PA) accumulation in heating and cooling berries altered the initial rates of PA accumulation, which was correlated strongly with the expression of core genes in the flavonoid pathway (Cohen et al. 2012).

The jasmonates, which include jasmonic acid and its methyl ester methyl jasmonates (MJ), play a central role in regulating the biosynthesis of many secondary metabolites, including flavonoids, and also are signaling molecules in environmental stresses. The expression of several phenylpropanoid-related genes, primarily those associated with anthocyanin biosynthesis, was increased under MJ treatment, which correlated with an increased accumulation of anthocyanin (Horbowicz et al. 2011). The growth hormone cytokinin also increased rutin accumulation by threefold in developing *Cucurbita* cotyledons.

Concentration of biomolecules also affects the rutin content in various plants. For example, the concentration of rutin was enhanced twofold in *Ficus deltoidea* by culture enriched with either glucose or fructose (Ong et al. 2011). However, not much information is available on biosynthesis and accumulation of flavonoids in relation to growth and development of different stages in a particular plant species.

9.4.1 Transcriptional Regulation of Flavonoid Biosynthesis

Coordinate transcriptional control of biosynthetic genes has arisen as a major mechanism dictating the final levels of secondary metabolites in plant cells. This regulation is achieved by specific transcription factors. Plant flavonoid accumulation is controlled by transcriptional regulation of the genes that encode the biosynthetic enzymes, in which the R2R3-MYB transcription factor is a key factor (Hichri et al. 2011).

Regulation of structural gene expression appears tightly organized in a spatial and temporal way during plant development, and is orchestrated by a ternary complex involving transcription factors from the R2R3-MYB, basic helix-loop-helix (bHLH), and WD40 classes (Xu et al. 2014). Overexpression of MYB134 in poplar resulted in transcriptional activation of the full proanthocyanidins biosynthetic pathway and a significant plant-wide increase in proanthocyanidins levels (Mellway et al. 2009). In tobacco plants, overexpressing the *Arabidopsis* PAP1 gene (encoding an MYB transcription factor), causes an accumulation of anthocyanin pigments and other flavonoids/phenylpropanoids, and in tomato, expression of maize transcription factors LC and C1 leads to a large increase in the level of kaempferol (Liu et al. 2016; Verhoeyen et al. 2002).

9.5 Biological Functions of Flavonoids

9.5.1 Distribution and Function in Plants

Besides their common feature to contribute to plant color, different flavonoids have distinct biological role depending upon site of accumulation, structural characteristics, and the pattern of glycosylation. There are evidences, from a number of experiments, that flavonoids in leaves deposited either in the epidermal cells or in the waxy upper leaf surface, provide protection from the potential damage of UV radiation. Physiological experiments in petunia flowers indicate that the flavonol glycosides secreted in plant pollen play a role in the reception of the pollen on the plant stigma and are required for successful reproduction. In addition, it has been discovered that the flavones of legume roots have an essential signaling role in the process of infection by *Rhizobium* bacteria and hence in the establishment of nitrogen fixation in these plants (Lee and Hirsch 2006). Flavonoids also exhibit a wide range of biological properties, including antibacterial, insecticidal, and chemoprotective activities. In one study, it has been proved that rutin, a flavonoid, is involved in tolerance mechanism of plants against salinity by regulating K^+ retention and Na^+ exclusion from leaf mesophyll cells (Ismail et al. 2016).

9.5.2 Studies in Animals and Human

Flavonoids occur widely in the fruits and vegetables that make up the human diet, and it has been estimated that a person consumes at least 1 g of flavonoid daily in diet. A lot of clinical studies and medications based on flavonoids are beneficial to human health, particularly for compounds such as quercetin and isoflavonoids, which have been linked to anticancer benefits and stilbenes that are believed to contribute to lowering of heart disease (de Pascual-Teresa et al. 2010).

Flavonoids are also major components of many plant drugs (e.g., *Ginkgo* leaves), and it is possible that they contribute to the curative properties. Some flavonoids exert a wide variety of biological actions, such as antioxidant (Heim et al. 2002), anti-inflammatory (Lopez-Lazaro 2009), anticarcinogenic (Nair et al. 2006), vasodilatory, and hepatoprotective on human system (Haid et al. 2012).

9.5.2.1 Antioxidant Activity

Oxidative stress is defined as an imbalance among free radicals and reactive oxygen species (ROS). The accumulation of unrepaired damaged products and ROS may be critical to the development of cancer, atherosclerosis, diabetes, and chronic inflammation (Zhang and Tsao 2016). Food sources rich in antioxidants are protective against a variety of diseases. Antioxidants and dietary fiber are thought to be the principal nutrients responsible for these protective effects. As a dietary component, flavonoids are thought to have health-promoting properties due to their high antioxidant capacity in both in vivo and in vitro systems (Ross and Kasum 2002).

Several in vitro studies have publicized that the flavonoids, including flavonols, flavones, isoflavones, flavanols, and anthocyanidins, possess antioxidant activity (Dai and Mumper 2010). Studies have shown many flavonoids to be effective antioxidants in a wide range of chemical oxidation systems, demonstrated by their capability to scavenge peroxyl radicals, alkyl peroxyl radicals, superoxide hydroxyl radicals, and peroxynite in aqueous and organic environments (Halliwell et al. 2005).

Recent studies have suggested that flavonoids may exert antioxidant abilities through protection or enhancement of endogenous antioxidants. Numerous flavonoids, such as quercetin, myricetin, and kaempferol, have reported to decrease oxidative stress by inducing the expression of glutathione S-transferase (GST) enzyme (Sahu and Gray 1996). Intake of these compounds as supplement may increase the activity of reduced form of glutathione (GSH), thus better fighting oxidative stress. Some flavonoids, including quercetin, rutin, fisetin, and myricetin, have shown significant increases in GST-specific activity (Chirumbolo 2010). The antioxidant capacity of flavonoids depends on the structure and arrangement of functional groups surrounding the flavonol nucleus, which may directly affect glutathione metabolism, antioxidant capacity, or the ability to maintain low Ca^{2+} levels despite high levels of ROS (Nakayama et al. 2011). Studies have indicated that the aglycones, including quercetin, luteolin, myricetin, and kaempferol, have greater antioxidant capacity than do the conjugate flavonoids, such as quercetin-3-glucoside and rutin (Pereira et al. 2013).

In addition to the free-radical-scavenging properties, some flavonoids can chelate those transition heavy metal ions responsible for the generation of reactive oxygen species and therefore inhibit the initiation of the lipoxygenase reaction. Some evidence has suggested that flavonoids also have antioxidant capacity in nontransition-metal-dependent oxidation (Kattappagari et al. 2015). Flavonoids in combination with other antioxidants, including vitamins, are supposed to constrain lipid peroxidation in the phospholipid bilayer caused by reactive oxygen species (Tay et al. 2013).

9.5.2.2 Antitumor/Anticancer Activity

Flavonoids are powerful bioactive molecules that possess anticarcinogenic properties since they can interfere with the initiation, development, and progression of cancer by proliferation, invasion, angiogenesis, and metastasis of tumor cells. Flavonoids have emerged as potential chemopreventive candidates for cancer treatment due to their ability to induce apoptosis (Ren et al. 2003). Flavonoids may check various stages in the carcinogenesis process, namely tumor initiation, promotion, and progression by many molecular mechanisms of action. These mechanisms of action have been identified as including carcinogen inactivation, antiproliferation, cell cycle arrest, induction of apoptosis, inhibition of angiogenesis, antioxidation, and reversal of multidrug resistance, or a combination of these mechanisms.

Flavonoids also inhibit many enzymes that are the targets in anticancer treatment, by the activation of procarcinogen substrates to carcinogens which makes them putative anticancer substances (Galati and O'Brien 2004). Certain members of the flavonoids kaempferol and genistein possess antiproliferative effects in different cancer cell lines, and the tea polyphenols have strong antiproliferative capacity (Yang et al. 2000).

9.5.2.3 Antimicrobial Activity

In many plants, flavonoids are synthesized in response to attack of microbes in order to neutralize the harmful effects produced by them. Extracts of various plants rich in flavonoid content, e.g., chromolaena, hypericum, and capsella, have been reported to possess antibacterial activity. Flavonoids and their different classes, viz., flavones, flavone glycosides, isoflavones, flavanones, isoflavanones, flavonols, flavonol glycosides, and chalcones possess potent antibacterial activity against a wide range of microbes. The mechanism by which flavonoids inhibit bacterial growth includes inhibition of nucleic acid synthesis, inhibition of cytoplasmic membrane function, and inhibition of energy metabolism.

Naringenin and sophoraflavanone G show intensive antibacterial activity against methicillin-resistant *Staphylococcus aureus* (MRSA) and streptococci by altering the membrane fluidity in the hydrophilic and hydrophobic regions (Tsuchiya and Iinuma 2000). Flavonoids from *Lippia turbinate* also play an important role in developing the antibacterial activity against methicillin-resistant *Staphylococcus aureus*. Its extract shows synergistic effect against bacteria when combined with *Aloysia polystachya* extract (Aguado et al. 2016).

Flavonoids obtained from methanolic extract of *Asparagus racemosus*, *Ocimum sanctum*, *Cassia fistula*, *Citrus aurantifolia*, *Catharanthus roseus*, and *Polyalthia longifolia* were found to be active against *Staphylococcus aureus* (gram positive) and *Escherichia coli* (gram negative). Flavonoids from *Ocimum sanctum* and *Citrus aurantifolia* were found to be the most potent against the microorganisms (Kaur and Mondal 2014).

Antifungal activity has been displayed by a number of flavonoids. Flavonoids isolated from the peelings of tangerine orange show antifungal activity against *Deuterophoma tracheiphila* while nobiletin and langeritin display strong and weak antiviral activities, respectively. The first chlorine-containing flavonoid-type antifungal antibiotic, chlorflavonin, was produced from the strains of *Aspergillus candidus*. Quercetin and epigallocatechin gallate have been reported for various antimicrobial effects but very little is known about the antifungal activity. But it shows antifungal activity against *Candida tropicalis* when combined with flucanazole. *Candida tropicalis* is the most prevalent pathogenic yeast, resistant to fluconazole. The synergistic effect of these flavonoids against the fungus is by promoting apoptosis by exposure of phosphatidylserine in the plasma membrane and morphological changes, mitochondrial depolarization, intracellular accumulation of ROS, condensation, and DNA fragmentation (da Silva et al. 2014).

Flavonoids exhibit antiviral activity against a wide range of viruses such as herpes simplex virus, respiratory syncytial virus, parainfluenza virus, and adenovirus (Wang et al. 1998). They restrain the viral activity either by interacting with intracellular replication of viruses or by inhibiting the infectious properties of the viruses (Seo et al. 2016). In some cases, modified forms of flavonoids show more restrictive action on virus infectivity compared to their unmodified forms; for example, flavonoids in glycone form appear to be inhibitory on rotavirus infectivity compared to flavonoids in their aglycone form (Bae et al. 2000). Quercetin was reported to exhibit both anti-infective and antireplicative abilities. Flavonoids have been reported as an anti-HIV agent in various studies under in vitro conditions. They inhibit activity of different enzymes such as reverse transcriptase, RNA-directed DNA polymerase, integrase, and proteases involved in replication cycle of HIV (Kumar and Pandey 2014).

9.5.2.4 Anti-inflammatory Activity

Inflammation is associated with different human diseases and has been identified as a major threat. Therefore, reducing inflammation is a favorable strategy for treating several human diseases. Flavonoids have been reported to have anti-inflammatory property. Different molecular mechanisms have been suggested which may be the reason behind the anti-inflammatory activities of flavonoids, for instance, inhibition of NF-κB and activating protein-1 (AP-1), proinflammatory enzymes (e.g., cyclooxygenase-2, lipoxygenase) inhibition, activation of phase II-antioxidant detoxifying enzymes, inhibition of NO synthase enzyme, and activation of mitogen-activated protein kinase (MAPK), phase II-antioxidant detoxifying enzymes, protein kinase C and nuclear factor-erythroid 2-related factor 2 (Pan et al. 2010).

The mechanism underlying the anti-inflammatory activity of Nepalese propolis has been studied which shows the effects of flavonoids on the IL-33 signaling pathway. The four flavonoids 3', 4'-dihydroxy-4-methoxydalbergione, 4-methoxy-dalbergion, cearoin, and chrysin strikingly repressed the IL-33-induced mRNA expression of inflammatory genes including IL-6, TNFα, and IL-13 in bone marrow-derived mast cells (BMMC), along with IL-33-induced activation of nuclear factor κB (NF-κB) (Funakoshi-Tago et al. 2015).

In Chinese medicine, for the treatment of various disorders, the roots of *Sophora flavescens* have been used which contains flavonoids and alkaloids. So, an anti-inflammatory agent has been developed by preparing the alkaloid-free prenylated flavonoid-enriched fraction (PFS) from the rhizomes of *Sophora flavescens*. The mechanism by which PFS shows anti-inflammatory activity includes inhibition of proinflammatory molecules such as COX-2, iNOS and IL-6, thereby inhibiting chronic inflammatory response. Thus, it can be used for combating the various chronic inflammatory disorders such as rheumatoid arthritis (Jin et al. 2010).

It has been suggested that anti-inflammatory effects are associated with higher dietary anthocyanin and flavonol intake. It was also reported that lower inflammation is associated with higher anthocyanin intake while higher flavonol intake was associated with lower cytokine and oxidative stress biomarker concentrations (Kim et al. 1998).

9.5.2.5 Other Properties

Several flavonoids, such as naringenin, apigenin, rutin, quercetin and catechin, have been documented so far as having hepatoprotective activities. They prevent development of hepatic manifestations by different mechanisms, for example, by increasing hepatic GCLC expression due to increase in cAMP levels which ultimately decrease ROS levels in the liver. Anthocyanin cyanidin-3-O-β glucoside (C3G) acts by lowering hepatic lipid peroxidation and inhibiting the release of proinflammatory cytokines, thus protecting against the development of hepatic steatosis (Zhu et al. 2012). Another flavanoid silymarin has been reported for liver regeneration in damaged livers by stimulating enzymatic activity of DNA-dependent RNA polymerase I which results in DNA biosynthesis and cell proliferation (Sonnenbichler and Zetl 1985).

Antidiabetic property has been found to be correlated with different flavonoids such as quercetin. It is found to be responsible for the regeneration of the cells involved in insulin synthesis in pancreas in streptozotocin-induced diabetic rats which ultimately results in higher production of insulin. In another study, it was found that quercetin may play an important role in noninsulin-dependent diabetes as it enhances the insulin release by stimulating Ca^{2+} uptake from isolated islets cell (Hii and Howell 1985).

Flavonoid intake has been associated with prevention of a number of cardiovascular diseases including hypertension and atherosclerosis. In the development of cardiovascular diseases, endothelial dysfunction is the major event which can be prevented by having a flavonoid-rich diet. Flavonoids enhance the vasorelaxant process which reduces the arterial pressure eventually preventing endothelial dysfunction (Jeong et al. 2016).

Thrombotic diseases occur due to platelet aggregation which generates lipid peroxides and oxygen free radicals and results in decreased concentration of endothelial prostacyclin and nitrous oxide. Particularly, flavonols show antithrombotic property as they maintain proper concentration of endothelial prostacyclin and nitric oxide because they have the capacity to scavenge free radicals directly. Flavonoids inhibit activities of different pathways such as cyclooxygenase and lipoxygenase pathways and thus prevent thrombotic diseases (Brady 2013).

9.6 Conclusion

Flavonoids are the well-known phytochemicals found ubiquitously in different parts of plants. Flavonoids vary in their structure and exhibit different physical, chemical, and biological properties. These substances are responsible for major biological activities such as antioxidant, antitumor, anti-inflammatory, antimicrobial, and various other therapeutic properties. Due to their therapeutic properties, they can be used as a medicinal drug after validation as they have no side effects, have a relatively long life and are found in many foods.

References

Aguado MI, Dudik NH, Zamora CMP, Torres CA, NuÑez MB (2016) Antioxidant and antibacterial activities of hydroalcoholic extracts from *Aloysia polystachya griseb moldenke* and *Lippia turbinata griseb* (verbenaceae). Int J Pharm Pharm Sci 8(3):393–395

Bae EA, Han MJ, Lee M, Kim DH (2000) *In vitro* inhibitory effect of some flavonoids on rotavirus infectivity. Biol Pharm Bull 23(9):1122–1124

Barnes S (2010) The biochemistry, chemistry and physiology of the isoflavones in soybeans and their food products. Lymphat Res Biol 8(1):89–98

Beecher GR (2003) Overview of dietary flavonoids: nomenclature, occurrence and intake. J Nutr 133(10):3248S–3254S

Besseau S, Hoffmann L, Geoffroy P, Lapierre C, Pollet B, Legrand M (2007) Flavonoid accumulation in *Arabidopsis* repressed in lignin synthesis affects auxin transport and plant growth. Plant Cell 19(1):148–162

Brady DM (2013) Flavocoxid: a medicinal extract of *Scutellaria bicalensis* and *Acacia catechu* acts as a dual inhibitor of Cyclooxegenase and 5- Lipoxygenase to reduce inflammation: nutritional perspectives. J Council Nutr 36(3):28–37

Braidot E, Zancani M, Petrussa E et al (2008) Transport and accumulation of flavonoids in grapevine (*Vitis vinifera* L.) Plant Signal Behav 3(9):626–632

Chirumbolo S (2010) The role of quercetin, flavonols and flavones in modulating inflammatory cell function. Inflamm Allergy Drug Targets 9(4):263–285

Cohen SD, Tarara JM, Gambetta GA et al (2012) Impact of diurnal temperature variation on grape berry development, proanthocyanidin accumulation and the expression of flavonoid pathway genes. J Exp Bot 63(7):2655–2665 da Silva CR, de Andrade Neto JB, de Sousa Campos R, Figueiredo NS, Sampaio LS, Magalhães HIF, Cavalcanti BC, Gaspar DM, de Andrade GM, Lima ISP, de Barros Viana GS (2014) Synergistic effect of the flavonoid catechin, quercetin, or epigallocatechin gallate with fluconazole induces apoptosis in Candida tropicalis resistant to fluconazole. Antimicro Agents Chemother 58(3):1468–1478

Dai J, Mumper RJ (2010) Plant phenolics: extraction, analysis and their antioxidant and anticancer properties. Molecules 15(10):7313–7352

de Pascual-Teresa S, Moreno DA, García-Viguera C (2010) Flavanols and anthocyanins in cardiovascular health: a review of current evidence. Int J Mol Sci 11(4):1679–1703

Du H, Huang Y, Tang Y (2010) Genetic and metabolic engineering of isoflavonoid biosynthesis. Appl Microbiol Biotechnol 86(5):1293–1312

Dwyer JT, Peterson J (2013) Tea and flavonoids: where we are, where to go next. Am J Clin Nutr 98(6):1611S–1618S

Foti MC, Amorati R (2016) ROS and phenolic compounds. In: Reactive oxygen species in biology and human health. CRC Press, Boca Raton, pp 49–53

Friedman M (2014) Antibacterial, antiviral, and antifungal properties of wines and winery byproducts in relation to their flavonoid content. J Agric Food Chem 62(26):6025–6042

Funakoshi-Tago M, Okamoto K, Izumi R, Tago K et al (2015) Anti-inflammatory activity of flavonoids in Nepalese propolis is attributed to inhibition of the IL-33 signaling pathway. Int Immunopharmacol 25(1):189–198

Galati G, O'Brien PJ (2004) Potential toxicity of flavonoids and other dietary phenolics: significance for their chemopreventive and anticancer properties. Free Radic Biol Med 37(3):287–303

Ghasemzadeh A, Jaafar HZ, Karimi E, Ibrahim MH (2012) Combined effect of CO_2 enrichment and foliar application of salicylic acid on the production and antioxidant activities of anthocyanin, flavonoids and isoflavonoids from ginger. BMC Complement Altern Med 12(1):1

Gupta N, Chauhan RS, Pradhan JK (2014) Rutin: a bioactive flavonoids. In: Handbook of medicinal plants and their bioactive compounds. CRC Press, Boca Raton, pp 51–57

Gurib-Fakim A (2006) Medicinal plants: traditions of yesterday and drugs of tomorrow. Mol Asp Med 27(1):1–93

Haid S, Novodomská A, Gentzsch J, Grethe C et al (2012) A plant-derived flavonoid inhibits entry of all HCV genotypes into human hepatocytes. Gastroenterology 143(1):213–222

Halliwell B, Rafter J, Jenner A (2005) Health promotion by flavonoids, tocopherols, tocotrienols, and other phenols: direct or indirect effects? Antioxidant or not? Am J Clin Nutr 81(1):268S–276S

Heim KE, Tagliaferro AR, Bobilya DJ (2002) Flavonoid antioxidants: chemistry, metabolism and structure-activity relationships. J Nutr Biochem 13(10):572–584

Hii CST, Howell SL (1985) Effects of flavonoids on insulin secretion and $45Ca^{2+}$ handling in rat islets of Langerhans. J Endocrinol 107(1):1–8

Hichri I, Barrieu F, Bogs J, Kappel C, Delrot S, Lauvergeat V (2011) Recent advances in the transcriptional regulation of the flavonoid biosynthetic pathway. J Exp Bot 62(8):2465–2483

Horbowicz M, Wiczkowski W, Koczkodaj D, Saniewski M (2011) Effects of methyl jasmonate on accumulation of flavonoids in seedlings of common buckwheat (*Fagopyrum esculentum* Moench). Acta Biol Hung 62(3):265–278

Huang B, Guo J, Yi B, Yu X, Sun L, Chen W (2008) Heterologous production of secondary metabolites as pharmaceuticals in *Saccharomyces cerevisiae*. Biotechnol Lett 30(7):1121–1137

Ismail H, Shabala L, Tian Y et al (2016) Rutin, a flavonoid with antioxidant activity, improves plant salinity tolerance by regulating K^+ retention and Na^+ exclusion from leaf mesophyll in quinoa and broad beans. Funct Plant Biol 43(1):75–86

Jeong HS, Kim S, Hong SJ et al (2016) Black raspberry extract increased circulating endothelial progenitor cells and improved arterial stiffness in patients with metabolic syndrome: a randomized controlled trial. J Med Food 19(4):346–352

Jin JH, Kim JS, Kang SS, Son KH, Chang HW, Kim HP (2010) Anti-inflammatory and antiarthritic activity of total flavonoids of the roots of *Sophora flavescens*. J Ethnopharmacol 127(3):589–595

Karimi E, Jaafar HZ, Ghasemzadeh A, Ibrahim MH (2013) Light intensity effects on production and antioxidant activity of flavonoids and phenolic compounds in leaves, stems and roots of three varieties of *Labisia pumila* Benth. Aust J Crop Sci 7(7):1016

Kattappagari KK, Teja CR, Kommalapati RK et al (2015) Role of antioxidants in facilitating the body functions: a review. J Orofacial Sci 7(2):71

Kaur S, Mondal P (2014) Study of total phenolic and flavonoid content, antioxidant activity and antimicrobial properties of medicinal plants. J Microbiol Exp 1(1):00005

Kim HP, Mani I, Iversen L, Ziboh VA (1998) Effects of naturally-occurring flavonoids and biflavonoids on epidermal cyclooxygenase and lipoxygenase from guinea-pigs. Prostaglandins Leukot Essent Fat Acids 58(1):17–24

Kim YB, Kim JK, Uddin MR et al (2013) Metabolomics analysis and biosynthesis of rosmarinic acid in *Agastache rugosa* Kuntze treated with methyl jasmonate. PLoS One 8(5):e64199

Kumar S, Pandey AK (2014) Medicinal attributes of *Solanum xanthocarpum* fruit consumed by several tribal communities as food: an *in vitro* antioxidant, anticancer and anti HIV perspective. BMC Complem Altern Med 14(1):1

Lee A, Hirsch AM (2006) Signals and responses: choreographing the complex interaction between legumes and α-and β-rhizobia. Plant Signal Behav 1(4):161–168

Li H, Qiu J, Chen F, Lv X, Fu C, Zhao D, Zhao Q (2012) Molecular characterization and expression analysis of dihydroflavonol 4-reductase (DFR) gene in *Saussurea medusa*. Mol Biol Rep 39(3):2991–2999

Lin CH, Chang CY, Lee KR, Lin HJ, Chen TH, Wan L (2015) Flavones inhibit breast cancer proliferation through the Akt/FOXO3a signaling pathway. BMC Cancer 15(1):958

Liu C, Long J, Zhu K, Liu L et al (2016) Characterization of a citrus R2R3-MYB transcription factor that regulates the flavonol and hydroxycinnamic acid biosynthesis. Sci Rep 6:25352

Lopez-Lazaro M (2009) Distribution and biological activities of the flavonoid luteolin. Mini-Rev Med Chem 9(1):31–59

Mellway RD, Tran LT, Prouse MB, Campbell MM, Constabel CP (2009) The wound-, pathogen-, and ultraviolet B-responsive MYB134 gene encodes an R2R3 MYB transcription factor that regulates proanthocyanidin synthesis in poplar. Plant Physiol 150(2):924–941

Nair MP, Mahajan S, Reynolds JL, Aalinkeel R et al (2006) The flavonoid quercetin inhibits proinflammatory cytokine (tumor necrosis factor alpha) gene expression in normal peripheral blood mononuclear cells via modulation of the NF-κβ system. Clin Vaccine Immunol 13(3):319–328

Nakayama M, Aihara M, Chen YN et al (2011) Neuroprotective effects of flavonoids on hypoxia-, glutamate-, and oxidative stress–induced retinal ganglion cell death. Mol Vis 17:1784–1793

Ong SL, Ling APK, Poospooragi R, Moosa S (2011) Production of flavonoid compounds in cell cultures of *Ficus deltoidea* as influenced by medium composition. Int J Med Aromatic Plant 1(2):62–74

Pan MH, Lai CS, Ho CT (2010) Anti-inflammatory activity of natural dietary flavonoids. Food Funct 1(1):15–31

Paredes A, Alzuru M, Mendez J, Rodríguez-Ortega M (2003) Anti-Sindbis activity of flavanones hesperetin and naringenin. Biol Pharm Bull 26(1):108–109

Pereira RB, Sousa C, Costa A, Andrade PB, Valentão P (2013) Glutathione and the antioxidant potential of binary mixtures with flavonoids: synergisms and antagonisms. Molecules 18(8):8858–8872

Pietta P, Minoggio M, Bramati L (2003) Plant polyphenols: structure, occurrence and bioactivity. Stud Nat Prod Chem 28:257–312

Ren W, Qiao Z, Wang H, Zhu L, Zhang L (2003) Flavonoids: promising anticancer agents. Med Res Rev 23(4):519–534

Ross JA, Kasum CM (2002) Dietary flavonoids: bioavailability, metabolic effects, and safety. Annu Rev Nutr 22(1):19–34

Sahu SC, Gray GC (1996) Pro-oxidant activity of flavonoids: effects on glutathione and glutathione S-transferase in isolated rat liver nuclei. Cancer Lett 104(2):193–196

Seo DJ, Jeon SB, Oh H et al (2016) Comparison of the antiviral activity of flavonoids against murine norovirus and feline calicivirus. Food Control 60:25–30

Sonnenbichler J, Zetl I (1985) Biochemical effects of the flavonolignane silibinin on RNA, protein and DNA synthesis in rat livers. Prog Clin Biol Res 213:319–331

Tay WM, da Silva GF, Ming LJ (2013) Metal binding of flavonoids and their distinct inhibition mechanisms toward the oxidation activity of Cu2+–β-amyloid: not just serving as suicide antioxidants! Inorg Chem 52(2):679–690

Tripoli E, La Guardia M, Giammanco S, Di Majo D, Giammanco M (2007) Citrus flavonoids: molecular structure, biological activity and nutritional properties: a review. Food Chem 104(2):466–479

Tsuchiya H, Iinuma M (2000) Reduction of membrane fluidity by antibacterial sophoraflavanone G isolated from *Sophora exigua*. Phytomedicine 7(2):161–165

Verhoeyen ME, Bovy A, Collins G, Muir S et al (2002) Increasing antioxidant levels in tomatoes through modification of the flavonoid biosynthetic pathway. J Exp Bot 53(377):2099–2106

Wang HK, Xia Y, Yang ZY, Natschke SLM, Lee KH (1998) Recent advances in the discovery and development of flavonoids and their analogues as antitumor and anti-HIV agents. In: Flavonoids in the living system. Springer US, Boston, pp 191–225

Weng CJ, Yen GC (2012) Flavonoids, a ubiquitous dietary phenolic subclass, exert extensive *in vitro* anti-invasive and *in vivo* anti-metastatic activities. Cancer Metastasis Rev 31(1–2):323–351

Xu F, Ning Y, Zhang W, Liao Y, Li L, Cheng H, Cheng S (2014) An R2R3-MYB transcription factor as a negative regulator of the flavonoid biosynthesis pathway in *Ginkgo biloba*. Funct Integr Genom 14(1):177–189

Yang CS, Chung JY, Yang GY, Chhabra SK, Lee MJ (2000) Tea and tea polyphenols in cancer prevention. J Nutr 130(2):472S–478S

Yao Liu H et al (2004) Flavonoids in food and their health benefits. Plant Foods Hum Nutr 59(3):113–122

Zhang H, Tsao R (2016) Dietary polyphenols, oxidative stress and antioxidant and anti-inflammatory effects. Curr Opin Food Sci 8:33–42

Zhang Q, Zhao X, Qiu H (2013) Flavones and flavonols: phytochemistry and biochemistry. Natural products. Springer, Heidelberg, pp 1821–1847

Zhu W, Jia Q, Wang Y, Zhang Y, Xia M (2012) The anthocyanin cyanidin-3-O-β-glucoside, a flavonoid, increases hepatic glutathione synthesis and protects hepatocytes against reactive oxygen species during hyperglycemia: involvement of a cAMP–PKA-dependent signaling pathway. Free Radic Biol Med 52(2):314–327

Enzymatic Approaches for the Synthesis of High Fructose Syrup

10

R.S. Singh, K. Chauhan, and R.P. Singh

Abstract

Fructose is a monosaccharide widely used for food, pharmaceutical, and medical applications. Such ubiquity can be ascribed to its various superior technical properties over the conventional sugar, sucrose, and many beneficial roles in the human body. High fructose syrup (HFS) can be synthesized from starch or inulin using enzymatic/chemical methods. A well-known conventional approach is the hydrolysis of starch using amylolytic enzymes and subsequent isomerization of dextrose to fructose by glucose isomerase. The product yield by this method is only 42% and the product mixture also contains 50% dextrose and 8% other saccharides. HFS can also be produced from inulin by a single-step method using inulinases. In the single-step enzymatic method, inulinase acts sequentially on β-(2, 1) linkages of inulin to release the fructose units. By this method approximately 95% fructose yield can be obtained. Acid hydrolysis of inulin/starch is not recommended because it imparts color to the product and a few undesired products, like difructose anhydride, are also formed. In the present chapter, enzymatic approaches are described for the preparation of HFS from starch and inulin.

10.1 Introduction

Sugars have enhanced the enjoyment of food for thousands of years. Earlier, honey was the principal sweetener in the human diet, until it was replaced by sucrose. But, presently some value-added sweeteners in processed foods and beverages have

R.S. Singh (✉) • K. Chauhan • R.P. Singh
Carbohydrate and Protein Biotechnology Laboratory, Department of Biotechnology, Punjabi University, Patiala 147 002, Punjab, India
e-mail: rssingh11@lycos.com; rssbt@pbi.ac.in

S.K. Gahlawat et al. (eds.), *Plant Biotechnology: Recent Advancements and Developments*, DOI 10.1007/978-981-10-4732-9_10

become an important part of modern diet. One of these important sweeteners is high fructose syrup (HFS). The mid-to-late twentieth century was a turbulent period for sugar production due to weather instability. This has created a window of opportunity for milling industries to access plentiful and dependable raw materials for HFS production. In the 1960s, due to concurrent advances in refining, isomerization, and separation technologies, the production of high fructose corn syrup (HFCS) was possible with sweetness equivalent to that of sucrose. In the 1980s, HFCS had completely replaced conventional sugar in beverages like colas. Large-scale production of crystalline fructose was commenced by A. E Staley Manufacturing Company in 1987. Conventionally, HFS is prepared from starch by a multienzymatic method using α-amylase, glucoamylase, and glucose isomerase. HFS can also be prepared from inulin using a single-step enzymatic method. In the single-step enzymatic method, exoinulinase (EC 3.2.1.80, β-2-1-D-fructan fructohydrolase) acts sequentially on β-2, 1 bonds of inulin yielding fructose.

Fructose is the sweetest natural sugar (1.2–2.0 times sweeter than sucrose) and generally recognized as safe (GRAS) food ingredient. It is generally used as a safe sugar substitute in many countries. It is an idyllic sweetener as it imparts the same sweetness and texture to food as sucrose. It plays various beneficial roles in the human body (Table 10.1); consequently, it is used as a safe alternative sweetener by many food industries. Its important physiological roles in the human body includes metabolism which bypasses the glucose metabolic pathway and hence does not require insulin. It has low glycemic index and does not induce a sharp increase and fall in blood glucose level. Furthermore, it is more sensitive to insulin in non-insulin dependent diabetes mellitus (Wolfgang and Südzucker 2004). It is also independent of some common problems like cariogenicity, corpulence, and atherosclerosis which otherwise can occur on using the conventional sugar sucrose. It also accelerates ethanol metabolism (Ylikahari et al. 1972). Moreover, it also increases the intestinal absorption of iron by making iron chelate complex which certainly has more bioavailability than inorganic iron (Gupta et al. 1994).

Fructose possesses many technically superior and functional properties over the conventional sugar, sucrose (Table 10.1), due to which it has been accepted as an important sweetener by many industries. Fructose is highly soluble at low temperature, crystallizes lesser than sucrose and has a large freezing point depression. These properties are beneficial for the formulators in manufacturing ice creams and other desserts as they strongly influence the taste and texture of food products. Fructose creates higher osmotic pressure and lowers water activity in food items. Lower water activity is useful in maintaining microbial stability in food products without affecting moisture content which is further necessary for better texture and edibility. Fructose is an excellent humectant; it retains moisture for a longer period even if relative humidity is low. This property is important in imparting palatable texture and longer shelf-life to the food products (Hanover and White 1993). It also has antioxidant and hygroscopic properties and has the ability to retain anions (Pawan 1973). Moreover, it is a preferred choice in acidic foods over sucrose as higher

Table 10.1 Technical and functional superiorities of fructose over other sugars and its important physiological roles in the human body

Technical properties
High solubility and large freezing point depression
Crystallizes less
High osmotic pressure and low water activity, useful against microbial contamination
Humectancy
Antioxidant and hygroscopic properties
Preferred choice for acidic foods
Functional properties
Relative sweetness is almost identical to relative sweetness of sucrose
Easily synergizes with other sugars to enhance total sweetness of the food product
Enhances flavor profile and reduces cost
Physiological roles
Ideal nutritive sweetener with low energy content
Insulin-independent metabolism; therefore, used as sugar substitute for diabetics
Low glycemic index as compared to other sugars
Non-cariogenic, non-corpulent, and non-atherosclerotic
Increases bioavailability of iron by making iron chelate complex
Accelerates ethanol metabolism

temperature and acidic conditions cause breakdown of sucrose. The relative sweetness (99) of fructose is very close to the relative sweetness of sucrose (100), due to which it has emerged as a potential replacement for sucrose in many food applications. It also synergizes with other sugars to achieve a higher degree of sweetness without increasing the total level of sweeteners and hence reduces the final cost of the product. The sweetness of fructose is perceived and diminishes rapidly as compared to other sugars. This is significant in augmenting the other flavors present in a food product. Being a GRAS sweetener, the possibility of its use in food and beverage industries has also been extended to pharmaceutical industries. In pharmaceuticals, it is exclusively used in the formulation of tablets and solutions, injections, and infusions (Hanover and White 1993). Fructose-derived furans are integrated in the preparation of some ecofriendly chemicals like pyrroles, pyridazines, and diazepinones. These chemicals are valuable intermediates and products for pharmaceutical and agrochemical industries (Lichtenthaler et al. 2001).

10.2 Inulin: A Potent Substrate

Inulin, a polyfructan, was discovered by Rose in 1804. Since, then its presence in many plants has become apparent. The German plant physiologist Julius Sachs (1864) was a pioneer in fructan research and was able to detect the spherocrystals of inulin after ethanol precipitation. Inulin is present as a reserve carbohydrate in the roots and tubers of many plants such as *Helianthus tuberosus*, *Dahlia pinnata*, *Cichorium intybus*, and *Asparagus officinalis*, etc. (Table 10.2), representing more than 30,000 species. Most commonly consumed inulin-containing vegetables and fruits are Mexican turnip, onion, garlic, banana, leek, wild leek, rye, etc. (van Loo et al. 1995; Kaur and Gupta 2002; Zubaidah and Akhadiana 2013).

Mostly, inulin-containing plants are dicotyledonous belonging to *Asteraceae* and *Campanulaceae* families, but some monocotyledonous plants have also been reported from *Poaceae*, *Liliaceae*, and *Amaryllidaceae* families (Singh and Singh 2010). A model for the biosynthesis of inulin in *Helianthus tuberosus* was first proposed by Edelman and Jefford (1968). They reported inulin as a mixture of oligosaccharides and polysaccharides comprising fructose units with β-configuration at anomeric C_2. But, due to the presence of this configuration, inulin-type fructans are not digested by human intestinal enzymes as these enzymes have specificity for α-glycosidic bonds only (Roberfroid 2007).

Inulin is a heterodisperse molecule. It consists of variable number of fructose units linked by β-(2, 1)-D-fructosyl-fructose bonds usually terminated with one glucose unit linked through α-(1, 2) sucrose-type linkage (Fig. 10.1). Inulins ending with a glucose unit are called α-D-glucopyranosyl-[β-D-fructofuranosyl]$_{n-1}$-D-fructofuranosides (fructooligosaccharides), while those consisting only fructose units are termed as fructopyranosyl-[α-D-fructofuranosyl]$_{n-1}$-D-fructofuranosides (inulooligosaccharides) (Ronkart et al. 2007). The length of these fructose units can vary from 2 to 60 monomers. The degree of polymerization (DPn) of inulin largely depends upon the plant source and age, harvesting time, duration, and conditions of postharvest storage (Kruger 2002; Ronkart et al. 2006). Permethylation studies have shown 1–2% degree of polymerization (DPn) in native chicory inulin (De Leenheer and Hoebregs 1994). Commercially available inulin from various sources was reported to have much higher DPn than raw inulin (De Leenheer 2007). The bacterial inulin is highly branched and has been reported to have DPn in the range of 10,000–100,000 (Franck and De Leenheer 2005). Due to variation in DPn, weight average (DPw) of inulin also varies to a great extent. High-performance anion exchange chromatography (HPAEC) and other chromatographic techniques are usually employed for the determination of DPn and DPw of inulin (Timmermans et al. 1994). The ratio between DPn and DPw is generally used to determine the molecular weight distribution (polydispersity) in inulin-rich plants. Molecular weight of inulin generally varies from ±3500 to 5500 Da (Vandamme and Derycke 1983). DPn and polydispersity of inulin have great impact on its physicochemical and functional properties (Blecker et al. 2003). The short-chain molecules are more soluble and sweeter than native and long-chain inulin. Furthermore, they also contribute in improving the mouth feel or taste as they share almost similar properties

Table 10.2 Inulin-rich plants

Plant name	Inulin-rich part	Reference(s)
Agave americana	Lobes	Partida et al. (1998)
Agave fourcryodes	Lobes	Villegas-Silva et al. (2014)
Allium ampeloprasum var. *porrum*	Bulbs	van Loo et al. (1995)
Allium cepa	Bulbs	van Loo et al. (1995) and Ayyachamy et al. (2007)
Allium sativum	Bulbs	Mahmoud et al. (2011)
Arctium sp.	Roots	van Loo et al. (1995)
Asparagus officinalis	Root tubers	Gupta and Kaur (1997)
Asparagus racemosus	Root tubers	Gupta and Kaur (1997)
Camassia sp.	Bulbs	van Loo et al. (1995)
Cichorium intybus	Roots	van Loo et al. (1995) and Gupta and Kaur (1997)
Cynara cardunculus	Leaves-heart	De Leenheer (2007)
Cynara scolymus	Heads	Leroy et al. (2010)
Dahlia sp.	Root tubers	Zubaidah and Akhadiana (2013) and Rawat et al. (2015)
Dioscorea esculenta	Root tubers	Zubaidah and Akhadiana (2013)
Echinops ritro	Roots	Vergauwen et al. (2003)
Fagopyrum esculentum	Grains	Bonciu et al. (2012)
Helianthus annuus L.	Root tubers	Chi et al. (2011)
Helianthus tuberosus	Root tubers	Ertan et al. (2003) and Sarchami and Rehman (2015)
Hordeum vulgare	Grains	van Loo et al. (1995)
Lactuca sativa	Roots	Hendry and Wallace (1993)
Microseris lanceolata	Roots	van Loo et al. (1995)
Musa acuminate	Fruits	van Loo et al. (1995)
Pachyrhizus erosus	Root tubers	Zubaidah and Akhadiana (2013)
Polymnia sonchifolia	Root tubers	Braz de Oliveira et al. (2011)
Saussurea lappa	Roots	Kuniyal et al. (2005)
Scorzonera hispanica	Roots	Dolota and Dabrowska (2004)
Secale cereal	Grains	van Loo et al. (1995)
Smallanthus sonchifolius	Roots	De Leenheer (2007)
Taraxacum officinale	Leaves	van Loo et al. (1995)
Tragopogon sp.	Roots	Gupta and Kaur (1997)
Vernonia herbacea	Rhizome	De Carvalho and Dietrich (1993) and Pessoni et al. (1999)
Viguiera discolor	Root tubers	Isejima et al. (1991)

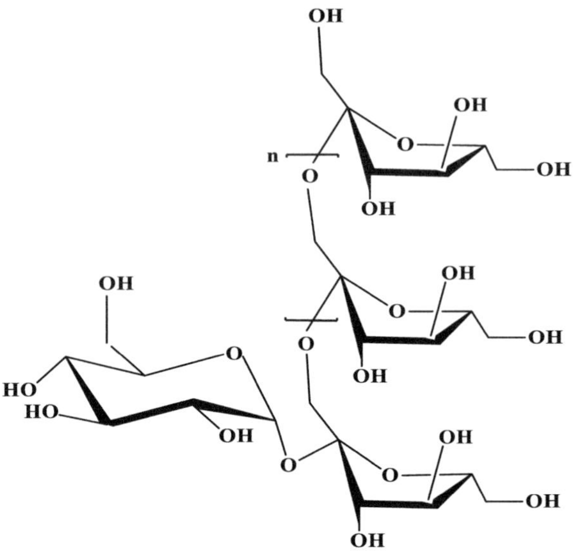

Fig. 10.1 Schematic structure of inulin

with other sugars also. High DPn represents good prebiotic potential of inulin (van de Wiele et al. 2007). Inulin possesses various health and food applications. The US Food and Drug Administration declared inulin as generally recognized as safe in 1992 (Kruger 2002). Some of its important medical applications include improvement in gastrointestinal system and kidney function, relief from constipation (Mensink et al. 2015), low caloric value (Roberfroid et al. 1993), prebiotic potential (Singh and Singh 2010), reduction in the chances of colon and breast cancer occurrence (Taper and Roberfroid 1999; Kato 2000), and most importantly reduction in blood sugar and cholesterol level (Hofer and Jenewein 1999). In pharmaceuticals, it is generally used as it increases the dissolution rate through solid dispersions (Visser et al. 2010), targets colon delivery (Imran et al. 2012), measures glomerulation filtration rate (Orlando et al. 1998), etc. In food industries, it is mainly used in various dairy products like milk, ice cream, jams, jellies, etc. In addition, it is also used as a low-calorie sweetener, sugar and fat substitute in many dairy products, and dietary fiber. It improves organoleptic properties and form gels to increase viscosity (Kaur and Gupta 2002).

Furthermore, after microcrystal formation it offers a creamy texture which gives a fat-like mouth feel (López-Molina et al. 2005). Recently, inulin has gained a great interest as a potential renewable source for the production of HFS (Ricca et al. 2007; Chi et al. 2011; Singh 2011), fructooligosaccharides (Singh and Singh 2010; Singh et al. 2016) and inulinases (Kango and Jain 2011; Treichel et al. 2012; Rawat et al. 2016). Inulin is a potent substrate for the production of fructooligosaccharides and HFS.

10.3 Action Pattern of Inulinases

Inulin-depolymerizing enzymes belong to glycosyl hydrolase (GH) family. Glycosyl hydrolase family includes the enzymes that hydrolyze the glycosidic linkages between two or more carbohydrates. On the basis of the catalytic action on inulin, these enzymes are categorized as transferases (GH91) and hydrolases (GH32). Inulin fructotransferase (inulin lyase, EC 4.2.2.18) produces difructose anhydride (DFA III) by exoacting intramolecular transfructosylation process (Saito and Tomita 2000), while the inulinases split β (2 → 1) glycosidic bonds of inulin by either exo- or endoaction (Fig. 10.2). Exoinulinase (β, 2-1-D-fructan fructohydrolase, EC 3.2.1.80) acts on β (2 → 1) bonds and depolymerizes inulin to produce fructose as the main product in addition to a small quantity of glucose (Singh et al. 2007a, b, 2008), while endoinulinase (β, 2 → 1-D-fructan fructanohydrolase, EC 3.2.1.7) acts randomly on inulin and produces a mixture of fructooligosaccharides having varied degree of polymerization (Singh and Singh 2014; Singh et al. 2016).

The prominent microbial inulinase producers are *Aspergillus* sp., *Penicillium* sp., *Kluyveromyces* sp., *Candida* sp., *Arthobactor* sp., *Xanthomonas* sp. and *Bacillus*

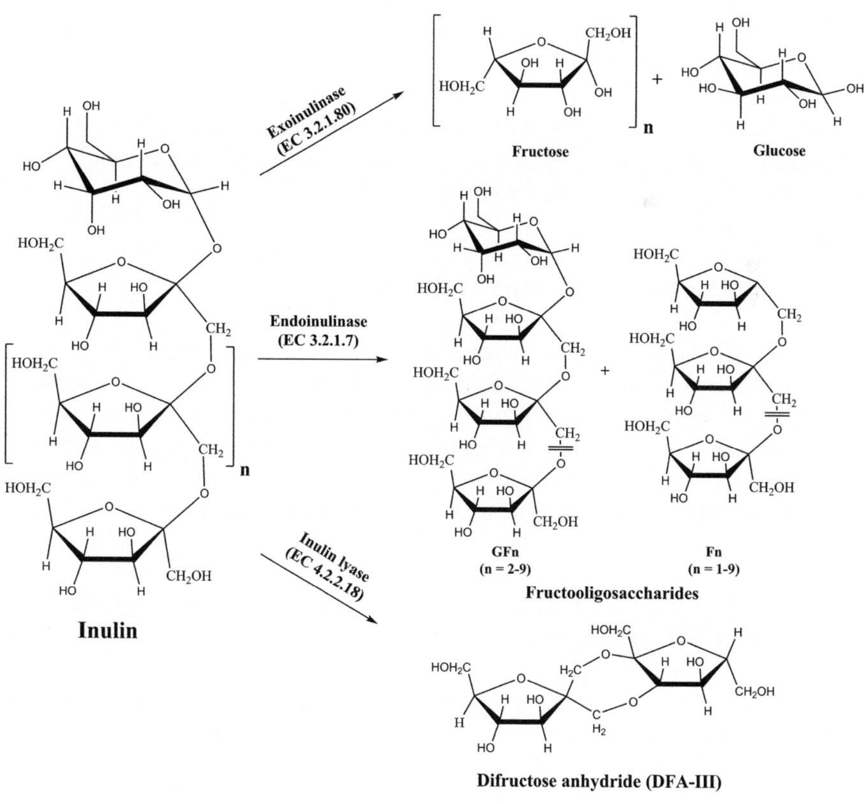

Fig. 10.2 Action pattern of inulinases on inulin

sp., etc. (Singh and Singh 2010; Kango and Jain 2011). Generally, the production of inulinases is inducible, but some reports are also available on the constitutive nature of the enzyme (Mukherjee and Sengupta 1985; Fontana et al. 1994; Cruz-guerrero et al. 1995; Saber and El-Naggar 2009). The exoacting or endoacting nature of the inulinase mainly depends upon the source of the enzyme. Inulinases from fungal cultures are generally exoacting (Pandey et al. 1999). But in some cases, fungal strains like *Aspergillus* sp. (Nakamura et al. 2001; Wang et al. 2003; Zhengyu et al. 2005), *Chrysosporium pannorum* (Xiao et al. 1988) and *Penicillium rugulosum* (Barthomeuf et al. 1991) and yeast strain *Cryptococcus aureus* G7a (Sheng et al. 2007) produce mixed forms of inulinases (both exo- and endoinulinases). Most of the microorganisms produce extracellular inulinases, but in some cases the enzyme localization is intracellular. Microbial species, such as *Fusarium oxysporum* (Gupta et al. 1990) and *Kluyveromyces* sp. (Guiraud et al. 1987; Rouwenhorst et al. 1988), have been reported as producing both extra- and intracellular inulinases. Majority of the microbial strains belonging to *Aspergillus niger* has been reported to produce only extracellular endoinulinases (Gern et al. 2001), while interestingly one strain of *Aspergillus niger* (*A. niger* strain 12) has been reported as producing both exo- and endoinulinases, intracellularly (Nakamura et al. 2001). Two forms of extracellular exoinulinases (EI & EII) have been reported from *Penicillium* sp. (Wenling et al. 1997). Some unique and interesting reports are also available on inulinases. Two intracellular (EI, EII) and one extracellular endoinulinase (Eexo) have been reported from *Kluyeromyces* sp. Y-85 (Wei et al. 1997). In an interesting study, five exoinulinases (Exo I, II, III, IV and V), three endoinulinases (Endo I, II and III), and one invertase (Inv) were isolated from a commercial enzyme preparation derived from *Aspergillus ficuum* (Ettalibi and Baratti 1987).

10.4 High Fructose Syrup Production

Conventionally HFS is produced from starch by a multienzymatic process, but it can also be prepared from inulin by a single-step method using inulinases. Chemical hydrolysis of starch/inulin can also be used for the production of HFS. However, chemical approach is generally not used due to some drawbacks like formation of difructose anhydrides and other colored products which lower fructose yield and demand an improved downstream processing and energy requirement.

10.4.1 Conventional Multienzymatic Method for the Preparation of HFS from Starch

Conventionally, high fructose syrup is produced by multienzymatic hydrolysis of corn starch, involving the action of three enzymes, namely α-amylase, glucoamylase, and glucose isomerase. The conventional production of HFS basically involves the following four steps: (a) liquefaction, (b) saccharification of polymer starch into its monomeric units (dextrose), (c) isomerization of dextrose to fructose, (d) and fractionation to enrich the final concentration of fructose by ion-exchange method. Besides these steps, crystallization of fructose is also carried out to obtain dry and crystalline fructose. Schematic representation of the conventional method is given in Fig. 10.3. The brief description of each step involved in conventional HFS production is given below.

10.4.1.1 Starch Liquefaction

Liquefaction and partial hydrolysis of starch is the first step in the production of HFS by the conventional method. In this step, starch obtained after wet milling of corn is hydrolyzed by thermostable α- or β-amylase (EC 3.2.1.1) into oligomers of intermediate molecular weight. Generally, α- or β-amylases are derived from thermostable bacterial strains like *Bacillus stearothermophilus*, *B. licheniformis*, and *B. subtilis* (Teague and Brumm 1992). *B. stearothermophilus* and *B. licheniformis* amylases are very thermostable and usually used at temperature above 100 °C, while *B. subtilis* enzyme is a little less thermostable and not used above 90 °C. Alpha-amylase randomly disrupts the backbone of starch containing the α-1, 4 glycosidic linkages producing soluble, low- and medium-molecular weight saccharides. Enzymatic liquefaction requires specific regulating parameters such as starch solids, pH, calcium level, reaction time, and temperature. Solid level is preferably maintained at 30–35% to achieve complete starch gelatinization. Starch slurry pH is normally controlled at 5.8–6.5. Higher pH increases byproduct formation, while lower pH causes inactivation of enzyme which correspondingly retards the product yield and increases refining cost. Calcium is used as a cofactor to enhance amylase thermostability. It is added in chloride and oxide form at a concentration of 100–200 ppm for *B. stearothermophilus* and *B. licheniformis* amylases, but at 300 ppm for *B. subtilis* amylase. Reaction time and temperature are generally balanced to provide optimal conditions for enzyme activity and starch gelatinization. Different processes like Enzyme-Heat-Enzyme (EHE), High Temperature (HT), and Low Temperature (LT) are used to carry out enzymatic liquefaction (Hebeda 1993).

10.4.1.2 Saccharification

Saccharification is carried out to convert partially hydrolyzed starch into dextrose using glucoamylase (EC 3.2.1.3). Glucoamylase hydrolyzes the di- and oligomeric products formed by the action of amylase by cleaving the α-1, 4 and α-1, 6 glycosidic linkages sequentially from the non-reducing end. Glucoamylase for commercial purposes is traditionally derived from filamentous fungi like *Aspergillus, Rhizopus*, etc. The type and number of linkages, size of the molecule, and the arrangement of α-1, 4 and α-1, 6 linkages affect the rate of hydrolysis by glucoamylase (Abdullah et al. 1963). Rate of hydrolysis increases with increase in the number of linkages. Approximately 95% of dextrose can be obtained after saccharification along with other sugars like maltose (1%), isomaltose (1.5%), and maltulose (0.2%). Both maltose and isomaltose are formed by reverse reaction catalyzed by glucoamylase. For commercial operation, minimizing this reversion is important in maximizing dextrose yield. This can be attained by adding a high dose of glucoamylase and reducing the reaction time so that the final hydrolysate can be processed rapidly

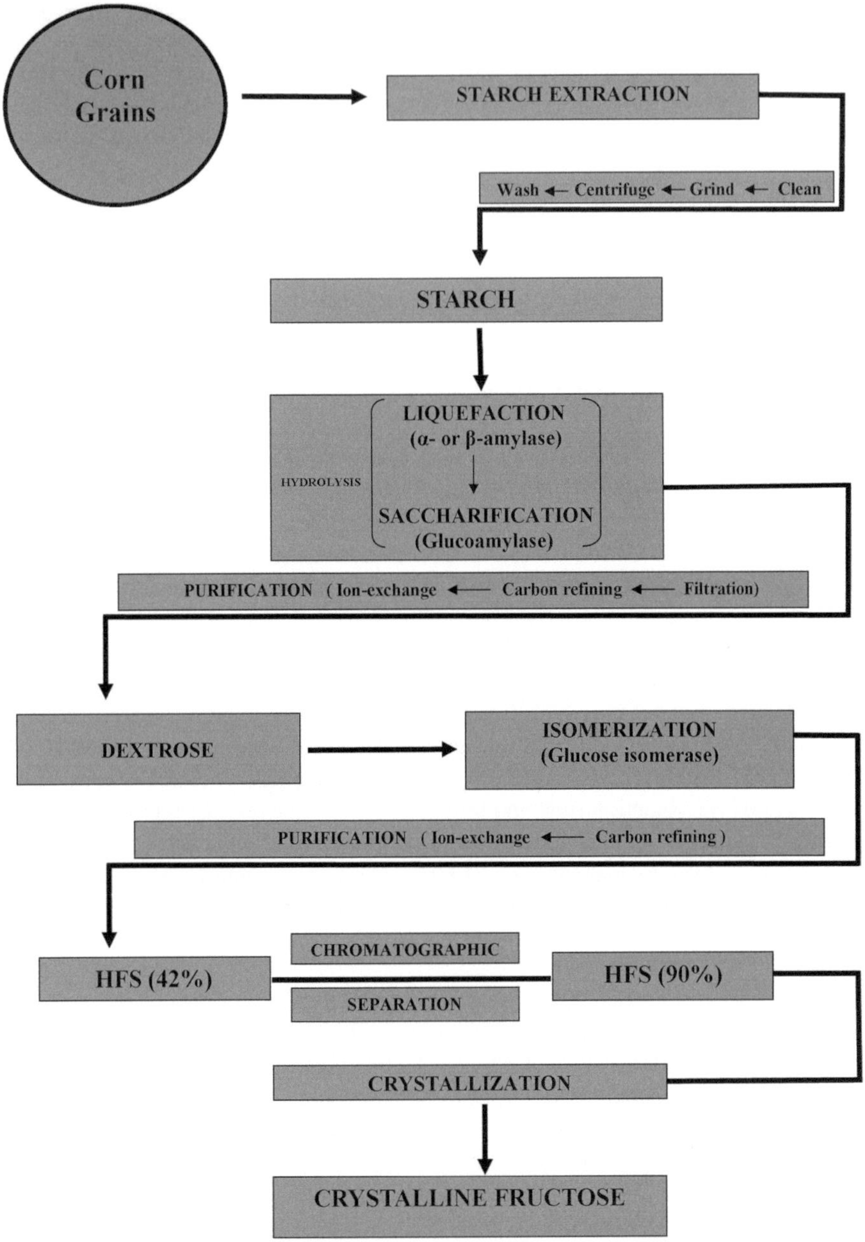

Fig. 10.3 Steps involved in conventional multienzymatic method for high fructose syrup production from starch

to prevent excessive reversion. Dextrose yield can also be increased on combining glucoamylase with either debranching enzyme from *B. acidopullulyticus* (Jensen and Norman 1984) or the enzyme exhibiting both amylase and transferase properties from *B. megaterium* (Hebeda et al. 1988). Saccharified hydrolysate is then clarified by vacuum filtration and refined by combining carbon- and ion-exchange technologies. The refined hydrolysate is processed to high dextrose corn syrup, crystalline or liquid dextrose, or as feed for high fructose syrup production.

10.4.1.3 Isomerization

Earlier, attempts were made to enzymatically isomerize dextrose into fructose, but these were hampered due to the expenses of regenerating the essential cofactors. Akabori et al. (1952) used glucose isomerase (EC 5.3.1.5) to convert dextrose into fructose without cofactor regeneration. Later, Takasaki and Tanabe (1971) improved the economics of enzyme catalysis by immobilizing glucose isomerase. An immobilized isomerase system has many benefits like easy passage of the substrate (dextrose) and the product (fructose) and reduction in reaction time due to the use of high enzyme concentration. Attempts were also made to carry out chemical-based isomerization that demonstrated the role of carbonates and diluted hydroxides in forming enediol intermediate for isomerizing dextrose into fructose. But, these catalysts produced unacceptable highly colored and flavored products with very low fructose yield. Moreover, fructose became more susceptible to degradation under such conditions. Therefore, for large scale-up operations, immobilized glucose isomerase (whole cell or soluble enzyme) from various sources like *Actinoplanes missouriensis*, *B. coagulans*, *Flavobacterium arborescens*, and *Streptomyces* emerged as a prompting tool to isomerize dextrose into fructose. Several parameters like feed purity, solid concentration, temperature, pH, magnesium content, and reaction time are carefully synchronized to optimize product yield and operational stability of the system (Hebeda 1993). Efficient refining by ion-exchange method is required to lower calcium level (below 3 ppm) and other impurities which are inhibitory to isomerase. Feed solid level is generally controlled at 45%, temperature 55–61 °C, and pH 7.5–8.2 to maintain the rate of diffusion and prevent any cross-microbial contamination. Moreover, these parameters are also important for sustaining long-term productivity. Maintaining dissolved oxygen up to a certain limit is detrimental to prevent possible oxidation of isomerase cysteine residues, which results in enzyme inactivation. Deaeration and addition of SO_2 (1–2 mM) as sodium bisulphite can generally reduce oxidation. Magnesium at 0.5–5 mM concentration is recommended as an enzyme cofactor and to counteract the adverse effects of calcium. Flow rate through the column enabling long reaction time is also important to maintain fructose level between 42 to 45%. Carbon- and ion-exchange treatments are used to refine 42% fructose product containing 51–54% dextrose to eliminate color, off-flavors, salts, and other impurities. A storage temperature of 27–32 °C is advised to prevent dextrose crystallization.

10.4.1.4 Fractionation

A higher content of fructose is required in fructose syrup so that it could match the sweetness of sucrose at amounts used as a beverage sweetener. Attempts were made to develop some chromatographic separation and blending technology using activated carbon or iron oxide to accomplish this requirement. But the separation efficiency of fructose: dextrose ratio using this method was very low and costly. Later, Mitsubishi Chemical Industries developed a technique to separate fructose from dextrose and other isomerized products on the basis of its affinity for calcium groups adsorbed on cation-exchange resin (Hanover and White 1993). Fructose is retained in the chromatographic column, while a stream containing predominantly non-fructose saccharides can be collected. This stream comprised of 80–90% dextrose and 5–10% fructose. Higher saccharide fraction is again recycled for saccharification and isomerization. Then, this high fraction of fructose contains approximately 80–90% fructose and 7–19% dextrose. In actual practice, simulated moving bed process is implemented for the continuous separation of enriched HFS and other saccharides.

Crystalline fructose is manufactured using 90% fructose product as raw material obtained after chromatographic separation. Aqueous or solvent-based systems are generally employed for the crystallization of fructose. But, today the use of aqueous solution for this purpose has become the most successful process. Batch and continuous systems have been incorporated with pre-programmed concentrating, cooling, warming, and recooling cycles for an efficient crystallization. The resultant crystals are harvested and washed free of impurities present on the surface of crystals for use in beverages (Hanover and White 1993).

10.4.2 HFS Production from Inulin Using Inulinases

Single-step hydrolysis of inulin using inulinases in a batch or continuous system is becoming a promising technology for HFS production. By this method as high as 95% fructose yield can be obtained.

10.4.2.1 HFS Production in a Batch System

Inulinases are an important class of industrial enzymes which play a vital role in catalyzing the single-step hydrolysis of inulin to produce HFS. Inulinases have been reported from various inulin-storing plant tissues but their quantity is very little (Pandey et al. 1999); hence, microbial sources have drawn much attention due to their many advantages like easy cultivation under controlled conditions and high enzyme yield. Inulinases target β-(2 $\rightarrow$ 1) linkages of inulin splitting off terminal fructosyl units, releasing fructose and glucose. The liberation of fructose is a single-chain mechanism and it occurs under a wide range of pH and temperature conditions. In literature, several reports are available on the hydrolysis of pure inulin or inulin-rich materials for HFS production using soluble enzyme and free whole cells (Table 10.3), and immobilized enzymes (Table 10.4). A lot of work has been reported on HFS production in batch system using inulinases from different sources

Table 10.3 Single-step enzymatic hydrolysis of inulin for high fructose syrup production using whole cells and soluble inulinase

Inulinase source	Substrate	Fructose productivity (g/L)	Reference(s)
Using whole cells			
Cladosporium cladosporioides	Jerusalem artichoke tuber extract	5	De Andrade et al. (1992)
	Jerusalem artichoke tuber extract	NS	Ferreira et al. (1991)
Saccharomyces cerevisiae[a]	Jerusalem artichoke tuber extract	≈83	Yu et al. (2011)
	Inulin	≈19	Brevnova et al. (1998)
	Inulin	≈55	Brevnova et al. (1998)
	Inulin	≈85	Brevnova et al. (1998)
Using soluble enzyme			
Aspergillus ficuum	Pure inulin (Chicory roots)	106.4	Mutanda et al. (2009)
A. niger	Artichoke tubers extract	NS	Cruz et al. (1998)
	Chicory tubers extract	NS	Cruz et al. (1998)
	Dahlia inulin extract	NS	Cruz et al. (1998)
	Pure inulin (Chicory roots)	37.5	Sirisansaneeyakul et al. (2007)
	Kuth (*Saussurea lappa*) root powder extract	35	Viswanathan and Kulkarni (1995)
A. niger and *Kluyveromyces marxianus* (mixed culture)	Jerusalem artichoke tubers extract	NS	Öngen-Baysal and Sukan (1996)
Candida guilliermondii	Pure inulin (Chicory roots)	35.3	Sirisansaneeyakul et al. (2007)
Fusarium oxysporum	Chicory root tubers extract	NS	Gupta et al. (1989)
K. marxianus	Agave juice	NS	García-Aguirre et al. (2009)
	Pure inulin	43.6	Singh et al. (2007a)
	Asparagus racemosus root extract	41.3	Singh et al. (2007a)
Kluyveromyces marxianus var. *Bulgaricus*	Pure inulin	653	Manzoni and Cavazzoni (1992)
		691	

(continued)

Table 10.3 (continued)

Inulinase source	Substrate	Fructose productivity (g/L)	Reference(s)
	Fresh Jerusalem artichoke tubers extract	773	Manzoni and Cavazzoni (1992)
		754	
	Dried Jerusalem artichoke tubers extract	839	Manzoni and Cavazzoni (1992)
		857	

NS: Not specified
[a]Recombinant cells of *Saccharomyces cerevisiae* were used for hydrolysis

like *Aspergillus* sp., *Penicillium* sp., *Kluyveromyces* sp., *Streptomyces* sp., etc. under wide range of temperature (28–60 °C) and pH (4.5–6.0). In a study, high yield of fructose has been obtained from Jerusalem artichoke tuber extract using inulinase from *Penicillium subrubescens* at 28 °C and pH 6.0 (Mansouri et al. 2013). An exoinulinase gene (rInuAGN25) from *Sphingobacterium* sp. G25 N cloned in *Escherichia coli* has shown 74.4% hydrolysis of Jerusalem artichoke tuber extract at room temperature (Zhou et al. 2014), while inulinase from *A. ficuum* has shown high fructose productivity at 50 °C and pH 5.0 (Mutanda et al. 2009). Singh et al. (2007a) also reported high temperature (55 °C) for the production of HFS from pure inulin and *Asparagus racemosus* root extract by inulinase from *K. marxianus*. Free whole cells of *Cladosporium cladosporioides* (Ferreira et al. 1991; De Andrade et al. 1992) and *Saccharomyces cerevisiae* (Brevnova et al. 1998; Yu et al. 2011) have also been reported for HFS production in a batch system. Yu et al. (2011) reported high fructose productivity from Jerusalem artichoke tuber extract by recombinant whole cells of *S. cerevisiae*.

Now-a-days, considerable attention has been paid toward research on various aspects of enzyme immobilization. Immobilized systems have a number of advantages over soluble precursors like development of production processes which assure no enzyme activity in end products, particularly relevant for food and feed sectors which contributes to improving operational performance, reuse of enzymes and continuous mode of operation, and economic feasibility. These advantageous features of immobilization have made the developed biocatalysts the best for HFS production. Various immobilized biocatalysts have been developed through covalent binding, encapsulation, cross-linking, and adsorption using different matrices/supports like chitin (Gill et al. 2006), chitosan beads (Yewale et al. 2013; Missau et al. 2014), cellulose (Gupta et al. 1992), alginate beads (Catana et al. 2005), agar gel (Bajpai and Margaritis 1985a), amberlite (Diaz et al. 2006), and montmorillonite (Coghetto et al. 2012) (Table 10.4). Some low-cost supports like activated carbon, diatomite, hen egg shell, amberlite, and porous silica have also been reported for the immobilization of inulinases (Paula et al. 2007). Santa et al. (2011) obtained maximum fructose productivity in a batch system by inulinase encapsulated in a

Table 10.4 Single-step enzymatic hydrolysis of inulin for high fructose syrup production using immobilized inulinase in a batch system

Inulinase source	Substrate	Immobilization matrix/support	Number of cycles	Fructose productivity (g/l)	Reference/s
Aspergillus ficuum	Jerusalem artichoke tubers extract	Chitin	NS	77.5	Kim and Rhee (1989)
Aspergillus sp.	Pure inulin (Chicory roots)	Sepabeads	NS	9.6	Ricca et al. (2010)
A. niger	Pure inulin (Chicory roots)	Sol-gel	29	47 ± 4	Santa et al. (2011)
	Pure inulin (Chicory roots)	Magnetic sol-gel particles	29	46 ± 2	Santa et al. (2011)
Fusarium oxysporum	Chicory root tubers extract	DEAE-Cellulose	3	NS	Gupta et al. (1992)
Kluyveromyces fragilis	Jerusalem artichoke tubers extract	2-Aminoethyl-cellulose	NS	34	Kim and Byun (1982)
Kluyveromyces marxianus	Jerusalem artichoke tubers extract	Agar gel	9	≈ 40	Bajpai and Margaritis (1985a)
Kluyveromyces marxianus[a]	Jerusalem artichoke tubers extract	Open pore gelatin matrix	10	42	Bajpai and Margaritis (1985b)
	Jerusalem artichoke tubers extract	Alginate beads	16	24	Parekh and Margaritis (1986)
	Pure inulin	Duolite A568	55	40.2	Singh et al. (2007b)
	Asparagus racemosus root extract	Duolite A568	55	39.2	Singh et al. (2007b)

NS: Not specified
[a]Immobilized whole cells were used for hydrolysis

sol-gel. In a number of studies, the developed immobilized biocatalyst has been recycled for 5–55 batches for HFS production (Parekh and Margaritis 1986; Singh et al. 2007b; Santa et al. 2011), which support operational and mechanical stability of the developed immobilized biocatalyst. Inulinase from *K. marxianus* was immobilized on Duolite A568 and recycled for 55 batches, and 82.6% hydrolysis of *Asparagus racemosus* root extract has been reported (Singh et al. 2007b). The developed biocatalyst can be a good candidate for large-scale applications.

10.4.2.2 HFS Production in a Continuous System

In a continuous flow system, the substrate is continuously fed into the reactor which emerges as a continuous stream of product from the outlet. Continuous operation possesses various advantages over batch operations like consistent quality of products due to controlled reaction parameters, less wastage, and lower storage requirement for both raw materials and products. Hence, the capital cost per ton of product is consequently lesser. Continuous reactors supposedly have superior heat transfer capacities as compared to batch reactors due to which higher reactant concentration can be added into the reactor. Different features like heat transfer capacity, temperature and pressure control, aspect ratio, flow rate, residence time, and substrate concentration play a vital role in a successful continuous operation. From a simulation outcome of a heterogeneous model, almost complete hydrolysis of inulin was attained practically at 100 °C by increasing the aspect ratio of the reactor (Abasaeed et al. 1999). Specifically, miniaturized and immobilized enzyme reactors have gained tremendous momentum for use in various bioprocesses. Continuous operations in miniaturized devices have several advantages like low requirement of manpower and consumables, automation, better heat and mass transfer, and probability of scaling out instead of scaling up. Ribeiro and Fernandes (2013) used a coated-wall mini reactor for inulin hydrolysis to produce fructose using immobilized inulinase, and a product yield of 80% was obtained after five days of continuous operation. Inulinases from several microbial sources like *Aspergillus niger, Kluyveromyces marxianus*, and *Debaryomyces phaffi* have been reported for HFS production in a continuous system. Fructose productivity from 2.3 to 614 g/L/h has been reported in a number of continuous systems (Table 10.5).

10.5 Conclusion

High fructose syrup has a broad range of food and health applications. It is a nutritive sweetener which contributes many useful physical and functional attributes to food and beverages. The present work outlines the basic development and techniques used in enzymatic approaches for the synthesis of high fructose syrup, conventionally from starch and single-step hydrolysis of inulin-using inulinases. Due to synchronized developments in refining, isomerization, and separation technologies, the production of HFS has become possible. A number of batch and continuous reactors employing free whole cells and soluble or immobilized inulinases have been developed for HFS production from inulin. Immobilized biocatalysts are considered advantageous as they provide more operational and mechanical stability to the system. Moreover, the biocatalysts can be recycled for concurrent batches. Further, process modeling and simulation approaches have not been included persistently for HFS production which otherwise could provide comprehensive understanding and prediction of phenomena in the bioreactors.

Table 10.5 Single-step enzymatic hydrolysis of inulin for high fructose syrup production using immobilized inulinase in a continuous system

Inulinase source	Substrate	Immobilization matrix/support	Fructose productivity (g/L/h)	Reference(s)
Aspergillus sp.	Pure inulin (Chicory roots)	Polyvinyl alcohol particles	146	Anes and Fernandes (2014)
	Pure inulin (Chicory roots)	Sepabeads	≈ 5.7	Ricca et al. (2010)
A. niger	Pure inulin (Dahlia tubers)	Amino-Cellulofine beads	397.7	Nakamura et al. (1995)
	Pure inulin (Chicory roots)	Silicone tube activated by APTES	NS	Ribeiro and Fernandes (2013)
	Kuth (*Saussurea lappa*) root powder	Chitosan beads	68	Yewale et al. (2013)
A. ficuum	Jerusalem artichoke tubers juice	Chitin	≈ 61	Kim and Rhee (1989)
A. fumigatus	Pure inulin	Chitin	2.7	Gill et al. (2006)
	Pure inulin	ConA-linked silica beads	3.4	Gill et al. (2006)
	Pure inulin	QAE-Sephadex	2.3	Gill et al. (2006)
Debaryomyces phaffii	Pure inulin	DEAE-Cellulose	NS	Guiraud et al. (1981)
Kluyveromyces sp.	Jerusalem artichoke tubers extract	Ionic polystyrene beads	199.66	Wenling et al. (1999)
Kluyveromyces fragilis	Jerusalem artichoke tubers extract	2-Aminoethyl-cellulose	NS	Kim and Byun (1982)
Kluyveromyces marxianus	Jerusalem artichoke tubers extract	Agar gel	31	Bajpai and Margaritis (1985a)
Kluyveromyces marxianus[a]	Jerusalem artichoke tubers extract	Open pore gelatin matrix	90	Bajpai and Margaritis (1985b)
	Pure inulin	Duolite A568	27.4	Singh et al. (2008)
	Asparagus racemosus root extract	Duolite A568	49.7	Singh et al. (2008)
Saccharomyces cerevisiae[b]	Pure inulin (Dahlia roots)	Amino-Cellulofine	614	Kim et al. (1997)
	Jerusalem artichoke tubers extract	Amino-Cellulofine	NS	Kim et al. (1997)

NS: Not specified
APTES: (3-aminopropyl) triethoxysilasne
[a]Immobilized whole cells were used for hydrolysis
[b]Recombinant exoinulinase from *S. cerevisiae* was used for hydrolysis

References

Abasaeed AE, Asif M, Fakeeha AH (1999) Zeolite-catalysed hydrolysis of inulin to fructose in a fluidized bed reactor. Bioprocess Eng 20:343–348

Abdullah M, Fleming ID, Taylor M, Whelan WJ (1963) Substrate specificity of the amyloglucosidase of *Aspergillus niger*. Biochem J 89:35–36

Akabori S, Nehara K, Muramatsu I (1952) Biochemical formation of tetrose, pentose and hexose. J Chem Soc Jpn 73:311

Anes J, Fernandes P (2014) Towards the continuous production of fructose syrups from inulin using inulinase entrapped in PVA-based particles. Biocatal Agric Biotechnol 3:296–302

Ayyachamy M, Khelawan K, Pillay D, Permaul K, Singh S (2007) Production of inulinase by *Xanthomonas campestris* pv. *phaseoli* using onion (*Allium cepa*) and garlic (*Allium sativum*) peels in solid state cultivation. Lett Appl Microbiol 45:439–444

Bajpai P, Margaritis A (1985a) Production of high fructose syrup from Jerusalem artichoke tubers using *Kluyveromyces marxianus* cells immobilized in agar gel. J Gen Appl Microbiol 31:305–311

Bajpai P, Margaritis A (1985b) Immobilization of *Kluyveromyces marxianus* cells containing inulinase activity in open pore gelatine matrix: 1. Preparation and enzymatic properties. Enzym Microb Technol 7:373–376

Barthomeuf C, Regerat F, Pourrat H (1991) Production of inulinase by a new mold of *Penicillum rugulosum*. J Ferment Bioeng 72:491–494

Blecker CS, Chevalier JP, Fougnies C, Van Herck JC, Deroanne C, Paquot M (2003) Characterisation of different inulin samples by DSC: influence of polymerisation degree on melting temperature. J Therm Anal Calorim 71:215–224

Bonciu CN, Constantin O, Bahrim G (2012) Increase in extracellular inulinase production for a new *Rhizoctonia* sp. strain by using buckwheat (*Fagopyrum esculentum*) flour as a single carbon source. Lett Appl Microbiol 55:195–201

Braz de Oliveira AJ, Correia Goncalves RA, Cantuaria Chierrito TP et al (2011) Structure and degree of polymerisation of fructooligosaccharides present in roots and leaves of *Stevia rebaudiana* (Bert.) Bertoni Food Chem 129:305–311

Brevnova EE, Kozlov DG, Efremov BD, Benevolensky SV (1998) Inulinase-secreting strain of *Saccharomyces cerevisiae* produces fructose. Biotechnol Bioeng 60:492–497

Catana R, Ferreira BS, Cabral JMS, Fernandes P (2005) Immobilization of inulinase for sucrose hydrolysis. Food Chem 91:517–520

Chi ZM, Zhang T, Cao TS, Liu XY, Cui W, Zhao CH (2011) Biotechnological potential of inulin for bioprocesses. Bioresour Technol 102:4295–4303

Coghetto CC, Scherer RP, Silva MF, Golunski S et al (2012) Natural montmorillonite as support for the immobilization of inulinase from *Kluyveromyces marxianus* NRRL Y-7571. Biocatalyst Agric Biotechnol 1:284–289

Cruz VD, Belote JG, Belline MZ, Cruz R (1998) Production and action pattern of inulinase from *Aspergillus niger*-245: hydrolysis of inulin from several sources. Rev Microbiol 29:301–306

Cruz-guerrero A, Garcia-Pena I, Barzana E et al (1995) *Kluyveromyces marxianus* CDBB-L-278: a wild inulinase hyperproducing strain. J Ferment Bioeng 80:159–163

De Andrade AVM, Ferreira MSS, Kennedy JF (1992) Selective fructose production by utilization of glucose liberated during the growth of *Cladosporium cladosporioides* on inulin or sucrose. Carbohydr Polym 18:59–62

De Carvalho MAM, Dietrich SMC (1993) Variation in fructan content in the underground organs of *Vernonia herbacea* (veil.) rusby at different phenological phases. New Phytol 123:735–740

De Leenheer L (2007) Production and use of inulin: industrial reality with a promising future. In: van Bekkum H, Röper H, Voragen F (eds) Carbohydrates as organic raw materials III. Wiley-VCH Verlag GmbH & Co. KGaA, Weinheim, pp 67–92

De Leenheer L, Hoebregs H (1994) Progress in elucidation of the composition of chicory inulin. Starch 46:193–196

Diaz EG, Catana R, Ferreira BS, Luque S et al (2006) Towards the development of a membrane reactor for enzymatic inulin hydrolysis. J Membr Sci 273:152–158
Dolota A, Dabrowska B (2004) Raw fibre and inulin content in roots of different Scorzonera cultivars (*Scorzonera hispanica* L.) depending on cultivation method. Folia Hortic 16:31–37
Edelman J, Jefford T (1968) The mechanism of fructosan metabolism in higher plants as exemplified in *Helianthus tuberosus*. New Phytol 67:517–531
Ertan F, Ekinci F, Aktac T (2003) Production of inulinases from *Penicillium spinulosum*, *Aspergillus parasiticus* NRRL 2999 and *Trichoderma viride*. Pak J Biol Sci 6:1332–1335
Ettalibi M, Baratti JC (1987) Purification, properties and comparison of invertase, exoinulinases and endoinulinases of *Aspergillus ficuum*. Appl Microbiol Biotechnol 26:13–20
Ferreira MSS, Andrade AVMD, Kennedy JF (1991) Properties of a thermostable non-specific fructofuranosidase produced by *Cladosporium cladosporioides* cells for hydrolysis of *Jerusalem artichoke* extract. Appl Biochem Biotechnol 31:1–9
Fontana JD, Baron M, Diniz ACP, Franco VC (1994) Microbial inulinase secretion using chemically modified inulins. Appl Biochem Biotechnol 45:257–268
Franck A, De Leenheer L (2005) Inulin. In: Steinbüchel A (ed) Biopolymers online. Wiley-VCH Verlag GmbH & Co. KGaA, Weinheim, pp 439–473
García-Aguirre M, Saenz-Alvaro VA, Rodri'Guez-Soto MA et al (2009) Strategy for biotechnological process design applied to the enzymatic hydrolysis of agave fructo-oligosaccharides to obtain fructose-rich syrups. J Agric Food Chem 57:10205–10210
Gern RMM, Furlan SA, Ninow JL, Jonas R (2001) Screening for microorganisms that produce only endoinulinase. Appl Microbiol Biotechnol 55:632–635
Gill PK, Manhas RK, Singh P (2006) Hydrolysis of inulin by immobilized thermostable extracellular exoinulinase from *Aspergillus fumigatus*. J Food Eng 76:369–375
Guiraud JP, Bourgi J, Stervinou M, Chaisse M, Galzy P (1987) Isolation of a respiratory-deficient *Kluyveromyces fragilis* mutant for the production of ethanol from *Jerusalem artichoke*. Biotechnol Bioeng 29:850–858
Guiraud JP, Demeulle S, Galzy P (1981) Inulin hydrolysis by the *Debaryomyces phaffii* inulinase immobilized on DEAE cellulose. Biotechnol Lett 3:683–688
Gupta AK, Kaur M, Kaur N, Singh R (1992) A comparison of properties of inulinases of *Fusarium oxysporum* immobilised on various supports. J Chem Technol Biotechnol 53:293–296
Gupta AK, Kaur N (1997) Fructan storing plants – a potential source of high fructose syrups. J Sci Ind Res 56:447–452
Gupta AK, Kaur N, Singh R (1989) Fructose and inulinase production from waste *Cichorium intybus* roots. Biol Wastes 29:73–77
Gupta AK, Rathore P, Kaur N, Singh R (1990) Production, thermal stability and immobilization of inulinase from *Fusarium oxysporum*. J Chem Technol Biotechnol 47:245–257
Gupta AK, Singh DP, Kaur N, Singh R (1994) Production, purification and immobilization of inulinase from *Kluyveromyces fragilis*. J Chem Technol Biotechnol 59:377–385
Hanover LM, White JS (1993) Manufacturing, composition and applications of fructose. Am J Clin Nutr 58:724S–732S
Hebeda RE (1993) Starches, sugars, and syrups. In: Nagodawithana T, Reed G (eds) Enzymes in food processing. Elsevier, London, pp 321–346
Hebeda RE, Styrlund CR, Teague WM (1988) A kinetic model of *Bacillus stearothermophilus* α-amylase under process conditions. Starch/Stärke 40:412–418
Hendry GAF, Wallace RK (1993) The origin, distribution, and evolutionary significance of fructans. In: Suzuki M, Chatterton NJ (eds) Science and technology of fructans. CRC Press, Boca Raton, pp 119–139
Hofer K, Jenewein D (1999) Enzymatic determination of inulin in food and dietary supplements. Eur Food Res Technol 209:423–427
Imran S, Gillis RB, Kok MS, Harding SE, Adams GG (2012) Application and use of inulin as a tool for therapeutic drug delivery. Biotechnol Genet Eng Rev 28:33–45

Isejima EM, Figueiredo-Ribeiro RCL, Zaidan LBP (1991) Fructan composition in adventitious tuberous roots of *Viguiera discolor* baker (asteraceae) as influenced by daylength. New Phytol 119:149–154

Jensen BF, Norman BE (1984) *Bacillus acidopullulyticus* pullulanase: application and regulatory aspects for use in the food industry. Process Biochem 19:129–134

Kango N, Jain SC (2011) Production and properties of microbial inulinases: recent advances. Food Biotechnol 25:165–212

Kato I (2000) Antitumour activity of lactic acid bacteria. In: Fuller R, Perdigon G (eds) Probiotics. Kluwer Academic Publishers, Dordrecht, pp 115–138

Kaur N, Gupta AK (2002) Applications of inulin and oligofructose in health and nutrition. J Biosci 27:703–714

Kim BW, Kim HW, Nam SW (1997) Continuous production of fructose-syrups from inulin by immobilized inulinase from recombinant *Saccharomyces cerevisiae*. Biotechnol Bioprocess Eng 2:90–93

Kim CH, Rhee SK (1989) Fructose production from *Jerusalem artichoke* by inulinase immobilized on chitin. Biotechnol Lett 11:201–206

Kim WY, Byun SM (1982) Hydrolysis of inulin from *Jerusalem artichoke* by inulinase immobilized on aminoethylcellulose. Enzym Microb Technol 4:239–244

Kruger CL (2002) Generally recognised as safe (GRAS) notification for Frutafit®. Silver Spring, MD, U.S. Food and Drug Administration, USA

Kuniyal C, Rawat Y, Oinam S, Kuniyal J, Vishvakarma S (2005) Kuth (*Saussurea lappa*) cultivation in the cold desert environment of the Lahaul valley, northwestern Himalaya, India: arising threats and needs to revive socio-economic values. Biodivers Conserv 14:1035–1045

Leroy G, Grongnet JF, Mabeau S, Le Corre D, Baty-Julien C (2010) Changes in inulin and soluble sugar concentration in artichokes (*Cynara scolymus* L.) during storage. J Sci Food Agric 90:1203–1209

Lichtenthaler FW, Brust A, Cuny E (2001) Sugar-derived building blocks. Part 26. Hydrophilic pyrroles, pyridazines and diazepinones from D-fructose and isomaltulose. Green Chem 3:201–209

López-Molina D, Navarro-Martínez MD, Rojas-Melgarejo F et al (2005) Molecular properties and prebiotic effect of inulin obtained from artichoke (*Cynara scolymus* L.) Phytochemistry 66:1476–1484

Mahmoud DAR, Mahdy ESME, Shousha WGH et al (2011) Raw garlic as a new substrate for inulinase production in comparison to dry garlic. Aust J Basic Appl Sci 5:453–462

Mansouri S, Houbraken J, Samson RA, Frisvad JC et al (2013) *Penicillium subrubescens*, a new species efficiently producing inulinase. A Van Leeuw 103:1343–1357

Manzoni M, Cavazzoni V (1992) Hydrolysis of topinambur (*Jerusalem artichoke*) fructans by extracellular inulinase of *Kluyveromyces marxianus var. bulgaricus*. J Chem Technol Biotechnol 54:311–315

Mensink MA, Frijlink HW, van der Maarschalk V, WLJ H (2015) Inulin, a flexible oligosaccharide I: review of its physicochemical characteristics. Carbohydr Polym 130:405–419

Missau J, Scheid AJ, Foletto EL, Jahn SL, Mazutti MA, Kuhn RC (2014) Immobilization of commercial inulinase on alginate-chitosan beads. Sustain Chem Process 2:13

Mukherjee K, Sengupta S (1985) The production of constitutive invertase and inulinase by the mushroom *Panaeolus papillonaceus* in submerged culture. Can J Microbiol 31:773–777

Mutanda T, Wilhelmi B, Whkteley CG (2009) Controlled production of fructose by an exoinulinase from *Aspergillus ficuum*. Appl Biochem Biotechnol 159:65–77

Nakamura T, Kuramori K, Zaita N, Akimoto H, Ohta K (2001) Purification and properties of intracellular exo- and endoinulinases from *Aspergillus niger* strain 12. Bull Faculty Agric Miyazaki Univ 48:49–58

Nakamura T, Ogata Y, Shitara A, Nakamura A, Ohta K (1995) Continuous production of fructose syrups from inulin by immobilized inulinase from *Aspergillus niger* mutant 817. J Ferment Bioeng 80:164–169

Öngen-Baysal G, Sukan SS (1996) Production of inulinase by mixed culture of *Aspergillus niger* and *Kluyveromyces marxianus*. Biotechnol Lett 18:1431–1434

Orlando R, Floreani M, Padrini R, Palatini P (1998) Determination of inulin clearance by bolus intravenous injection in healthy subjects and ascitic patients: equivalence of systemic and renal clearances as glomerular filtration markers. Brit J Clin Pharmacol 46:605–609

Pandey A, Soccol CR, Selvakumar P, Soccol VT, Krieger N, Fontana JD (1999) Recent developments in microbial inulinases, its production, properties and industrial applications. Appl Biochem Biotechnol 81:35–52

Parekh SR, Margaritis A (1986) Application of immobilized cells of *Kluyveromyces marxianus* for continuous hydrolysis to fructose of fructans in *Jerusalem artichoke* extracts. Int J Food Sci Technol 21:509–515

Partida VZ, Lopez AC, Gomez AJM (December 8, 1998) Method of producing fructose syrup from agave plants. US patent 5,84,6333

Paula FC, Cazetta ML, Monti R, Contiero J (2007) Screening of supports for *Kluyveromyces marxianus* var. *bulgaricus* inulinase immobilization. Curr Trends Biotechnol Pharm 1:34–40

Pawan GLS (1973) Fructose. In: Birch GG, Green LF (eds) Molecular structure and function of food carbohydrates. Applied Science Publishers, London, pp 65–80

Pessoni RAB, Figueiredo-Ribeiro RCL, Braga MR (1999) Extracellular inulinases from *Penicillium janczewskii*, a fungus isolated from the rhizosphere of *Vernonia herbacea* (Asteraceae). J Appl Microbiol 87:141–147

Rawat HK, Ganaie MA, Kango N (2015) Production of inulinase, fructosyltransferase and sucrose from fungi on low-value inulin-rich substrates and their use in generation of fructose and fructo-oligosaccharides. A Van Leeuw 107:799–811

Rawat HK, Soni H, Treichel H, Kango N (2016) Biotechnological potential of microbial inulinases: recent perspective. Crit Rev Food Sci Nutr. doi:10.1080/10408398.2016.1147419. (Epub ahead of print)

Ribeiro EM, Fernanades P (2013) Coated-wall mini reactor for inulin hydrolysis. Curr Biol 2:47–52

Ricca E, Calabrò V, Curcio S, Basso A, Gardossi L, Iorio G (2010) Fructose production by inulinase covalently immobilized on sepabeads in batch and fluidized bed bioreactor. Int J Mol Sci 11:1180–1189

Ricca E, Calabrò V, Curcio S, Iorio G (2007) The state of the art in the production of fructose from inulin enzymatic hydrolysis. Crit Rev Biotechnol 27:129–145

Roberfroid M, Gibson GR, Delzenne N (1993) Biochemistry of oligofructose, a non-digestible fructo-oligosaccharide: an approach to estimate its caloric value. Nutr Rev 151:137–146

Roberfroid MB (2007) Inulin-type fructans: functional food ingredients. J Nutr 137:2493S–2502S

Ronkart SN, Blecker CS, Fougnies C, Van Herck JC, Wouters J, Paquot M (2006) Determination of physical changes of inulin related to sorption isotherms: an X-ray diffraction, modulated differential scanning calorimetry and environmental scanning electron microscopy study. Carbohydr Polym 63:210–217

Ronkart SN, Blecker CS, Fourmanoir H et al (2007) Isolation and identification of inulooligosaccharides resulting from inulin hydrolysis. Anal Chim Acta 604:81–87

Rose V (1804) About a peculiar substance. Neues Algem Chem 3:217–219. (in German)

Rouwenhorst RJ, Visser VL, Vander AA, Scheffers WA, Dijken JP (1988) Production, distribution and kinetic properties of inulinase on continuous culture of *Kluyveromyces marxianus* CBS 6556. Appl Environ Microbiol 54:1131–1137

Saber W, El-Naggar N (2009) Optimization of fermentation conditions for the biosynthesis of inulinase by the new source; *Aspergillus tamarii* and hydrolysis of some inulin containing agro-wastes. Biotechnol 8:425–433

Saito K, Tomita F (2000) Difructose anhydrides: their mass production and physiological functions. Biosci Biotechnol Biochem 64:1321–1327

Santa GLM, Bernardino SMSA, Magalhães S et al (2011) From inulin to fructose syrups using sol-gel immobilized inulinase. Appl Biochem Biotechnol 165:1–12

Sarchami T, Rehmann L (2015) Optimizing acid hydrolysis of *Jerusalem artichoke*-derived inulin for fermentative butanol production. Biol Res 8:1148–1157

Sheng J, Chi Z, Li J, Gao L, Gong F (2007) Inulinases production by the marine yeast *Cryptococcus aureus* G7a and inulin hydrolysis by the crude inulinases. Process Biochem 42:805–811

Singh RS (2011) Enzymatic preparation of high fructose syrup from inulin. In: Panesar PS, Sharma HK, Sarkar BC (eds) Bioprocessing of foods. Asiatech Publishing Inc, New Delhi, pp 77–98

Singh RS, Dhaliwal R, Puri M (2007a) Partial purification and characterization of exoinulinase from *Kluyveromyces marxianus* YS-1 for preparation of high-fructose syrup. J Microbiol Biotechnol 17:733–738

Singh RS, Dhaliwal R, Puri M (2007b) Production of high fructose syrup from *Asparagus* inulin using immobilized exoinulinase from *Kluyveromyces marxianus* YS-1. J Ind Microbiol Biotechnol 34:649–655

Singh RS, Dhaliwal R, Puri M (2008) Development of a stable continuous flow immobilized enzyme reactor for the hydrolysis of inulin. J Ind Microbiol Biotechnol 35:777–782

Singh RS, Singh RP (2010) Production of fructooligosaccharides from inulin by endoinulinases and their prebiotic potential. Food Technol Biotechnol 48:435–450

Singh RS, Singh RP (2014) Response surface optimization of endoinulinase production from a cost effective substrate by *Bacillus safensis* AS-08 for hydrolysis of inulin. Biocatal Agric Biotechnol 3:365–372

Singh RS, Singh RP, Kennedy JF (2016) Recent insights in enzymatic synthesis of fructooligosaccharides from inulin. Int J Biol Macromol 85:565–572

Sirisansaneeyakul S, Worawuthiyanan N, Vanichsriratana W, Srinophakun P, Chisti Y (2007) Production of fructose from inulin using mixed inulinases from *Aspergillus niger* and *Candida guilliermondii*. World J Microbiol Biotechnol 23:543–552

Takasaki Y, Tanabe O (October 26, 1971) Enzyme method for converting glucose in glucose syrups to fructose. US patent 3,616,221

Taper HS, Roberfroid M (1999) Influence of inulin and oligofructose on breast cancer and tumor growth. J Nutr (Suppl) 129:1488–1491

Teague WM, Brumm PJ (1992) Commercial enzymes for starch hydrolysis products. In: Schenck FW, Hebeda RE (eds) Starch hydrolysis products: worldwide technology, production and applications. VCH Publishers, New York, pp 45–77

Timmermans JW, van Leeuwen MB, Tournois H, de Wit D, Vliegenthart JFG (1994) Quantitative analysis of the molecular weight distribution of inulin by means of anion exchange HPLC with pulsed amperometric detection. J Carbohydr Chem 13:881–888

Treichel H, Oliveria D, Lerin L, Astolfi V et al (2012) A review on the production and partial characterization of microbial inulinases. Global J Biochem 3:7

van de Wiele T, Boon N, Possemiers S, Jacobs H, Verstraete W (2007) Inulin-type fructans of longer degree of polymerization exert more pronounced *in vitro* prebiotic effects. J Appl Microbiol 102:452–460

van Loo L, Coussement P, de Leenheer L, Horbregs H, Smith G (1995) On the presence of inulin and oligofructose as natural ingredients in the Western diet. Crit Rev Food Sci Nutr 35:525–552

Vandamme EJ, Derycke DG (1983) Microbial inulinases: fermentation process, properties, and applications. Adv Appl Microbiol 29:139–174

Vergauwen R, Van Laere A, Van den Ende W (2003) Properties of fructan: fructan 1-Fructosyltransferases from chicory and globe thistle, two asteracean plants storing greatly different types of inulin. Pl Physiol 133:391–401

Villegas-Silva PA, Toledano-Thompson T, Canto-Canché BB et al (2014) Hydrolysis of *Agave fourcroydes* Lemaire (henequen) leaf juice and fermentation with *Kluyveromyces marxianus* for ethanol production. BMC Biotechnol 14:14

Visser MR, Baert L, Klooster GV, Schueller L et al (2010) Inulin solid dispersion technology to improve the absorption of the BCS class IV drug TMC240. Eur J Pharm Biopharm 74:233–238

Viswanathan P, Kulkarni PR (1995) Properties and application of inulinase obtained by fermentation of costus (*Saussurea lappa*) root powder with *Aspergillus niger*. Nahrung 39:288–294

Wang J, Zhengyu J, Bo J, Xueming X (2003) Separation and identification of exo- and endoinulinases from *Aspergillus ficuum*. Curr Microbiol 47:109–112

Wei W, Yu X, Dai Y, Zheng J, Xie Z (1997) Purification and properties of inulinase from *Kluyveromyces sp.* Y-85. Wei Sheng Wu Xue Bao 37:443–448. (Article in Chinese)

Wenling W, Le Huiying WW, Shiyuan W (1999) Continuous preparation of fructose syrups from Jerusalem artichoke tuber using immobilized intracellular inulinase from *Kluyveromyces* sp. Y-85. Process Biochem 34:643–646

Wenling W, Zhaong X, Qian Z, Xinsheng Z (1997) Purification and properties of *Penicillium* extracellular inulinase. Ind Microbiol 27:8–12

Wolfgang W, Südzucker AGM (2004) Fructose. In: Hubert P (ed) Electronic ullmann's encyclopedia of industrial chemistry, Wiley-VCH GmbH & Co. KGaA, Weinheim, doi:10.1002/14356007.a12_047.pub2

Xiao R, Tanida M, Takao S (1988) Inulinase from *Chrysosporium pannorum*. J Ferment Technol 66:553–558

Yewale T, Singhal RS, Vaidya AA (2013) Immobilization of inulinase from *Aspergillus niger* NCIM 945 on chitosan and its application in continuous inulin hydrolysis. Biocatal & Agric Biotechnol 2:96–101

Ylikahari RH, Kahonen MT, Hassinen I (1972) Modification of metabolic effects of ethanol by fructose. Acta Medica Scand Suppl 542:141–150

Yu J, Jiang J, Ji W, Li Y, Liu J (2011) Glucose-free fructose production from *Jerusalem artichoke* using a recombinant inulinase-secreting *Saccharomyces cerevisiae* strain. Biotechnol Lett 33:147–152

Zhengyu J, Jing W, Bo J, Xueming X (2005) Production of inulooligosaccharides by endoinulinase from *Aspergillus ficuum*. Food Res Int 38:301–308

Zhou J, Gao Y, Zhang R, Mo M et al (2014) A novel low-temperature-active exo-inulinase identified based on molecular-activity strategy from *Sphingobacterium* sp. GN25 isolated from feces of *Grus nigricollis*. Process Biochem 49:1656–1663

Zubaidah E, Akhadiana W (2013) Comparative study of inulin extracts from dahlia, yam, and gembili tubers as prebiotic. Food Nutr Sci 4:8–12

Starch Nanoparticles: Their Preparation and Applications

11

Kawaljit Singh Sandhu and Vikash Nain

Abstract

The word "nano" originates from the Greek term "dwarf." Nanoparticles are generally recognized as those with a particle size below 100 nm which enables novel applications and benefits. Materials and devices with nanostructures have different properties from the materials they are made. Starch is a widely available, abundant biopolymer produced in plants by the process of photosynthesis and stored in the form of granules. The starch granules are in microsize which can be broken down into nanosize using different nanotechnological methods. These nano-sized particles are safe for human health and are increasingly utilized in food and nonfood industries. Nanotechnology, in general, is making the most significant progresses in biomedical application, including novel drug delivery practices in which starch nanoparticles can be used as drug carriers. Other potential applications of starch nanoparticles include their use in waste water treatment where they can replace the costly activated carbon as adsorbents. It has also attracted interest as a packaging material due to its better barrier properties. Their addition in natural rubber matrix improved barrier and mechanical properties and thus they are used commercially in tire making. Recently starch nanoparticles are found to have applications as fat replacers and emulsion stabilizers. Due to growing interest in starch nanoparticles, in the last decade, various methods have been developed for their synthesis. Acid hydrolysis, regeneration, and physical methods are widely used for the preparation of starch nanoparticles.

K.S. Sandhu (✉) • V. Nain
Department of Food Science and Technology, Chaudhary Devi Lal University, Sirsa 125 055, Haryana, India
e-mail: kawsandhu@rediffmail.com; kawsandhu5@gmail.com

S.K. Gahlawat et al. (eds.), *Plant Biotechnology: Recent Advancements and Developments*, DOI 10.1007/978-981-10-4732-9_11

11.1 Introduction

Starch is a widely available renewable biopolymer that is produced by photosynthesis in plants and a major source of energy for humans. It is made up of two components: amylose, a linear or slightly branched (1 → 4)-α-D-glucose units, and amylopectin, an extremely branched molecule in which glucose is linked with α-(1 → 4) and α-(1 → 6) linkages (Paris et al. 1999). Starch granules are made up of nano-sized blocklets having starch chains in semicrystalline organization (Oates 1997). Various methods have been used by researchers to study the starch structure, such as nuclear magnetic resonance (NMR) spectroscopy, microfocus X-ray diffractions (Waigh et al. 1997), and microfocus small-angle X-ray scattering (Waigh et al. 1999; Suzuki et al.1997). The results from these methods revealed that starch granules possess a structure of alternatively arranged crystalline and semicrystalline layers. Gallant et al. (1997) reported blocklets in starches, with those possessing B- and C-type crystalline structures having larger sizes as compared to A-type crystalline starches (Gallant et al. 1997). Amylopectin has been reported as the main element of the crystalline region in starch granules (Jenkins and Donald 1995; Perez et al. 2009); its packing forms a double helical structure (O'Sullivan and Perez 1999; Putaux et al. 2000).

Starch nanoparticles can be easily synthesized by acid or enzymatic hydrolysis of native starch. Mild acidic hydrolysis by HCl or H_2SO_4 results in the separation of crystalline regions. It is reported that at low temperatures (below gelatinization temperature), acid molecules prefer to hydrolyze amorphous regions of the granules (Robin et al. 1974; Wang et al. 2003). Starch origin and the method of nanoparticle formation affect the morphologies and crystal structures of the nanoparticles. Starch nanoparticles are safe for human health and have increasing utilization in food and non-food industrial applications (Chakraborty et al. 2005; LeCorre et al. 2010).

11.2 Starch Nanocrystals and Starch Nanoparticles

Starch nanocrystals (SNCs) are crystalline platelets formed by the breakdown of starch granules by hydrolysis. Recent studies by various workers report SNCs of different sizes: 40-70 nm for potato starch (Chen et al. 2008a); 30–80 nm (Chen et al. 2008b) and 60–150 nm (Yu et al. 2008 and Zheng et al. 2009) for pea starch; and 50 (Garcia et al. 2009) and 70–100 nm (Namazi and Dadkhah 2010) for waxy maize starch. Putaux et al. (2003) acid hydrolyzed waxy maize starch and reported morphological dimensions of nanoparticles with length and width in the range 20–40 nm and 15–30 nm, respectively.

To the best of our knowledge, there is no information regarding the industrial production of SNCs. Most of the patents on SNCs described either the alteration of SNCs or their utilization as nanocomposites. The degree of hydrolysis decides the formation of starch crystallites, starch nanocrystals, microcrystalline starch, or hydrolyzed starches.

Starch nanocrystals or nanoparticles are usually produced with the aim to be used as filler in polymeric matrices to improve their mechanical and/or barrier properties. Different methods have been described by different workers to formulate starch nanoparticles: (1) by precipitation (Ma et al. 2008; Tan et al. 2009), (2) by combining complex formation and enzymatic hydrolysis (Kim and Lim 2009), and (3) by microfluidization (Liu et al. 2009). Due to differences in methodologies, the resulting starch nanoparticles have different characteristics and crystallinities.

Ma et al. (2008) described the precipitation method for the production of starch nanoparticles by precipitating drop wise the gelatinized starch in ethanol. The suspension was centrifuged to remove solvent and then air-dried at 50 °C. Hebeish et al. (2014) proposed an amended solvent replacement technique which involves the application of a hydrophilic nonionic surfactant. A concentration of 20% surfactant resulted in the production of smaller size starch nanoparticles (SNPs) as compared to those without surfactants.

Lim and Kim (2008, 2009) filed two Korean patents on starch nanoparticles preparation. The first claims the formation of starch/alcohol complex nanoparticles (Kim and Lim 2009) and the second described the manufacturing of enzymatically hydrolyzed nanoparticles (Kim et al. 2008). Liu et al. (2009) also described the method of starch nanoparticles formation using microfluidizer. They used high amylose starch (5% slurry) which was passed through microfluidizer several times to achieve a gelatinized suspension of self-stable SNPs.

11.3 Preparation of Starch Nanocrystals/Nanoparticles

Starch nanocrystals or nanoparticles can be synthesized mainly by approaches described below:

1. Hydrolysis (acid and/or enzymatic hydrolysis, or in combination with other techniques).
2. Regeneration/ Nanoprecipitation.
3. Physical methods of SNP development (reactive extrusion, high-pressure homogenization, using gamma radiation, and ultrasonication).

11.3.1 Hydrolysis

11.3.1.1 Acid Hydrolysis

Depending on the botanical source of starch and isolation procedures used, the starch nanoparticles of diverse sizes and forms can be synthesized. Morphological properties of starch nanoparticles depend on the botanical source of starch, its crystallinity, and percentage of amylose and amylopectin (Gerard et al. 2002; Jayakody and Hoover 2002). The procedure for the synthesis of starch nanocrystals by acid hydrolysis is shown in Fig. 11.1.

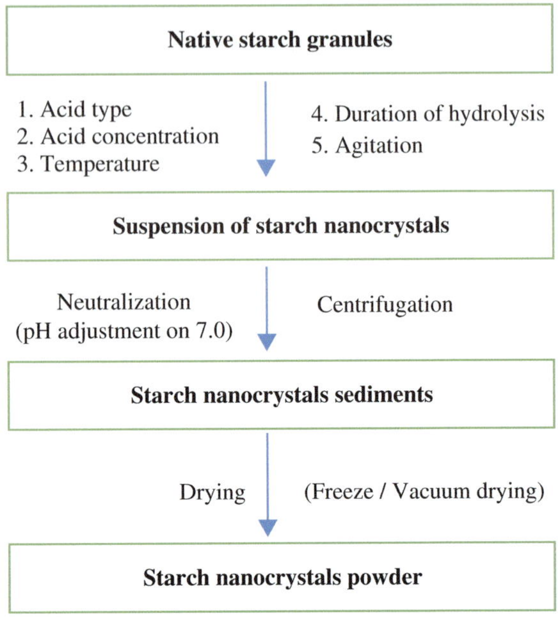

Fig. 11.1 Procedure for the preparation of starch nanocrystals by acid hydrolysis

The main objective of employing low temperature conditions during hydrolysis may be to prevent starch gelatinization and the breakdown of crystalline structure. Type of acid and hydrolysis period have been reported to affect acid hydrolysis. When HCl was used for hydrolysis it required more than 15 days, whereas it took 5–7 days when hydrolysis was carried out with H_2SO_4 (Gerard et al. 2002; Jayakody and Hoover 2002).

Angellier et al. (2004) optimized the hydrolysis conditions for starch nanocrystals formation using response surface methodology (RSM) and reported the effect of different types of acids on morphology and yield of nanocrystals. Hydrolysis using HCl resulted in the formation of separate starch nanoplatelets, the prolonged treatment time and poor yield which lower their potential to be used as nanofiller in nanomaterials, while hydrolysis using H_2SO_4 has been widely accepted to produce starch nanocrystals at a comparatively large scale. RSM was used to optimize the process conditions as: 14.69% starch slurry in 3.16 M H_2SO_4 was stirred at 100 rpm at 40 °C for 5 days. The nanoparticles formed had similar morphological characteristics as that attained by HCl hydrolysis for 6 weeks, but the yield was significantly improved (from 0.5 wt% to 15.7 wt%) while the hydrolysis time was reduced to 5 days. Various conditions of hydrolysis influence the size and yield of starch nanocrystals. Table 11.1 shows the effect of acid concentration, type of acids, temperature, and duration on the structural characteristics of starch nanoparticles from different botanical sources as reported by different workers.

The impact of type of acid and starch on hydrolysis kinetics has been observed by various workers, and they concluded that hydrolysis kinetics is a two-stage

Table 11.1 Effect of acid type, concentration, temperature, and duration of hydrolysis on structural characteristics of SNCs

Type of starch	Concentration of acid for hydrolysis	Temperature (°C)	Duration (in Days)	Structural parameters	Characterization	References
Waxy corn	2.2 N HCl	36	40	L 20–40 nm W 15–30 nm	TEM	Putaux et al. (2003)
Waxy corn	3.16 M H_2SO_4	40	5	T 5–7 nm L 40–60 nm W 15–30 nm	TEM	Angellier et al. (2004) and Angellier et al. (2005)
Waxy corn	3.16 M H_2SO_4	40	5	T 6–8 nm L 70–100 nm	TEM	Namazi and Dadkhah (2010)
Corn	2.87 M H_2SO_4	45	7	L 50 nm	SEM	Song et al. (2011)
Pea	3.16 M H_2SO_4	40	5	L 60–150 nm W 15–30 nm	TEM	Zheng et al. (2009)
Pea	3.16 M H_2SO_4	40	5	L 30–80 nm	TEM	Chen et al. (2008b)
Potato	3.16 M H_2SO_4	40	5	L 40–70 nm W 10–20 nm	TEM	Chen et al. (2008a)

process: the first stage includes faster hydrolysis, whereas the second stage is slower, as the hydrolysis of more dense crystalline regions takes more time (Biliaderis et al. 1981; Singh and Ali 1987; Singh and Ali 2000; Jayakody and Hoover 2002; Singh and Ali 2008). It has been observed that using H_2SO_4 hydrolysis method, the distribution and stability in aqueous solution increase due to the formation of sulfate groups at the surface; however, the thermal stability of the nanocrystals decreases. This limitation can be overcome using a small quantity of ammonia (0.5%) supplemented to H_2SO_4 starch slurry before centrifugation (Chang et al. 2009). This improved the heat stability of the starch nanocrystals so that they can withstand chemical modifications. Thus, starch nanocrystals are prepared so as to maintain the crystalline structure while absolutely eliminating the amorphous regions. It can also be prepared by separating the nanocrystalline lamellae to obtain the separated, small degree of self-aggregated platelet starch nanocrystals. The disadvantages associated with this method are poor yield, lengthy treatment duration, and involvement of acid.

11.3.1.2 Enzymatic Hydrolysis

To best of our knowledge, scarce literature is available that reports the synthesis of SNCs by sole utilization of enzymes for hydrolysis. An effort has been made by Kim et al. (2008). They utilized α- amylase enzyme to hydrolyze the native starch at 37 °C for 24 h at 150 rpm to produce SNCs of 500 nm size. The amorphous regions in the starch granules could be specifically separated by α-amylase hydrolysis. The resultant particles were unstable and swell rapidly by mechanical energy provided by ultrasonication, thus forming starch aggregates. They pointed toward the requirement of precise controls during enzymatic hydrolysis. Foresti et al. (2014) described the effect of α-amylase action on waxy maize starches and reported two-phase hydrolysis kinetics during hydrolysis.

11.3.1.3 Combined Enzymatic and Acid Hydrolysis

For the development of SNPs, a combination of acid and enzymatic hydrolysis has been used. Amylase has been used to minimize the duration of acid hydrolysis. Among α and β amylases, the latter was found more efficient in producing microporous starch having intact crystallinity. This enzyme pretreatment created routes for the acid to reach into the granules, which results in easy hydrolysis in amorphous regions. This combined hydrolysis also followed the two-stage kinetic profile. A pretreatment of 2 h before acid hydrolysis reduces the hydrolysis time from 24 h to 6 h for the same degree of hydrolysis. LeCorre et al. (2012) reported a reduction in hydrolysis time from 120 to 45 h after 2 h of pretreatment by enzymes.

11.3.1.4 Combination of Acid Hydrolysis and Ultrasonication

A combination method using acid hydrolysis and ultrasonication has been reported for SNP development from waxy maize starches (Kim et al. 2013). When acid hydrolysis was used alone, it took several days to develop SNPs. However, combination of acid hydrolysis (for 2 days) and subsequent ultrasonication for 3 min at 60% vibration amplitude produced high yield of nanoparticles. The crystalline behavior was disrupted due to ultrasonication with the resultant crystallinity below 30%. To overcome this problem, the process was carried out at a low temperature (4 °C). The combination method resulted in the complete conversion of microparticles to nanoparticles with a high yield of SNPs (up to 78%).

11.3.2 Regeneration/Nanoprecipitation

Regeneration method is usually employed for the synthesis of amorphous SNPs. In this process, the starch is gelatinized before regeneration step which is mostly carried out by solvent precipitation. Kim and Lim (2009) used regeneration method to prepare SNCs. Complex formation technique was used for nanocrystal formation, followed by selective enzymatic hydrolysis to maintain crystallinity. A suspension of high amylose maize starch in 90% di-methyl sulfoxide (DMSO) was heated in a water bath at boiling temperature followed by stirring for 24 h at 35 °C. A glass filtration apparatus with two compartments was chosen for filtration with the upper compartment filled with an aliquot of starch and the bottom compartment having n-butanol. Then this assembly was placed in a hot air oven at 70 °C for 6 days. The precipitate was recovered by centrifugation of this suspension and washed thrice in n-butanol. TEM images of starch-butanol complex showed the polygonal platelets with a length less than 100 nm. Enzymatic hydrolysis with α-amylase for 60 min removed the matrix around the platelets, and the sphere- or oval-shaped nanoparticles were observed as bunches with a relatively uniform size (10–20 nm diameter). Since most of starch was hydrolyzed in this process, the yield of nanoparticles was extremely low.

Chin et al. (2011) synthesized starch nanoparticles using nanoprecipitation technique by precipitating starch solution in absolute ethanol. NaOH/urea (NU) mixture was used as solvent to dissolve starch. A part of suspension was gently mixed and stirred in a known amount of absolute ethanol. This mixture was then centrifuged, washed with ethanol, and regenerated SNPs were collected. Ma et al. (2008) described the synthesis of SNPs by the precipitation of pregelatinized starch followed by cross-linking. Ethanol was used as a precipitant and SNPs having dimensions of 50–100 nm were formed.

A less time-consuming innovative technique which is a combination of enzymatic hydrolysis and recrystallization was presented by Sun et al. (2014). They enzymatically hydrolyzed the pregelatinized starch by pullulanase and then kept it for 12 h for retrogradation. This resulted in the formation of nanoparticles during the recrystallization process. This approach was time-saving and produced higher yield in comparison to the ordinary acid hydrolysis process for the development of SNCs.

11.3.3 Physical Methods for SNP/SNC Development

11.3.3.1 Reactive Extrusion

Starch extrusion has been widely studied by many researchers, particularly in food industry. The process includes exposing starch to high pressure, temperature, and shear. These process conditions bring changes in physicochemical and structural characteristics. Lai and Kokini (1990) reported incomplete gelatinization of starch granules due to limited water availability during extrusion. In 2000, the first patent for preparing starch nanoparticles via reactive extrusion was filed by Giezen et al. (2000). They used twin screw extruder, and high yields of nanoparticles were obtained with a size less than 400 nm. LeCorre et al. (2010) termed the starch nanoparticles prepared by reactive extrusion as "regenerated starch." Similarly, Song et al. (2011) used reactive extrusion method for starch nanoparticle preparation using a cross-linking agent, and the size achieved was around 160 nm.

11.3.3.2 High-Pressure Homogenization

In 2009, a simple and environmental friendly means for preparing starch nanoparticles using high pressure homogenization method was proposed by Liu and others. This technique is based on continuously passing the homogenized starch slurry through microfabricated membranes of a specially designed microfluidizer. 5% starch suspension was allowed to pass under pressure which resulted in particle size reduction from around 5 µm to 10–20 nm (Liu et al. 2009). With a combination of high-pressure homogenization technique with miniemulsion cross-linking, Shi et al. (2011) prepared starch nanoparticles. Although this technique is ecofriendly, it can be employed only for low concentration starch suspension and productivity was low. Recent studies confirmed that the combination of heat moisture treatment (0.075 M, for 8 h) and homogenization (60 min) produced nanoparticles with diameters of about 50 nm and a high yield of up to 80% (Park et al. 2016; Kim et al. 2017).

11.3.3.3 Using Gamma Radiation

SNPs can be produced by applying gamma radiations, and these radiations were found to cause cleavage of the glycosidic linkage in starch at a rapid rate which leads to the fragmentation of large molecules. It can produce free radicals which are capable of breaking chemical bonds in starch, resulting in the production of small molecules known as dextrins (Yu and Wang 2007). A dose of 20 KGy (at a rate of 14 KGy/h) of gamma radiation was applied to cassava and waxy maize starches which resulted in the production of SNPs of 20–30 nm size (Lamanna et al. 2013). SNP obtained by this process were found to be more susceptible to thermal degradation as compared to their counterpart native starches. It indicates that thermal degradation begins on the surface of SNPs (García et al. 2012). Like other physical treatments, SNPs by gamma radiation displayed amorphous X-ray diffraction patterns. It may be due to thermal treatment given before gamma radiation which was required to attain stable dispersions.

11.3.3.4 Ultrasonication

Ultrasound refers to the sound waves with frequencies in the range 15–20 kHz. This frequency range is above the hearing range of human beings. Piezoelectric or magnetostrictive transducers are generally used to create high-energy vibrations. This energy upon magnification can be utilized by transferring it to a probe which is in direct contact with the fluid. Recently, this approach is utilized by Haaj et al. (2013) using waxy maize starch suspension for preparing SNPs. Ultrasonication in a water bath for 75 min at 8 °C was used. SEM observations revealed that SNPs of limiting size (30–100 nm) were obtained through this process. The effect of ultrasonication on starch has been reported to depend on various parameters which include sonication power, thermal conditions, treatment time, starch concentration, and the botanical origin of starch (Zuo et al. 2009).

11.4 Physicochemical Properties of Starch Nanoparticles

11.4.1 Morphological and Geometrical Dimensions

The size of starch nanoparticles varies with the botanical sources and extracting techniques used. Starch nanoparticles of different shapes have been obtained mainly from corn, pea, and potato starches by different workers. Starch nanoparticles' size varied with the botanical sources of starch hydrolyzed and depends on the concentration and type of acid used for hydrolysis. Starch nanocrystals produced from different sources have been morphologically studied by using TEM or SEM. Various workers have reported their length, breadth, and thickness in the range 20–200, 10–130, and 5–10 nm, respectively (Gallant et al. 1997; Angellier-Coussy et al. 2009; LeCorre et al. 2011; Kim et al. 2012; Liu et al. 2016; Patel et al. 2016). The process employed for the synthesis of SNPs/SNCs also influences the geometric and morphological characteristics of the resulting nanoparticles. In comparison to SNCs, SNPs were found to have smaller sizes. SNCs had the platelet-like morphology and possesses a negative surface charge due to sulfuric acid used in hydrolysis (Haaj et al. 2016). It was observed that nanoparticles produced from B- and C-type starches were larger than those obtained from A-type starches, which may be due to their greater resistance to acid hydrolysis (LeCorre et al. 2011). It was also observed that the size of nanoparticles was influenced by amylose content; higher the amylose content, larger the nanoparticles produced.

11.4.2 X-ray Diffraction

Starch from different biopolymers has semi-crystalline structure and different degrees of crystallinity (15–45%). Native starch granules have crystalline regions which can be studied by X-ray diffraction patterns (XRD) (Imberty and Perez 1988). Starch nanoparticles synthesized by various methods retain their crystallinity. Starch nanoparticles from waxy maize and nanoparticles revealed their A-type crystalline patterns (Vodalkar and Thakore 2010). Starch nanoparticles prepared by enzymatic hydrolysis revealed their crystallinity which confirmed that α-amylolysis occurred mainly in the amorphous region of the complex, without changing their crystalline structures (Kim and Lim 2009). Disappearance of small peaks and broadening of other peaks of X-ray diffraction were observed during the process of homogenization and ultrasonication used in SNC formation (Haaz et al. 2013; Liu et al. 2009; Patel et al. 2016).

The packing of amylose and amylopectin chains results in the crystalline structure of starches. The degree of crystallinity (ratio between mass of crystalline domains and the total mass of nanocrystals) of polysaccharide nanocrystals is less than theoretical value of 100% due to incomplete removal of amorphous regions. It has been reported that depending on the botanical sources, the crystallinity values of starch nanocrystals vary in the range 45–50% (LeCorre et al. 2011).

11.4.3 Thermal Transition Properties

Differential scanning calorimeter (DSC) has been extensively used to study the thermal behavior of starches. The change in heat flow behavior is observed which has first-order transition (melting) and second-order transition (glass transition) for polymers. The thermal transition behavior of starch nanoparticles synthesized by acid hydrolysis has been studied by many researchers. They observed that acid hydrolysis has a significant effect on the melting temperature and enthalpy of native starches. Onset temperature (T_o) of starch nanoparticles produced by acid hydrolysis was lower from their native counterparts whereas peak (T_p), conclusion of melting temperatures (T_c) and melting enthalpy (ΔH) were higher for nanoparticles as compared to their native counterpart starches. It indicates that the cleavage of starch chains could lead to its rearrangements which increase crystallinity. A lower glass transition temperature (T_g) for starch nanoparticles was observed as compared to native starches (Kim and Lim 2009; Kim et al. 2012; Shi et al. 2012a). These researchers also studied the effect of annealing on T_g of starch nanoparticles and concluded that annealing led to the lowering of T_g.

11.4.4 Rheological Properties of Starch Nanoparticles

Rheological behavior of SNCs in aqueous suspension is an important parameter to be assessed in order to determine its industrial applications. SNC suspension exhibited a shear thinning behavior in the examined range of shear rate (Shi et al. 2012a). Although no correlation among viscosity and particle size, thickness, surface area of SNCs was found. Rheological properties of spray and vacuum freeze dried starch nanoparticles were investigated by various workers (Shi et al. 2012a, b, 2013). They also studied the effect of NaCl addition on the rheological behavior of spray and vacuum freeze dried starch nanoparticle suspensions. The parameters studied were continuous shear, temperature sweep, frequency sweep and creep recovery tests. Pasting properties of starch nanoparticles have also been reported by some researchers with different observations (Shi et al. 2012a, b). Some starch nanoparticles showed their pasting behavior similar to their counterpart native starches. Although some variation in gelatinization temperature was observed but pasting behavior/patterns remain the same. On the contrary, our study (unpublished) revealed that starch nanoparticles made from potato by acid hydrolysis lost their pasting behavior during pasting studies. Generally native starch slurry upon heating shows an increase in viscosity during pasting studies with characteristic peak and setback viscosities. But starch nanoparticles studied by our group revealed no increase in viscosity during pasting studies and the characteristic curve appears flattened probably due to loss of starch granules integrity happening during the process of nanoparticles formation. The characteristic pasting curve of the starch nanoparticles in our study is shown in Fig. 11.2.

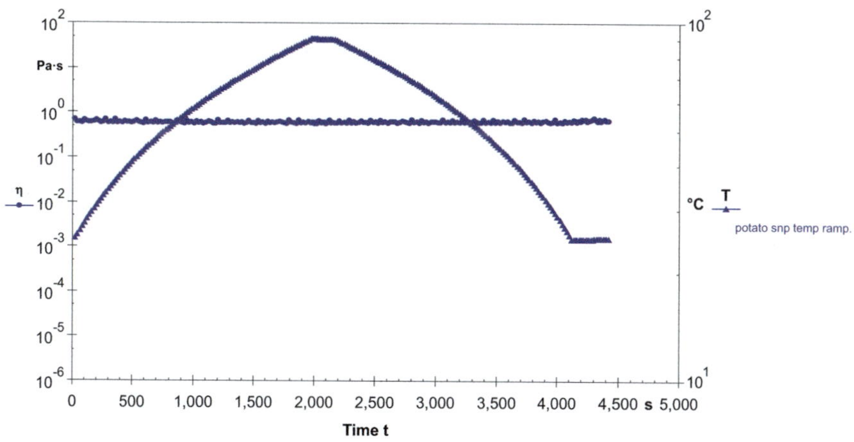

Fig. 11.2 Pasting behavior of potato SNPs

11.4.5 Digestibility of Starch Nanoparticles

In vitro digestion behavior of native and debranched waxy maize starches and starch nanoparticles were studied by Cai et al. (2010). Native waxy maize starch was found to have a small amount of resistant starch content. Resistant starch (RS) content increased with increase in debranching period. During initial 4 h, RS content increased at low rate but after this period it significantly increased from 4–24 h. After 24 h, debranching resulted in 71.4% RS.

The influence of pullulanase concentration, duration of hydrolysis and storage temperature on SDS content of thermally processed rice starches was described by Guraya et al. (2001). Hydrolysis kinetics of starch was significantly affected by degree of debranching and percentage of amylose present in starch. Debranching resulted in the development of smaller amylose chains that aggregate into more compactly arranged crystalline arrays upon cooling. These compactly arranged crystals showed more resistant to digestion process.

11.5 Application of Starch Nanoparticles

11.5.1 Carrier in Drug Delivery

Due to exclusive capabilities of polymeric nanoparticles in drug delivery systems their use has received significant attention. Nanodrug delivery systems are defined as suspension or solid particles having size of 10–1000 nm. Nanoparticle matrix is the medium on which the drug is either loaded, entrapped, encapsulated or bound (Kommareddy et al. 2005; Lee and Kim 2005). Different properties of SNPs including its non-toxicity, decomposability, and ability to carry drugs make it promising candidate for carrier in drug delivery system.

The possibility of starch nanoparticles as a transdermal drug delivery system (TDDS) was investigated by Santander-Ortega and their team in 2010. The problem encountered in drug delivery by such system was due to barrier properties of skin which decreased effective drug distribution and therefore needs to be addressed. Nanoparticles have been revealed to enable drug delivery without disturbing the skin's integrity. The process employed to formulate the nanoparticles was emulsification. Nanodrug delivery approaches have numerous benefits, such as better encapsulation efficiency, effective defense to biochemical or enzymatic degradation and exclusive capability to produce a controlled release (Soma et al. 1999). Their application is getting a huge consideration because of improved hydrophilicity, biocompatibility and biodegradability (Elvira et al. 2002). Modified starch nanoparticles were used for indomethacin carrier (Simi and Abraham 2007). It was observed that modified starch nanoparticles had an improved loading capacity and in vitro release behavior for indomethacin drug (Brecher et al. 1997; Dufresne et al. 2006). Recently in vitro and in vivo studies proved the possible application of SNCs for skin use. The SNCs were gentle approach to increase skin penetration of lipophilic bioactive compounds with better acceptance by human subjects (Marto et al. 2016). Starch nanoparticles enabled effective dispersal of drug in aqueous media, permitting controlled release of it in both basic and acidic mediums. Because of small size, these starch nanoparticles gave a huge opportunity for its applications in drug delivery (Ayadi et al. 2016).

11.5.2 Nano Composite in Food Packaging Industry

The crucial function of nanotechnology in food packaging is considered as the largest commercial utilization in food sector (Chaudhary et al. 2010). In recent years, there is more concern about research and innovation in the food packaging materials ranging from films, carbon nanotubes, to waxy nano-coating for some foods. The use of nanoparticles might help in the production of new food packaging materials with better properties to increase shelf life (Chaudhary et al. 2008). Besides antimicrobial characteristics, nanoparticles can be used as carrier to distribute antioxidants, enzymes, flavors, anti-browning agents and other materials to extend shelf life, even after opening (Cha and Chinnan 2004; LaCoste et al. 2005; Weiss et al. 2006). Starch-based colloidal coating filling with antimicrobial nanoparticles enabled protection and food packaging has been reported by Boumans (2003). SNPs have gained huge attention as a packaging material due to its better barrier properties. Barrier properties in packaging material primarily focus on water vapor permeability and oxygen transfer rate. Recent studies have reported that incorporation of SNPs from various botanical sources to films resulted in decreased water transfer rate. An addition of 30–40% SNPs of waxy maize to sorbitol-pullulan films showed a significant decrease in water vapor permeability (Kristo and Biliaderis 2007, Goyal and Goyal 2012). The incorporation of maize SNCs to amaranth protein improved the general behavior of protein films as the resultant films had decreased water vapor permeability, and increased mechanical properties as compared to films prepared by proteins alone (Condes et al. 2015). Physicochemical properties of films from pea starches were improved when potato SNPs were added at different level. SNPs added films had higher tensile strength and lower water vapor permeability (Jiang et al. 2016).

11.5.3 Natural Polymers as Matrix

Starch nanoparticles have been used for developing nanocomposites along with natural rubber latex. With these nanoparticles acting as reinforcing phase, they have shown a uniform dispersion in rubber matrix which is most desired to achieve good mechanical properties. Incorporation of SNPs in rubber based polymer decreased water vapor and oxygen permeability significantly (Angellier et al. 2005). Their addition also enhanced the mechanical properties of bio composites. In manufacturing of rubber tires, carbon block and silica is traditionally used. In order to enhance better mechanical properties, tires are made jointly by companies Novamont and Goodyear using corn starch nanoparticles (Materne et al. 2000 European patent no EP0995775A1).

11.5.4 Adsorbents in Wastewater Treatment

Starch nanoparticles after chemical modifications can be utilized as adsorbents for removing aromatic and organic contaminants from water. The activated carbon is traditionally used for wastewater treatment but its use is limited because of its high cost. Use of synthetic polymer can be the solution to this limitation. During the past decade an increasing attention is being given to the renewable and cost effective alternate. Biopolymers have been found to be most suitable for this purpose. Recently, for removal of aromatic organic compounds from wastewater, polystyrene-modified cellulose nano whiskers has been studied. Alila et al. (2011) was the first to report the potential use of SNPs as adsorbents after modifying it by grafting with stearate. Nanoparticles from starches have shown improved efficiency and capability due to enhanced surface area as compared to their native counterparts. The chemical modification of SNPs (grafting, crosslinking etc.) has further increased the efficiency of adsorption capacity. Adsorption capacity of SNPs grafted with stearate increased from 150 and 900 μmol g^{-1} when 2-naphthol and nitrobenzene, respectively were used as compared to much lower adsorption of 50 and 40 μmol g^{-1} using unmodified starch (Chern and Chien 2003; Orr et al. 2004; Kamegawa et al. 2005; Sharma et al. 2007, 2008; Morandi et al. 2009; Alila et al. 2011). It is assumed that Van der Waals interactions between grafted chain and organic molecules is the main driving force behind the adsorption. Planer structure of aromatic organic solute also favors the interaction of the molecules inside the domain formed in grafted SNPs.

11.5.5 Fat Replacers

In an era of health concern and developed taste buds high calorie dense fat is one main concern of food technologists. So replacement of fat without disturbing the mouth feel is a burning issue. Food technologists are therefore looking for fat substitute which can replace the fat and provide the mouth feel identical to fats. SNPs have also found their use in this field as they can be used as fat replacers. Fat replacers are the substance which mimic the characteristics of fat without providing the calories or provide fewer calories. Starches and their modified versions can sometimes be used as fat replacers. Particle size of starch is the main factor which decides both fat like taste and mouth feel (Daniel and Whistler 1990; Jane et al. 1992). Mechanical disintegration and chemical depolymerization are the most commonly used modification for producing fat substitute starch. When SNPs are mixed with other constituents, their blend forms a smooth texture which is similar to fat. Further, addition of SNPs as fat replacers also reduces the calories of food. However, till date there has been no study reported on the role of SNPs as fat replacers.

11.5.6 Emulsion Stabilizer

Various kinds of emulsion stabilizers are needed in food processing and its preservation. Presently various kind of particles viz. hydrophobic silica, carbon nanotubes, latex, bacterial cellulose nanocrystals and microfibrillated cellulose have been used as emulsion stabilizers in foods. In recent years, SNPs also have found their place as emulsion stabilizers. It was observed that by mixing a little amount of SNP suspension, a stable emulsion of water and paraffin could be achieved. SNP addition at a concentration higher than 0.02% by weight stabilized the suspension for more than 2 months. Though, the phase separation occurred when suspension was heated at 80 °C for 2 h (Andresen and Stenius 2007; Kalashnikova et al. 2011; Li et al. 2012). SNPs may also be used as emulsion stabilizers in cosmetics and pharmaceuticals in addition to food system.

11.5.7 Binder in Papermaking and Paper Coating

In paper-making cooked starch is widely used as an additive. The adsorption of starch on paper matrix is limited due to saturation of cellulose substrates. Other limitation of starch usage is the increase in viscosity of starch paste upon cooking. SNPs can be used to overcome this problem as they have low viscosity than their native counterpart starches. The performance of binders has been found to improve when SNPs were used in place of cooked starch. Viscosity has an inversely proportional relationship with binding capacity in paper making process. SNPs have reduced viscosity as compared to cooked starch, however, they have increased binding capacity (Lee et al. 2015). SNPs applications in different fields are summarized in Table 11.2.

Table 11.2 Application of starch nanoparticles in various fields

Application field	Roles of SNPs/SNCs	References
Food application		
Emulsion stabilizer	Stabilize oil in water emulsion	Li et al. (2012)
Fat replacer	Improve taste and mouth feel with reduced calories	Jane et al. (1992) and Daniel and Whistler (1990)
Non-food applications		
Drug delivery	Regulated drug release system	Santander-Ortega et al. (2010), Elvira et al. (2002), Simi and Abraham (2007), Brecher et al. (1997) and Dufresne et al. (2006)
Composite filler	Reinforcement of polymeric matrix	Angellier et al. (2005)
Waste water treatment	Cost-effective adsorbent	Alila et al. (2011)
Binder in paper-making	Decreased paste viscosity, high binding capacity	Lee et al. (2015)
Packaging components	Improved oxygen and water vapor barrier	Kristo and Biliaderis (2007), Ma et al. (2008) and Garcia et al. (2009)

11.6 Starch Nanocrystal Toxicity

Public is concerned about the use of nanoparticles due to their release in environment during manufacturing, transportation, use and disposal and therefore their influence on surroundings and on human health need to be study. Thus, sustainable usage could not be established without addressing the concern of starch nanoparticles potential toxicity. Direct exposure of consumers to nanoparticles used in food industry poses a serious problem to human health. Some nanoparticles of other polymers are supposed to interact with protein and enzymes leading to induction of oxidative stress (Hajipour et al. 2012). Unfortunately, there are limited studies on potential toxicity of nanoparticles to humans. Therefore, risk assessment studies to show their adverse effect on human health and biocompatibility of starch nanoparticles should be critically examined.

11.7 Conclusion

Surveys reported in the literature indicate that during previous decade a huge consideration has been focused on biopolymers and their nanoparticles. Biopolymer nanoparticles gain this significant amount of interest due to their availability in nature, inexpensive, renewability, biocompatibility, biodegradability, and nontoxicity. The growing technical and industrial attention for SNPs has resulted in the development of various approaches for synthesis of small-sized starch composites. Innovative processing approaches should be used in order to increase the applications of starch nanocrystals in nanomaterials. A simple, economic, and high yielding method should be developed for SNPs production at commercial/industrial level. The problem of aggregation of starch nanoparticles should also be addressed to increase its utilization in different fields. SNPs may be used as a unique component for regulation of physicochemical and rheological properties of processed foods in food industries. The functional modifications of SNPs under mild conditions may greatly improve its application in future.

References

Alila S, Aloulou F, Thielemans W et al (2011) Sorption potential of modified nanocrystals for the removal of aromatic organic pollutant from aqueous solution. Ind Crop Prod 33(2):350–357

Andresen M, Stenius P (2007) Water-in-oil emulsions stabilized by hydrophobized microfibrillated cellulose. J Dispers Sci Technol 28(6):837–844

Angellier H, Choisnard L, Molina-Boisseau S et al (2004) Optimization of the preparation of aqueous suspensions of waxy maize starch nanocrystals using a response surface methodology. Biomacromolecules 5(4):1545–1551

Angellier H, Molina-Boisseau S, Dufresne A (2005) Mechanical properties of waxy maize starch nanocrystal reinforced natural rubber. Macromolecules 38(22):9161–9170

Angellier-Coussy H, Putaux JL, Molina-Boisseau S et al (2009) The molecular structure of waxy maize starch nanocrystals. Carbohydr Polym 344(12):1558–1566

Ayadi F, Bayer IS, Marras S et al (2016) Synthesis of water dispersed nanoparticles from different polysaccharides and their application in drug release. Carbohydr Polym 136:282–291

Biliaderis CG, Grant DR, Vose JR (1981) Structural characterization of legume starches. II Studies on acid-treated starches. Cereal Chem 58(6):502–507

Boumans H (2003) Release on command: Bioswitch, in leads in life science. TNO Nutr Food Zeist 22:4–5

Brecher ME, Owen HG, Bandarenko N (1997) Alternatives to albumin: starch replacement for plasma exchange. J Clin Apher 12(3):146–153

Cai L, Shi YC, Rong L et al (2010) Debranching and crystallization of waxy maize starch in relation to enzyme digestibility. Carbohydr Polym 81(2):385–393

Cha D, Chinnan M (2004) Biopolymer-based antimicrobial packaging: a review. Crit Rev Food Sci Nutr 44(4):223–237

Chakraborty S, Sahoo B, Teraoka I et al (2005) Solution properties of starch nanoparticles in water and DMSO as studied by dynamic light scattering. Carbohydr Polym 60(4):475–481

Chang PR, Ai F, Chen Y et al (2009) Effects of starch nanocrystal-graft-polycaprolactone on mechanical properties of waterborne polyurethane-based nanocomposites. J Appl Polym Sci 111(2):619–627

Chaudhary Q, Scotter M, Blackburn J et al (2008) Applications and implications of nanotechnologies for the food sector. Food Addit Contam A 25(3):241–258

Chaudhary Q, Castle L, Watkins R (2010) Nanotechnologies in food. Royal Society of Chemistry Publishers Cambridge, UK

Chen G, Wei M, Chen J et al (2008a) Simultaneous reinforcing and toughening: new nanocomposites of waterborne polyurethane filled with low loading level of starch nanocrystals. Polymer 49(7):1860–1870

Chen Y, Cao X, Chang PR et al (2008b) Comparative study on the films of poly(vinyl alcohol)/pea starch nanocrystals and poly(vinyl alcohol)/native pea starch. Carbohydr Polym 73(1):8–17

Chern JM, Chien YW (2003) Competitive adsorption of benzoic acid and p -nitrophenol onto activated carbon: isotherm and breakthrough curves. Water Res 37(10):2347–2356

Chin SF, Pang SC, Tay SH (2011) Size controlled synthesis of starch nanoparticles by a simple nanoprecipitation method. Carbohydr Polym 86(4):1817–1819

Condes MC, Anonn MC, Mauri AN et al (2015) Amaranth protein films reinforced with maize starch nanocrystals. Food Hydrocoll 47:146–157
Daniel JR, Whistler RL (1990) Fatty sensory qualities of polysachharides (Abstr). CFW 35(825):321–330
Dufresne A, Thielemans W, Leonard M et al (2006) Starch nanocrystals with long chain surface modifications. Langmuir 22(10):4804–4810
Elvira C, Mano JF, San RJ et al (2002) Starch-based biodegradable hydrogels with potential biomedical applications as drug delivery systems. Biomaterials 23(9):1955–1966
Foresti ML, Williams MDP, Martinez-Garcia R et al (2014) Analysis of a preferential action of α-amylase from *B. licheniformis* towards amorphous regions of waxy maize starch. Carbohydr Polym 102:80–87
Gallant DJ, Bouchet B, Baldwin PM (1997) Microscopy of starch: evidence of a new level of granule organization. Carbohydr Polym 32(3):177–191
Garcia NL, Ribba L, Dufresne A et al (2009) Physico-mechanical properties of biodegradable starch nanocomposites. Macromol Mater Eng 294(3):169–177
García NL, Lamanna M, Accorso ND et al (2012) Biodegradable materials from grafting of modified PLA onto starch nanocrystals. Polym Degrad Stab 97(10):2021–2026
Gerard C, Colonna P, Buléon A et al (2002) Order in maize mutant starches revealed by mild acid hydrolysis. Carbohydr Polym 48(2):131–141
Giezen FE, Jongboom ROJ, Feil H, Gotlieb KF, Boersma A (2000) WO 2000069916 A1, 20001123 (Patent)
Goyal S, Goyal GK (2012) Nanotechnology in food packaging- a critical review. Russ J Agric Soc Econ Sci 10(10):14–24
Guraya HS, James C, Champagne ET (2001) Effect of enzyme concentration and storage temperature on the formation of slowly digestible starch from cooked debranched rice starch. Starch/Stärke 53(3–4):131–139
Haaj SB, Magnin A, Pétrier C et al (2013) Starch nanoparticles formation via high power ultrasonication. Carbohydr Polym 92(2):1625–1632
Haaj SB, Thielemans W, Magnin A et al (2016) Starch nanocrystals and starch nanoparticles from waxy maize as nanoreinforcement: a comparative study. Carbohydr Polym 143:310–317
Hajipour MJ, Fromm KM, Ashkarran AA et al (2012) Antibacterial properties of nanoparticles. Trends Biotechnol 30(10):599–611
Hebeish A, El-Rafie MH, MA EL-S et al (2014) Ultra-fine characteristics of starch nanoparticles prepared using native starch with and without surfactant. J Inorg Organomet Polym Mater 24(3):515–524
Imberty A, Perez S (1988) A revisit to the three-dimensional structure of B-type starch. Biopolymers 27(8):1205–1221
Jane J, Shen L, Wang L et al (1992) Preparation and properties of small-particle corn starch. Cereal Chem 69(3):280–283
Jayakody L, Hoover R (2002) The effect of lintnerization on cereal starch granules. Food Res Int 35(7):665–680
Jenkins PJ, Donald AM (1995) The influence of amylose on starch granule structure. J Biol Macromol 17(6):315–321
Jiang S, Liu C, Wang X et al (2016) Physicochemical properties of starch nanocomposite films enhanced by self-assembled potato starch nanoparticles. LWT Food Sci Technol 69:251–257
Kalashnikova I, Bizot H, Cathala B et al (2011) New pickering emulsions stabilized by bacterial cellulose nanocrystals. Langmuir 27(12):7471–7479
Kamegawa K, Nishikubo K, Kodama M et al (2005) Aqueous-phase adsorption of aromatic compounds on water-soluble nanographite. Colloids Surf A Physicochem Eng Asp 254(1):31–35
Kim JY, Lim ST (2009) Preparation of nano-sized starch particles by complex formation with n-butanol. Carbohydr Polym 76(1):110–116
Kim JY, Park DJ, Lim ST (2008) Fragmentation of waxy rice starch granules by enzymatic hydrolysis. Cereal Chem 85(2):182–187

Kim HY, Lee HJ, Kim JY et al (2012) Characterization of nanoparticles prepared by acid hydrolysis of various starches. Starch/Stärke 64(5):367–373
Kim HY, Park DJ, Kim JY et al (2013) Preparation of crystalline starch nanoparticles using cold acid hydrolysis and ultrasonication. Carbohydr Polym 98(1):295–301
Kim JH, Park DH, Kim JY (2017) Food hydrocolloids effect of heat-moisture treatment under mildly acidic condition on fragmentation of waxy maize starch granules into nanoparticles. Food Hydrocoll 63:59–66
Kommareddy S, Tiwari S, Amiji M (2005) Long-circulating nano-vectors for tumor-selective gene delivery. Technol Cancer Res Treat 4(6):615–625
Kristo E, Biliaderis C (2007) Physical properties of starch nanocrystal-reinforced pullulan films. Carbohydr Polym 68(1):146–158
LaCoste A, Schaich K, Zumbrunnen D et al (2005) Advanced controlled release packaging through smart blending. Packag Technol Sci 18(2):77–87
Lai LS, Kokini JL (1990) The effect of extrusion operating conditions on the on-line apparent viscosity of 98% amylopectin (Amioca) and 70% amylose (Hylon 7) corn starches during extrusion. J Rheol 34(8):1245–1266
Lamanna M, Morales N, Garacia NL et al (2013) Development and characterization of starch nanoparticles by gamma radiation: potential application as starch matrix filler. Carbohydr Polym 97(1):90–97
LeCorre D, Bras J, Dufresne A (2010) Starch Nanoparticles : a review. Biomacromolecules 11(5):1139–1153
LeCorre D, Bras J, Dufresne A (2011) Influence of botanic origin and amylose content on the morphology of starch nanocrystals. J Nanopart Res 13(12):7193–7208
LeCorre D, Vahanian E, Dufresne A et al (2012) Enzymatic pretreatment for preparing starch nanocrystals. Biomacromolecules 13(1):132–137
Lee M, Kim S (2005) Polyethylene glycol-conjugated copolymers for plasmid DNA delivery. Pharm Res 22(1):1–10
Lee YK, Kim SG, Won JM et al (2015) Application technology of environmental-friendly starch based biobinder and synthesized binder in order to substitute SB latex for paper coating(1)-application of substitute binder for pre-coating later. J Korea TAPPI 47(5):134–140
Li C, Sun P, Yang C (2012) Emulsion stabilized by starch nanocrystals. Starch-Starke 64(6):497–502
Lim ST, Kim JY (2008) CAPLUS AN 2008:1490656 (Patent)
Lim ST, Kim JY (2009) CAPLUS AN 2009:679477 (Patent)
Liu D, Wu Q, Chen H et al (2009) Transitional properties of starch colloid with particle size reduction from micro- to nanometer. J Colloid Interface Sci 339(1):117–124
Liu C, Qin Y, Li X et al (2016) Preparation and characterization of starch nanoparticles via self-assembly at moderate temperature. Int J Biol Macromol 84:354–360
Ma X, Jian R, Chang PR et al (2008) Fabrication and characterization of citric acid-modified starch nanoparticles/plasticized-starch composites. Biomacromolecules 9(11):3314–3320
Marto J, Gouveia LF, Gonçalves LM (2016) A quality by design (QbD) approach on starch-based nanocapsules: a promising platform for topical drug delivery. Colloids Surf B Biointerfaces 143:177–185
Materne T, Corvasce F, Leitz P (2000) European patent no. EP0995775A1, 2000 (Patent)
Morandi G, Heath L, Thielemans W (2009) Cellulose nanocrystals grafted with polystyrene chains through surface-initiated atom transfer radical polymerization (SI-ATRP). Langmuir 25(14):8280–8286
Namazi H, Dadkhah A (2010) Convenient method for preparation of hydrophobically modified starch nanocrystals with using fatty acids. Carbohydr Polym 79(3):731–737
O'Sullivan AC, Perez S (1999) The relationship between internal chain length of amylopectin and crystallinity in starch. Biopolymers 50(4):381–390
Oates CG (1997) Towards an understanding of starch granule structure and hydrolysis. Trends Food Sci Technol 8(11):375–382

Orr PT, Jones GJ, Hamilton GR (2004) Removal of saxitoxins from drinking water by granular activated carbon, ozone and hydrogen peroxide-implications for compliance with the Australian drinking water guidelines. Water Res 38(20):4455–4461

Paris M, Bizot H, Emery J et al (1999) Crystallinity and structuring role of water in native and recrystallized starches by 13C CP-MAS NMR spectroscopy.1: Spectral decomposition. Carbohydr Polym 39(4):327–339

Park EY, Kim MJ, Cho ML et al (2016) Production of starch nanoparticles using normal maize starch via heat-moisture treatment under mildly acidic conditions and homogenization. Carbohydr Polym 151:274–282

Patel CM, Chakraborty M, Murthy ZVP (2016) Fast and scalable preparation of starch nanoparticles by stirred media milling. Adv Powder Technol 27:1287-1294 (http://dx.doi.org/10.1016/j.apt.2016.04.021)

Perez S, Baldwin PM, Gallant DJ (2009) Structural features of starch granules I, 3rd edn. Starch: Chemistry and Technology. ISBN:978-0-12-746275-2

Putaux JL, Buléon A, Chanzy H (2000) Network formation in dilute amylose and amylopectin studied by TEM. Macromolecules 33(17):6416–6422

Putaux JL, Molina-Boisseau S, Momaur T et al (2003) Platelet nanocrystals resulting from the disruption of waxy maize starch granules by acid hydrolysis. Biomacromolecules 4(5):1198–1202

Robin JP, Mercier C, Charbonniere R et al (1974) Lintnerized starches, gel filtration and enzymatic studies of insoluble residues from prolonged acid treatment of potato starch. Cereal Chem 51:389–406

Santander-Ortega MJ, Stauner T, Loretz B et al (2010) Nanoparticles made from novel starch derivatives for transdermal drug delivery. J Control Release 141(1):85–92

Sharma YC, Uma SSN et al (2007) Fly ash for the removal of Mn(II) from aqueous solutions and wastewaters. Chem Eng J 132(1):319–323

Sharma YC, Singh B, Agrawal A et al (2008) Removal of chromium by riverbed sand from water and wastewater: effect of important parameters. J Hazard Mater 151(2):789–793

Shi A, Li D, Wang L et al (2011) Preparation of starch-based nanoparticles through high-pressure homogenization and miniemulsion cross-linking: influence of various process parameters on particle size and stability. Carbohydr Polym 83(4):1604–1610

Shi A, Li D, Wang L et al (2012a) The effect of NaCl on the rheological properties of suspensions containing spray dried starch nanoparticles. Carbohydr Polym 90(4):1530–1537

Shi A, Li D, Wang L et al (2012b) Rheological properties of suspensions containing cross-linked starch nanoparticles prepared by spray and vacuum freeze drying methods. Carbohydr Polym 90(4):1732–1738

Shi A, Wang L, Li D et al (2013) Suspension of vacuum-freeze dried starch nanoparticles: influence of NaCl on their rheological properties. Carbohydr Polym 94(2):782–790

Simi CK, Abraham TE (2007) Hydrophobic grafted and cross-linked starch nanoparticles for drug delivery. Bioprocess Biosyst Eng 30(3):173–180

Singh V, Ali SZ (1987) Comparative acid modification of various starches. Starch/Stärke 39(11):402–405

Singh V, Ali SZ (2000) Acid degradation of starch. The effect of acid and starch type. Carbohydr Polym 41(2):191–195

Singh V, Ali SZ (2008) Properties of starches modified by different acids. Int J Food Prop 11(3):495–507

Soma CE, Dubernet C, Barratt G et al (1999) Ability of doxorubicin-loaded nanoparticles to overcome multidrug resistance of tumor cells after capture by macrophages. Pharma Res 16(11):1710–1716

Song D, Thio YS, Deng Y (2011) Starch nanoparticle formation via reactive extrusion and related mechanism study. Carbohydr Polym 85(1):208–214

Sun Q, Gong M, Li Y et al (2014) Effect of retrogradation time on preparation and characterization of proso millet starch nanoparticles. Carbohydr Polym 111:133–138

Suzuki T, Chiba A, Yano T (1997) Interpretation of small angle X-ray scattering from starch on the basis of fractals. Carbohydr Polym 34(4):357–363
Tan Y, Xu K, Li L et al (2009) Fabrication of size-controlled starch-based nanospheres by nanoprecipitation. ACS Appl Mater Interfaces 1(4):956–959
Vodalkar M, Thakore S (2010) Isocyanate crosslinked reactive starch nanoparticles for thrmo-responsive conducting application. Carbohydr Polym 345(16):2354–2360
Waigh TA, Hopkinson I, Donald AM (1997) Analysis of the native structure of starch granules with X-ray microfocus diffraction. Macromolecules 30(13):3813–3820
Waigh TA, Donald AM, Heidelbach F et al (1999) Analysis of the native structure of starch granules with small angle X-ray microfocus scattering. Biopolymers 49(1):91–105
Wang YJ, Truong VD, Wang L (2003) Structures and rheological properties of corn starch as affected by acid hydrolysis. Carbohydr Polym 52(3):327–333
Weiss J, Takhistov P, McClements DJ (2006) Functional materials in food nanotechnology. J Food Sci 71(9):R107–R116
Yu Y, Wang J (2007) Effect of γ-ray irradiation on starch granule structure and physicochemical properties of rice. Food Res Int 40(2):297–303
Yu J, Ai F, Dufresne A et al (2008) Structure and mechanical properties of poly(lactic acid) filled with (starch nanocrystal)-graft-poly(ε-caprolactone). Macromol Mater Eng 293(9):763–770
Zheng H, Al F, Chang PR et al (2009) Structure and properties of starch nanocrystal-reinforced soy protein plastics. Polym Compos 30(4):474–480
Zuo JY, Knoerzer K, Mawson R et al (2009) The pasting properties of sonicated waxy rice starch suspensions. Ultrason Sonochem 16(4):462–468

Recent Updates on Molecular Biotechnological Intervention in Isabgol

12

Sundeep Jaglan, Rakesh Yadav, Priyanka Siwach, and Namita Singh

Abstract

Isabgol is a very important native medicinal plant and export commodity of India, which is the largest producer. At this stage, it becomes very important to study the genetic diversity within isabgol gene pool for proper maintenance, conservation, validation, and effective utilization. There are rare reports of DNA fingerprinting of *Plantago ovata* genotypes for determining genetic diversity or varietal identification. In this chapter, we discuss about the role of DNA markers especially random amplified polymorphic DNA (RAPD) and inter-simple sequence repeats (ISSRs) for genetic diversity study for useful crops by taking isabgol as a case study. No other reports of recent markers are available for isabgol. Genetic variation in *Plantago* species is of prime significance for focusing on the constructive breeding of improved varieties/cultivars with added value and durable resistance against diseases.

S. Jaglan
Department of Bio and Nano Technology, Guru Jambheshwar University of Science and Technology, Hisar 125001, Haryana, India

Indian Institute of Integrative Medicine (IIIM), Canal Road, Jammu 180001, Jammu, India

R. Yadav • N. Singh (✉)
Department of Bio and Nano Technology, Guru Jambheshwar University of Science and Technology, Hisar 125001, Haryana, India
e-mail: namitasingh71@gmail.com

P. Siwach
Department of Biotechnology, Chaudhary Devi Lal University, Sirsa 125055, Haryana, India

S.K. Gahlawat et al. (eds.), *Plant Biotechnology: Recent Advancements and Developments*, DOI 10.1007/978-981-10-4732-9_12

12.1 Introduction

Isabgol or blond *psyllium* belongs to genus *Plantago* among the three genera comprising family Plantaginaceae, and the other two are *Littorella* Bergius and *Bougueria* Decne. Out of the ~300 species of *Plantago*, only two species, namely, *P. psyllium* and *P. ovata* (French *psyllium*), are cultivated for their seed husk having wide applications in pharmaceutical and cosmetic industries. Swelling qualities and colorlessness of the finer seed husks of *P. ovata* were the good reasons which displaced French psyllium from the world market.

The diploid chromosome number of *P. ovata* is eight, and the chromosome size varies from 2.5 to 2.9 μm as obtained in Feulgen-stained preparations. The haploid set of chromosome comprises two nonnucleolar and two nucleolar chromosomes which were also identified through FISH and C-banding pattern (Dhar et al. 2002).

Seeds of *Plantago ovata* enclose a plantiose sugar (French et al. 1953), fatty oil, mucilage, a pharmacologically inactive glucoside, namely, aucubin ($C_{13}H_{19}O_8H_2O$), and a large quantity of albuminous matter (Chopra 1930; Khorana et al. 1958). Cold water extracts of seeds contain polysaccharides made up of 20% uronic anhydride, 52% pentosan, and 18% methylpentosan (Laidlaw and Percival 1949). Hot water extracts of seeds contain polysaccharides, i.e., P-I and P-II. Upon hydrolysis, P-I yields D-xylose, L-arabinose, an aldobiuronic acid called 2D galacturonoside-L-rhamnose, and an insoluble residue. P-II yields D-xylose, L-arabinose, and an insoluble residue. The seed oil has two oxygenated fatty acids, namely, 9-hydroxyacetadec-cis-12-enoic acid and 9-oxoacetadec-cis-12-enoic acid (Jamal et al. 1987). The mucilage contained in the husk swells by soaking the seeds in water (Hyde 1970; Karimzadeh and Omidbaigi 2004).

The species under genus *Plantago* are grown for their immense medicinal and other economic values. India holds monopoly in both production and trade of psyllium in the world. The seeds of *P. major*, *P. ovata*, and *P. psyllium* are consumed as laxatives (Wasicky 1961). *Plantago ovata* seed husks mixed with gum acacia are used as ice cream stabilizer (Upadhyay et al. 1978). *Plantago lanceolata* and *P. major* have been utilized as external antirheumatic herbal tonic in the conventional medicine of Italy (Khorana et al. 1958). Biologically active polysaccharides have been isolated from *P. major* (Patel et al. 1979). Cholesterol-lowering properties of the husk of *Plantago ovata* have been well established (Turley et al. 1994). Plantamajoside, a caffeic acid sugar ester from *P. major*, possesses antibacterial activity (Ravn and Brimer 1988). Seeds of psyllium contain 17–19% proteins (Patel et al. 1979). The amino acids extracted from psyllium seeds by paper chromatography include valine, glycine, alanine, cystine, glutamic acid, lysine, tyrosine, leucine, and serine. The seed husks produce 85% polysaccharide (single species) and 15% nonpolysaccharides by giving mild alkali treatment. The mucilage contained in psyllium seeds is an arabinosyl (galactosyluronic acid) rhamnosyxylan (Sandu et al. 1981). It is comparable to methyl cellulose and is better than sodium carboxymethyl cellulose and sodium alginate in suspending properties (Mithal and Gupta 1965).

Many of the problems associated with plant breeding program in the past were based on the phenotypic estimation of an agronomic trait such as environmental effects or qualitative inheritance which can now be eliminated by DNA-based molecular marker diagnostic approaches. These markers are used to obtain information indirectly about the genetics of traits of interest in different genotypes. In the last two decades, a rapid progress has been made toward the development and application of molecular marker technology in plant genome analysis. It has facilitated the analysis and cataloging of available germplasm, variability and DNA fingerprinting studies, phylogenetic analysis, synteny mapping, the development of marker-based gene tagging, and marker-assisted selection (MAS) of desirable genotypes. It is now possible to dissect complex physiological traits such as salt tolerance by using improved methods of identifying and measuring the physiological components (Yeo et al. 1990) and molecular markers (Jansen and Stam 1994; Flowers et al. 2000).

Several different kinds of molecular markers are now available, namely, restriction fragment length polymorphism (RFLP), PCR-based markers such as random amplified polymorphic DNA (RAPD), inter-simple sequence repeats (ISSRs), short sequence repeats (SSRs), sequence-characterized amplified regions (SCARs), inter-simple sequence repeat amplification (ISA), sequence-tagged sites (STSs), amplicon length polymorphisms (ALPs), and amplified fragment length polymorphisms (AFLPs).

RAPD markers are dominant and detect variation in both coding and noncoding regions of genome. It has several advantages as this technique is simple and fast; requires a small quantity of template DNA and no prior sequence information; involves no radioactivity or Southern blotting and hybridization; has high sensitivity, automation, and ease of generating primers; and can be applied at any stage of plant growth. RAPD analysis is technically simple, robust, and provides an approach to characterize the different plant genotypes, thus helps to estimate genetic diversity, which will be very useful in future crop improvement program of plant breeders.

ISSR refers to amplification of DNA fragments near an amplifiable distance in between two similar microsatellite repeat areas arranged in opposite direction. This tool employs microsatellites as primers in a single primer PCR reaction that targets several genomic loci for amplifying mainly ISSRs of discrete sizes. The microsatellite used as primers for amplification can be dinucleotide, trinucleotide, tetranucleotide, or pentanucleotide. The primers employed can be either unanchored (Meyer et al. 1993; Gupta et al. 1994; Wu et al. 1994) or often anchored at 3′ or 5′ end with one to four degenerate bases extended into the bordering sequences (Zietkiewicz et al. 1994). Unlike RAPD primers (10 mers), ISSRs use longer primers of 15–30 nucleotides, which help the successive use of higher annealing temperature that leads to higher inflexibility. The annealing temperature ranges from 45 to 65 °C, which depends on the GC content of the primer employed. The amplified DNA bands usually range between 200 and 2000 bp in length and amenable to resolve by both agarose and polyacrylamide gel electrophoresis.

Although codominant segregation has been obtained in certain cases (Wu et al. 1994; Akagi et al. 1996; Wang et al. 1998; Sankar and Moore 2001), ISSR segregation takes place mostly as dominant markers (Gupta et al. 1994; Tsumura et al. 1996; Ratnaparkhe et al. 1995; Wang et al. 1998). Sometimes like RAPD, DNA fragments with the same size and mobility are amplified from nonhomologous regions (Sanchez et al. 1996). Additionally, the vigorousness, informativeness, and cost effectiveness make ISSR more applicable and a marker of choice.

India is the biggest producer and exporter of isabgol. At this stage it becomes very important to study the genetic diversity within isabgol gene pool for proper maintenance, conservation, validation, and effective utilization. Till now there are few reports of DNA fingerprinting of *Plantago ovata* genotypes for determining genetic diversity or varietal identification by using RAPD and ISSR (Das and Raychaudhuri 2003; Vahabi et al. 2008; Singh et al. 2009). Samantaray et al. (2010) assess genetic relationships in seven species of *Plantago* and evaluate the genetic variance within populations of *P. ovata* (Forsk.), *P. indica* (L.), *P. arenaria* (Waldst.), *P. psyllium* (Linn.), *P. lanceolata* (Linn.) *P. serraria* (Linn.), and *P. coronopus* (Linn.) by using RAPD markers. This chapter includes the detailed study on DNA fingerprinting of superior isabgol varieties for maintaining the germplasm at genetic level so that it could be applied for future improvement in the breeding program of *Plantago ovata.*

12.2 Marker Technology in Molecular Biology

The advancement of techniques with due time led the recombinant DNA technology and molecular biology to open novel opportunities for the genetic tailoring of crop plants. Genetic manipulation further leads to genetic tailoring of crop species to improve their economic traits such as yield and resistance to biotic and abiotic stresses, and thus achieve the final goal of breeding. The selection of individuals or plants with desirable traits is the most fundamental and important step in a plant breeding program. The traditional approach used by breeders is the phenotypic selection where phenotypic or morphological agronomic traits such as plant height and grain yields are kept in mind.

Exploration and exploitation of polymorphism in organisms either prokaryotic or eukaryotic represents one of the most significant developments in biological sciences, so the analysis of genetic diversity and relatedness between or within different species, population, and individuals is a major and central task. The advent of molecular markers has revolutionized the entire scenario of various disciplines such as taxonomy, phylogeny, ecology, genetics, and plant breeding (Weising et al. 1995).

12.3 Molecular Markers or DNA Markers

Molecular markers or DNA markers appeared over the last three decades as a new class of markers which have completely revolutionized the entire area of molecular biology. Although biochemical markers particularly isozymes are still frequently used in many laboratories due to simplicity and low cost, yet DNA markers are more preferred due to the high information content and high quality. They exhibit unique merits over the phenotypic and biochemical markers. They look directly into the basic level of variation which gives insight into the genetic makeup, screen the whole genome, and uncover variations in both coding and noncoding regions and thus present high level of polymorphism. It is highly important in plants as it is the only marker technique to calculate organelle genome diversity as many traits are regulated by cytoplasmic inheritance (mitochondria and chloroplast). Study of chloroplast genome is valuable in phylogenetics at taxonomic level because these genomes are conserved in comparison with mitochondrial and nuclear genomes.

An ideal DNA marker should bear some desirable characters as follows:

1. A DNA marker should be frequently and evenly spread across the genome.
2. It must be highly polymorphic as it represents polymorphism which is assessed for genetic diversity analysis.
3. The DNA sequences of all organisms are indeterminate to management practices or environmental conditions.
4. Codominant inheritance.
5. Easy and rapid assay.
6. Easy accessibility.
7. Easy sharing of data between researchers and laboratories.
8. High reproducibility.

It is enormously tough to discover a molecular marker which would meet all desirable measures. Diverse molecular approaches are available that reveal polymorphism at DNA level. A marker system can be known that fulfill at least a few of the desired characteristics depending upon the nature of study to be carried out (Weising et al. 1995).

12.3.1 Random Amplified Polymorphic DNA (RAPD) Markers

Random amplified polymorphic DNA (RAPD) study, a PCR-based molecular marker technique, was first developed originally by Williams et al. (1990), Welsh and McClelland (1990). These markers are formed as a result of PCR amplification (Saiki et al. 1985, 1988) of random genomic DNA sequences with single primers (generally ten nucleotides long) of arbitrary sequence. The primers are generated with at least 60% G + C content to ensure efficient annealing such that the sequences are not capable of internal pairing that can produce PCR artifacts. The primers target complexes that are employed as substrates for Taq DNA polymerase to copy the genomic sequence at 3′ end of the primer. This procedure yields a distinct set of amplified DNA products (amplicons) that represent target sequence flanked by antiparallel-oriented primer binding sites within an amplifiable distance. Amplified products can be separated by electrophoresis on agarose or polyacrylamide gels and visualized by staining with ethidium bromide or by silver staining methods. RAPD are usually dominant markers with polymorphisms between individuals defined as the presence or absence of a particular RAPD band (Staub et al. 1996). There are distinct advantages of RAPD over conventional RFLP technology which includes (Waugh and Powell 1992):

- Requirement of a small amount of genomic DNA (25–100 ng per reaction) compared to 5–10 μg for RFLP analysis.
- An ethidium-bromide-based detection system.
- Many primers can be screened on a single PCR run (Gale and Witcombe 1992).
- RAPD may provide markers in regions of genomic DNA inaccessible to RFLP analysis due to the presence of repetitive DNA sequence (Williams et al. 1990).

RAPD markers have been proved to be four to six times more efficient than screening for the same polymorphisms using RFLP technology and over tenfold more efficient in terms of time and labor involved. RAPD overcomes many of the technical limitations of RFLP analysis. It requires neither previous knowledge of any genomic sequence such as general PCR nor tedious procedure such as of RFLP analysis (Tingey and del-Tufo 1993). RAPD is a useful genetic marker for a variety of eukaryotic organisms including bacteria, fungi plants, and animals (Welsh and McClelland 1990).

12.3.1.1 Applications of RAPDs in Plant Breeding for Making Genetic Maps and Gene Introgression

Molecular markers offer an excellent prospect for developing saturated genetic maps and to integrate molecular, genetic, and cytological maps. Comprehensive molecular maps have become available in maize (Weber and Helentjaris 1989), rice (Causse et al. 1994, Kurata et al. 1994), and several other crop plants. Over 1000 markers have been mapped on the tomato genetic map (Tanksley 1993).

RAPD markers can be used in gene mapping, gene tagging phylogeny, and evaluation studies in crops such as soybean (Zhang and Gali 1997) and chickpea (Banerjee et al. 2001). Studies also analyzed the association of three morphological traits linked with RAPD markers. These markers can also be used to demonstrate that dihaploids produced after interspecific pollination of the tetraploid *Solanum tuberosum* cv. Pentland crown with *Solanum phureja* dihaploid inducer clones were not of parthenogenetic origin, thus detecting gene introgression in potato (Waugh et al. 1992). RAPD markers have been mapped on all 12 chromosomes from DNAs of cultivars Nipponbare (*Japonica*) and Kasalath (*Indica*) and of F_2 population generated by single cross of these parents using 60 random primers (Monna et al. 1994). The loci of RAPD markers were determined on to the frame work of RFLP linkage map.

12.3.1.1.1 RAPD Mapping of Quantitative Trait Loci (QTLs)

An individual locus regulating a quantitative character is known as polygene or quantitative trait locus (QTL). The objective of QTL mapping is to discover DNA diagnostic markers for a particular phenotype in a study so that marker-assisted selection (MAS) can be employed. For most quantitative traits, high-density RAPD maps provide an opportunity to identify, measure, and map their effects by detecting QTLs. Methodologies are available to detect and map QTL involved in sampling strategies, different populations, types of population, and threshold levels for detecting QTL and missing data (Paterson et al. 1990; Stuber et al. 1992; Tanksley 1993; Zheng 1994). One method to work out the map position of a QTL is by revealing a set of near-isogenic lines (NILs) which vary for markers flanking the QTL (Paterson et al. 1990, Kaeppler et al. 1993). This method can spot out a small region of the genome linked with a quantitative trait, thus defining more accurately the map position for a specific QTL. Backcross introgression has been the most frequently used approach to develop NILs for QTL studies in *Teosinte glume* (Dorweilar et al. 1993), soybean (Muchlbauer et al. 1991), and tomato (Paterson et al. 1990). The backcross method was employed to introgress a small fragment of donor genome imparting a specific allele for QTL into a recipient genome with a discrete allele.

RAPD markers help in studying quantitative trait loci (QTLs) in two diverse strains of pigeon peas and the F_1 and F_2 progeny. But the polymorphism index among parents was very low. F_1 was transitional between two parents, but F_2 exhibited little variation, signifying that both parents were distinctive morphologically but were found to have little genetic variation at DNA level (Tyagi 1997). Similarly, RAPD markers can also be employed for investigating QTLs in chickpea (*Cicer arietinum*); out of seven morphological characters studied for linkage with the associated markers in the segregating plants, four QTLs were identified for the character of leaf length and three QTLs each for the character of erect plant habit and leaf width (Banerjee et al. 2001).

12.3.1.1.2 Studies on Genetic Diversity and DNA Fingerprinting

The maintenance and evaluation of germplasm collections is a major feature of many international agricultural research centers (Williams 1989). The extent of the analysis and allocation of genetic diversity in crop plants are important in figuring out the evolutionary relationships among accessions and to sample genetic resources in an efficient manner for conservation and breeding function.

In the sequence of the development of a molecular toolkit for studying genetic diversity among huge plant germplasm collections, RAPD technology has been employed to genotypes of rice obtained from International Rice Research Institute (Virk et al. 1995). An RAPD protocol was optimized for studying diversity in germplasm collections and also identified duplicated accessions in germplasm collections. Various broccoli and cauliflower cultivars (*Brassica oleracea* L.) can be discriminated by banding patterns of RAPD marker. The markers produced by two and three primers were adequate to identify each of the broccoli and cauliflower genotypes. Devos and Gale (1992) optimized RAPD protocol in wheat with respect to DNA concentration, Mg^{+2} concentration, Taq DNA polymerase concentration, and denaturing temperature, and concluded that though RAPD was a sensitive process, it has many applications in the analysis of genotypes (Hu and Quiros 1991).

Similarly, genetic distances, discrimination among cultivars, and phylogenetic relationship were studied in other crops. Polymorphism can be established among various species of genus *Hordeum* by RAPD analysis utilizing numerous arbitrary primers (De Bustos et al. 1998), between various coffee species and *Coffea arabica* genotypes (Orozco-Castillo et al. 1994), various cultivars of *Gossypium hirsutum* (Multani and Lyon 1995), apple cultivars (Koller et al. 1993), papaya (Stiles et al. 1993), mungbean (Santalla et al. 1998), *Glycine soja* (Lee 1998), *Phaseolus lunatus* (Fofana et al. 1997), *Trifolium incarnatum* (Steiner et al. 1998), cowpea (Mignouna et al. 1998), *Phaseolus vulgaris* (Duarte et al. 1999; Beebe et al. 2000), *Medicago polymorpha* L. (Paredes et al. 2002), Malawian cowpea (*Vigna unguiculata*; Nkongolo 2003), *Cyamopsis tetragonoloba* (Punia et al. 2009a, b; Kumar et al. 2015) *P. ovata* (Vala et al. 2011; Rohilla et al. 2012; Kaswan et al. 2013; Kumar et al. 2014), and more recently, Kour et al. 2016 deduced genetic distances among different accessions of *P. ovata* and different species of *Plantago* by employing RAPD primers.

Based upon data, dendrogram can be constructed showing the relationship between cultivars. Ancestry study identified via a minor set of RAPD primers established the associations among selected plant introductions (PIs) and major North American soybean (*Glycine max*) ancestors. Two hundred and eighty-one polymorphic RAPD fragments assessed on 35 ancestors and PIs were selected for reproducibility and diversity (Thompson et al. 1998).

Recognition of cytoplasmic male fertile and male sterile lines in pigeon pea has been achieved by Souframanien et al. (2003). Phylogenetic tree based on the similarity index exhibited that significant genetic variation existed between CMS lines, two putative R lines, and wild species.

Establishing the identity of a given crop variety has assumed a great importance for protecting plant breeders and farmer rights particularly in developing countries. All unambiguous characteristic patterns of crop varieties can be obtained using DNA markers, and the technique has been termed as DNA fingerprinting. Jeffery et al. (1985) developed the technique of DNA fingerprinting in human beings using hypervariable minisatellite DNA. Dallas (1988) demonstrated the usefulness of DNA fingerprinting technique for cultivar identification for the first time in rice using human minisatellite DNA probes. Smith and Smith (1992) have reviewed the utility of DNA fingerprinting for characterizing the genetic diversity and relatedness in plant populations.

Besides helping in varietal identification, the fact that DNA fingerprinting has numerous other applications in crop improvement has been reported by Smith and Smith (1992). They are devoted for studying genetic diversity within a taxa; evolutionary relationship; assessment of genetic purity of inbred lines and varieties; selection of recurrent parental genome in backcross and segregating generations, thereby decreasing the number of generations required for breeding pure-line varieties; paternity analysis in perennials and tree crops; identification of hybrids and maternal seedlings in plants with polyembryony, e.g., citrus and mango; monitoring genetic stability of germplasm conserved in the form of seeds and tissues; the detection of somaclonal variants; and screening of duplicate accessions in gene banks, etc. Hence DNA fingerprinting plays an important role in plant variety protection, registration, certification, and patents.

12.3.2 Inter-Simple Sequence Repeat (ISSR) Marker

The first report of using ISSR markers was available in 1994, which was on studies of cultivated genotypes and demonstrated the hypervariable nature of ISSR markers. For testing the utility of the tool in natural populations, an identified hybrid complex of four species of *Penstemon* having the availability of three additional molecular data sets was reexamined. The results evidently confirmed the use of ISSR markers in connection with queries of hybridization and diploid hybrid speciation. It is virtually identical to RAPD techniques except that ISSR primer sequences have higher annealing temperatures and are nonrandom designed from microsatellite regions which make them more suitable than RAPD markers. ISSR markers have huge potential for examining natural populations because of the published, unpublished, and in-progress studies conducted so far by using these markers. The primer-annealing sites are disseminated squarely all over the genome so that the primer will anneal to two sites orientated on antiparallel DNA strands. The section between the two primers will be amplified through PCR, if these are within an appropriate distance from one another. The section would not be amplified if structural rearrangements of the chromosomes had occurred, divergence exists at the primer binding sites, or binding site was lost. ISSRs are potentially used for natural populations to analyze hybridization issues. These studies advocate the wider application of the technique for plant families. The ISSR marker has its specific advantages over other molecular markers including RAPD, SSR, and AFLP.

Following are few advantages of ISSR markers:

1. Greater stability
2. Prior sequence information not required
3. Acquiescent to laboratory level
4. Simple and rapid operation
5. Plenty of genomic information
6. Exhibit high polymorphism
7. No radioactivity usage is required

ISSR marker infiltrates variation in several microsatellite areas widely dispersed in the various genomes and outwits the challenge of characterizing individual loci. Microsatellites are brief, hypervariable stretches of DNA which are expressed as diverse variants among different species and within populations. Especially they bear mono-, di- or trinucleotide repeats, e.g., AA… or AG… CAG… that have four to ten repeat units side by side. ISSRs are specifically targeting the di- and trinucleotide repeat types of microsatellite which are characteristic of the nuclear genome.

The ISSR primers are applied to deduce the variation in a DNA sample having any highly variable microsatellite sequences and an arbitrary pair of nucleotide bases at the 3′ (rear) end. Single primer is used at a time for finding variation among DNA samples in small PCR reactions where the primer effectively locates two microsatellite regions within an amplifiable distance far on the DNA strands of the sample and the PCR reaction will produce a band of a specific molecular weight for that "locus" on behalf of the prevailing stretch of DNA between the microsatellites. Usually such "paired" microsatellite areas exist in a particular DNA sample several times, so one obtains many bands generated in the reaction for a sample.

The application of ISSRs has been verified in numerous studies such as gene tagging (Ammiraju et al. 2001; Marczewski et al. 2002), detection of somaclonal variation (Leroy et al. 2000), cultivar identification (Nagaraju et al. 2002), phylogenetic studies (Paris et al. 2003; Han and Wang 2010), genetic stability (Yuan et al. 2009; Zhang et al. 2010; Lata et al. 2010), hybrid identification (Lin et al. 2010), determining production quality (Tamhankar et al. 2009; Wu et al. 2010), and genetic relatedness (Rajwade et al. 2010). ISSRs have been used for assessing genetic diversity studies in *Malus* sp. (Goulão and Oliveira 2001), *Pisum sativum* (Baranger et al. 2004), *Morus alba* (Awasthi et al. 2004), *Cicer* sp. (Souframanien and Gopalakrishna 2004), *Tribulus terrestris* (Sarwat et al. 2008), *Lobelia* sp. (Geleta and Bryngelsson 2009), *Cynodon* (Gulsen al. 2009), wheat (Carvalho et al. 2009; Thomas and Bebeli 2010), *Dioscorea alata* (Wu et al. 2009), *Punica granatum* (Narzary et al. 2010), *Salvia miltiorrhiza* (Song et al. 2010), and *Panax ginseng* (Reunova et al. 2010). They have also been employed for the analysis of repeat motifs in mungbean (Singh et al. 2000) and for establishing the population molecular identification and genetic structure of medicinal plant *Changium smyrnioides*, *Chuanmin shenviolaceum* (Qiu et al. 2004), and *Jatropha* (Senthil Kumar et al. 2009). ISSRs are highly informative as compared to RAPDs (Korbin et al. 2002; Galvan et al. 2003) and exhibit more polymorphic information content (PIC) than AFLPs (Blair et al. 1999).

12.3.2.1 ISSR Markers in Genetic Diversity Analysis

Genetic diversity designates any variation in the nucleotides, genes, chromosomes, or whole genomes of the organism. Higher scale of genetic diversity exists in non-cultivated medicinal plants across the geographic scale, which can be assessed by ISSR marker. Genetic diversity study is among the major application of ISSR marker, and it is also used for medicinal plant study. ISSR markers were employed for analyzing genetic diversity of *Swertia chirayita* cultivars collected from the temperate Himalayan region of India. Allied species of *S. chirayita* were also involved in the study as outliers. ISSR study revealed high polymorphism (~98.7%) representing a high genetic diversity among the genotypes assayed (Joshi et al. 2007). Genetic diversity in wild populations of *Glycyrrhiza uralensis* Fisch was detected by using various screened and optimized ISSR primers (Yao et al. 2008). The comparative genetic diversity in *Humulus lupulus* was undertaken using AFLP, RAPD, ISSR, and STS molecular markers. All molecular approaches precisely distinguish all tested genotypes, excluding Osvald's clones (only distinguishable by AFLP;

Josef 2001). Geographic differentiation and genetic diversity of disjunct *Psychotria ipecacuanha* (Rubiaceae) was studied by ISSR marker system. Larger number of amplification bands obtained by ISSR primers was polymorphic (Ana et al. 2009). Hybrid performance and genetic diversity of *Benincasa hispida* inbred lines were studied by ISSR and RAPD markers. Out of 42 primers tested, all produced polymorphic and reproducible bands except only 4 which produced monomorphic bands. Out of 26, 11 marker bands were polymorphic, which were produced by five ISSR primers. Genetic diversity of cashew germplasm has been studied by using RAPD and ISSR markers (Thimmappaiah et al. 2009). Selective amplification of microsatellite polymorphic loci (SAMPL), AFLP, RAPD, and ISSR markers help for the uncovering genetic polymorphism in *Tribulus terrestris* medicinal plant. The samples were collected from diverse geographic locations of India. Assays of AFLP, ISSR, SAMPL, and RAPD markers each interprets polymorphic nature of the population (Sarwat et al. 2008). Genetic characterization in *Cyamopsis tetragonoloba* has been efficiently undertaken by ISSR marker (Punia et al. 2009a, b). In a separate study, assessment of *Mactra veneriformis* using ISSR markers not only reveals high polymorphism but also overcomes limitation of allozyme and RAPD technique (Hou et al. 2006). ISSR marker can be used to estimate genetic diversity within and among various populations of *Rhodiola chrysanthemifolia* (Xia et al. 2007). Genetic diversity in Indian bitter gourd (*Momordica charantia* L.) was achieved using ISSR and RAPD markers. Seventy-six amplicons were polymorphic out of total 208 amplicons while assessing 38 *M. charantia* accessions with 29 RAPD primers, whereas 125 bands were produced of which 94 were polymorphic by 15 ISSR primers in the accessions studied (Behera et al. 2008). The genetic diversity study of *Vanilla planifolia* was undertaken by employing ISSR primers. Ten out of 11 ISSR primers amplified 108 marker bands out of which 93 were polymorphic (86.11%) (Verma et al. 2009). Molecular markers can resolve interspecific and intraspecific genetic variability in *Plantago ovata* (Kotwal et al. 2013).

12.3.2.2 ISSR Markers in Authentication of Plants

Since several important medicinal plants have been adulterated naturally or artificially, these can also be authenticated to know correct botanical identification for obtaining better-quality herbal drugs by using ISSR markers. Identification of adulteration is thereby serving improvement in quality control. Use of ISSR markers in pharmacognosy is not new. Correct identification of most popular mushroom *Flammulina velutipes* was undertaken by using strain-specific sequence-characterized amplified region (SCAR) markers where eight primers selected out of 20 amplified 104 distinct and stable bands among which 81 were polymorphic (Hongyan et al. 2008). Microsatellite markers which were able to amplify the corresponding microsatellite loci from five different spp. of *Eucalyptous*, namely, *E. grandis*, *E. nitens*, *E. globulus*, *E. camaldulensis*, and *E. urophylla*, were developed for their genetic study because RAPD results were not reproducible. DNA analysis of disputed chili samples (*Capsicum annum*) was achieved by employing FISSR-PCR and ISSR-PCR markers. Fifteen ISSR (seven di- and eight trinucleotide) anchored primers amplified a total of 212 and 288 bands respectively. Four out of eight trinucleotide primers and five out of nine dinucleotide primers clearly differentiated all the four disputed chili samples. The FISSR-PCR assay developed a total of 566 bands by employing three tri- and one dinucleotide primers. These four primers were highly reliable to identify and discriminate all the four disputed samples clearly (Kumar et al. 2001).

12.3.2.3 ISSR Markers in Identification of Plants

ISSR markers are very important for correct botanical identification of plants. They can undoubtedly distinguish inter- and intraspecies distinction. There are several case studies including ISSR markers which are employed for cultivar or species identification. A successful example of identification by means of ISSR markers was of Mediterranean *Dentex dentex* (Sparidae) and *Diplodus* spp., where eight primers amplified a total of 97 fragments out of which 95 referred by genotype to a distinct species (97.9%) were polymorphic in nature (Casu et al. 2009). RAPD, SRAP (sequence-related amplified polymorphism), and ISSR were used for genetic diversity study and identification of 35 elite late-bolting radish genotypes. Detected by 35 RAPD, 17 SRAP, and 22 ISSR primer combinations, the proportions of polymorphic bands obtained were 85.44%, 85.41%, and 85.2% respectively, and the mean genetic similarity coefficients among pairs of cultivars were 0.781, 0.764, and 0.787 respectively. Each molecular marker system could efficiently distinguish all the cultivars. The polymorphism level in tomato (*Lycopersicon esculentum*) was analyzed by employing ISSR-PCR markers. Five tomato species, *Lycopersicon esculentum*, *Lycopersicon humboldtii*, *Lycopersicon pennellii*, *Lycopersicon cheesmanii*, *Lycopersicon esculentum* and *Lycopersicon hirsutum*, and two substitution lines WSL 6 and IL 6–3 were studied with 14 ISSR-PCR primers. Among them, nine primers were individually capable to differentiate all tomato species (Tikunov et al. 2003). Twelve accessions in eight related *Solanum* species and eight cultivars of eggplant (*Solanum melongena*) were analyzed by ISSR for evaluating the phylogenetic relationships and identifying cultivars. Thirty-four out of the 100 primers tested produced a total of 552 polymorphic bands, and the percentage of polymorphisms obtained was quite high: ~99.1% (Isshiki et al. 2008).

12.3.2.4 ISSR in Germplasm Authentication

Over the last several decades, the impact of research on clonal plants has been globally cherished, and immense improvement has been made in studying the morphology, evolution, ecology, and physiology of clonal plants. As on today, high interest in clonal plants has focused on studying the influence of their pattern of development on growth, capturing resources, and responding to environmental variation in space and time. Even though allozyme study has long been employed to detect clones and to reveal their population genetics, it often underrates genetic polymorphism and has a restricted ability to distinguish individuals genetically. Recently, several PCR-based molecular markers such as RAPD, ISSR, and SSR have been widely employed to study population genetic structure and clonal diversity which cope with the limitations associated with allozyme markers. Genetic fidelity of micropropagated Gerbera (*Gerbera jamesonii*) clones was performed by employing ISSR markers. Out of 15, 12 ISSR primers exhibited monomorphic behavior among in vitro raised clones and the control plant. However, polymorphic bands were obtained with three ISSR primers (Bhatia et al. 2009). Tung tree (*Vernicia fordii*) is an essential oil-rich woody plant. For determining the genetic diversity, breeding method, and germplasm resource in tung tree, ISSR was employed to study the cultivars in China. Ninety bands were polymorphic out of 110 bands amplified with 12 primers. Both PCA and UPGMA cluster showed distinct genetic correlation among the 64 tung cultivars (Li et al. 2008). Genetic fidelity study of 20 commercial varieties and 20 exotic germplasm lines of the two cultivated species (*Corchorus capsularis* and *C. olitorius*) and two wild relatives of jute (*C. trilocularis* and *C. aestuans*) was undertaken by employing ISSR, sequence-tagged microsatellite site (STMS), and RAPD markers. The 22 RAPD and four ISSR primers used in the study gave 100% and 98.44% polymorphism respectively among all the species, while the polymorphism index was considerably low within a species (Roy et al. 2006). RAPD markers have also been used to validate the clonal purity of tissue culture raised plants in *Lilium* sp. (Yadav et al. 2014) and *Acacia auriculiformis* (Yadav et al. 2015). In the coming years, there can be many applications to this field.

12.3.2.5 ISSR Markers in Plants Genotyping

Genetic structure of a population usually determines nonrandom distribution of genetic variation. Geographic distance is commonly considered as an important factor affecting both gene flow and genetic structure because distance limits the movements of propagules, gametes, and individuals that alter the spatial distributions of genes. The relation between distance scales and genetic variation has been analyzed for populations of several organisms, including a few species of seaweed. Numerous DNA markers such as AFLP, RAPD, ISSR, and SSR have been employed in the analysis of the population genetics. Besides this, these markers can be helpful for correlating plant chemotypes and genotypes, which is very useful for evaluating the quality of medicinal species. Twelve individual cultivars chosen from Juniperus populations, varieties, and species were studied by using ITS sequences, ISSRs, RAPDs, and leaf volatile terpenoids. These data sets can be employed at diverse organizational levels: specific, intraspecific, and interspecific (Adams et al. 2003). ISSR study was employed to reveal genetic variations of 184 haploid and diploid samples obtained from nine North Atlantic *Chondrus crispus* population. Twenty-two out of 50 primers were screened, and 163 loci were obtained in genetic diversity study. Genetic diversity was variable among populations where percentage of polymorphic bands ranged from 27.0 to 55.8% (Wang et al. 2008). Genetic linkage map of cucumber (*Cucumis sativus* L.) consisting of 33 RAPDs, 116 SRAPs, 9 SCARs, 11 SSRs, 3 ISSRs, and 1 STS was constructed by 130 F2 progeny originated from a narrow cross between line S06 (greenhouse European type) and line S94 (Northern China open-field type) (Yuan et al. 2009). To retrieve the genetic integrity of the nuclear, chloroplast, and mitochondrial genomes among the hardened regenerants of *Coffea arabica*, DNA markers (RFLP, ISSR, RAPD) were used for sampling different regions of the genome. A total 480 genetic loci obtained from 16 nuclear, chloroplast, and mitochondrial gene probes along with 38 RAPD, 9 restriction enzyme digests, and 17 SSR primers were screened in 27 somatic embryo-derived plants and a control. Among all, 44 loci were found to be polymorphic. A relatively low polymorphism (4.36%) was obtained in the nuclear genome, whereas polymorphism in the mitochondrial genome (41%) was in higher range (Rani et al. 2000).

12.4 Limitations of DNA-Based Molecular Markers

1. The main demerit of RAPDs is low reproducibility factor (Schierwater et al. 1993), and consequently highly optimized experimental protocols are needed due to their sensitivity to the PCR conditions. Short arbitrary primers are used for the analysis that are capable of amplifying DNA bands in several organisms; hence, RAPD analyses generally need purified, high-molecular-weight DNA to evade contamination of samples. In general, the inborn problems of reproducibility lead to unsuitability of RAPD markers for comparison or transference of results among researchers working in similar species. RAPD markers are not locus specific as that of the other multilocus techniques due to which its banding profile cannot be inferred in terms of loci and alleles (dominance of markers), and similar-sized PCR products may not be homologous. But RAPD markers are ease to carry out by various laboratories, but reproducibility was not obtained to a reasonable level (Jones et al. 1997), and therefore, the method cannot be preferred for regular identifications.
2. Similarly ISSR is also a multilocus technique; drawback includes the probable nonhomology of comparable-sized fragments. Furthermore, ISSRs can also show reproducibility problems.

12.5 Conclusion

Agriculture is a highly important occupation in India as around 70% of the population rely on it. A remarkable research in traditional breeding for many decades has made this country self-contained in many respects. Conversely, the ever-increasing population has troubled food security in India, and efforts have been initiated to combine modern biotechnology approaches in traditional breeding for improving the essential crops such as wheat, rice, legumes, and medicinal plants of immense values.

Although significant progress has been made in other legumes such as chickpea, cowpea, mungbean, pigeon pea, and soybean, and other cereal and vegetable crops such as rice, maize, and wheat using various DNA-based molecular markers as reviewed above, there has been limited report on DNA-based molecular marker study in *Plantago ovata.* Numerous efforts have been made through conventional breeding methods to ascertain genetic variance among isabgol genotypes, but the assessment of genetic diversity, genetic relatedness, and cataloging at molecular level has not been reported so far. No reports are available for isabgol molecular markers other than RAPD, SSR, and ISSR.

The maintenance and evaluation of germplasm collections is one of the main priorities of the international agricultural researches. The extent of analysis and distribution of genetic diversity in crop plants are of paramount importance in understanding the evolutionary relationship between accessions and to catalog genetic resources in a more systemic fashion for breeding and conservation purposes. Phylogenetic relationships illustrate the evolutionary process and biological conservation and also provide information for the selection of parents in conventional breeding.

Thus, DNA-based molecular markers (RAPD and ISSR) enable for characterizing the superior genotypes of *Plantago ovata* using and creating data base, which would be very useful for future isabgol breeding and improvement programs, so that it could lead to boost the production and export capabilities for isabgol.

References

Adams RP, Andrea E, Schwarzbach R, Pandey N (2003) The concordance of terpenoid, ISSR and RAPD markers and ITS sequence data sets among genotypes: an example from *Juniperus*. Biochem Syst Ecol 31:375–387

Akagi H, Yokozeki Y, Inagaki A, Nakamura A, Fujimura T (1996) A codominant DNA marker closely linked to the rice nuclear restorer gene, *Rf-1*, identified with inter-SSR fingerprinting. Genome 39:1205–1209

Ammiraju JSS, Dholakia BB, Santra DK, Singh H et al (2001) Identification of inter simple sequence repeat (ISSR) markers associated with seed size in wheat. Theor Appl Genet 102:726–732

Ana A, Bandini R, Luiz Orlando de O, Bruna AV, Silva RS (2009) Genetic diversity and geographic differentiation of disjunct Atlantic and Amazonian populations of *Psychotriai pecacuanha* (Rubiaceae). Genetica 136:57–67

Awasthi AK, Nagaraja GM, Naik GV et al (2004) Genetic diversity and relationships in mulberry (genus *Morus*) as revealed by RAPD and ISSR marker assays. BMC Genet 5:100–104

Banerjee H, Pai RA, Moss JP, Sharma RP (2001) Use of random amplified polymorphic DNA markers for mapping the chickpea genome. Biol Plant 44:195–202

Baranger A, Aubert G, Arnau G, Lainé AL et al (2004) Genetic diversity within *Pisum sativum* using protein and PCR-based markers. Theor Appl Genet 108:1309–1321

Beebe S, Skroch PW, Tohme J et al (2000) Structure of genetic diversity among common bean landraces of Middle American origin based on correspondence analysis of RAPD. Crop Sci 40:264–273

Behera TK, Singh AK, Staub JE (2008) Comparative analysis of genetic diversity in Indian bitter gourd (*Momordica charantia* L.) using RAPD and ISSR markers for developing crop improvement strategies. Sci Hortic 115:209–217

Bhatia R, Singh KP, Jhang T, Sharma TR (2009) Assessment of clonal fidelity of micropropagated gerbera plants by ISSR markers. Sci Hortic 119:208–211

Blair MW, Panaud O, McCouch SR (1999) Inter-simple sequence repeat (ISSR) amplification for analysis of microsatellite motif frequency and fingerprinting in rice (*Oryza sativa* L). Theor Appl Genet 98:780–792

Carvalho A, Lima-Brito J, Ma ÃÃB, Guedes-Pinto H (2009) Genetic diversity and variation among botanical varieties of old Portuguese wheat cultivars revealed by ISSR assays. Biochem Genet 47:276–294

Casu M, Tiziana L, Marco CG, Alberto R, Antonio P (2009) Identification of Mediterranean Diplodus spp and *Dentex dentex* (Sparidae) by means of DNA Inter-Simple Sequence Repeat (ISSR) markers. J Exp Mar Biol Ecol 368:147–152

Causse MA, Fulton TM, Cho YG, Alin SN et al (1994) Saturated molecular map of the rice genome based on an interspecific backcross population. Genetics 138:1251–1274

Chopra RN (1930) *Plantago ovata* (Isbaghul) in chronic diarrhoeas and dysenteries. Ind Med Gaz 65:428–433

Dallas JF (1988) Detection of DNA fingerprints of cultivated rice by hybridization with a human minisatellite DNA probe. Proc Natl Acad Sci U S A 85:6831–6835

Das M, Raychaudhuri SS (2003) Estimation of genetic variability in *Plantago ovata* cultivars. Biol Plant 47(3):459–462

De Bustos, Casanova A, Soler C, Jouve N (1998) RAPD variation in wild populations of four species of the genus *Hordeum* (Poaceae). Theor Appl Genet 96:101–111

Devos KM, Gale MD (1992) The use of random amplified polymorphic DNA markers in wheat. Theor Appl Genet 84:567–572

Dhar MK, Kaul S, Friebe B, Gill BS (2002) Chromosome identification in *Plantago ovata* Forsk through C-banding and FISH. Curr Sci 83(2):150–152

Dorweilar J, Stec A, Kermiele J, Doebley J (1993) *Teosinte glune* architecture I: a genetic locus controlling a key step in maize evaluation. Science 262:233–235

Duarte JM, Santos JB, Melo LC (1999) Genetic divergence among common bean cultivars from different races based on RAPD markers. Genet Mol Biol 22:419–426

Flowers TJ, Koyama ML, Flowers SA et al (2000) QTL: their place in engineering tolerance of rice to salinity. J Exp Bot 51:99–106

Fofana B, Vekemans X, Jardin P, Baudoin JP (1997) Genetic diversity in lima bean (*Phaseolus lunatus* L.) as revealed by RAPD markers. Euphytica 95:157–165

French D, Wild GM, Young, James WJ (1953) Constitution of plantiose. J Am Chem Soc 75:709–712

Gale MD, Witcombe JR (1992) DNA markers and marker-mediated applications in plant breeding with particular reference to pearl millet breeding. In: Moss JP (ed) Biotechnology and crop improvement in Asia, ICRISAT, Patancheru, AP 502324, India, pp 323–332

Galvan MZ, Bornet B, Balatti PA, Branchard M (2003) Inter simple sequence repeat (ISSR) markers as a tool for the assessment of both genetic diversity and gene pool origin in common bean (*Phaseolus vulgaris* L). Euphytica 132:297–301

Geleta M, Bryngelsson T (2009) Inter simple sequence repeat (ISSR) based analysis of genetic diversity of *Lobelia rhynchopetalum* (Campanulaceae). Hereditas 146:122–130

Goulão L, Oliveira CM (2001) Molecular characterization of cultivars of apple (*Malus×domestica* Borkh) using microsatellite (SSR, ISSR) markers. Euphytica 122:81–89

Gulsen O, Sever-Mutlu S, Mutlu N, Tuna M et al (2009) Polyploidy creates higher diversity among *Cynodon* accessions as assessed by molecular markers. Theor Appl Genet 118:1309–1319

Gupta M, Chyi YS, Romero-Severson J, Owen JL (1994) Amplification of DNA markers from evolutionarily diverse genomes using single primers of simple-sequence repeats. Theor Appl Genet 89:998–1006

Han Y, Wang HY (2010) Genetic diversity and phylogenetic relationships of two closely related northeast China *Vicia* species revealed with RAPD and ISSR markers. Biochem Genet 48:385–401

Hongyan S, Wang L, Liu L, Chi X, Zhang Y (2008) Use of inter-simple sequence repeats markers to develop strain-specific SCAR markers for *Flammulina velutipes*. J Appl Genet 49(3):233–235

Hou L, Hongli L, Zou L et al (2006) Genetic characterizations of *Mactra veneriformis* (Bivalve) along the Chinese coast using ISSR-PCR markers. Aquaculture 261:865–871

Hu J, Quiros CF (1991) Identification of broccoli and cauliflower cultivars with RAPD markers. Plant Cell Rep 10:505–511

Hyde BB (1970) Mucilage producing cells in the seed coat of *Plantago ovata*: development fine structure. Am J Bot 57(10):1197–1206

Isshiki S, Iwata N, Mizanur M, Khan R (2008) ISSR variations in eggplant (*Solanum melongena* L.) and related Solanum species. Sci Hortic 117:186–190

Jamal SI, Ahmad R, Agrawal M, Ahamed OSM (1987) A novel oxo fatty acid in Plantago ovata seed oil. Phytochem 26(11):3067–3069

Jansen RC, Stam P (1994) High resolution of quantitative traits into multiple loci via interval mapping. Genetics 136:1447–1455

Jeffery AJ, Wilson V, Thein SL (1985) Hypervariable minisatellite regions in human DNA. Nature 314:67–72

Jones SD, Wipff JK, Montgomery PM (1997) Vascular plants of Texas: a comprehensive checklist including synonymy, bibliography, and index. University of Texas Press, Austin

Josef P (2001) Comparison of RAPD, STS, ISSR and AFLP molecular methods used for assessment of genetic diversity in hop (*Humulus lupulus* L.) Euphytica 121:9–18

Joshi AK, Kumari M, Singh VP, Reddy CM, Kumar S, Rane J, Chand R (2007) Stay green trait: variation, inheritance and its association with spot blotch resistance in spring wheat (Triticum aestivum L.) Euphytica 153:59–71

Kaeppler SM, Phillips RL, Kim TS (1993) Use of near isogenic lines derived by backcrossing or selfing to map qualitative traits. Theor Appl Genet 87:233–237

Karimzadeh G, Omidbaigi R (2004) Growth and seed characteristics of Isabgol (*Plantago ovata* Forsk) as influenced by some environmental factors. J Agric Sci Technol 6:103–110

Kaswan V, Joshi A, Maloo SR (2013) Assessment of genetic diversity in Isabgol (*Plantago ovata* Forsk.) using 19 random amplified polymorphic DNA (RAPD) and intersimple sequence repeat (ISSR) markers for developing 20 crop improvement strategies. Afr J Biotechnol 12(23):3622–3635

Khorana ML, Prabhu VG, Rao MRR (1958) Pharmacology of an alcoholic extract of *Plantago ovata*. Indian J Pharm 20:3–6

Koller B, Lehmann A, McDermott JM, Gessler C (1993) Identification of apple cultivars using RAPD markers. Theor Appl Genet 85:901–904

Korbin M, Kuras A, Zurawicz E (2002) Fruit plant germplasm characterization using molecular markers generated in RAPD and ISSR-PCR. Cell Mol Biol Lett 7:785–794

Kotwal S, Dhar MK, Kour B et al (2013) Molecular markers unravel intraspecific and interspecific genetic variability in *Plantago ovata* and some of its wild allies. J Genet 92:293–298

Kour B, Kotwal S, Dhar MK, Kaul S (2016) Genetic diversity analysis in *Plantago ovata* and some of its wild allies using RAPD markers. Russ Agric Sci 42(1):37–41

Kumar LD, Kathirvel M, Rao GV, Nagaraju J (2001) DNA profiling of disputed chilli samples (*Capsicum annum*) using ISSR-PCR and FISS-RPCR marker assays. Forensic Sci Int 116:63–68

Kumar M, Fougat RS, Sharma AK et al (2014) Phenotypic and molecular characterization of selected species of *Plantago* with emphasis on *Plantago ovata*. Aust J Crop Sci 8(12):1639–1647

Kumar S, Joshi UN, Saini ML et al (2015) Biomolecular characterization of guar [*Cyamopsis tetragonoloba* (L.)Taub.] genotypes along with wild species *C. serrata* and *C. senegalensis*. Plant Syst Evol 301(4):1249–1262

Kurata N, Nagamura Y, Miyao A et al (1994) Progress in cDNA analysis and construction of genetic and physical maps in RGP Presented in 4th International Workshop on Rice Molecular Biology 1–2 August, Science University of Tokyo, Tokyo, Japan

Laidlaw RA, Percival EGV (1949) Studies on seed mucilage. Part III. Examination of polysaccharide extracted from the seeds of Plantago ovata Forsk. J Chem Soc:1600–1607

Lata H, Chandra S, Techen N, Khan IA, ElSohly MA (2010) Assessment of the genetic stability of micropropagated plants of *Cannabis sativa* by ISSR markers. Planta Med 76:97–100

Lee SK (1998) Identification of DNA polymorphisms in the field bean *(Glycine soja* S) using RAPD markers. J Korean Soc Grassland Sci 18:143–150

Leroy XJ, Leon K, Branchard M (2000) Plant genomic instability detected by microsatellite-primers. Eur J Biotechnol 3(2):354–360

Li P, Xiaoping Z, Yicun C et al (2008) Genetic diversity and germplasm resource research on tung tree (*Vernicia fordii*) cultivars, investigated by inter-simple sequence repeats. Afr J Biotechnol 7(8):1054–1059
Lin XC, Lou YF, Liu J et al (2010) Crossbreeding of *Phyllostachys* species (Poaceae) and identification of their hybrids using ISSR markers. Genet Mol Res 9:1398–1404
Marczewski W, Hennig J, Gebhardt C (2002) The potato virus S resistance gene Ns maps to potato chromosome VIII. Theor Appl Genet 105:564–567
Meyer W, Mitchell TG, Freedman EZ, Vilgalys R (1993) Hybridization probes for conventional DNA fingerprinting used as single primers in the polymerase chain reaction to distinguish strains of *Cryptococcus neoformans*. J Clin Microbiol 31:2274–2280
Mignouna HD, Ng NQ, Ikea J, Thottapilly G (1998) Genetic diversity in cowpea as revealed by random amplified polymorphic DNA. J Genet Breed 52:151–159
Mithal BM, Gupta VD (1965) Suspending properties of *Plantago ovata* seed husk (Ispaghula) seed husk. Ind J Pharma 27(12):331–334
Monna L, Miyao A, Inoue T et al (1994) Determination of RAPD markers in rice and their conversion into sequence tagged sites (STSs) and STS-specific primers. DNA Res 1:139–148
Muchlbauer GJ, Staswick PE, Specht JE et al (1991) RFLP mapping using near isogenic lines in the soybean. Theor Appl Genet 81:189–198
Multani DS, Lyon BR (1995) Genetic fingerprinting of Australian cotton cultivars with RAPD markers. Genome 38:1005–1008
Nagaraju J, Kathirvel M, Kumar RR et al (2002) Genetic analysis of traditional and evolved Basmati and non-Basmati rice varieties by using fluorescence-based ISSR-PCR and SSR markers. Proc Natl Acad Sci U S A 99:5836–5841
Narzary D, Rana TS, Ranade SA (2010) Genetic diversity in inter-simple sequence repeat profiles across natural populations of Indian pomegranate (*Punica granatum* L.) Plant Biol 12:806–813
Nkongolo KK (2003) Genetic characterization of Malawian cowpea (*Vigna unguiculata* L.) Walp landraces: diversity and gene flow among accessions. Euphytica 129:219–228
Orozco-Castillo C, Chalmers KJ, Waugh R, Powell W (1994) Detection of genetic diversity and selective gene introgression in coffee using RAPD markers. Theor Appl Genet 87:934–940
Paredes M, Becerra V, Rojo C et al (2002) Ecotypic differentiation of *Medicago polymorpha* L. along an environmental gradient in central Chile. Euphytica 123:431–439
Paris HS, Yonash N, Portnoy V et al (2003) Assessment of genetic relationships in *Cucurbita pepo* (cucurbitaceae) using DNA markers. Theor Appl Genet 106:971–978
Patel RB, Rana NG, Patel MR, Dhyani HK, Chauhan HV (1979) Chromatographic screening of proteins of *Plantago ovata* Forsk. Ind J Pharm Sci 41(6):249
Paterson AH, de Verna JW, Lanini B, Tanksley SD (1990) Fine mapping of quantitative trait loci using selected overlapping recombinant chromosomes, in an interspecies cross of tomato. Genetics 124:735–742
Punia A, Arora P, Yadav R, Chaudhury A (2009a) Optimization and inference of PCR conditions for genetic variability studies of commercially important cluster bean varieties by RAPD analysis. Asian Pac J Mol Biol Biotechnol 17(17):33–38
Punia A, Yadav R, Arora P, Chaudhury A (2009b) Molecular and morphophysiological characterization of superior cluster bean (*Cymopsis tetragonoloba*) varieties. J Crop Sci Biotechnol 12(3):143–148
Qiu YX, Fu CX, Wu FJ (2004) Analysis of population genetic structure and molecular identification of *Changium smyrnioides* and *Chuanmin shenviolaceum* with ISSR marker. Zhong Yao Cai 27:164–169
Rajwade AV, Arora RS, Kadoo NY et al (2010) Relatedness of Indian flax genotypes (*Linumu sitatissimum* L.): an inter-simple sequence repeat (ISSR) primer assay. Mol Biotechnol 45:161–170

Rani VK, Singh P, Shiran B et al (2000) Evidence for new nuclear and mitochondrial genome organizations among high frequency somatic embryogenesis derived plants of allotetraploid *Coffea arabica* L. (Rubiaceae). Plant Cell Rep 19:1013–1020

Ratnaparkhe MB, Gupta VS, Venmurthy MR, Ranjekar PK (1995) Genetic fingerprinting of pigeonpea [*Cajanus cajan* (L.) Mill sp.] and its wild relatives using RAPD markers. Theor Appl Genet 91:893–898

Ravn H, Brimer L (1988) Structure and antibacterial activity of plantamajo side, a caffeic acid sugar ester from *Plantago major subsp. major*. Phytochemistry 27(11):3433–3437

Reunova GD, Kats IL, Muzarok TI, Zhuravlev IN (2010) Polymorphism of RAPD, ISSR and AFLP markers of the *Panax ginseng* CA Meyer (*Araliaceae*) genome. Genetika 46:1057–1066

Rohilla AK, Mukesh SA, Boora KS (2012) Genetic diversity analysis of the medicinal herb *Plantago ovate* (Forsk.) Afr J Biotechnol 11(92):15835–15842

Roy A, Bandyopadhyay AK, Mahapatra SK, Ghosh NK et al (2006) Evaluation of genetic diversity in jute (*Corchorus* species) using STMS, ISSR and RAPD markers. Plant Breed 125:292–297

Saiki RK, Scharf S, Faloona F et al (1985) Enzymatic amplification of β-globin genomic sequences and restriction site analysis for diagnostics of sickle cell anaemia. Science 230:1350–1353

Saiki RK, Gelfoud DH, Stoffel S et al (1988) Primer directed enzymatic amplification of DNA with a thermostable DNA polymerase. Science 239:487–491

Samantaray S, Geetha KA, Hidayath KP, Maiti S (2010) Identification of RAPD markers linked to sex determination in guggal [*Commiphora wightii* (Arnott.)] Bhandari. Plant Biotechnol Rep 4:95

Sanchez MP, Davila JA, Loarce Y, Ferrer E (1996) Simple sequence repeat primers used in polymerase chain reaction amplifications to study genetic diversity in barley. Genome 39:112–117

Sandu SS, Hudson GJ, Kennedy JF (1981) The gel nature and structure of the carbohydrate of Ispaghula husk ex *Plantago ovata* Forsk. Carbohydr Res 93(2):247–259

Sankar AA, Moore GA (2001) Evaluation of inter-simple sequence repeat analysis for mapping in Citrus and extension of genetic linkage map. Theor Appl Genet 102:206–214

Santalla M, Power JB, Davey MR (1998) Genetic diversity in mungbean germplasm revealed by RAPD markers. Plant Breed 117:473–478

Sarwat M, Das S, Srivastava PS (2008) Analysis of genetic diversity through AFLP, SAMPL, ISSR and RAPD markers in *Tribulus terrestris*, a medicinal Herb Plant. Cell Rep 27:519–528

Schierwater B, Ender A, Streit B (1993) The use of RAPD markers for molecular phylogenies. Verhdtzool Ges 86(1):58

Senthil Kumar R, Parthiban KT, Govinda RM (2009) Molecular characterization of Jatropha genetic resources through inter-simple sequence repeat (ISSR) markers. Mol Biol Rep 36:951–956

Singh BM, Sharma KD, Katoch M, Guleria S, Sharma TR (2000) Molecular analysis of variability in *Podophyllum hexandrum* Royle an endangered medicinal herb of northwestern Himalaya. PGR Newslett 124:57–61

Singh N, Lal RK, Shasany AK (2009) Phenotypic and RAPD diversity among 80 germplasm accessions of the medicinal plant isabgol (*Plantago ovata*, Plantaginaceae). Genet Mol Res 8(3):1273–1284

Smith JSC, Smith DS (1992) Fingerprinting crop varieties. Adv Agron 47:85–140

Song Z, Li X, Wang H, Wang J (2010) Genetic diversity and population structure of *Salvia miltiorrhizabge* in China revealed by ISSR and SRAP. Genetica 138:241–249

Souframanien J, Gopalakrishna T (2004) A comparative analysis of genetic diversity in blackgram genotypes using RAPD and ISSR markers. Theor Appl Genet 109:687–1693

Souframanien J, Manjaya JG, Krishna TG, Pawar SE (2003) Random amplified polymorphic DNA analysis of cytoplasmic male sterile and male fertile pigeonpea (*Cajanus cajan* (L.) Millsp). Euphytica 129:293–299

Staub JE, Serquen FC, Gupta M (1996) Genetic markers, map construction and their application in plant breeding. Hortic Sci 31:729–740

Steiner JJ, Piccioni E, Falcinelli M, Liston A (1998) Germplasm diversity among cultivars and the NPGS crimson clover collection. Crop Sci 38:263–271

Stiles JI, Lemme C, Sondur S, Morshid MB, Manshardt R (1993) Using randomly amplified polymorphic DNA for evaluating genetic relationship among papaya cultivars. Theor Appl Genet 85:697–701

Stuber CW, Lincoln SE, Wolf DW, Helentjaris T, Lander E (1992) Identification of genetic factors contributing to heterosis in a hybrid from two elite maize inbred lines using molecular markers. Genetics 132:823–839

Tamhankar S, Ghate V, Raut A, Rajput B (2009) Molecular profiling of "Chirayat" using Inter Simple Sequence Repeat (ISSR) markers. Planta Med 75:1266–1270

Tanksley SD (1993) Mapping polygenes. Annu Rev Genet 27:205–233

Thimmappaiah WG, Santhosh D, Shobha GS, Melwyn (2009) Assessment of genetic diversity in cashew germplasm using RAPD and ISSR markers. Sci Hortic 120:411–417

Thomas KG, Bebeli PJ (2010) Genetic diversity of Greek *Aegilops* species using different types of nuclear genome markers. Mol Phylogenet Evol 56:951–961

Thompson JA, Nelson RL, Vodkin LO (1998) Identification of diverse soybean germplasm using RAPD markers. Crop Sci 38:1362–1368

Tikunov YM, Khrustaleva LI, Karlov GI (2003) Application of ISSR markers in the genus, Lycopersicon. Euphytica 131:71–80

Tingey SV, del-Tufo JP (1993) Genetic analysis with random amplified polymorphic DNA markers. Plant Physiol 101:349–352

Tsumura Y, Ohba K, Strauss SH (1996) Diversity and inheritance of inter-simple sequence repeat polymorphisms in Douglas-fir (*Pseudotsugamenziesii*) and sugi (*Cryptomeria japonica*). Theor Appl Genet 92:40–45

Turley SD, Daggy BP, Dietschy JM (1994) Psyllium augments the cholesterol-lowering action of cholestyramine in hamsters by enhancing sterol loss from the liver. Gastroenterology 107(2):444–452

Tyagi AP (1997) Investigation of quantitative trait loci (QTLs) in two diverse strains of pigeonpea (*Cajanus cajan*) using RAPD markers. J S Pac Agric 4:92–98

Upadhyay KG, Patel AR, Vyas SH (1978) Evaluation of isabgul (*psyllium*) husk and gum acacia as ice cream stabilizers. Gujarat Agric Univ Res J 4:45–50

Vahabi AA, Lotifi A, Solouki M, Bahrami S (2008) Molecular and morphological markers for the evaluation of diversity between *Plantago ovata* in Iran. Biotechnology 7(4):702–709

Vala AG, Fougat RS, Jadeja GC (2011) Genetic diversity of *Plantago ovata* Forsk through RAPD markers. Electronic J Plant Breed 2(4):592–596

Verma PC, Chakrabarty D, Jena SN et al (2009) The extent of genetic diversity among *Vanilla* species: comparative results for RAPD and ISSR. Ind Crop Prod 29:581–589

Virk PS, Ford-Lloyd BV, Jackson MT, Nehbury HJ (1995) Use of RAPD for the study of diversity within plant germ collections. Heredity 74:170–179

Wang G, Mahalingan R, Knap HT (1998) (C-A) and (GA) anchored simple sequence repeats (ASSRs) generated polymorphism in soybean, *Glycine max* (L.) Merr. Theor Appl Genet 96:1086–1096

Wang X, Zhao F, Hu Z et al (2008) Inter-simple sequence repeat (ISSR) analysis of genetic variation of *Chondrus crispus* populations from North Atlantic. Aquat Bot 88:154–159

Wasicky R (1961) Investigations of the seeds of *Plantago ovata, P. psyllium* and *P. major* variety cruneta as laxatives. Plant Med 9:232–244

Waugh R, Powell W (1992) Using RAPD markers for crop improvement. Trends Biotechnol 10:186–191

Waugh R, Baird E, Powell W (1992) The use of RAPD markers for the detection of gene introgression in potato. Plant Cell Rep 11:466–469

Weber D, Helentjaris T (1989) Mapping RFLP loci in maize using B-A translocations. Genetics 2:583–590

Weising K, Nybom H, Wolff K, Meyer W (1995) DNA fingerprinting in plants and fungi (ed.) CRC Press, Boca Raton, pp 1–3

Welsh J, McClelland M (1990) Fingerprinting genomes using PCR with arbitrary primers. Nucleic Acids Res 18:7213–7218

Williams TJ (1989) Plant germplasm preservation: a global perspective. In: Knutson L, Stoner AK (eds) Biotic diversity and germplasm preservation, global imperatives. Kluwer Academic Publishers, Amsterdam, pp 81–96

Williams JGK, Kubelik AR, Livak KJ et al (1990) DNA polymorphisms amplified by arbitrary primers are useful as genetic markers. Nucleic Acids Res 18:6531–6535

Wu K, Jones R, Dannaeberger L, Scolnik PA (1994) Detection of microsatellite polymorphisms without cloning. Nucleic Acids Res 22:3257–3258

Wu Z, Leng C, Tao Z, Wei Y, Jiang C (2009) Genetic diversity of *Dioscorea alata* based on ISSR analysis. Zhongguo Zhong Yao Za Zhi 34:3017–3020

Wu Y, Shi HM, Bao Z, Wang MY et al (2010) Application of molecular markers in predicting production quality of cultivated *Cistanchede serticola*. Biol Pharm Bull 33:334–339

Xia T, Chen S, Chen S, Zhang D et al (2007) ISSR analysis of genetic diversity of the Qinghai-Tibet Plateau endemic *Rhodiola chrysanthemifolia* (Crassulaceae). Biochem Syst Ecol 35:209–214

Yadav R, Yadav N, Pal M, Goutam U (2014) Multiple shoot proliferation, bulblet induction and evaluation of genetic stability in Asiatic hybrid lily (*Lilium* sp.) Indian J Plant Physiol 18(4):354–359

Yadav R, Yadav N, Kumar S (2015) An improved micropropagation and assessment of genetic fidelity in multipurpose medicinal tree, *Acacia auriculiformis*. Proceedings of the National Academy of Sciences, India - Section B: Biological Sciences. doi:10.1007/s40011-015-0550-9

Yao HYZ, Chen DF, Chen JK, Zhou TS (2008) ISSR primer screening and preliminary evaluation of genetic diversity in wild populations of *Glycyrrhiza uralensis*. Biol Plant 52(1):117–120

Yeo AR, Yeo ME, Folowers SA, Flowers TJ (1990) Screening of rice (*Oryza sativa* L.) genotypes for physiological characters contributing to salinity resistance, and their relationship to overall performance. Theor Appl Genet 79:377–384

Yuan XF, Dai ZH, Wang XD, Zhao B (2009) Assessment of genetic stability in tissue-cultured products and seedlings of *Saussurea involucrata* by RAPD and ISSR markers. Biotechnol Lett 31:1279–1287

Zhang ZY, Gali JY (1997) Utilization of RAPD in soybean germplasm resources and genetic linkage studies. Soybean Sci 16:60–65

Zhang F, Lv Y, Dong H, Guo S (2010) Analysis of genetic stability through inter-simple sequence repeats molecular markers in micropropagated plantlets of *Anoectochilus formosanus* Hayata, a medicinal plant. Biol Pharm Bull 33:384–388

Zheng ZB (1994) Precision mapping of quantitative trait loci. Genetics 136:1457–1468

Zietkiewicz E, Rafalski A, Labuda D (1994) Genome fingerprinting by simple sequence repeats (SSR)-anchored PCR amplification. Genomics 20:176–183

Flavonoids: A Nutraceutical and Its Role as Anti-inflammatory and Anticancer Agent

13

Nidhi Saini, S.K. Gahlawat, and Viney Lather

Abstract

Medicinal plants nowadays are the richest source of various medicinal agents which possess many significant therapeutic and biological properties. They contain secondary metabolites, which do not directly take part in the growth of plants but are protective against different diseases. As microorganisms are developing resistance against various diseases, so herbal formulations are widely prepared and used for the treatment of disease. Various phytochemicals are present in plants like flavonoids, alkaloids, sterols, tannins, glycosides, and so on. They are present in various parts of the plant (roots, shoots, leaves, flowers, fruit, etc.) and in different concentrations. These phytochemicals play significant roles such as photoprotectant, disease-protective agents, antioxidants, anti-inflammatory agents, antimetastatic agents, and so on. Flavonoids, one of the important phytoconstituents present in plants, are present in fruits and vegetables in different quantities particularly in green tea and wine. They are categorized/classified into six groups, namely, flavonols, flavanols (catechins), flavanones, flavones, anthocyanins, and isoflavonoids. They effectively protect the body from major diseases such as diabetes mellitus, cancer, chlotis, and inflammation. They also boost up the immune system. This chapter highlights the diversity of flavonoids, their classification, and their role as anticancer and anti-inflammatory agent.

N. Saini (✉) • S.K. Gahlawat
Department of Biotechnology, Chaudhary Devi Lal University, Sirsa 125 055, Haryana, India
e-mail: nid0029@gmail.com

V. Lather
JCDM College of Pharmacy, Sirsa 125 055, Haryana, India

S.K. Gahlawat et al. (eds.), *Plant Biotechnology: Recent Advancements and Developments*, DOI 10.1007/978-981-10-4732-9_13

13.1 Introduction

Flavonoids are renowned bioactive polyphenols having less molecular weight. They are a group of secondary metabolites found aplenty in plant kingdom and are universally prominent due to their pharmacological and physiological roles and health benefits (Nijveldt et al. 2001; Hendriks et al. 2003; Gattuso et al. 2007; Nathiya et al. 2014; Calderon-Montano et al. 2011). About 4000 miscellany flavonoids are familiar and are recognized in plant kingdom. They manage the luring colors in flowers, fruits, leaves, and so on; anthocyanin (based on delphinidin) is obtained from plant's petal for the formation of blue color. These polyphenols play a remarkable role in plant coloration and as defensive agents from various superficial agents. In consequence of the antioxidant, anti-inflammatory, antiallergenic, antimicrobial, and anticancer properties of flavonoids, there is an elevating curiosity in flavonoids in the recent times (Janicijevic et al. 2007). Flavonoids exceptionally reduce the immedicable diseases in humans and are freely established in plant food products. In plants, they react as antioxidants (direct and indirect), anti-inflammatory, immunomodulatory agents and are prospective photoprotectors as they absorb the harmful light, that is, UV rays (Justyna et al. 2014; Janicijevic et al. 2007). Also from previous times, many publications reported flavonoids with many therapeutic activities like antimicrobial, anti-inflammatory, antiulcer, antiallergic, hepatoprotective, anticancer, and so on. Flavonoids have vasodilatory and gastroprotective actions, which cause hindrance in human platelet accumulation; this is also communicated in many studies (Gattuso et al. 2007; Danihelova et al. 2012; Hendriks et al. 2003; Tapas et al. 2008). Environmental and biochemical properties of flavonoids helps in the communication of plants (with various micro-organisms and environmental stress). (Liu 2002; Saewan and Jimtaisong 2013; Ghasemzadeh and Ghasemzadeh 2011).

13.2 Nomenclature and Classification of Flavonoids

Flavonoids are made up of C_6-C_3-C_6 carbon framework including two benzene rings A and B along with heterocyclic pyrane ring C as shown in Fig. 13.1. The structure is thus called benzo-γ-pyrane structure (Kumar and Pandey 2013).

All the therapeutic activities and chemical behavior of flavonoids are structure dependent and owe this behavior to the various structures of flavonoids. Structural variation in flavonoids is due to the pattern of substitution: degree of hydroxylation, methoxylation, prenylation, glycosylation, and degree of polymerization (Kumar

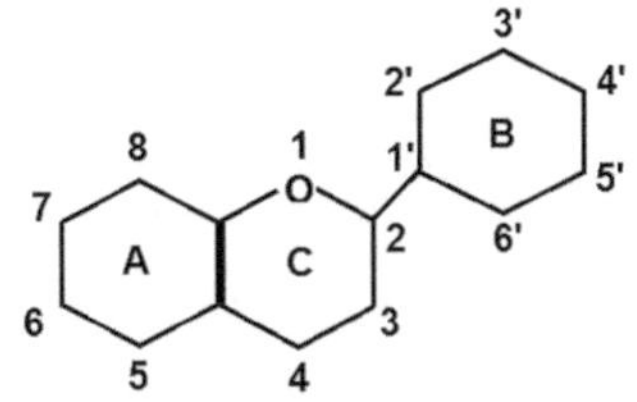

Fig. 13.1 General structure of flavonoids (Nathiya et al. 2014)

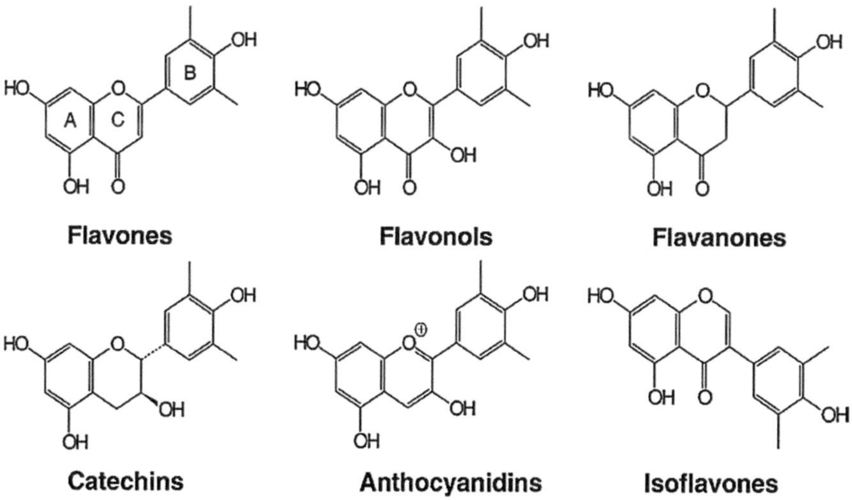

Fig. 13.2 Chemical structures of flavonoid family (Hollman 2004)

and Pandey 2013; Prasain et al. 2004). Flavonoids are broadly divided into two classes: anthocyanins (glycosylated derived from anthocyanidin, found in eye-catching fruits and flowers) and anthoxanthins (a category of uncolored compounds). Further, anthoxanthins are divided into five classes, namely, flavonols, flavones, flavanols (catechins), flavanones, and isoflavonoids (Ozcan et al. 2014) as shown in Fig. 13.2.

The structure of flavones includes a phenyl ring (B) as a substituent at the 2-position of pyrone ring (C) with a C2-C3 double bond and a C4-oxo function. It is found in cereals and herbs. Major flavonoids include apigenin, chrysin, and atherosclerosis. It is beneficial against osteoporosis (bone disease), juvenile diabetes, and certain cancers. Flavanol's structure is similar to that of flavones with the addition of an –OH, that is, alcoholic group at the third position of pyrone ring (C). It is found in abundance in vegetables like onion, kale (leaf cabbage), broccoli, and tomato, and in fruits like apple, grape, berries, and tea. The greenest leaf has the highest amount of flavonols found in it. Major flavonoids include quercetin, kaempferol, myricetin, and isorhamnetin. Another name for flavanonols is dihydroflavonols. Its structure resembles the structure of flavonols but positions second and third of pyrone ring (C) do not possess a double bond between them. A phenyl group is found at carbon's third position of the C ring (pyrone ring) in the structure of isoflavones. It is found in edible parts of the legume plants like soybeans, black beans, soya, string beans, bengal gram, that is, chickpeas, and so on and is favorable in various heart diseases, cancers, and so on. Another name of flavanols is catechins and flavan-3-ols. C2 and C3 do not carry a double bond, whereas Ring C has no C4 carbonyl. C3 hydroxylation results in two chiral centers (onto C2, C3 molecule), and four diastereoisomers are attainable. The chiral centers known as catechin and epicatechin (EC) have two sterioisomers, i.e., (+) Catechin, (−) Catechin and

(+) EC and (−) EC respectively. Chief source of flavanols are fruits (like grape's skin, apple, blueberries) and various teas (e.g., black tea, green tea). The sugar moiety in anthocyanins is mainly attached at the third position on C-ring and on A-ring; it is present at the fifth and seventh positions. The B-ring glycosylation at the third, fourth, and fifth positions is very rarely observed. The extent of aromatic or aliphatic acids acylates the sugar moieties by using most common acylating agents such as cinnamic acids. They are abundantly found in berries (like strawberries and raspberries), grapes (red and green), and cherries with dark blue or red colors. Cyanidin-3-glucoside, delphinidin-3-glucoside, petunidin, quercetin-3-glu, peonidin, quercetin, pelargonidin, pelargonidin-3-arabinose, cyaniding, delphinidin and malvidin are the major flavonoids found (Saewan and Jimtaisong 2013; Goncalves et al. 2015; Nijveldt et al. 2001).

13.3 Flavonoids as Promising Nutraceutical

Dr. Stephen L. DeFelice, chairperson of the Foundation of Innovation Medicine (FIM), Crawford, New Jersey, coined the term "nutraceutical" in 1989. "Nutrition" in conjugation with "pharmaceutical" creates "nutraceutical." "Nutrition" stands for a nourishing food and "pharmaceutical" stands for a medical drug and both words together make "nutraceutical" which is a nonpoisonous food or any food product that provides health and medical benefits, including the prevention and treatment of diseases (Tapas et al. 2008; Ganesh et al. 2015; Gupta et al. 2013) as shown in Fig. 13.3.

Flavonoids are phenolic substances (hydroxylated), synthesized within plants which work against microbial infections (Dixon et al. 1983). Various active components of flavonoids present in the food products are under obligation for their nutrition and also a remedy for some lethal and serious diseases in humans. Flavonoids in diet are an important source of antioxidants as their antioxidant activity depends upon the arrangement of functional groups around the nuclear structure (Kumar and Pandey 2013). Some flavonoids, namely, apigenin, rutin, catechin, naringenin, quercetin, and venoruton are reported for their hepatoprotective activities (Tapas et al. 2008). Flavonoid-rich plant extracts have significant in vitro antimicrobial activity in response to microbial infection (Kumar and Pandey 2013). Several flavonoids including apigenin, galangin, flavone and flavonol glycosides, isoflavones,

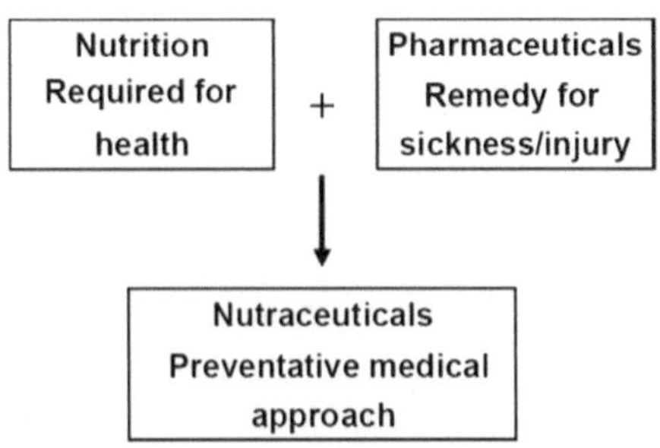

Fig. 13.3 Concept of nutraceutical (Gupta et al. 2013)

flavanones, and chalcones have been reported to possess potent antibacterial activity (Cushnie and Lamb 2005). Flavonoids, such as quercetin and iso catechin, showed potent anti-inflammatory activity (Pan et al. 2010). Many other reports are also present showing their beneficial therapeutic activities. Hippocrates, known as the father of Western medicine, stated, "Let food be the medicine," and flavonoids are demonstrating themselves to be nutritious medicines for human health (Chintale et al. 2013). But it is important to take the bioavailability of flavonoids into consideration.

Bioavailability can be commonly defined as the proportion of the nutrient which is digested, absorbed, and metabolized through normal pathways by human body (D'Archivio et al. 2007). From ancient times, plant kingdom has abundant flavonoids which have been used in folk medicine around the world (Ren et al. 2003). Daily human diet of flavonoid present in the food like apple, onion, wine, spices, and herbs varies from 23 mg/day to more than 500 mg/day (Comalada et al. 2005). In our daily diet, the flavonoid known as luteolin is required in low amounts (less than 1 mg/day) (Seelinger et al. 2008). It is reported that intake of flavonoid is a few hundreds of mg/day in normal human diet. The flavonol quercetin (5, 7, 3′, 4′-hydroxyflavanol) is a polyphenolic compound which is important in our diet with an average intake of 16 mg (Graefe et al. 2001). The flavonoid epicatechin when taken at 10 mg/kg reduces lesion severity in both acute and chronic models of colitis (Vasconcelos et al. 2012). Fisetin is another flavonoid which when taken at 10 mg/kg body weight cures diabetes mellitus (Durga et al. 2014).

13.4 Flavonoids as Anticancer Agent

Cancer is a major threat and dreadful disease which occurs due to the proliferation of abnormal cells (Dantu et al. 2012). When a usual cell changes into a cancer cell, it undergoes behavioral changes. Therefore, it is classified on the basis of the type of cell that gets infected (Merina et al. 2012). From the olden times, the conventional drugs and therapies, like radiotherapy, chemotherapy, and so on, are commonly prescribed to cancer patients (Charoensin 2014). But resistance of the various cancer types to these therapies is a major problem (Plengsuriyakarn et al. 2012). To overcome this problem, researchers are now finding new effective anticancer compounds with negligible side effects. Some metabolites like flavonoids from various plant sources play an important part in leading to new pharmaceutical compounds for antitumor drugs (Zandi et al. 2010).

Flavonoids significantly act against cancer as they have many mechanisms which are responsible for anticancerous activity (Ren et al. 2003) as shown in Fig. 13.4. Some important processes of the anticancer activity are as follows:

The mechanism involves the synergy of flavonoids with phase I metabolizing enzymes resulting in arousal of various procarcinogens to reactive intermediates. An example of phase I metabolizing enzyme is cytochrome P450. These reactive intermediates combine with cellular neutrophils and result in cancer induction. Flavonoids inhibit the activities of certain isozymes of enzymes like P450 isozymes (CYP1A1 and CYP1A2) making them protective against cancer and hence they are

Fig. 13.4 Pathway of carcinogenesis and the potential benefits of flavonoids (Durga et al. 2014)

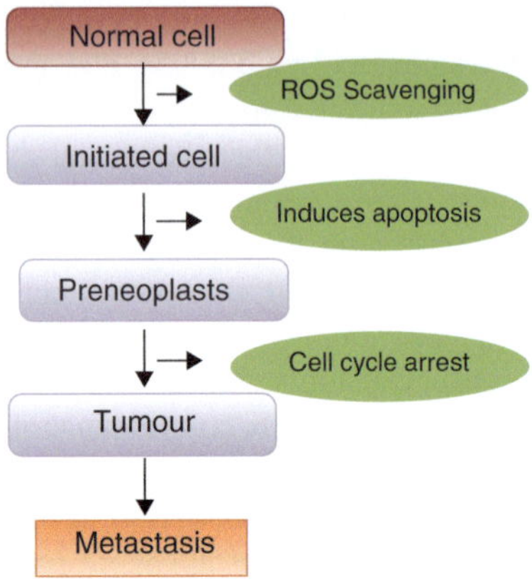

known as anticancerous agents (Ren et al. 2003; Korobkova and Sukala 2012). Also, flavonoids and their metabolites inhibit or stimulate the various signaling cascades {phosphoinositide 3-kinase (PI 3-kinase), Akt/protein kinase B (Akt/PKB), tyrosine kinases, protein kinase C (PKC), and mitogen activated protein kinase (MAP kinase)}, which affects the cellular functioning by altering the phosphorylation state and modulating gene expression of target molecules (Williams et al. 2004). Induction of phase II metabolizing enzymes like glutathione-S-transferase, quinone reductase, and UDP-glucuronyl transferase results in the removal of carcinogens from the body and thus protecting the body from cancer (Ren et al. 2003; Korobkova and Sukala 2012). Inhibition of various enzymes (xanthine oxidase, aldose reductase, Ca + 2-ATPase, phosphodiesterase, cycloxygenase, lipoxygenase, etc.) and hormones (estrogens, androgens, and thyroid) by various flavonoids also inhibit cancer formation (Narayana et al. 2001). Breast cancer is a serious threat to women worldwide. Flavonoids have a significant role in breast cancer as they are potent aromatase inhibitors. Estradiol is an endogenous estrogen, synthesized from androgens by aromatase (cytochrome P450 enzyme complex). Inhibition of aromatase reduces the growth stimulatory effects of estrogens in hormone-dependent breast cancer (Ren et al. 2003). Formation of reactive oxygen species (ROS) is one of the major reasons for tumor formation. They act as catalyst in tumor formation and progression. Flavonoids inhibit the metabolizing enzymes of pro-oxidant enzymes (which activate the tumors), such as cyclooxygenases (COX) and lipoxygenases (LOX), resulting in tumor cell proliferation inhibition (Ren et al. 2003; Durga et al. 2014).

Enzyme inhibition is another mechanism for anticancerous activity in which certain enzymes like inducible nitric oxide synthase (iNOS) and cyclooxygenase (COX-2) as well as some mediators of the inflammatory process like reactive C protein or adhesion molecules are involved (Garcia-Mediavilla et al. 2007). Signal transduction enzymes like protein tyrosine kinase (PTK), protein kinase C (PKC), and phosphoinositide 3-kinases (PIP3) are involved in the regulation of cell proliferation which is inhibited by flavonoids. Also, flavonoids such as apigenin, kaempferol, luteolin, and so on, alter the checkpoints in cell-cycle progression, resulting in anticancerous activity (Ren et al. 2003).

13.5 Flavonoids as Anti-inflammatory Agent

Inflammation in the body is due to its protective mechanism. Inflammation occurs due to many reasons like physical trauma, virulent stimuli by chemical agents, heat, antigen-antibody reaction, and microbial effect. It occurs to remove or limit the spread of foreign harmful agents from the host (Soni et al. 2014). To manage pain and inflammation, various drugs (narcotics or non-narcotics and corticosteroids) are given, for example, opioids, salicylates, hydrocortisone, and so on. It is well documented that these drugs for anti-inflammation have well-known toxic side effects, for example, nonsteroidal anti-inflammatory drugs (NSAIDs) cause intestinal tract ulcers, erosions of the stomach lining and intestinal tract (Neha et al. 2013). Thus, there is still need to develop newer and safer anti-inflammatory drugs from natural sources in order to overcome these problems (Singh and Swapnil 2011).

Flavonoids are well-known phytoconstituents that act as anti-inflammatory agents. It acts via various routes/mechanisms like COX, ins, cytokines, matrix metalloproteinases, and nuclear factor-κB as shown in Fig. 13.5. When a flavonoid

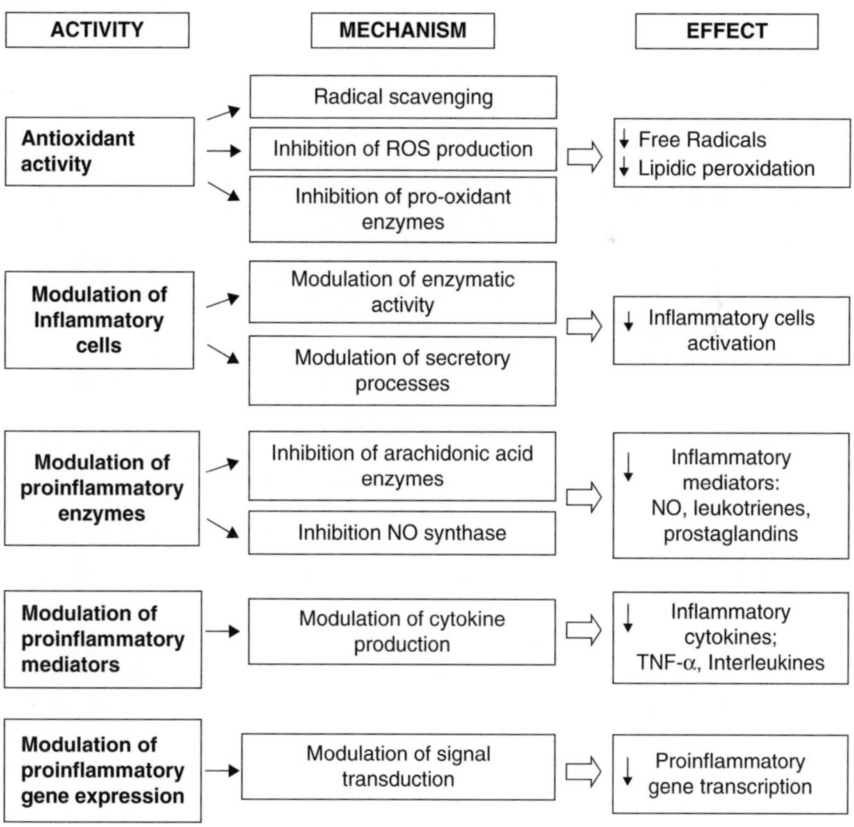

Fig. 13.5 Various mechanisms showing anti-inflammatory activity (Kim et al. 2004)

compound structure has catechol or guaiacol-like B ring then it shows anti-inflammatory activity. Flavonoids modulate cellular action of mast cells, macrophages, lymphocytes, neutrophils, and so on, which are known as inflammation-related cells. In some cases, for example, they inhibit histamine release (from mast cells) plus T-cell proliferation (Kim et al. 2004).

Several mechanisms explain the anti-inflammatory activity of flavonoids. The first mechanism involves antioxidative and radical scavenging activities. In this mechanism, the impairment of the production of active oxygen intermediates by the neutrophils and other phagocytes by flavonoids contribute significantly to the anti-inflammatory activity of these compounds (Middleton et al. 2000). Production of free radicals (ROS) is life threatening and causes serious damage to body cells and tissues. Some major effects on the body include nucleic acid's or carbohydrates' oxidative damage, membrane lipid peroxidation, in proteins—the oxidation of sulfhydryl and other susceptible groups. The free radicals entice the various inflammatory mediators which accord to inflammatory response and tissue damage (Garcia-Lafuente et al. 2009). The change in the activities of arachidonic acid metabolism enzymes (lipoxygenase, phospholipase A2, and cyclooxygenase) and nitric oxide synthase is also done by certain flavonoids. Inhibition of these biological catalysts minimizes the assembly of significant inflammation mediators like arachidonic acid, leukotrienes, prostaglandins, and nitric oxide (Kim et al. 2004). Arachidonic acid metabolism is inhibited effectively by flavonoids. It takes place by the inhibition of gene expression of COX-1 (cyclooxygenase 1) and COX-2 (cyclooxygenase 2) enzymes (Raja et al. 2016). Another modulation of the generation of other proinflammatory molecules is also the reason for anti-inflammatory activity.

Inflammatory diseases are concerned with various cytokinins, particularly tumor necrosis factor-α (TNF-α) and IL-1 in addition to COX-2/iNOS. These cytokinins are the leading contributors to chronic inflammatory disorders which can be successfully reduced by using SAIDs viz., prednisolone, dexamethasone (Kim et al. 2004). Inhibition of the synthesis and various activities of different proinflammatory mediators (eicosanoids, cytokines, adhesion molecules, and C-reactive protein) are also responsible for the anti-inflammatory actions of flavonoids in vitro or in cellular models (Hemalata and Shiva Kameshwari 2015). Diverse documentation has assisted the proposal that definite flavonoids are known as modulators of proinflammatory gene expression, thus leading to the weakening of the inflammatory response. COX-2 discriminatory inhibitors are claimed to possess anti-inflammatory activity and are regularly being developed to manufacture the safer anti-inflammatory drugs. In numerous studies, inhibition of COX-2 activity is rarely reported but the effect on suppression of COX-2 expression is documented.

Various flavonoids and (−)-epigallocatechin-3-gallate were studied by Liang et al. (1999) on activities of analytical cyclooxygenase (COX-2) and nitric oxide synthase (iNOS). It was found that apigenin was the dominant inhibitor of transcriptional activation of both COX-2 and iNOS. Particularly, apigenin and kaempferol were the active inhibitors of transcriptional activation of iNOS. Western and northern blot analyses also revealed that apigenin extremely blocked the protein and mRNA expression of COX-2 and iNOS in LPS-activated macrophages. Flavonoids inhibit the development of the cancer-like induced granuloma (Mohammed et al. 2014). Myricetin and delphinidin possess marked anti-inflammatory effects. Another biflavonoid known as amentoflavone (biapigenin) has antiulcerogenic properties (in rats and guinea pigs). Such properties appeared to be of interest with respect to the adverse effect of gastric ulceration, which develops commonly in subjects taking anti-inflammatory drugs (Middleton et al. 2000). Table 13.1 below shows the structure and anticarcinogenic and anti-inflammatory activities of some important flavonoids.

13.6 Rich Food Sources of Flavonoids

Most fruits and some legumes contain catechins in the range 4.5–610 mg/kg but preparation and processing of food may decrease flavonoid levels depending on the methods used. For example, orange juices contains 81–200 mg/L soluble flavanones after processing whereas before processing it is 206–644 mg/L (Kumar and Pandey 2013). Onion contains major flavonol quercetin with relatively high bioavailability between dietary flavonoid components. The fruits and vegetables diet has beneficial effects on the treatment of gastric cancer (Woo et al. 2014). Available data from population studies show that red wine (rich in procyanidins) consumption has greater cardiovascular benefits than similar alcohol intake from beer due to the high flavonoid content of wine (Hodgson and Croft 2006). Some sources that contain various flavonoids in plenty include wine and bilberry (anthocyanins), apples and tea (flavones), all citrus fruits (flavanones), and soya products (isoflavones) as described in Table 13.2. Berries, parsley, legumes, and green tea are also some of the good flavonoid sources (Pal and Verma 2013).

13.7 Conclusion

Flavonoids are one of the secondary metabolites that have potential to play an important role in food and medical industry. Due to its high nutritional and medicinal value they are considered as significant nutraceuticals. Also they are effective in various cancers like gastric tumors, lung cancer, breast cancer, prostate cancer, and so on. Inflammation is correlated with cancer and so it can also be controlled by using flavonoids in diet. Taking flavonoids in appropriate doses proves that it acts as a drug against many major health issues like high blood pressure, cholitis, and nervous system disorders. Due to its high antioxidant property, it is also used in cosmetic industry to make beauty products. Therefore, overall studies on the flavonoids point toward the development of nutraceutical which are able to cure major diseases with no side effects.

Table 13.1 Some important flavonoids with anticarcinogenic and anti-inflammatory activities

S. No.	Flavonoid	Structure	Role as anticancerous agent	Role as anti-inflammatory agent	References
1.	Apigenin	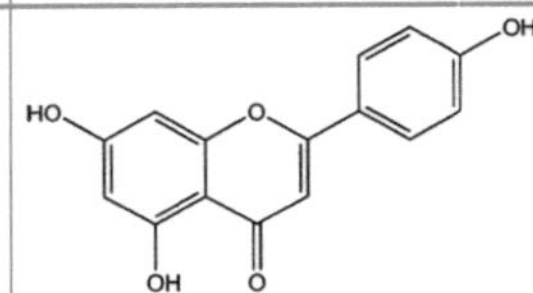	It inhibits proliferation of human bladder cancer T-24 cells via blocking cell-cycle progression and inducing apoptosis	Apigenin inhibits LPS-induced inflammatory response through multiple mechanisms in macrophages	Zhang et al. (2014), Shi et al. (2015), Campbell et al. (2004), Escande et al. (2013) and Saewan and Jimtaisong (2013)
2.	Cynidin-3-glucoside	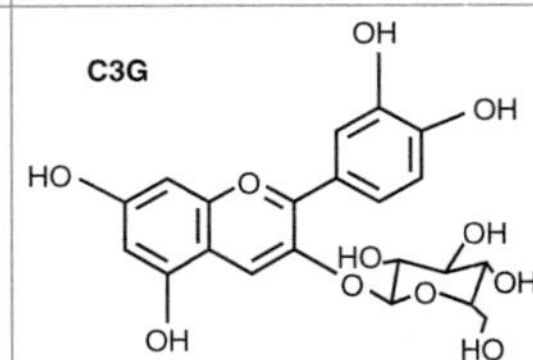 C3G	Treatment with peonidin 3-glucoside or cyanidin 3-glucoside resulted in a strong inhibitory effect on cell growth via G2/M arrest. Cyanidin 3-glucoside or peonidin 3-glucoside also induced caspase-3 activation, chromatin condensation, and cell death.	C3G and its metabolites suppressed the production of the proinflammatory cytokines, TNF-α, and IL-1β, and the inflammatory mediators, NO and prostaglandin E2 (PGE_2), as well as the gene expression of nitric oxide synthase (iNOS) and cyclooxygenase-2 (COX-2)	Serra et al. (2013), Chen et al. (2005) and Min et al. (2010)
3.	(−)-Epicatechin	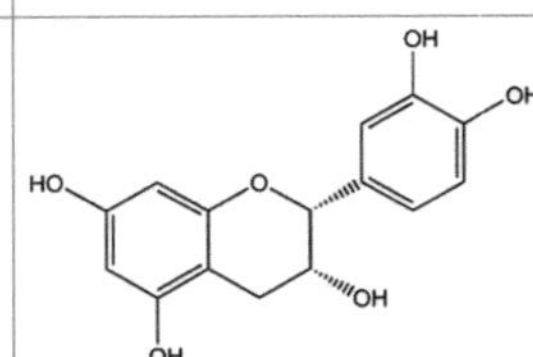	It protects cellular damage by inhibiting DNA damage and oxidation of LDL and has ability to scavenge free radicals.	To make better overproduction of proinflammatory cytokines and mediators, reduces the activity of NF-κB and AP-1 and the subsequent formation of peroxynitrite with NO and reactive oxygen species	Campbell et al. (2004), Vasconcelos et al. (2012) and Tipoe et al. (2007)

4.	(−)-Epigallocatechin gallate	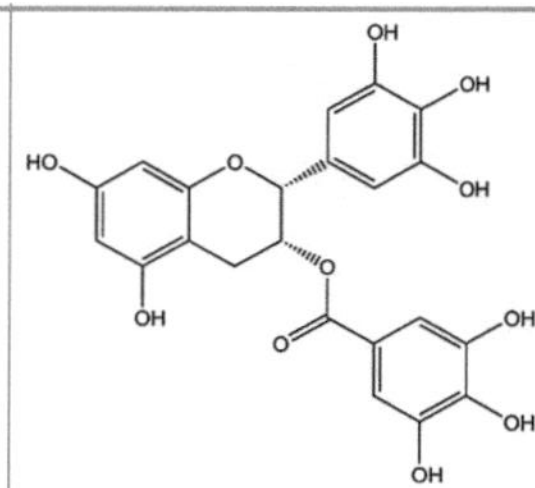	The green tea polyphenol (−)-epigallocatechin-3-gallate blocks nuclear factor-κB activation by inhibiting IκB kinase activity in the intestinal epithelial cell line IEC-6	Generation of hydrogen peroxide primarily contributes to the induction of Fe(II)-dependent apoptosis in Jurkat cells by (−)-epigallocatechin gallate	Campbell et al. (2004), Vasconcelos et al. (2012), Tipoe et al. (2007) and Nakagawa et al. (2004)
5.	Genistien (4′,5,7-trihydroxyisoflavone)	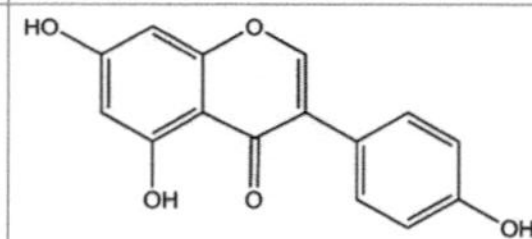	It induces apoptosis in many cells	Genistein may attenuate proinflammatory pathways through inhibition of cytokine-induced overexpression of proinflammatory mediators and inflammatory reactions in human brain microvascular endothelial cells	Campbell et al. (2004) and Lee and Lee (2008)
6.	Kaempferol	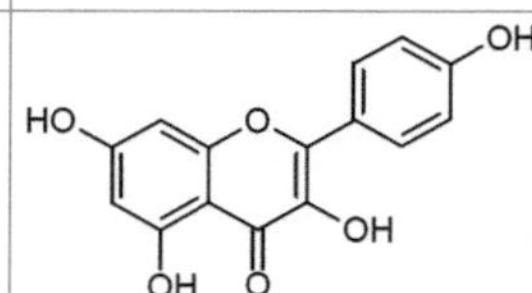	Effective in ovarian cancer cells by reducing vascular endothelial growth factor (VEGF) expression in them. Also it promotes apoptosis by enhancing the effect of cisplatin through downregulation of cMyc	Inhibition of inducible nitric oxide synthase, cyclooxygenase-2 and reactive C-protein, and downregulation of the nuclear factor-κB pathway result in anti-inflammation	Agrawal (2011), Batra and Sharma (2013) and Garcia-Mediavilla et al. (2007)

(continued)

Table 13.1 (continued)

S. No.	Flavonoid	Structure	Role as anticancerous agent	Role as anti-inflammatory agent	References
7.	Quercetin(3,5,7,3′,4′-pentahydroxyflavone)	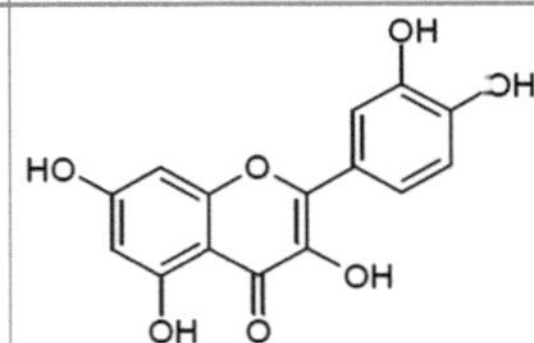	Apoptosis and metastasis makes the quercetin, a potential anticancer agent. Quercetin and intratumoral doxorubicin injection together result in increased immune responses in breast tumors. Quercetin inhibits tumor growth and induces apoptosis in various tumor cells, e.g., human cervical cancer, prostate cancer, epidermal growth factor receptor-overexpressing oral cancer, osteosarcoma, and so on. It also possesses antiproliferative activity against CRC by persuading cell cycle arrest in the G2/M phase, cytotoxicity activity, cell differentiation, apoptosis, and attenuation of cell migration		Agrawal (2011), Pal and Verma (2013), Maalik et al. (2014), Patel et al. (2011), Song et al. (2014) and Koosha et al. (2016)

Table 13.2 Various food sources with active components of flavonoids present

S. No.	Food sources	Flavonoids present	Active component	References
1.	Dark chocolate	Proanthocyanidins	Epicatechin	Rodriguez-Mateos et al. (2014) and Pal and Verma (2013)
2.	Green tea	Flavan-3-ols	(−)-epigallocatechin, (+)-gallocatechin, (−) epicatechin-3-*O*-gallate and (−)-epigallocatechin-3-*O*-gallate	Rodriguez-Mateos et al. (2014)
3.	Citrus peels	Flavonones	Naringen, herpesitin	Durga et al. (2014)
4.	Tea, onion	Flavonol	Quercetin, rutin	Durga et al. (2014)
5.	Red wine, tea, fruits	Catechins	Catechins, epicatechins	Durga et al. (2014)
6.	Beans	Isoflavones	Daidzen, glycindin	Durga et al. (2014)
7.	Citrus fruit, parsley	Flavones	Apigenin, luteolin, quercetin	Durga et al. (2014)
8.	Grapes, apple skin, celery, berries, olives, grapes, tea	Anthocyanidins	Cyanidin, delphinidin, malvidin	Durga et al. (2014)
9.	Herbs; vegetable	Flavones	Apigenin; luteolin	Hodgson and Croft (2006)

References

Agrawal AD (2011) Pharmacological activities of flavonoids: a review. Int J Pharm Sci Nanotech 4(2):1394–1398

Batra P, Sharma AK (2013) Anti-cancer potential of flavonoids: recent trends and future perspectives. Biotech 3(6):439–459

Calderon-Montano JM, Burgos-Moron E, Perez-Guerrero C, Lepez-Lazaro M (2011) A review on the dietary flavonoid kaempferol. Mini Rev Med Chem 11(4):298–344

Campbell EL, Chebib M, Johnston GA (2004) The dietary flavonoids apigenin and (−)-epigallocatechin gallate enhance the positive modulation by diazepam of the activation by GABA of recombinant GABA A receptors. Biochem Pharmacol 68(8):1631–1638

Charoensin S (2014) Antioxidant and anticancer activities of *Moringa oleifera* leaves. J Med Plant Res 8(7):318–325

Chen PN, Chu SC, Chiou HL, Chiang CL et al (2005) Cyanidin 3-glucoside and peonidin 3-glucoside inhibit tumor cell growth and induce apoptosis *in vitro* and suppress tumor growth *in vivo*. Nutr Cancer 53(2):232–243

Chintale Ashwini G, Kadam Vaishali S, Sakhare Ram S et al (2013) Role of nutraceuticals in various diseases: a comprehensive review. Int J Res Pharma Chem 3:290–299

Comalada M, Camuesco D, Sierra S, Ballester I et al (2005) *In vivo* quercitrin anti-inflammatory effect involves release of quercetin, which inhibits inflammation through down-regulation of the NF-κB pathway. Eur J Immunol 35:584–592

Cushnie TPT, Lamb AJ (2005) Antimicrobial activity of flavonoids. Int J Antimicrob Agents 26(5):343–356

D'Archivio M, Filesi C, Benedetto RD, Gargiulo R et al (2007) Polyphenols, dietary sources and bioavailability. Ann 1st Super Sanita 43(4):348–361
Danihelova M, Viskupicova J, Sturdik E (2012) Lipophilization of flavonoids for their food, therapeutic and cosmetic applications. Acta Chimica Slovaca 5(1):59–69
Dantu AS, Shankarguru P, Ramya DD, Vedha HB (2012) Evaluation of *in vitro* anticancer activity of hydroalcoholic extract of *Tabernaemontana divaricata*. Asian J Pharm Res Health Care 5(3):59–61
Dixon RA, Dey PM, Lamb CJ (1983) Phytoalexins: enzymology and molecular biology. Adv Enzymol Relat Areas Mol Biol 55:1–136
Durga M, Nathiya S, Devasena T (2014) Multifarious actions of dietary flavonoids – implications in cancer and cataract. Int J Pharm Biol Sci 5(2):404–416
Escande C, Nin V, Price NL, Capellini V et al (2013) Flavonoid Apigenin is an inhibitor of the NAD+ ase CD38 implications for cellular NAD^+ metabolism, protein acetylation, and treatment of metabolic syndrome. Diabetes 62(4):1084–1093
Ganesh GNK, Ramachandran A, Suresh KR et al (2015) Nutraceuticals – a regulatory review. Int J Drug Regul Aff 3(2):22–29
Garcia-Lafuente A, Guillamon E, Villares A et al (2009) Flavonoids as anti-inflammatory agents: implications in cancer and cardiovascular disease. Inflamm Res 58(9):537–552
Garcia-Mediavilla V, Crespo I, Collado PS, Esteller A, Sanchez-Campos S, Tunon MJ, Gonzalez-Gallego J (2007) The anti-inflammatory flavones quercetin and kaempferol cause inhibition of inducible nitric oxide synthase, cyclooxygenase-2 and reactive C-protein, and down-regulation of the nuclear factor-κB pathway in Chang Liver cells. Eur J Pharmacol 557(2):221–229
Gattuso G, Barreca D, Gargiulli C, Leuzzi U, Caristi C (2007) Flavonoid composition of citrus juices. Molecules 12(8):1641–1673
Ghasemzadeh A, Ghasemzadeh N (2011) Flavonoids and phenolic acids: role and biochemical activity in plants and human. J Med Plant Res 5(31):6697–6703
Goncalves MPJ, Yanez L, Morales C, Navarro M et al (2015) Isolation and characterization of phenolic compounds and Anthocyanins from Murta (*Ugni molinae* Turcz.) fruits. Assessment of antioxidant and antibacterial activity. Molecules 20:5698–5713
Graefe EU, Wittig J, Mueller S, Riethling AK et al (2001) Pharmacokinetics and bioavailability of quercetin glycosides in humans. J Clin Pharmacol 41(5):492–499
Gupta SK, Yadav SK, Patil SMM (2013) Nutraceutical: a bright scope and Oppourtunity of Indian healthcare market. Int J Res Dev Pharm Life Sci 2(4):478–481
Hemalata SK, Shiva Kameshwari MN (2015) Anti inflammatory activity in *Urginea wightii*, (Lakxmin) Kunth, Hyacinthaceae. Int J Pure Appl Biosci 3(3):249–253
Hendriks JJ, de Vries HE, van der Pol SM et al (2003) Flavonoids inhibit myelin phagocytosis by macrophages: a structure–activity relationship study. Biochem Pharmacol 65(5):877–885
Hodgson JM, Croft KD (2006) Dietary flavonoids: effects on endothelial function and blood pressure. J Sci Food Agric 86(15):2492–2498
Hollman PC (2004) Absorption, bioavailability and metabolism of flavonoids. Pharma Biol 42(sup1):74–83
Janicijevic J, Tosic S, Mitrovic T (2007) Flavonoids in plants. In: 9th symposium on flora of Southeastern Serbia and neighbouring regions. Proc Nis (Serbia), pp 153–156
Justyna M, Kamil K, Anna K (2014) Flavonoids as important molecules of plant interaction with the environment. Molecules 19:16240–16265
Kim HP, Son KH, Chang HW, Kang SS (2004) Anti-inflammatory plant flavonoids and cellular action mechanisms. J Pharmacol Sci 3:229–245
Koosha S, Alshawsh MA, Looi CY et al (2016) An association map on the effect of flavonoids on the signaling pathways in colorectal cancer. Int J Med Sci 13(5):374
Korobkova EA, Sukala L (2012) Chemistry & biology interface. Chem Biol 2(2):59–75
Kumar S, Pandey AK (2013) Chemistry and biological activities of flavonoids: an overview. Sci World J 2013:1. 162750

Lee YW, Lee WH (2008) Protective effects of genistein on proinflammatory pathways in human brain microvascular endothelial cells. J Nutr Biochem 19(12):819–825

Liang YC, Huang YT, Tsai SH, Lin-Shiau SY, Chen CF, Lin JK (1999) Suppression of inducible cyclooxygenase and inducible nitric oxide synthase by apigenin and related flavonoids in mouse macrophages. Carcinogenesis 20(10):1945–1952

Liu RH (2002) Health benefits of dietary flavonoids: Flavonols and flavones. New York Fruit Q 10:21–23

Maalik A, Khan FA, Mumtaz A, Mehmood A et al (2014) Pharmacological applications of quercetin and its derivatives: a short review. Trop J Pharm Res 13(9):1561–1566

Merina N, Chandra KJ, Jibon K (2012) Medicinal plants with potential anticancer activities: a review. Int Res J Pharma 3(6):26–30

Middleton E, Kandaswami C, Theoharides TC (2000) The effects of plant flavonoids on mammalian cells: implications for inflammation, heart disease and cancer. Pharmacol Rev 52(4):673–751

Min SW, Ryu SN, Kim DH (2010) Anti-inflammatory effects of black rice, cyanidin-3-O-β-D--glycoside, and its metabolites, cyanidin and protocatechuic acid. Int Immunopharmacol 10(8):959–966

Mohammed MS, Osman WJ, Garelnabi EA, Osman Z et al (2014) Secondary metabolites as anti-inflammatory agents. J Pharm Sci Pharmacol 3(4):275–285

Nakagawa H, Hasumi K, Woo JT, Nagai K, Wachi M (2004) Generation of hydrogen peroxide primarily contributes to the induction of Fe (II)-dependent apoptosis in Jurkat cells by (−)-epigallocatechin gallate. Carcinogenesis 25(9):1567–1574

Narayana KR, Reddy MS, Chaluvadi MR, Krishna DR (2001) Bioflavonoids classification, pharmacological, biochemical effects and therapeutic potential. Indian J Pharmacol 33(1):2–16

Nathiya S, Durga M, Thiyagarajan D (2014) Quercetin, encapsulated Quercetin and its application-a review. Int J Pharm Pharm Sci 6(10):20–26

Neha MPV, Suganthi V, Gowri S (2013) Evaluation of anti-inflammatory activity in ethanolic extract of *Coriandrum sativum* L. using carrageenan induced paw oedema in albino rats. Der Pharma Chemica 5(2):139–143

Nijveldt RJ, Van Nood EL, Van Hoorn DE et al (2001) Flavonoids: a review of probable mechanisms of action and potential applications. Am J Clin Nutr 74(4):418–425

Ozcan T, Akpinar-Bayizit A, Yilmaz-Ersan L, Delikanli B (2014) Phenolics in human health. Int J Chem Eng Appl 5(5):393

Pal DK, Verma P (2013) Flavonoids: a powerful and abundant source of antioxidants. Int J Pharma Pharm Sci 5(3):95–98

Pan MH, Lai CS, Ho CT (2010) Anti-inflammatory activity of natural dietary flavonoids. Food Funct 1(1):15–31

Patel JR, Choubisa B, Dholakiya BZ (2011) Plant derived compounds having activity against P388 and L1210 leukemia cells. J Chem Sci 1:1–16

Plengsuriyakarn T, Viyanant V, Eursitthichai V et al (2012) Cytotoxicity, toxicity and anticancer activity of *Zingiber officinale* Roscoe against cholangiocarcinoma. Asian Pac J Cancer Prev 13(9):4597–4606

Prasain JK, Wang CC, Barnes S (2004) Mass spectrometric methods for the determination of flavonoids in biological samples. Free Radic Biol Med 37(9):1324–1350

Raja S, Ramya I, Ravindranadh K (2016) A review on protective role of Phytoconstituents against isoproterenol induced myocardial necrosis. J Pharmacogn Phytochem Res 8(5):848–864

Ren W, Qiao Z, Wang H, Zhu L, Zhang L (2003) Flavonoids: promising anticancer agents. Med Res Rev 23(4):519–534

Rodriguez-Mateos A, Vauzour D, Krueger CG et al (2014) Bioavailability, bioactivity and impact on health of dietary flavonoids and related compounds: an update. Arch Toxicol 88(10):1803–1853

Saewan N, Jimtaisong A (2013) Photoprotection of natural flavonoids. J Appl Pharm Sci 3(09):129–141

Seelinger G, Merfort I, Wolfle U, Schempp CM (2008) Anti-carcinogenic effects of the flavonoid luteolin. Molecules 13(10):2628–2651

Serra D, Paixao J, Nunes C, Dinis TC, Almeida LM (2013) Cyanidin-3-glucoside suppresses cytokine-induced inflammatory response in human intestinal cells: comparison with 5-aminosalicylic acid. PLoS One 8(9):e73001

Shi MD, Shiao CK, Lee YC, Shih YW (2015) Apigenin, a dietary flavonoid, inhibits proliferation of human bladder cancer T-24 cells via blocking cell cycle progression and inducing apoptosis. Cancer Cell Int 15(1):1

Singh S, Swapnil SKV (2011) Antibacterial properties of alkaloid rich fractions obtained from various parts of *Prosopis juliflora*. Int J Pharma Sci Res 2:114–120

Song YH, Sun H, Zhang AH, Yan GL, Han Y, Wang XJ (2014) Plant-derived natural products as leads to anti-cancer drugs. J Med Plant Herb Ther Res 2:6–15

Soni RK, Irchhaiya R, Dixixt V et al (2014) Anti-inflammatory activity of *Kirganelia reticulata* (Poir). Baill. root by carrageenan-induced rat paw oedema model Group. Int J Pharm Pharm Sci 36(60):82–38

Tapas AR, Sakarkar DM, Kakde RB (2008) Flavonoids as nutraceuticals: a review. Trop J Pharma Res 7:1089–1099

Tipoe GL, Leung TM, Hung MW, Fung ML (2007) Green tea polyphenols as an anti-oxidant and antiinflammatory agent for cardiovascular protection. Cardiovasc Haematol Disord-Drug Targets (Formerly Curr Drug Targets-Cardiovasc Hematol Disord) 7(2):135–144

Vasconcelos PCDP, Seito LN, Di Stasi LC et al (2012) Epicatechin used in the treatment of intestinal inflammatory disease: an analysis by experimental models. Evid Based Complement Alternat Med. Article ID 508902 1–12

Williams RJ, Spencer JP, Rice-Evans C (2004) Flavonoids: antioxidants or signalling molecules. Free Radic Biol Med 36(7):838–849

Woo HD, Lee J, Choi IJ, Kim CG et al (2014) Dietary flavonoids and gastric cancer risk in a Korean population. Forum Nutr 6(11):4961–4973

Zandi K, Ahmadzadeh S, Tajbakhsh S et al (2010) Anticancer activity of *Sargassum oligocystum* water extract against human cancer cell lines. Eur Rev Med Pharmacol Sci 14(8):669–673

Zhang X, Wang G, Gurley EC, Zhou H (2014) Flavonoid apigenin inhibits lipopolysaccharide-induced inflammatory response through multiple mechanisms in macrophages. PLoS One 9(9):e107072

Recent Advances in Biodegradable Films, Coatings and Their Applications

14

Kawaljit Singh Sandhu, Loveleen Sharma, Charanjiv Singh, and Anil Kumar Siroha

Abstract

Biodegradable films have appeared as an alternative for synthetic plastics. These films have generated huge attention in recent years as they are beneficial over synthetic films. These films increase the shelf life of foods because they limit migration of moisture and gases. Biomaterials such as polysaccharides, proteins and lipids are generally chosen for the formation of biodegradable films. Functional properties of biodegradable films and coatings can be improved by adding compounds with antimicrobial and antioxidant properties. Biodegradable materials can be used in variety of food products such as dairy products, ready-to-eat meals, meat products, beverages, frozen products, dehydrated products and fruits and vegetables. Different protein-based biodegradable films from different plant sources such as soy, zein, wheat protein, peanut protein, cotton seed, pumpkin, sesame, pea, mung bean, castor bean, faba bean, kidney bean, casein, whey protein have been discussed in detail. Films prepared from cellulose, starch, pectin, and seaweeds have also been discussed.

K.S. Sandhu (✉) • L. Sharma • A.K. Siroha
Department of Food Science and Technology, Chaudhary Devi Lal University, Sirsa 125 055, Haryana, India
e-mail: kawsandhu@rediffmail.com

C. Singh
Department of Food Engineering and Technology, Sant Longowal Institute of Engineering and Technology, Longowal 148 106, Punjab, India

S.K. Gahlawat et al. (eds.), *Plant Biotechnology: Recent Advancements and Developments*, DOI 10.1007/978-981-10-4732-9_14

14.1 Introduction

Biodegradable films and coatings have enticed a constant research over the last two decades as they can be potentially used in food packaging (Pavlath and Orts 2009). These films may be considered as edible or non-edible depending upon their composition. Biodegradable/edible films are thin layers of material prepared from biodegradable polymers which provide a barrier to water, gases and solute movement for the food (Park 1999). It may be applied as a wrapping material after being formed separately or may be used as a coating. Edibility of film or coating depends on the legality, safety, acceptability and nutritional value of components used. Biodegradable films have appeared as a substitute for plastic films and they are receiving substantial attention.

14.1.1 History of Biodegradable/Edible Films and Coatings

China was the first country that used coatings in the twelfth century for citrus (Hadenberg 1967). It was later used in England for improving the storage life of meat products (Labuza and Contrereas-Medellin 1981). Coatings have been used as casings by using collagen for sausages (Becker 1938, 1939), to add shine and prevent water loss from fruits and vegetables (Baldwin et al. 1995) and as a sugary coating on chocolates and confectionaries (Biquet and Labuza 1988). Gelatin has been used as a coating material for meat products (Antoniewski et al. 2007). Cellulose, starch derivatives, pectin derivatives, seaweed extracts, exudate gums and chitosan have been used as oxygen, aroma and oil barriers for different food products (Krochta and Mulder-Johnston 1997; Krochta 2002; Kester and Fennema 1986).

14.2 Uses of Biodegradable/Edible Films and Coatings

Films are structures which are formed and then applied (Guilbert et al. 1996). Edible films and coatings come under the category of GRAS as they can be consumed with the product (Park et al. 1994; Krochta and Mulder-Johnston 1997). Biodegradable films and coatings are widely used for fruits and vegetables, to reduce moisture loss, prevent weight loss, and improve appearance and to reduce gas transfer rates (Shah et al. 2016; Aquino et al. 2015; Talens et al. 2012; Ribeiro et al. 2007; Rojas-Grau et al. 2007; Bai et al. 2003; Baldwin et al. 1995). Fruit and vegetable cells use oxygen and produce carbon dioxide, during respiration. Also, fruits produce ethylene, responsible for ripening and senescence (Wills et al. 1981). Shelf life of fruits and vegetables can be increased by providing high carbon dioxide and low oxygen storage conditions, which decelerate respiration rates, resulting in decreased production of ethylene and slow ripening process. Biodegradable films and coatings can have the same effect as controlled and modified storages, depending upon their formulations. Water loss has been prevented by hydrophobic coatings (Baldwin et al. 1997). Organoleptic properties of packaged foods are improved by applying edible films provided they contain various components (flavourings, colourings, sweeteners). Films and coatings may operate as carriers for many functional ingredients and to enhance the storage life of foods by inhibiting microbial deterioration, rancidity, enzymatic browning, off flavours, etc. These ingredients include antioxidants, nutraceuticals, spices and colourants which enhance the functional properties of packaging material by adding innovative or additional functions (Salmieri and Lacroix 2006; Tapia et al. 2008).

14.3 Components of Biodegradable/Edible Films and Coatings

The major components of biodegradable films and coatings are polymers such as proteins and polysaccharides (Li and Barth 1998; Guilbert et al. 1996; Park et al. 1994). Sometimes waxes and resins as coating agents are also used (Baldwin et al. 1995). Renewable sources are used to produce biodegradable or edible ingredients, thus they degenerate faster than synthetic materials. Apart from polymers and waxes, other ingredients also have a prime role in the making of biodegradable materials. To improve the flexibility and strength of films and viscosity of coatings, glycerol, propylene glycol or sorbitol are generally used. The bulk of film-forming solutions or coatings are formed of solvent, usually aqueous alcohol (Bai et al. 2003; Ribeiro et al. 2007).

14.3.1 Protein-Based Biodegradable Films and Coatings

The intrinsic characteristics of proteins make them exceptional material for the formation of films and coatings. Proteins structure can be changed to ameliorate the protein configuration which results in film formation. Formulation, formation method and modification methods arbitrate, whether the films are biodegradable or edible (Krochta 2002). The potential of biodegradable/edible films can be determined by mechanical properties, water vapour permeability and oxygen permeability. Protein film formation involves disulphide bonds, hydrophobic interfaces and hydrogen bonds. Different factors such as physical and enzymatic treatments affect the properties of films (Park et al. 2002). Different animal or plant sources are used as film-forming materials. Proteins from plant origin such as those from soybean, corn, wheat, peanut, cottonseed, pumpkin seeds, sesame, pea, mung bean, castor bean, faba bean, kidney bean and animal-based proteins will be considered in this chapter.

14.3.1.1 Soy Protein Films and Coatings

Proteins from soy protein isolate (SPI) ranged from 200 to 600 kDa. Globulins (~90%) are the main soy proteins, and they are classified according to their sedimentation coefficients (Cho and Rhee 2004). A two-step process including protein denaturation by heat and then surface dehydration is used to prepare films from soy protein isolate. Intermolecular interactions in unfolded protein, bond through leading to network formation during drying (Gennadios and Weller 1991; Cho and Rhee

2004). Soy protein films are flexible and transparent when plasticizer is used but they have moisture barrier properties. Concentration of protein and plasticizer are the major contributors to the film properties. Films formed with 6, 7, 8 and 9% (w/w) soy protein isolate and 40, 60 and 70% (w/w) glycerol at alkaline pH 10 presented different water vapour permeability (WVP), thermal and wettability properties (Kokoszka et al. 2010). Synergistic effect of both glycerol and protein content on WVP has been studied by Cho and Rhee (2002). On increasing glycerol and protein concentrations, WVP of the films was increased. The lowest SPI (6%) and glycerol content (40%) exhibited better barrier properties of films. Glycerol, sorbitol and 1:1 ratio of glycerol and sorbitol affect the elongation at break and tensile strength of films. Increased plasticizer concentration resulted in increased EB and decreased TS. Films with glycerol exhibited flexibility at low RH conditions compared to sorbitol.

A study by Kim et al. (2002) on soy protein films revealed that heat curing at selected temperatures and pressures affect the moisture content, WVP, TS, E, colour (L*, a*, b*, ΔE) TSM of films. SPI films were cured at 60, 72.5, or 85 °C and at 101.3, 81.32, or 61.32 kPa for 24 h. Temperature and pressure synergistically and individually exhibited significant effects on moisture content of soy protein films. Moisture content decreased with increase in temperature. All heat-cured soy protein films showed lower WVP values as compared to unheated films. WVP values of soy protein films, however, were reported to be higher than whey protein and sesame protein films (Table 14.1). TS of heat-cured films were higher than unheated films. SPI film heated at 85 °C showed the highest TS compared to films cured at 60 and 72.5 °C. Drying conditions also affect properties of SPI films. Denavi et al. (2009) studied the effect of drying conditions on physical properties of SPI films by using response surface methodology (RSM). The films were dried under relative humidity (24, 30, 45, 60 and 66%) and air temperature (34, 40, 55, 70 and 76 °C). Both drying temperature and relative humidity showed significant effect on the TS and solubility of films. Development methods also exhibit effect on soy protein films. Guerrero et al. (2010) reported that thermal properties were not influenced by the processing method employed. Antimicrobial properties of soy protein films can also be improved by adding different antimicrobials (Table 14.1). Sivarooban et al. (2008) prepared SPI film, and the best quality of film was prepared with grape seed extract (GSE 1%), nisin 10,000 IU/g, and ethylene diamine tetra acetic acid (EDTA 0.16%). The film was able to reduce *Listeria monocytogenes* populations by 2.9log CFU/ml, while the population of *Escherichia coli* O157:H7 and *Salmonella typhimurium* were reduced by 1.8 and 0.6 log CFU/ml, respectively. SPI-coated sweet cherries show prolonged shelf life and reduced weight loss compared to uncoated samples. SPI-coated cherries exhibited increased resistance to moisture loss with increased concentration of protein (Lim et al. 2011).

14.3.1.2 Zein Films and Coatings

Zein protein from maize or corn (*Zea mays*) has been reported to produce biodegradable plastics (Biswas et al. 2009; Tihminlioglu et al. 2010). Zein films are generally prepared in alcohol or aqueous solutions of alcohol. Development of films involved hydrogen and hydrophobic bonds, while disulphide bonds participate little due to the

Table 14.1 Water vapour permeability (WVP) and tensile strength of biodegradable films from different sources

Source	WVP (gmm/kPa/h/m^2)	TS (MPa)	References
Soy protein	4.5	5.4	Lee et al. (2014)
Soy protein	2.4	4.1	Denavi et al. (2009)
Zein protein + chitosan	5.4	2.5	Cheng et al. (2015)
Zein	1.5	–	Gaona-Sánchez et al. (2015)
Wheat gluten	1.7	4.9	Ansorena et al. (2016)
Wheat gluten	2.1	1.9–4.2	Türe et al. (2012)
Peanut	1.6	0.77	Li et al. (2015)
Peanut (compression moulded)	–	5.7	Reddy et al. (2013)
Pea protein	–	7.0	Kowalczyk et al. (2014)
Pea protein	1.08	–	Kowalczyk and Baraniak (2011)
Pumpkin oil cake protein	–	6.56	Popovic et al. (2012)
Pumpkin oil cake protein	–	68.0	Popovic et al. (2011)
Castor bean protein	0.44–0.61	3.1–6.5	Chambi et al. (2014)
Castor bean protein	–	3.9–5.5	Makishi et al. (2013)
Kidney protein	3.32	6.2	Ma et al. (2013)
Mung bean protein	1.04	6.31	Bourtoom (2008)
Faba bean protein	1.5	4.5	Saremnezhad et al. (2011)
Whey protein	–	2.5	Perez-Gago et al. (2005)
Whey protein	0.82	1.0	Ramos et al. (2013)
Collagen	–	1.7	Oechsle et al. (2016)
Fish gelatin	–	3.42	Nur Hanani et al. (2012)
Beef gelatin	–	4.04	Nur Hanani et al. (2012)
Pork gelatin	–	4.46	Nur Hanani et al. (2012)
Pectin	1.47	26.07	Lorevice et al. (2016)
Cellulose + glycerol	1.18	17.62	Sayanjali et al. (2011)
Cellulose + PEG	12.0	20.66	Sayanjali et al. (2011)
Alginate + Silver nanoparticles	1.05–1.51	40–67	Shankar et al. (2016)
Carrageenan + Silver nanoparticles	0.94–1.29	56–68	Rhim and Wang (2014)
Chitosan	12.0	78	Pastor et al. (2013)

low content of cysteine (Gennadios et al. 1994). Zein films are extremely brittle; therefore, they require plasticizers to enhance their flexibility. Plasticizers normally used for preparation of zein film are glycerol, polyethylene glycol, polypropylene glycol and different fatty acids. Temperature, RH conditions, additives and different film formation methods affect the properties of zein films. Good water vapour barrier properties and volatile components were observed for zein films (Ozcalik and Tihminlioglu 2013). Brittleness is the major problem of zein films, and incorporation of plasticizers in film-forming solutions reduces the brittleness (Lawton 2002; Paramawati et al. 2001; Tillekeratne and Easteal 2000). Wang and Padua (2006) observed the effect of temperature on WVP of zein films. WVP decreased for films prepared at 4 °C compared to those prepared at 25 °C. Lim et al. (2011) observed the effect of glycerol on oxygen permeability, microstructure and molecular mobility of amorphous zein films. Molecular mobility was observed by using erythrosin B (Ery B) phosphorescence in the temperature range from 0 to 100 °C.

Electrospraying, an alternative method to film formation, is generally applied for the production of food nanoparticles, nanofibres and nanocapsules. Zein films were developed by electrospraying method and their structural, thermal and barrier properties were evaluated (Gaona-Sánchez et al. 2015). Films prepared by electrospraying method showed lighter yellow colour and more transparency than those prepared by casting methods. There were no significant differences observed for WVP values of zein films prepared from casting and electrospraying methods. T_g values were observed to be higher in case of electro spraying formulated films. Moradi et al. (2016) prepared zein films infused with 1% *monolaurin* (ML) and 2 or 3% *Zataria multiflora Boiss.* (ZEO) essential oil and studied antioxidant and antimicrobial activities in minced beef. Zein films were prepared using casting methods. Films from ZEO had a higher phenolic acid than plain zein films. Antimicrobial properties were observed to be increased by the addition of ZEO. Bai et al. (2003) coated apples to improve shelf life. Zein protein was used to coat apples as an alternative to other shiny coatings by dissolving zein in aqueous alcohol solution with propylene glycol. Gloss levels of apple surface varied due to different ratios of zein and propylene glycol when compared to control and shellac-coated samples. Permeability to CO_2, O_2 and WVP was significantly affected by concentration of zein. O_2 level was decreased when concentration of zein was increased in coating formulations. An optimum coating of 10% zein and 10% propylene glycol was applied to apples to maintain the overall fruit quality.

14.3.1.3 Wheat Protein Edible Films and Coatings

Wheat gluten is an elastic and cohesive protein that is left over after the removal of starch from wheat flour dough. Wheat protein is classified, according to its solubility, into four primary fractions: albumins, globulins, gliadins and glutenin (Haard and Chism 1996). Up to 47% and 34% of glutenin and gliadin, respectively, are present in wheat flour (Kinsella 1982). Gliadin is a single-chained peptide with intramolecular disulphide bonds (Kinsella 1982) which provides strength, elasticity and film formation. Strength of the protein matrix can be analysed by the disulphide bonds present in glutenin and gliadin. Shearing force also affects the gluten protein matrix, as it causes proteins to unfold, weakens hydrogens bonds and rearranges the proteins to form parallel fibres, which results in increased elasticity and stretchability of films. Gluten, therefore, is considered to show both plastic and elastic behaviours. Film formation from wheat gluten has been widely studied by various workers (Gennadios and Weller 1990; Gennadios et al. 1993; Park and Chinnan 1995; Guilbert et al. 2002; Tanada-Palmu and Grosso 2005; Marcuzzo et al. 2010; Türe et al. 2012; Zuo et al. 2009). Compression moulding is another technique used to formulate wheat gluten films. Zuo et al. (2009) compared the wheat gluten/methylcellulose bilayer casted films with compression-moulded composite films. The properties were reported to be superior for casted films. Blending with methylcellulose also improved the mechanical, morphological and barrier properties of films. Different treatments also affect the properties of wheat gluten films.

Effect of ultrasound treatment (UT) on wheat gluten films was studied. UT interestingly improves the appearance of gluten films. Marcuzzo et al. (2010) developed films under acidic conditions. SE-HPLC was used to determine the distribution of gliadin and glutenin.

In biodegradable films, various types of fibres such as leaf and wood fibres (Averous et al. 2001), flax and ramie fibres (Wollerdorfer and Bander 1998) and wood pulp (De carvalho et al. 2002) have been found to be useful in increasing their strength. Spelt and wheat bran fibres with glycerol were used to improve the mechanical properties of wheat gluten films by using RSM (Mastromatteo et al. 2008). Elastic modulus of films was increased with increased concentration of bran and reduced amount of glycerol. Türe et al. (2012) developed gluten films containing potassium sorbate by compression moulding method and showed antimicrobial activity against *A. niger* and *F. incarnatum*. Microbial resistance of films was observed up to 1 week when films were left in agar solution. In addition to antimicrobial properties, potassium sorbate also improved the dynamic, mechanical, permeable and morphological properties of wheat gluten films. Active food packaging has been prepared by using different compounds, comprising enzymes, spices and organic acids (Martínez et al. 2013; Tharanathan 2003). Terpenoids and phenolic are chief compounds present in essential oils (EOs) isolated from plants (Burt 2004; Lamber et al. 2001).

Ansorena et al. (2016) used thermoplastic process to develop films. The analysis was done for thermal, structural, mechanical, antimicrobial and antioxidant properties of the formulations. Increased concentration of thyme oils resulted in films with increased antimicrobial activity. Coatings from wheat gluten were used to prolong the shelf life of refrigerated strawberries. Coatings were formed directly on the surface of fruits and the fruit quality was scrutinized by weight loss, firmness, total soluble solids, titratable acidity, reducing sugars and sensory evaluation. Coatings had a significant effect on the quality of fruit as compared to control (Tanada-Palmu and Grosso 2005).

14.3.1.4 Peanut Protein Films and Coatings

About 75% of the total peanut proteins are arachin and conarachin, and they are reported to have good thermal stability. Therefore, peanut protein acts as a good source to prepare films (Sun et al. 2014) and formation of active peptides (Ji et al. 2014). First peanut protein-lipid films were developed on the surface of heated peanut milk (Wu and Bates 1973; Aboagye and Stanley 1985). The tensile strength of these films was lower compared to films formed by casting of peanut protein isolate solutions. Liu et al. (2004) scrutinized the physical and mechanical properties of peanut protein films. Different physical and chemical treatments were implied on the film-forming solutions.

Reddy et al. (2012) prepared crosslinked films from peanut protein isolated from peanut meal. Crosslinked peanut protein films had good wet and dry tensile strength properties but poor biocompatibility compared to films made from crosslinked films. Reddy et al. (2013) also developed thermoplastic peanut protein films by compression moulding method. Peanut protein was isolated and compression moulded into films, and they exhibited better tensile strength properties than casted and crosslinked films. Compared to solution casting method compression moulding was more convenient, environment friendly and they had better properties. Physical modifications play an important role in the improvement of film properties. Li et al. (2015) studied the properties of peanut protein isolate prepared by glycation method using gum arabic for 3, 6 and 9 days. Effect of glycation on the structural and physical properties of glycated films was studied. As the glycation increased, tensile strength of peanut protein glycated films decreased while WVP and elongation increased.

14.3.1.5 Plant Protein Films of Limited Availability

14.3.1.5.1 Cottonseed Protein Films and Coatings

Cottonseed kernel consists of 30 to 40% (w/w) proteins. The proteins are mainly comprised of globulins (60%), albumins (30%), gluteins (0.5%) and prolamins (8.6%) (Saroso 1989). First cottonseed films were prepared by steeping grains in water at 90 °C (Wu and Bates 1973). These films exhibited poor mechanical properties which might be reason for their limited usage. Marquié et al. (1995) prepared cotton seed protein films directly from cotton flour by casting method. The proteins were solubilized by using appropriate pH and temperature conditions and insoluble components were removed by centrifugation. The supernatant collected was homogenized and casted. Plasticizer was added to supernatant prior to casting to obtain films with good mechanical properties. Films were prepared under different conditions to optimize the processing condition of films from cotton seed protein films (Marquie and Guilbert 2002). Marquie (2001) also crosslinked cotton seed protein by formaldehyde, glutaraldehyde and glyoxal to enhance the mechanical properties of films. HPLC was used to analyse the crosslinking effects of chemicals on cotton seed protein. Crosslinking could deliver useful chemical tools for modifying the mechanical properties of cottonseed protein films.

14.3.1.5.2 Pumpkin Seed Protein Films

Pumpkin seeds are among the richest sources of protein (25 to 35%). Austria, Hungary, Slovenia and Serbia are the major regions producing pumpkin seed oil and pumpkin seed oil cake (PuOC). Films were prepared from pumpkin oil seed cake and the effect of pH and temperature was studied. The highest TS (68.1 MPa) and elongation to break (EB) (36.6%) were recorded when the pH was 12 and temperature was 90 °C (Popovic et al. 2011). Mechanical properties were observed to be higher for pumpkin oil seed films as compared to films from other protein sources (Table 14.1). pH and heating temperature had no significant effect on PuOC films. Influence of pH and temperature on morphology was determined by scanning electron microscopy (SEM). Popovic et al. (2012) developed biodegradable films as a potential for gas barrier properties from pumpkin seed protein isolate. Protein was isolated from PuOC and films were developed at different pH (2–12) and glycerol concentrations (0.3–0.6 g glycerol/g PuOC protein isolate). Films formed at alkaline pH exhibited better mechanical and barrier properties while those prepared at pH 4–6 were not formed which might be due to insolubility of protein. Films showed excellent barrier properties for N_2, O_2, CO_2 and air and exhibited antioxidant properties. Good barrier and antioxidant properties of pumpkin seed protein films make them a potential source of coatings for food products.

14.3.1.5.3 Sesame Protein Films

Sesame meal has high possibilities of use as a protein source or as an ingredient in the food industry. Sesame meal (protein content of 40–60%) and isolated protein have particularly high contents of methionine (2.4–4.0%) and total sulphur-containing amino acids (3.8–5.5%) (Johnson et al. 1979). Sesame protein has low solubility that limits its food applications in its native form. Physicochemical properties of sesame protein concentrate (Inyang and Iduh 1996; Zaghloul and Prakash 2002), isolate (Khalid et al. 2003; Gandhi and Srivastava 2007; Onsaard et al. 2010; Sharma et al. 2016) have been studied. Films prepared from sesame protein isolate exhibited better barrier properties than soy, peanut, whey and mung bean protein films (Table 14.1). First sesame-protein-based films developed were reported to be nanocomposite films with addition of nanoclay at different levels. Lee et al. (2014) prepared nanocomposite sesame protein isolate/nanoclay biodegradable films. Montmorillonite and cloisite nanoclays were used in different concentrations to prepare films. Incorporation of nanoclays enhanced the mechanical and physical properties of films. Then, Sharma and Singh (2016a) optimized the process conditions for sesame protein isolate films using RSM. Protein concentration, pH, temperature and plasticizer significantly affected the characteristics of sesame protein edible films (Table 14.1). Thermal and water barrier properties of films were better when compared with films prepared from other sources. Overall, the optimized film prepared with 9% protein, pH 12, 90 °C and 10% glycerol concentration showed better properties. However, this film was less transparent and dark in colour. Further, to improve mechanical and barrier properties and transparency of films, composite films were developed by utilizing gum rosin (GR) and sesame protein isolate at different concentrations. Composite films from sesame protein isolate and GR were developed and the effect of GR addition on the mechanical, thermal, physical, optical and morphological properties was scrutinized. Addition of GR significantly enhanced the properties of films as compared to control developed from only sesame protein isolate. Transparency, mechanical, WVP and thermal properties of films were enhanced (Sharma and Singh 2016b).

14.3.1.5.4 Pea Protein Films

Dried peas mainly contain carbohydrates (35%), proteins (27%), fibres (27%) and a very low amount of lipids (Boulter 1983). Protein fraction has not been fully utilized, whilst the fibre and starch fraction of peas are widely used in different food industries. Pea protein mainly contains globulins, and it is composed of legumin, vicilin and convicilin (Boulter 1983). Albumins constitute 13 to 14% of the total proteins. Edible films developed from denatured pea protein concentrate (PPC) solution showed good elasticity and strength. TS decreased while elongation and WVP increased when plasticizer (glycerol) concentration in film was increased. PPC/glycerol composition of 70/30 and 60/40 have been reported to form very tough and stretchable films (Choi and Han 2001). The physical and mechanical properties of pea protein films were similar to those of other protein films. Film formation methods and heat denaturation also affect the properties of pea protein biodegradable films. Choi and Han (2002) developed pea protein films by providing different heat treatments. Film-forming solutions were heated at 90 °C for 5, 10, 20, 30, 40 and 50 min for heat denaturation of proteins. Heat denaturation treatment increased strength and transparency but reduces wettability of films when compared to those developed from non-heated film-forming solutions of pea proteins.

Kowalczyk and Baraniak (2011) studied the effect of pH, plasticizer and heat on film preparing solutions for the preparation of biodegradable films from pea proteins. They studied the effects of different concentrations of sorbitol, glycerol, pH and varying time and temperature of film-forming solution (FFS) on properties of films. Kowalczyk et al. (2014) reported that films plasticized with glycerol showed significantly higher WVP, lower MC and lower solubility, in comparison with glycerol-plasticized films. Mechanical properties were evaluated for pea protein films plasticized with glycerol and sorbitol at different concentrations. Films plasticized with sorbitol exhibited better mechanical properties than films plasticized with glycerol. However, increased concentration of both plasticizers exhibited decreased tensile strength and increased elongation at break.

14.3.1.5.5 Mung Bean Protein Films

Approximately 25–30% protein is present in whole seed of mung bean (Magee 1996). Mung bean is also widely used in the manufacture of noodles. Protein is a by-product thrown away along with wash water during recovery of the starch from mung bean. Mung bean possesses desirable nutritional and functional properties and acts as a potential source of proteins. Due to this reason, mung bean protein is generally accepted as a food additive in the form of a protein isolate (Coffman and Garciaj 2007). Bourtoom (2008) prepared edible films from mung bean proteins and RSM was used to study the effect of pH, temperature and heating time of the film solution on various film characteristics. Heating time showed less significant effect compared to pH and temperature. The tensile strength was the highest at pH 9 but elongation of film was reportedly low. WVP, solubility and moisture content were the lowest for films prepared at pH 9 and comparable to films from other sources. Films were dark yellowish in colour and darkness increased with increasing pH of film-forming solutions. Nasriyanti et al. (2009) prepared films from mung bean protein isolates and water vapour permeability (WVP) and tensile strength of biodegradable film from different sources were investigated. To study the effect of pH on film-forming properties, alkaline pH (8, 10 and 12) was maintained for film-forming solutions. WVP decreased with increasing pH and tensile strength was found the highest at pH 12. The protein film (pH 8) had a higher moisture sorption isotherm than those made at pH 10 and 12.

14.3.1.5.6 Castor Bean Protein Films

Castor bean cake has a protein content of 28–43%. It is the major by-product during castor oil production. Due to its toxicity it cannot be utilized as animal feed but it can be used as a soil fertilizer or as a filler (Anandan et al. 2005; Godoy et al. 2009). Makishi et al. (2013) developed biodegradable films from castor protein crosslinked by glutaraldehyde and glyoxal. Protein was extracted by alkaline solubilization (pH 12) at 50 °C and freeze dried. Films were developed by removal of water from film-forming solutions by using 2, 4, 6, 8 and 10% (w/w) protein concentration and 1 g of glyoxal or glutaraldehyde/100 g of protein as crosslinking agents. Solubility, microstructure, colour, opacity, gloss and mechanical properties of films were studied by puncture and tensile strength tests. Films prepared with 6% protein exhibited the best properties. Chambi et al. (2014) also studied the effect of crosslinking on films. Glyoxal crosslinked films exhibited better properties compared to films crosslinked with glutaraldehyde. Further, the properties of castor bean protein films have been improved by increasing pH conditions (10, 11 and 12) during extraction of protein from castor bean de-oiled cake. Sufficiently cohesive films were obtained from these proteins. The pH at which the protein was extracted affected the tensile strength, elongation, Young's modulus, WVP and hydrophobicity of films. Also, films were developed from castor bean proteins by reinforced cellulose fibres for use in agriculture, as bags for planting seedlings. The main aim of study was to scrutinize the effect of fibre concentration of opacity, mechanical, WVP, solubility, morphological, chemical and thermal properties of developed films. Cellulose fibres are reported to improve the tensile properties of films, and scanning electron microscopy revealed that these fibres completely dissolve in film matrix (Oliveira et al. 2015).

14.3.1.5.7 Faba Bean Protein Films

Faba bean protein contains up to 35% protein (dry basis) depending on its cultivar. It is a popular legume food in many countries and is cultivated in large areas. The protein fractions of faba bean were highly soluble at acidic and alkaline pH and exhibited only a narrow range of insolubility at pH 4.5. Saremnezhad et al. (2011) formulated the films from faba bean protein that was isolated from its seed flour. The effects of pH, and plasticizer and protein concentration were studied on the mechanical and barrier properties of films. Plasticizer did not affect the WVP values for faba bean protein films, whereas solubility and elongation were increased and tensile strength decreased by increased concentration of the plasticizer. The lowest WVP and highest tensile strength were observed for the films prepared at pH 12. The film properties were comparable to those of other protein sources. Further effect of pH on faba bean protein films was observed by Montalvo-Paquini et al. (2013). Faba bean protein films were developed by casting method by using 3% (w/w) of faba bean protein concentrate and 50% (w/w of faba bean protein concentrate) glycerol as a plasticizer. At alkaline pH conditions, the films showed the highest tensile strength and puncture strength as alkaline conditions promoted denaturation of protein.

14.3.1.5.8 Kidney Bean Protein Films

The protein content of kidney bean (*Phaseolus vulgaris* L.) varies from 20 to 30% (Sathe 2002) and reported to have good functional properties (Kimura et al. 2008; Tang 2008). Composite films, by the fabrication of chitosan with kidney bean protein isolate (KPI), were developed and characterized (Ma et al. 2013). Films developed from KPI at neutral pH exhibited poor mechanical properties. Therefore, acidic pH conditions were used to prepare composite films from kidney bean protein isolate-chitosan (KPI/CH). Tensile and thermo-mechanical properties and microstructure of the composite films showed that films were less rigid and much more flexible. Surface hydrophobicity of films was enhanced by incorporation of chitosan. Further properties of KPI/CH films were enhanced by preparing films by ultrasonic pre-treatment. Ultrasound generator was used for preparing film-forming solution. Ultrasound pre-treated films showed a stiff behaviour, whereas the composite films were much more flexible. Conformational changes might have occurred due to ultrasonic treatment which leads to the visible effect on tensile properties (Fan et al. 2014).

14.3.1.6 Casein Protein Films and Coatings

Casein is the main protein in milk, and it is secreted in the mammary glands and found nowhere else, which makes this protein unique from other proteins (Fox and McSweeny 1998). Some amount of secondary and tertiary structures is present in casein due to high proline content (Fox and McSweeny 1998). Caseins directly form films with aqueous solutions without any treatment due to their random-coil nature and capability of hydrogen bonding. Hydrophobic interactions, electrostatic and extensive hydrogen bonds facilitate the formation of films from casein (McHugh and Krochta 1994). Films solubility in water was reportedly higher compared to other protein sources (Vachon et al. 2000). Casein films appeared to exhibit similar moisture barrier behaviours as exhibited by whey protein or soy protein films (Table 14.1). Films were developed from different casein products such as sodium caseinate, calcium caseinate, and magnesium caseinate or rennet casein (Chen 2002). Magnesium and calcium caseinate films showed improved WVP values than sodium caseinate films. Sodium caseinate films treated with calcium chloride at pH 9.5 induce crosslinking effect and reduced the WVP of films (Avena-Bustillos and Krochta 1993). UV radiations were also used to modify the properties of sodium caseinate films. Schou et al. (2005) observed the effect of plasticizer on the properties of casein. Increased plasticizer concentration increased the WVP of films. Atarés et al. (2010) developed biodegradable films from sodium caseinate after incorporation of cinnamon and ginger essential oils to enhance the microbial and antioxidant activity. Lipid addition does not affect the mechanical properties of films, while WVP was observed to be slightly reduced. Cinnamon oil affected the optical characteristics of films as compared to ginger oil. Stearic acid was used as an additive in sodium caseinate films and physical and mechanical properties were characterized. Effects of the ratio of stearic acid and sodium caseinate to water on the WVP and mechanical properties of the prepared films were determined. Higher amount of stearic acid was found responsible for increased elastic modulus and consistency coefficient. WVP values were reduced due to the addition of stearic acid (Rezvani et al. 2013). Hsan-tsao gum (HG) was incorporated in casein and prepared films were characterized for physical, structural and antioxidant properties of films. Addition of HG improved the mechanical and barrier properties of casein films. NMR analysis exhibited that HG changed the mobility of water molecules in casein films (Yang et al. 2015).

14.3.1.7 Whey Protein Films and Coatings

Whey protein is the remaining portion in milk serum after cheese manufacturing (Morr and Ha 1993). Whey proteins are divided into two categories (whey protein concentrates and whey protein isolates) depending upon their industrial utilization. Heat denaturation of whey protein is necessary for the formation of films. Heating improves the three-dimensional structure and exposed the internal SH and hydrophobic groups (Shimada and Cheftel 1998). Perez-Gago et al. (2005) studied the effect of whey protein and hydroxyl propyl methyl cellulose based edible coatings on the shelf life of fresh-cut apples. Apples were coated with composite coatings and their loss in weight, colour and browning index were determined. Higher L* value and lower a*, b* values indicated that whey protein coatings exerted anti-browning effect. Weight loss did not reduce in coated samples which might be due to higher relative humidity. Whey protein films exhibited antimicrobial properties when incorporated with oregano, rosemary and garlic essential oils. The films with garlic essential oil at 3 and 4% levels showed more antimicrobial properties (Seydim and Sarikus 2006).

Ramos et al. (2012) investigated the antimicrobial activity of whey protein films induced with various antimicrobial agents. Chitooligosaccharide and lactic acid exhibited the antimicrobial effect against gram –ve and gram +ve bacteria. Coated cheese surface was more resistant to bacterial growth while less effective against yeast and mould. Further, effect of whey protein purity and glycerol content was studied by Ramos et al. (2013). Films were developed from whey protein concentrate and whey protein isolate and glycerol at three different levels. Films were characterized and molecular structure, mechanical, thermal and barrier were analysed. Whey protein films were observed to have good barrier and mechanical properties when compared to soy, zein and wheat protein films. Films plasticized at 40 and 50% glycerol exhibited overall better properties. Thermal properties were determined by differential scanning calorimetry (DSC) and thermal gravimetric analysis (TGA), which revealed that whey proteins exhibited excellent thermal properties and were comparable to synthetic films.

14.3.1.8 Collagen and Gelatin Protein Films and Coatings

Collagen is a protein that consists of connective tissues, skins and tendons. Gelatin is prepared from collagen by partial hydrolysis. Collagen films are commercially used as casings for sausages (Balian and Bowes 1977). The collagen films have excellent gas barrier properties comparable to synthetic films, but no data is available about the water barrier properties of collagen films. Collagen films also exhibited good mechanical properties. Simelane and Ustunol (2005) utilize collagen in the manufacture of polish sausage and studied the effect on mechanical properties. O'Sullivan et al. (2006) studies the difference in mechanical strength between collagen films from different fish species. Oechsle et al. (2016) prepared extruded collagen films with the addition of co-gelling proteins. Results revealed that co-gelling proteins have higher consistency index. Gelatin biodegradable films are poor water barriers but act as good gas barriers. Plasticized films from gelatin are clear, flexible and strong.

Bradbury and Martin (1952) were the first to develop cold and hot casted films. More TS was observed for solutions dried at 20 °C and heated at 60 °C. Guilbert et al. (1996) found the effect of crosslinking/denaturation of gelatin with calcium ions on WVP of films. Gelatin from tuna fish showed the ability to act as antioxidant carriers in two different murta ecotype leaves (Gomez-Guillen et al. 2007). Pérez-Mateos et al. (2009) formulated biodegradable films from cod gelatin and sunflower blends, and these films had improved hydrophobic properties and reduced water vapour transfer rates. Gómez-Estaca et al. (2010) studied the effect of essential oils of clove, fennel, cypress, lavender, thyme, pine and rosemary which were added to gelatin-chitosan blends to develop biodegradable films with antimicrobial activity. Films showed antimicrobial activities against six selected microorganisms. Cloves exhibited the highest antimicrobial activity, followed by lavender and rosemary. Nur Hanani et al. (2012) manufactured films from beef, pork and fish gelatin sources and the mechanical properties were characterized. Mechanical properties of gelatin films, WVP, oxygen permeability, oil permeability and aqueous solubility of films were evaluated. Tensile strength and puncture strength were observed to be higher for films with higher concentration of gelatin. Guo et al. (2014) determined the crosslinking effect of periodate oxidation of xanthan gum on biodegradable films from gelatin protein. Oxidized xanthan gum with different levels of aldehyde content was prepared and used as a crosslinking agent for gelatin films. Xanthan gum exhibited drastic improvement in the moisture content, soluble matter and water vapour permeability.

14.3.2 Polysaccharide-Based Edible Films and Coatings

Polysaccharide films are made up of starch, alginate, cellulose ethers, chitosan, carrageenan, or pectin and have film-forming abilities with solvents. Films from polysaccharides impart excellent gas barrier properties due to the presence of the polymer chains (Baldwin et al. 1995). During the past years, demand and utilization of polysaccharides have increased for making biodegradable films and edible coatings especially in the field of dairy, bakery, meat products, fruits and vegetables. Starch and cellulose derivatives are widely used, and other sources such as gums and pectin are also gaining importance.

14.3.2.1 Cellulose and Derivatives

Cellulose materials have gained much attention in the last decades as a prospective replacement for synthetic polymers, because of their biodegradable character that can be suitably used for different applications. Regenerated cellulose films like cellophane, due to their poor water vapour permeability, have not been used for food applications compared to their synthetic counterparts, while transparent cellulose films showed good gas barrier properties at dry conditions and are comparable to dense polyethylene (HDPE) and low-density polyethylene (LDPE) films. Main cellulose and cellulose derivatives used for biodegradable/edible film developments are microcrystalline cellulose, nanocrystalline cellulose, microfibrillated and nanofibrillated cellulose, methylcellulose (MC), hydroxypropyl methylcellulose (HPMC), hydroxypropyl cellulose (HPC) and carboxymethyl cellulose. Microcrystalline or nanocrystalline cellulose is prepared by acid hydrolysis which disrupts the reactive crystalline regions as fine crystals. Microfibril and nanofibril cellulose is prepared by passing the native cellulose under high shear force and pressure which changes cellulose into microfibrillar or nanofibrillar fragments. Also carboxy, methoxy and hydroxypropyl substitutions affect the physical and chemical properties of cellulose.

Su et al. (2010) developed biodegradable films from carboxymethyl cellulose and soy protein isolate. Edible films from carboxymethyl cellulose (CMC) and soy protein isolate (SPI), plasticized with glycerol were prepared and their properties were studied. Glass transition showed single T_g value, indicating both CMC and SPI form single phase with good blending properties. Composite edible films based on hydroxypropyl methylcellulose reinforced with microcrystalline cellulose were prepared and the mechanical and moisture barrier properties were investigated. Three different sizes of microcrystalline cellulose (MCC) were incorporated in HPMC solutions and identical MCC particles were coated with lipids and then incorporated in HPMC solutions. Effects of non-lipid MCC and lipid-coated MCC were then compared. MCC-incorporated HPMC films showed 40% reduction in WVP and lipid-coated MCC showed 50% reduction in WVP, compared to plain HPMC films. Both types of MCC particles improved the tensile properties of HPMC films (Bilbao-Sáinz et al. 2010).

Fresh pistachios were coated with films having carboxymethyl cellulose and potassium sorbate and film properties were evaluated (Sayanjali et al. 2011). Antimicrobial effects were studied against *Aspergillus* species by using agar diffusion assay. Tensile strength of films decreased with increasing concentration of sorbate, and WVP values increased with increased concentration of sorbate.

14.3.2.2 Starches and Derivatives

Starch is made up of amylose and amylopectin. Amylose is linear molecule with α-(1–4)-D linkages of glycosidic units whereas amylopectin is a branched macromolecule having both α-(1–4) and α-(1–6) linkages. Starch is relatively cheap, and it is widely used in food materials as a functional ingredient. Tubers, cereal grains and legume seeds are rich sources of starch. Film-forming ability and mechanical properties of aqueous native maize starches were evaluated by Palviainen et al. (2001). Films from maize starch with high amylose content formed stronger films compared to those from maize with less amylose. Waxy varieties of corn and rice which contains high amylopectin have poor film-forming properties. Films formed from amylose are tasteless, colourless and non-toxic. Garg and Jana (2007) studied the properties starch films prepared from crosslinked and glycerol-modified corn starch using epichlorohydrin as a crosslinking agent. Corn starch, crosslinked starch and glycerol-modified starch in LDPE were blended to form different films. Mechanical properties of films were characterized for tensile strength, burst strength, elongation and melt flow index. Blending starch on LDPE decreased the tensile strength and increased the burst strength. However, films with crosslinked starch showed better mechanical properties than native starch blended films. Reddy and Yang (2010) crosslinked corn starch films with citric acid to increase their tensile strength and thermal stability.

Corn starch varying in amylose content was used to prepare edible films. It was observed that starch with high amylose content was difficult to process by extrusion, which was overcome by increasing process temperature and moisture content (Li et al. 2011). Saberi et al. (2016) optimized the properties of edible films made from pea starch and guar gum. The effect of processing variables on various properties was studied.

14.3.2.3 Pectin and Derivatives

Pectin, a component of all the cell walls, is composed of (1–4) α-D-galacturonic acid units with single (1–2) α-L-rhamnose residues (Ridley et al. 2001). Properties of high and low methoxy pectin films were improved by incorporating chitosan nanoparticles and the mechanical, thermal, barrier and structural properties of these films were analysed (Lorevice et al. 2016). FTIR results confirmed the blending of both components. The other results exhibited that pectin/chitosan nanocomposite films had improved mechanical properties compared to films from native pectin. Recently, films from pomegranate peel pectin with addition of nanoclay were also developed. Films were prepared from pectin extracted from pomegranate peels, with different contents of montmorillonite (MMT) as a nanoreinforcement material. Tensile strength and modulus of films were improved by MMT, when added up to 6 wt%; further addition of MMT to 8 wt%, however, reduced the strength of films.

14.3.2.4 Seaweed Extracts

14.3.2.4.1 Alginates

Alginate is low cost, non-toxic and biodegradable, which makes it suitable for film-forming properties (Vu and Won 2013). Various functional and film-forming properties have been studied by different workers (Dhanapal et al. 2012; Zactiti and Kieckbusch 2006). Alginate dissolves in this solution and extracted as sodium alginate. Then alginate is separated by precipitation, as alginic acid or calcium alginate (McHugh 2003). Rojas-Graü et al. (2007) developed the apple puree-alginate coatings as a carrier of antimicrobials agents to enhance the shelf life of fresh-cut apples. Lemongrass, oregano oil and vanillin were incorporated in apple puree-alginate edible coatings as antimicrobial agents and their effect was determined. Growth of psychrophilic aerobes, yeasts and moulds were significantly inhibited by all antimicrobial coatings. Alginate-based edible coatings with antibrowning agents were developed to maintain the sensory properties of fresh-cut pears (Oms-Oliu et al. 2008). N-acetylcysteine and glutathione were used as antibrowning agents. Both antibrowning agents inhibited the browning of fresh-cut pears for 2 weeks and also reduced the microbial growth. Increased vitamin C content and polyphenols were observed in the case of coated fresh-cut pears compared to non-coated sample. Azarakhsh et al. (2012) optimized the alginate-gellan-based edible coatings for fresh-cut pineapples using RSM. The results showed good fitting of response variables using RSM models. Recently, Shankar et al. (2016) prepared alginate/silver composite films and studied the effect of types of silver particles. Films were prepared by reinforcing different nanoparticles and the effect of the silver particles was studied for optical, mechanical, water barrier and antimicrobial properties. Strong antimicrobial activity against two food pathogenic bacteria *Escherichia coli* and *Listeria monocytogenes* was observed for all composite films.

14.3.2.4.2 Carrageenan Edible Films and Coatings

Carrageenans have high potentiality for film formations as they are water-soluble. Different types of carrageenans were extracted from different seaweeds. Numbers of sulphate ester groups, together with the anhydrogalactose bridges, are responsible for carrageenan functionality. Lee et al. (2003) reported applications of carrageenan as an edible films and coatings on different kinds of food products. Formation of three-dimensional network by polysaccharide double helices with gelation mechanism occurs during drying in the film-forming mechanism. Karbowiak et al. (2006) evaluated wetting properties on the surface of carrageenan-based edible films. Contact angle measurements were used to study surface properties. Alves et al. (2010) evaluated the barrier properties of biodegradable composite films from carrageenan and pectin blends and mica flakes. Significant decrease in WVP, O_2, CO_2, was observed by the addition of mica flakes in composite films. Films were prepared from different ratios of carrageenan and locust bean gum, and TGA, X-RD, FTIR and DMA techniques were used to highlight the interactions between the two polysaccharides. Blending of films improved the overall properties of films, including tensile strength, when compared to native films from both sources (Martins et al. 2012). Rhim and Wang (2014) prepared nanocomposite films from carrageenan by reinforcing nanoclay and silver nanoparticles. They observed increased tensile strength (TS) and decreased WVP of the nanocomposite films. Shankar et al. (2015) studied the antimicrobial and other properties of nanocomposite films reinforced with chitin nanofibrils. Increase in tensile strength of carrageenan/chitin nanocomposite films was observed with the reinforcement at 5% level and elongation at break, transparency and WVP value were slightly decreased. Nanocomposite films from carrageenan/chitin films showed strong antimicrobial activity against *Listeria monocytogenes* (a gram-positive food-borne pathogen).

14.3.2.5 Chitosan

Chitosan made from chitin has been substantially used as coating applications due to its film-forming properties. Films made from chitosan are not soluble in water, and they had good tensile strength. Abugoch et al. (2011) prepared and characterized chitosan-blended quinoa protein edible films. Films were developed by solution casting method and characterized by FTIR, SEM, thermal properties, and mechanical and barrier properties. Structural and thermal properties exhibited clear

Fig. 14.1 Biodegradable films from sesame protein isolate (**a**, **b**) and corn starch (**c**)

interaction among both components and showed enhanced properties of composite films compared to native films. Ripening process of banana was delayed by applying chitosan-glycerol-based coating. Optimization of concentration of chitosan and glycerol for coating of Berangan banana (*Musa sapientum* cv. *Berangan*) were carried out by using RSM. Chitosan concentration showed significant ($P < 0.1$) effect on variables studied excluding TSS (Jafarizadeh Malmiri et al. 2011). Pastor et al. (2013) evaluated the physical and antioxidant properties of chitosan- and methylcellulose-based films containing resveratrol. Resveratrol reduces the transparency, gloss and stretchability of films. WVP was slightly improved and no antimicrobial activity of films was observed. However, antioxidant property of film was enhanced with increased concentration of resveratrol. Recently, tomatoes were supplemented with iodine by applying iodine-doped chitosan edible films (Limchoowong et al. 2016). Tomatoes were coated with an iodide-doped chitosan (CT-I) solution with an edible film (1.5 μm). Films were characterized by FTIR and SEM. Antioxidant activity of tomatoes was not affected by iodide supplementation (Fig. 14.1).

14.4 Applications of Biodegradable/Edible Films and Coatings

Synthetic films are not biodegradable, so they pose environmental problems. In recent years people are becoming more aware towards the safety of environment and the adverse effect of non-biodegradable films on the environment. This has resulted in the use of biodegradable films which are generally made from agriculture feedstock and food processing industry wastes. Complete shifting from synthetic to biodegradable films seems not possible now but BIOPACKAGING will be the future. Protein and polysaccharide films are regarded as most multifaceted due to their large number of different applications. These act as potential carriers of antimicrobials and antioxidants which help to increase the shelf life of meat products, such as sausages, fillets or beef patties, by preventing their oxidation and inhibiting microbial growth. These are also used for supplementation of various food products by coating applications. Moreover, these can be used to improve the moisture and oxygen barriers to increase the shelf life of fruits and vegetables.

Mechanical properties of crosslinked films are comparable to synthetic films and can be substituted for light-weighted materials such as pouches, bags and as primary packaging films.

14.5 Conclusion

Synthetic materials used for making packaging films pose serious threat to environment and these materials may be substituted with films made from biodegradable materials. Protein-, cellulose-, starch- and pectin-based films can be produced conveniently. These films show properties comparable to synthetic films. Chemical or enzymatic modification of biomaterials can be done for making them suitable for diverse applications. At present, for packaging applications, synthetic films are primarily used, but the use of biodegradable films have gained momentum. Lots of research needs to be done by biotechnologists, food technologists, chemists, polymer technologists, environmental scientists, and microbiologists for making biodegradable films more commercially viable.

References

Aboagye Y, Stanley DW (1985) Texturization of peanut proteins by surface film formation. Influence of process parameters on film forming properties. Can Inst Food Sci Technol J 18:12–20

Abugoch LE, Tapia C, Villamán MC et al (2011) Characterization of quinoa protein-chitosan blend edible films. Food Hydrocoll 25:879–886

Alves VD, Costa N, Coelhoso IM (2010) Barrier properties of biodegradable composite films based on kappa-carrageenan/pectin blends and mica flakes. Carbohydr Polym 79:269–276

Anandan S, Kumar GKA, Ghosh J et al (2005) Effect of different physical and chemical treatments on detoxification of ricin in castor cake. Animal Feed Sci Technol 126:159–168

Ansorena MR, Francisco Z, Norma EM (2016) Active wheat gluten films obtained by thermoplastic processing. LWT – Food Sci Technol 69:47–54

Antoniewski MN, Barringer SA, Kinipe CL et al (2007) Effect of gelatin coating on the shelf life of fresh meat. J Food Sci 72:382–387

Aquino de AB, Blank AF, Santana de LCL (2015) Impact of edible chitosan–cassava starch coatings enriched with *Lippia gracilis* Schauer genotype mixtures on the shelf life of guavas (*Psidium guajava* L.) during storage at room temperature. Food Chem 171:108–116

Atarés L, Bonilla J, Chiralt A (2010) Characterization of sodium caseinate-based edible films incorporated with cinnamon or ginger essential oils. J Food Eng 100:678–687

Avena-Bustillos RJ, Krochta JM (1993) Water vapour permeability of caseinate-based edible films as affected by pH, calcium crosslinking and lipid content. J Food Sci 58:904–907

Averous L, Fringant C, Moro L (2001) Plasticized starch-cellulose interactions in polysaccharide composites. Polymer 42:571–578

Azarakhsh N, Osman A, Ghazali HM et al (2012) Optimization of alginate and gellan-based edible coating formulations for fresh-cut pineapples. Int Food Res J 19:279–285

Bai J, Alleyne V, Hagenmaier RD et al (2003) Formulation of zein coatings for apples (*Malus domestica* Borkh). Postharvest Biol Technol 28:259–268

Baldwin EA, Nisperos-Carriedo MO, Baker RA (1995) Use of edible coatings to preserve quality of lightly (and slightly) processed products. Crit Rev Food Sci Nutr 35:509–524

Baldwin EA, Nisperos-Carriedo MO, Hagenmaier RD et al (1997) Using lipids in coatings for food products. Food Technol 51:56–61
Balian G, Bowes JH (1977) The structure and properties of collagen. In: Ward AG, Courts A (eds) The science and technology of gelatin. Academic Press, New York, pp 1–30
Becker OW (1938) Sausage casing. US patent 2:115:607
Becker OW (1939) Method of and apparatus for making artificial sausage casings. US Patent 2:161:908
Bilbao-Sáinz C, Avena-Bustillos RJ, Wood DF et al (2010) Composite edible films based on hydroxypropyl methylcellulose reinforced with microcrystalline cellulose nanoparticles. J Agric Food Chem 58:3753–3760
Biquet B, Labuza TP (1988) Evalutation of the moisture permeability characterstics of chocolate films as an edible moisture barrier. J Food Sci 53:989–998
Biswas A, Selling GW, Woods KK et al (2009) Surface modification of zein films. Indus Crops Prod 30:168–171
Boulter D (1983) Regulation of storage protein synthesis and deposition in developing pea seeds. In: Peterson MS, Johnson AH (eds) Encyclopedia of food science. Martinus Nijhoff Publishers, The Hague, pp 221–224
Bourtoom T (2008) Factors affecting the properties of edible film prepared from mung bean proteins. Int Food Res J 15:167–180
Bradbury E, Martin C (1952) The effect of the temperature of preservation on the mechanical properties and structure of gelatin films. Proc Roy Soc A 214:183–192
Burt S (2004) Essential oils: their antibacterial properties and potential applications in foods: a review. Int J Food Microb 94:223–253
Chambi H, Roseli L, Gisele M et al (2014) Protein extracted from Castor bean (*Ricinus communis* L.) cake in high pH results in films with improved physical properties. Indus Crops Prod 61:217–224
Chen H (2002) Formation and properties of casein films and coatings. In: Gennadious A (ed) Protein based films and coatings. CRC Press, Boca Raton, pp 181–211
Cheng SY, Wang BJ, Weng YM (2015) Antioxidant and antimicrobial edible zein/chitosan composite films fabricated by incorporation of phenolic compounds and dicarboxylic acids. LWT-Food Sci Technol 63:115-121
Cho SY, Rhee C (2002) Sorption characteristics of soy protein films and their relation to mechanical properties. LWT – Food Sci Technol 35:151–157
Cho SY, Rhee C (2004) Mechanical properties and water vapor permeability of edible films made from fractionated soy protein with ultrafilteration. LWT Food Sci Technol 37:833–839
Choi W, Han JH (2001) Physical and mechanical properties of pea-protein-based edible films. J Food Sci 66:319–322
Choi WS, Han JH (2002) Film-forming mechanism and heat denaturation effects on the physical and chemical properties of pea-protein-isolate edible films. J Food Sci 67:1399–1406
Coffman CW, Garciaj VV (2007) Functional properties and amino acid content of a protein isolate from mung bean flour. Int J Food Sci Technol 12:473–484
De Carvalho AJF, Curvelo AAS, Agnelli JAM (2002) Wood pulp reinforced thermoplastic starch composites. Int J Polym Mater 51:647–660
Denavi G, Tapia-Blácido DR, Añón MC et al (2009) Effects of drying conditions on some physical properties of soy protein films. J Food Eng 90:341–349
Dhanapal A, Sasikala P, Rajamani L et al (2012) Edible films from polysaccharides. Food Sci Qualt Managm 3:9–18
Fan J, Ma W, Liu G et al (2014) Food hydrocolloids preparation and characterization of kidney bean protein isolate (KPI) chitosan (CH) composite films prepared by ultrasonic pretreatment. Food Hydrocoll 36:60–69
Fox PF, McSweeny PLH (1998) Dairy chemistry and biochemistry. Blackie Academic and Professional, London

Gandhi AP, Srivastava J (2007) Studies on the production of protein isolates from defatted sesame seed (*Sesamum indicum*) flour and their nutritional profile. ASEAN Food J 14:175–180

Gaona-Sánchez VA, Calderón-Domínguez G, Morales-Sánchez E et al (2015) Preparation and characterisation of Zein films obtained by Electrospraying. Food Hydrocoll 49:1–10

Garg S, Jana AK (2007) Studies on the properties and characteristics of starch-LDPE blend films using cross-linked, glycerol modified, cross-linked and glycerol modified starch. Eur Polym J 43:3976–3987

Gennadios A, Weller CL (1990) Edible films and coatings from wheat and corn proteins. Food Technol 44:63–69

Gennadios A, Weller CL (1991) Edible films and coating from soy milk and soy proteins. Cereal Food World 36:1004–1009

Gennadios A, Weller CL, Testin RF (1993) Temperature effect on oxygen permeability of edible protein-based films. J Food Sci 58:212–219

Gennadios A, Brandenburg AH, Park JW et al (1994) Water vapour permeability of wheat gluten and soy protein isolate films. Ind Crop Prod 2:189–195

Godoy MG, Gulbrra MLE, Maciel FM et al (2009) Use of a low cost methodolgy for biodesonification of castor bean waste and lipase production. Enz Microb Technol 31:317–322

Gómez-Estaca J, López de LA, López-Caballero ME et al (2010) Biodegradable gelatin–chitosan films incorporated with essential oils as antimicrobial agents for fish preservation. Food Microbiol 27:889–896

Gomez-Guillen MC, Ihl M, Bifani V et al (2007) Edible films made from tuna fish gelatin with antioxidant extracts of two different murta ecotypes leaves (*Ugni molinae Turcz*). Food Hydrocoll 21:1133–1143

Guerrero P, Retegi A, Gabilondo N et al (2010) Mechanical and thermal properties of soy protein films processed by casting and compression. J Food Eng 100:145–151

Guilbert S, Gontard N, Gorris LGM (1996) Prolongation of the shelf life of perishable food products using biodegradable films and coatings. LWT Food Sci Technol 29:10–17

Guilbert S, Gontard N, Morel MH et al (2002) Formation and properties of wheat glute films and coatings. In: Gennadious A (ed) Protein based films and coatings. CRC Press, Boca Raton, pp 43–67

Guo J, Ge L, Li X et al (2014) Periodate oxidation of xanthan gum and its crosslinking effects on gelatin-based edible films. Food Hydrocoll 39:243–250

Haard NF, Chism GW (1996) Characteristics of edible plant tissue. In: Fennema O (ed) Food chemistry. Marcel Dekker, New York, p 1067

Hadenberg RE (1967) Wax and related coatings for horticulture products. In: Agricultural Research Service Bulletin. U.S Department of Agriculture, Washington, DC, pp 15–51

Inyang UE, Iduh AO (1996) Influence of pH and salt concentration on protein solubility, emulsifying and foaming properties of sesame protein concentrate. J Am Oil Chem Soc 73:1663–1667

Jafarizadeh Malmiri H, Osman A, Tan CP et al (2011) Development of an edible coating based on chitosan-glycerol to delay "Berangan" banana (*Musa sapientum* cv. *Berangan*) ripening process. Int Food Res J 18:989–997

Ji N, Sun C, Zhao Y, Xiong L, Sun Q (2014) Purification and indentification of antioxidant peptides from peanut protein isolate hydrolysates using UHR-Q-TOF mass spectrometer. Food Chem 161:148–154

Johnson LA, Suleiman TM, Lusas EW (1979) Sesame protein: a review and prospectus. J Am Oil Chem Soc 56:463–468

Karbowiak T, Debeaufort F, Champion D, Voilley A (2006) Wetting properties at the surface of iota-carrageenan-based edible films. J Colloid Interface Sci 294:400–410

Kester JJ, Fennema OR (1986) Edible films and coatings: a review. Food Technol 40:47–59

Khalid EK, Babiker EE, Eltinay AH (2003) Solubility and functional properties of sesame proteins as influenced by pH and/or salt concentration. Food Chem 82:361–366

Kim KM, Weller CL, Hanna MA et al (2002) Heat curing of soy protein films at selected temperatures and pressures. LWT – Food Sci Technol 35:140–145

Kimura A, Fukuda T, Zhang M et al (2008) Comparison of physicochemical properties of 7S and 11S globulins from pea, Fava bean, cowpea, and French bean with those of soybean and French bean 7S globulin exhibits excellent properties. J Agri Food Chem 56:10273–10279

Kinsella JE (1982) Relationships between structure and functional properties of food proteins. In: Fon, Codon (eds) Food proteins. Springer, New York, pp 51–104

Kokoszka S, Debeaufort F, Hambleton A et al (2010) Protein and glycerol contents affect physicochemical properties of soy protein isolate-based edible films. Innov Food Sci Emerg Technol 11:503–510

Kowalczyk D, Baraniak B (2011) Effects of plasticizers, pH and heating of film-forming solution on the properties of pea protein isolate films. J Food Eng 105:295–305

Kowalczyk D, Waldemar G et al (2014) A study on the mechanical properties of pea protein isolate films. J Food Process Preserv 38:1726–1736

Krochta JM (2002) Proteins as raw materials for films and coatings: definations, current status and oppertunities. In: Gennadious A (ed) Protein based films and coatings. CRC press, Boca Raton, pp 1–41

Krochta JM, Mulder-Johnston CD (1997) Edible and biodegradable polymer films: challenges and opportunities. Food Technol 51:61–74

Labuza TP, Contrereas-Medellin R (1981) Prediction of moisture protection requirements for foods. Cereal Food World 26:335–340

Lamber RJW, Skandamis PN, Coote PJ et al (2001) A study of the minimum inhibitory concentration and mode of action of oregano essential oil. J Appl Microbiol 91:453–462

Lawton JW (2002) Zein: a history of processing and use. Cereal Chem 79:1–18

Lee JY, Park HJ, Lee CY et al (2003) Extending shelf life of minimally processed apples with edible coatings and antibrowning agents. LWT- Food Technol 36:323–329

Lee JH, Song NB, Jo WS et al (2014) Effects of nano-clay type and content on the physical properties of sesame seed meal protein composite films. Int J Food Sci Technol 49:1869–1875

Li P, Barth MM (1998) Impact of edible coating on nutritional and physiological changes in lightly processed carrots. Postharv Bio Technol 14:51–60

Li M, Liu P, Zou W et al (2011) Extrusion processing and characterization of edible starch films with different amylose contents. J Food Eng 106:95–101

Li C, Zhu W, Xue H et al (2015) Physical and structural properties of peanut protein isolate-gum Arabic films prepared by various Glycation time. Food Hydrocoll 43:322–328

Lim R, Stathopoulos CE, Golding JB (2011) Effect of edible coatings on some quality characteristics of sweet cherries. Int Food Res J 18:1237–1241

Limchoowong N, Sricharoen P, Techawongstien S et al (2016) An iodine supplementation of tomato fruits coated with an edible film of the iodide-doped chitosan. Food Chem 200:223–229

Liu CC, Tellez-Garay AM, Castell-Perez ME et al (2004) Physical and mechanical properties of peanut protein films. LWT-Food Science and Technology 37:731–738

Lorevice MV, Otoni CG, de MR M (2016) Chitosan nanoparticles on the improvement of thermal, barrier, and mechanical properties of high- and low-methyl pectin films. Food Hydrocoll 52:732–740

Ma W, Tang C, Yang X et al (2013) Food hydrocolloids fabrication and characterization of kidney bean (*Phaseolus vulgaris* L.) protein isolate e chitosan composite films at acidic pH. Food Hydrocoll 31:237–247

Magee E (1996) Eat well for healthy menopouse. John Wiley and Sons, New York

Makishi GLA, Lacerda RS, Bittante AMQB et al (2013) Films based on castor bean (*Ricinus communis* L.) proteins crosslinked with glutaraldehyde and glyoxal. Ind Crop Prod 50:375–382

Marcuzzo E, Peressini D, Debeaufort F et al (2010) Effect of ultrasound treatment on properties of gluten-based film. Innov Food Sci Emerg Technol 11:451–457

Marquié C (2001) Chemical reactions in cottonseed protein cross-linking by formaldehyde, glutaraldehyde, and glyoxal for the formation of protein films with enhanced mechanical properties. J Agric Food Chem 49:4676–4681

Marquie C, Guilbert S (2002) Formation and properties of cottonseed protein films and coatings. In: Protein based films and coatings. CRC press, Boca Raton, pp 139–158

Marquie C, Aymard C, Cuq JL et al (1995) Biodegradable packaging made from cottonseed flour: formation and improvement by chemical treatments with gossypol, formaldehyde and glutaraldehyde. J Agric Food Chem 43:2762–2767

Martínez I, Partal P, García-Morales M (2013) Development of protein-based bioplastics with antimicrobial activity by thermo-mechanical processing. J Food Engin 117:247–254

Martins JT, Cerqueira MA, Bourbon AI et al (2012) Synergistic effects between κ-carrageenan and locust bean gum on physicochemical properties of edible films made thereof. Food Hydrocoll 29:280–289

Mastromatteo M, Chillo S, Buonocore GG et al (2008) Effects of spelt and wheat bran on the performances of wheat gluten films. J Food Engin 88:202–212

McHugh DJ (2003) A guide to the seaweed industry. FAO fisheries technical paper, Rome Italy, pp 1–105

McHugh TH, Krochta M (1994) Sorbitol vs glycerol-plasticize whey protein edible films: integrated oxygen permeability and tensile property evaluation. J Agri Food Chem 42:841–845

Montalvo-Paquini C, Rangel-Marron M, Palou E et al (2013) Effect of pH on physical properties of edible films from faba bean protein. Recent Adv Chem Eng Biochem Comput Chem:29–34

Moradi M, Tajik H, Rohani SMR et al (2016) Antioxidant and antimicrobial effects of zein edible film impregnated with *Zataria multiflora* Boiss essential oil and monolaurin. LWT-Food Sci Technol 72:37–43

Morr CV, Ha EYW (1993) Whey protein concentrates and isolates: processing and functional properties. Crit Rev Food Sci Nutr 33:431–476

Nasriyanti MN, Aziah AAN, Ahmad R (2009) Comparison of properties of mung bean (*Vigna radiata*) protein isolate films at different pH. J Trop Agric Fd Sc 37(2):179–185

Nur Hanani ZA, Roos YH, Kerry JP (2012) Use of beef, pork and fish gelatin sources in the manufacture of films and assessment of their composition and mechanical properties. Food Hydrocoll 29:144–151

O'Sullivan A, Shaw NB, Murphy SC et al (2006) Extraction of collagen from fish skins and its use in the manufacture of biopolymer films. J Aqua Food Pro Technol 15:21–32

Oechsle AM, Häupler M, Weigel F et al (2016) Modulation of extruded collagen films by the addition of co-gelling proteins. J Food Engin 171:164–173

Oliveira TG, Makishi GLA, Chambi HNM et al (2015) Cellulose fiber reinforced biodegradable films based on proteins extracted from castor bean (*Ricinus communis* L.) cake. Ind Crop Prod 67:355–363

Oms-Oliu G, Soliva-Fortuny R, Martín-Belloso O (2008) Edible coatings with antibrowning agents to maintain sensory quality and antioxidant properties of fresh-cut pears. Postharvest Biol Technol 50:87–94

Onsaard E, Pomsamud P, Audtum P (2010) Functional properties of sesame protein concentrate from sesame meal. Asian J Food Agri Ind 3:420–431

Ozcalik O, Tihminlioglu F (2013) Barrier properties of corn zein nanocomposite coated polypropylene films for food packaging applications. J Food Engin 114:505–513

Palviainen P, Heinamaki J, Myllarinen P et al (2001) Corn starches as film formers in aqueous-based film coating. Pharmac Developm Technol 6:351–359

Paramawati R, Yoshino T, Isobe S (2001) Properties of plasticized-zein film as affected by plasticizer treatments. Food Sci Technol Res 7:191–194

Park HJ (1999) Development of advanced edible coatings for fruits. Trends Food Sci Technol 10:254–260

Park HJ, Chinnan MS (1995) Gas and water vapor barrier properties of edible films from protein and cellulosic materials. J Food Eng 25:497–507

Park HJ, Chinnan MS, Shewfelt RL (1994) Edible coating effects on storage life and quality of tomatoes. J Food Sci 59:568–570

Park SK, Hettiarachchy ZYJ, Gennadios A (2002) Formation and properties of soy protein films and coatings. In: Protein based films and coatings. CRC press, Boca Raton, pp 305–327
Pastor C, Sánchez-González L, Chiralt A et al (2013) Physical and antioxidant properties of chitosan and methylcellulose based films containing resveratrol. Food Hydrocoll 30:272–280
Pavlath AE, Orts W (2009) Edible films and coatings: why, what, and how? In: Embuscado ME, Huber KC (eds) Edible films and coatings for food applications. Springer, New York, pp 1–23
Perez-Gago MB, Serra M, Alonso M et al (2005) Effect of whey protein- and hydroxypropyl methylcellulose-based edible composite coatings on color change of fresh-cut apples. Postharvest Biol Technol 36:77–85
Pérez-Mateos M, Montero P, Gómez-Guillén MC (2009) Formulation and stability of biodegradable films made from cod gelatin and sunflower oil blends. Food Hydrocoll 23:53–61
Popovic S, Peričin D, Vaštag Ž et al (2011) Evaluation of edible film-forming ability of pumpkin oil cake; effect of pH and temperature. Food Hydrocoll 25:470–476
Popovic S, Peričin D, Vaštag Ž et al (2012) Pumpkin oil cake protein isolate films as potential gas barrier coating. J Food Engin 110:374–379
Ramos ÓL, Santos AC, Leão MV et al (2012) Antimicrobial activity of edible coatings prepared from whey protein isolate and formulated with various antimicrobial agents. Int Dairy J 25:132–141
Ramos ÓL, Reinas I, Silva SI et al (2013) Effect of whey protein purity and glycerol content upon physical properties of edible films manufactured therefrom. Food Hydrocoll 30:110–122
Reddy N, Yang Y (2010) Citric acid cross-linking of starch films. Food Chem 118:702–711
Reddy N, Jiang Q, Yang Y (2012) Preparation and properties of peanut protein films crosslinked with citric acid. Ind Crop Prod 39:26–30
Reddy N, Chen L, Yang Y (2013) Thermoplastic films from peanut proteins extracted from peanut meal. Ind Crop Prod 43:159–164
Rezvani E, Schleining G, Sümen G et al (2013) Assessment of physical and mechanical properties of sodium caseinate and stearic acid based film-forming emulsions and edible films. J Food Engin 116:598–605
Rhim J, Wang L (2014) Applied clay science preparation and characterization of carrageenan-based nanocomposite films reinforced with clay mineral and silver nanoparticles. Appl Clay Sci 97-98:174–181
Ribeiro C, Vicente A, Teixeira JA et al (2007) Optimization of edible coating composition to retard strawberry fruit senescence. Postharv Bio Technol 44:63–70
Ridley BL, O'Neill MA, Mohnen D (2001) Pectins: structure, biosynthesis, and oilgogalacturonide related signaling. Phytochemistry 57:929–967
Rojas-Grau MA, Raybaudi-Massilia RM, Soliva-Fortuny RC et al (2007) Apple puree-alginate edible coating as carrier of antimicrobial agents to prolong shelf-life of fresh-cut apples. Postharvest Biol Technol 45:254–264
Saberi B, Thakur R, Vuong QV et al (2016) Optimization of physical and optical properties of biodegradable edible films based on pea starch and guar gum. Ind Crop Prod 86:342–352
Salmieri S, Lacroix M (2006) Physiochemical properties of alginate /polycaprolactone-based films containing essential oils. J Agri Food Chem 54:10205–10214
Saremnezhad S, Azizi MH, Barzegar M et al (2011) Properties of a new edible film made of faba bean protein isolate. J Agri Sci Technol 13:181–192
Saroso, B (1989) Chemical properties of protein in cottonseed kernels. Industr Crops Res J 1:60–65
Sathe SK (2002) Dry bean protein functionality. Crit Rev Biotechnol 22:175–223
Sayanjali S, Ghanbarzadeh B, Ghiassifar S (2011) Evaluation of antimicrobial and physical properties of edible film based on carboxymethyl cellulose containing potassium sorbate on some mycotoxigenic *Aspergillus* species in fresh pistachios. LWT – Food Sci Technol 44:1133–1138
Schou M, Longares A, Montesinos-Herrero C et al (2005) Properties of edible sodium caseinante films and their applications as food wrapping. LWT- Food Sci Technol 38:605–610

Seydim AC, Sarikus G (2006) Antimicrobial activity of whey protein based edible films incorporated with oregano, rosemary and garlic essential oils. Food Res Int 39:639–644

Shah NN, Vishwasrao C, Singhal RS et al (2016) *n*-Octenyl succinylation of pullulan: effect on its physico-mechanical and thermal properties and application as an edible coating on fruits. Food Hydrocoll 55:178–188

Shankar S, Reddy JP, Rhim JW et al (2015) Preparation, characterization, and antimicrobial activity of chitin nanofibrils reinforced carrageenan nanocomposite films. Carbohydr Polym 117:468–475

Shankar S, Wang LF, Rhim JW (2016) Preparations and characterization of alginate/silver composite films: effect of types of silver particles. Carbohydr Polym 146:208–216

Sharma L, Singh C (2016a) Sesame protein based edible films: development and characterization. Food Hydrocoll 61:139–147

Sharma L, Singh C (2016b) Composite film developed from the blends of sesame protein isolate and gum rosin and their properties thereof. Polym Compos 16:101–113

Sharma L, Singh C, Sharma HK (2016) Assessment of functionality of sesame meal and sesame protein isolate from Indian cultivar. J Food Measure Charac 1:1–7

Shimada K, Cheftel JC (1998) Sulfydryl group disulphide bond interchange during heat induced gelation of whey protein isolate. J Agric Food Chem 37:161–168

Simelane S, Ustunol Z (2005) Mechanical properties of heat-cured whey protein-based edible films compared with collagen casings under sausage manufacturing conditons. J Food Sci 70:131–134

Sivarooban T, Hettiarachchy NS, Johnson MG (2008) Physical and antimicrobial properties of grape seed extract, nisin, and EDTA incorporated soy protein edible films. Food Res Int 41:781–785

Su JF, Huang Z, Yuan XY et al (2010) Structure and properties of carboxymethyl cellulose/ soy protein isolate blend edible films crosslinked by Millard reactions. Carbohydr Polym 79:145–153

Sun Q, Sun CX, Xiong L (2014) Mechanical: barrier and morphological properties of pea starch and peanut protein isolate blend films. Carb Hyd Polym 98:630–637

Talens P, Perez-Masia R, Fabra MJ et al (2012) Application of edible coatings to partially dehydrated pineapple for use in fruit–cereal products. J Food Engin 112:86–93

Tanada-Palmu PS, Grosso CRF (2005) Effect of edible wheat gluten-based films and coatings on refrigerated strawberry (*Fragaria ananassa*) quality. Postharvest Biol Technol 36:199–208

Tang CH (2008) Thermal denaturation and gelation of vicilin-rich protein isolates from three *Phaseolus* legumes: a comparative study. LWT – Food Sci Technol 41:1380–1388

Tapia MS, Rojas-Grau MA, Carmona A et al (2008) Use of alginate and gellan-based coatings for improving barrier, texture and nutritional properties of fresh-cut papaya. Food Hydrol 22:1493–1503

Tharanathan RN (2003) Biodegradable films and composite coatings: past, present and future. Trends Food Sci Technol 14:71–78

Tihminlioglu F, Atik İD, Özen B (2010) Water vapor and oxygen-barrier performance of corn–zein coated polypropylene films. J Food Engin 96:342–347

Tillekeratne M, Easteal AJ (2000) Modification of zein films by incorporation of poly (ethylene glycol)s. Polym Int 49:127–134

Türe H, Gällstedt M, Hedenqvist MS (2012) Antimicrobial compression-moulded wheat gluten films containing potassium sorbate. Food Res Int 45:109–115

Vachon C, Yu HL, Yefsah R et al (2000) Mechanical and structural properties of milk protein edible films cross-linked by heating and γ-irradiation. J Agri Food Chem 48:3202–3209

Vu CHT, Won K (2013) Novel water-resistant UV-activated oxygen indicator for intelligent food packaging. Food Chem 140:52–56

Wang Y, Padua GW (2006) Water barrier properties of zein oleic acid films. Cereal Chem 83:331–334

Wills RH, Lee TH, Graham D et al (1981) Postharvest, an introduction to the physiology and handling of fruits and vegetables. AVI, Westport

Wollerdorfer M, Bander H (1998) Influence of natural fibres on the mechanical properties of biodegradable polymers. Indus Crops Prod 8:105–112

Wu LC, Bates RP (1973) Influence of ingredients upon edible protein-lipid film characterstics. J Food Sci 37:36–39

Yang H, Wen XL, Guo SG et al (2015) Physical, antioxidant and structural characterization of blend films based on hsian-tsao gum (HG) and casein (CAS). Carbohydr Polym 134:222–229

Zactiti EM, Kieckbusch TG (2006) Potassium sorbate permeability in biodegradable alginate films: effect of the antimicrobial agent concentration and crosslinking degree. J Food Engr 77:462–467

Zaghloul M, Prakash V (2002) Effect of succinylation on the functional and physicochemical properties of α-globulin, the major protein fraction from *Sesamum indicum* L. Nahrung/Food 46:364–369

Zuo M, Song Y, Zheng Q (2009) Preparation and properties of wheat gluten/methylcellulose binary blend film casting from aqueous ammonia: a comparison with compression molded composites. J Food Engin 91:415–422

Bioplastics: A Sustainable Approach Toward Healthier Environment

15

Kiran Nehra, Pragati Jamdagni, and Priyanka Lathwal

Abstract

An uncontrolled usage of conventional nonbiodegradable plastics over a long period of time has resulted in large-scale accumulation of waste and release of greenhouse gases in the environment, thus giving rise to severe environmental problems. More recently, a widespread global concern toward these environmental issues has provided a major impetus to the search for suitable biodegradable substitutes which may have minimal side effects on the environment, but still retain the properties and the ability to be used in applications similar to the existing conventional synthetic polymers. Biodegradable plastics or bioplastics may be defined as polymers which can be completely converted by living organisms, usually microorganisms, to carbon dioxide, water, and humic material. Among the various biopolymers produced by microorganisms, polyhydroxyalkanoates (PHAs) have generated a specific interest among the biotechnologists because of their thermoplastic characteristics and resemblance to synthetic polymers to a large extent, thus holding a potential for replacing conventional nonbiodegradable plastics. Two forms of PHA, that is, polyhydroxybutyrate (PHB) homopolymer and 3HB-3HV (-3-hydroxybutyrate-3-hydroxyvalerate) copolymer, are popularly used for commercial production of bioplastics. However, PHB is the best characterized form of PHA, found to be produced naturally by several microorganisms as energy storage molecule in response to stressful conditions. The present review highlights several aspects of these potential biopolymers including their history, basic structure and properties, genetic background of their biosynthesis, and finally their applications in different areas.

K. Nehra (✉) • P. Jamdagni • P. Lathwal
Department of Biotechnology, Deenbandhu Chhotu Ram University of Science and Technology, Murthal 131 039, Sonipat, Haryana, India
e-mail: nehrakiran@gmail.com

S.K. Gahlawat et al. (eds.), *Plant Biotechnology: Recent Advancements and Developments*, DOI 10.1007/978-981-10-4732-9_15

15.1 Introduction

Synthetic plastics, considered as one of the major discoveries in the history of mankind, are being extensively used throughout the world, right from the making of bottles to automobile parts, and also from preparing simple plastic carry bags to complex surgical and medical implants (Sharma and Dhingra 2016). Current turnover of global plastic industry is $1 trillion/year and GDP is 3.7% (Arun et al. 2006). These plastics have high molecular weight and their molecules are tightly bonded together. This property makes these compounds nonbiodegradable and is responsible for their long-term accumulation in the environment (Sharmila et al. 2011). The current techniques for disposal of plastics include recycling or incineration; but recycling is a very long and nonefficient process, and incineration is difficult, expensive, and dangerous, resulting in the release of several toxins. These conventional plastics, therefore, either persist in the environment for years, choking our ecosystem, or their degradation processes make the environment highly toxic. More recently, the world has been becoming more and more cognizant about these hazardous effects of plastics on the environment. But, since plastics are involved in our everyday lives and life cannot be thought of without plastics, production of biodegradable plastics for a healthier and sustainable environment has become the need of the hour. A sustainable approach regarding the matter could be found in natural alternatives for plastics, that is, bioplastics. Bioplastics can be defined as plastics synthesized from renewable resources such as plants or microbial species (Soam et al. 2012).

Bioplastics belong to a family of compounds called polyhydroxyalkanoates (PHAs). The PHAs include several polymeric esters such as polyhydroxybutyrates (PHB), polyhydroxybutyrate co-hydroxyvalerates (PHBV), polyhydroxybutyrate co-hydroxyhexanoate (PHBHx), and polyhydroxybutyrate co-hydroxyoctanoate (PHBO), however, PHB being the most common type of PHA. These polyesters are produced as storage compounds by bacteria within a cell during stressful conditions when excessive amount of carbon and limited amount of nitrogen or phosphorous is present (Ishizaki et al. 2001). During starvation period, bacteria that contain PHA as storage molecule are able to survive better in comparison to bacteria which do not produce PHA.

These biopolymers are much safer than conventional plastics as they are synthesized, as well as degraded naturally with no toxin emission, and are also biocompatible in the human system (Wani et al. 2016). Increased production of bioplastics can significantly reduce carbon dioxide emissions, curtail plastic waste generation, and decrease consumption of fossil fuels. However, one of the main hurdles in the widespread usage of bioplastics is that their production cost is several times higher as compared to that of conventional plastics. This high cost of industrial production and recovery of bioplastics is presently not able to compete with traditional ways of synthetic plastic production. Hence, there is still a need for the discovery of polymers which in addition to being completely biodegradable are cost-effective as well. Alternately, another possible solution to this issue appears in the form of genetic engineering of natural producers for enhanced as well as directed production of biopolymers that may be of use in various applications. This could eventually help overcome the existing shortcomings of the bioplastics produced from microorganisms.

Bioplastics, therefore, is an exciting new field, offering research challenges to the biotechnologists and genetic engineers for the development of successful recombinants capable of cost-effective production of bioplastics.

Taking all these things into account, the current review is an attempt to brief up the remarkable journey of bioplastics: from their discovery as storage granules in the humble soil bacterium *Bacillus megaterium* as early as in the early twentieth century to the designing of the present-day highly complex genetically engineered microorganisms and plants for the production of PHBs.

15.2 History of Bioplastics

Conventionally, plastics are predominantly manufactured from crude oil. But, more recently, the problems associated with solid waste management of conventional plastics have generated a heightened interest in the development of biodegradable plastics at a global level (Anderson and Dawes 1990). Although the last few decades have witnessed a boom in the research related to biodegradable plastics, bioplastics are not new to this century. They were first derived from cellulose (a derivative of wood pulp) by a British chemist, as early as in 1850s. This early discovery was then followed by Henry Ford in the early twentieth century when he tried to develop soy-based plastics. After a lag period of about half a century, the interest in biodegradable plastics was renewed during the oil crisis in 1970s. This ushered in the era of development of biopolymers using monomers from natural renewable resources.

Different sources can be used for the production of bioplastics, but biopolymers produced by microorganisms are the most popular. The several biopolymers produced by microorganisms, polyhydroxyalkanoates (PHAs) stored in the form of lipid granules in bacterial cells, are unique due to their plastic-like properties. The presence of these lipid-like inclusions was initially observed in *Azotobacter chroococcum* almost a century ago. However, the chemical nature of these inclusion bodies as poly-(3)-hydroxybutyric acid was first established by Lemoigne in 1925 while studying the bacterium *Bacillus megaterium*. After its initial discovery in 1920s, the following 30 years saw the expansion of PHB (the most popular form of PHA) research on several other *Bacillus* species. The studies included detection, cell-content estimation methods, and the various culture parameters affecting PHB accumulation inside *Bacillus* cells (Macrae and Wilkinson 1958a). It was observed that PHB accumulation in *B. megaterium* was higher when the glucose-to-nitrogen source ratio was kept high in the medium (Macrae and Wilkinson 1958b). In 1970s and 1980s, reports regarding the significance of polyhydroxyalkanoates, other than PHBs, started emerging. In 1974, Wallen and Rohwedder reported the identification of poly-3-hydroxyvalerate and poly-3-hydroxyhexanoate. De Smet et al. (1983) observed that *Pseudomonas oleovorans* could synthesize and accumulate poly-3-hydroxyoctanoate and small amounts of poly-3-hydroxyhexanoates when cultivated on n-octane. PHB was found to have limited applications due to its brittle nature; however, incorporation of these other hydroxyalkanoates (HAs) significantly enhanced the biopolymer properties. The search for alternate HAs led to the identification of an astounding number of approximately 150 different hydroxyalkanoates (Steinbuchel and Valentin 1995) produced by more than 300 different microorganisms (Lee 1996b).

By the 1990s, the research on PHA shifted to the molecular level involving development of techniques for characterization and cloning of genes involved in PHA biosynthesis. PHA biosynthesis genes of *Ralstonia eutropha* were cloned and shown to be functionally active in *E. coli* (Schubert et al. 1988; Slater et al. 1988). By the end of the twentieth century, 40 different PHA synthase genes (gene responsible for the synthesis of one of the key enzyme, PHA synthase, in the PHA biosynthetic pathway) from more than 35 different bacteria had been cloned (Sudesh et al. 2000). The beginning of the twenty-first century has been marked by a further encroachment of this field into protein engineering, wherein the research studies are targeted at the manipulation of PHA synthase and other enzymes of the PHB biosynthetic pathway for enabling the production of environment-friendly and cost-efficient biopolymers (Sudesh et al. 2000). Metabolic engineering is being experimented to enable the introduction of new metabolic pathways which might result in enhanced PHA accumulating capacity, or may produce novel PHAs. These strategies involve the introduction of heterologous PHA biosynthetic pathway from PHA-producing bacteria into others which normally do not produce PHA, but possess metabolic advantages over these. More recently, PHA production from economically cheap substrates using recombinant *E. coli* has also gained much attention. PHA production using cheese whey, a coproduct of dairy industry, as a sole carbon substrate has been reported (Solaiman et al. 2006).

The introduction of molecular techniques has thus boosted the biopolymer research and has enabled the understanding of various regulations in the in vivo synthesis of PHA to produce it in an efficient way. Using genetic engineering techniques, recombination involving PHA synthesis genes of required characters has been undertaken to obtain specific functions. The isolation and cloning of genes involved in PHA biosynthesis has resulted in heterologous expression of the cloned PHA genes in suitable host organisms and in metabolic engineering for their higher and safer productivity.

15.3 Polyhydroxyalkanoates

A number of biopolymers such as polynucleotides, polyamides, polyphenols, polythioesters, polyanhydrides, polyoxoesters (including PHAs), and many others have been proposed to offer an alternative to conventional plastics (Steinbuchel 2001). More particularly, PHAs belonging to the group of polyoxoesters are considered the most promising alternative for modern polymers and have, therefore, become an emerging topic for research (Antipov et al. 2006). In recent years, PHAs have received greater attention because they have been found to have a high potential in household, medical, industrial, and other applications (Albuquerque et al. 2007). The PHA family includes several polymeric esters such as poly-3-hydroxybutyrate (PHB), poly-(3-hydroxybutyrate-co-3-hydroxyvalerate) (PHBV), polylactic acid (PLA), polycaprolactone (PCL), polyethylene succinate (PES), polybutylene succinate-co-adipate (PBSA), and their combination polymers. However, PHB and PHBV are the most commonly used PHAs.

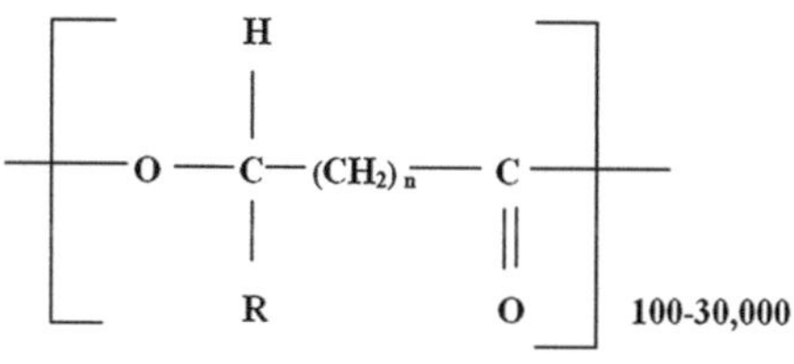

Fig. 15.1 General structure of PHA (Lee 1996a)

15.3.1 Polyhydroxyalkanoates: Structure and Classification

Polyhydroxyalkanoates are linear polyesters composed of saturated or unsaturated hydroxy fatty acid monomers in which the carboxy group and the hydroxyl group of two different monomers form an ester bond (Endres and Raths 2011). They mainly consist of two to six hydroxy fatty acids having different groups such as alkyl, cyano, epoxy, ether, acyl, halogen, aryl, ester, ether, and acid groups. Figure 15.1 depicts the general structure of PHAs (Steinbuchel 1991; Lee 1996a). The nomenclature of different forms of PHAs depends upon the number of (CH_2) units (n), and the type of side group (R) attached in the common general structure of the PHA (Endres and Raths 2011).

PHAs have been classified on the basis of composition of their monomer units into two broad groups: (i) short chain length (scl) polyhydroxyalkanoates, with a chain length of 3–5 carbons, and (ii) medium chain length (mcl) polyhydroxyalkanoates with a chain length of 6–14 carbon atoms (Madison and Huisman 1999; Kunasundari and Sudesh 2011). In addition to the common scl- and mcl-PHAs, copolymers or blends of these two forms of PHA, commonly referred to as scl-co-mcl polymers (Kang et al. 2001; Lee et al. 2000) consisting of 3–14 carbons, and long chain length (lcl-PHAs) having 16 or more carbon atoms have also been reported. Approximately 150 different monomeric units that are polymerized to yield PHAs have been identified till date (Steinbuchel 2001; Steinbuchel and Valentin 1995).

Short chain length PHAs includes mainly 3-hydroxybutyrate and 3-hydroxyvalerate; however, other monomer units which are oxidized at positions other than the third carbon such as 4-hydroxybutyrate, 4-hydroxyvalerate, and 5-hydroxyvalerate also fall into this group. Medium chain length PHAs include 3-hydroxyhexanoate, 3-hydroxyheptanoate, 3-hydroxyoctanoate, and 3-hydroxydecanoate. Substrate specificity of PHA synthases, one of the major enzymes of the PHB biosynthetic pathway, determines the accumulation of either scl-PHAs or mcl-PHAs (Anderson et al. 1992). PHA synthase of *Alcaligenes eutrophus* can polymerize 3-hydroxyalkanoic acids consisting of six or lesser number of carbon atoms only; however, *Pseudomonads* are responsible for accumulation of mcl-PHAs (Witholt and Kessler 1999) as their PHA synthases accept 3-hydroxyalkanoic acids of 6–14 carbon atoms. Metabolic pathways involved in the synthesis of scl- and mcl-PHAs are also different (Tappel et al. 2014). Thus, the different classes of PHA accumulation by microorganisms are dictated by the organism's metabolic background, its PHA biosynthesis function, and genetic composition (Solaiman and Ashby 2005).

15.3.2 Polyhydroxyalkanoates: Properties

PHAs are optically active and easily biodegradable thermoplastics having material properties quite similar to those of conventional plastics such as polypropylene. However, physical properties of these biopolymers vary considerably from material to material. The chemical structure, monomer composition, and incorporation of different functional groups in the basic structure of PHA are important factors in determining the physical and material properties of different PHAs (Hori et al. 1994; Casini et al. 1997). But, more generally, the bacterial PHAs exhibit an average molecular mass of up to 4.0×10^6 Da and a melting point temperature around 180 °C (Marchessault and Yu 2004; Tsz-Chun et al. 2005).

PHB, the most popular scl-PHA, is optically pure, 100% stereospecific with D (-) configuration and is isotactic in nature. It is highly crystalline with crystallinity ranging between 55 and 88% (Lee 1996a). Some of the mechanical properties of PHB are similar to polypropylene, but it is relatively stiff and brittle due to the proximity of glass transition range to ambient temperature (Antipov et al. 2006). Due to low thermal stability and excessive brittleness, PHB has less permeability than polypropylene and polyethylene (Matsusaki et al. 2000). While PHB is the most researched version of PHA, other variants such as copolymers of PHB are also generating interest as they not only possess major characteristics of PHB but are also less brittle. Although the basic helical crystalline structure of PHB homopolymer is retained in its various copolymers (Padermshoke et al. 2004), incorporation of a second monomer unit such as with 3-hydroxyvalerate in PHB leads to the formation of a copolymer, namely, poly-(3-hydroxyvalerate-co-3-hydroxyvalerate) (PHBV) which is not only flexible but also easily degradable (Shah 2014). Moreover, the decomposition temperature for PHBV is also increased from 246·3 °C (PHB homopolymer) to 260·4 °C, implementing the role of valerate in increasing the thermal stability.

In contrast to scl-PHA, the mcl-PHAs are generally elastomers having low melting point and are highly amorphous. In these biopolymers, the crystallinity is hampered with the incorporation of side chains into mcl-PHA, although saturated PHA-mcl are able to crystallize due to their isotactic configuration. But, in general, PHA-mcl is semicrystalline in nature and show decreased melting point and a low tensile strength. They, however, offer an additional advantage of high elongation for breakage and are a potential candidate for use as biodegradable rubber. Table 15.1 gives a comparative account of properties of PHB (homopolymer), PHBV, PHB4B (scl-copolymers), and PHBHx (mcl-copolymer) with that of polypropylene (PP).

These values and comparisons indicate that bioplastics are well capable of replacing the conventional polymeric plastics. Besides these favorable physical and chemical properties, another very important property of all PHAs which has been responsible for creating an immense interest in research on their large-scale production is their complete biodegradability.

Table 15.1 Comparison of physical properties of different PHAs and polypropylene (PP)

Parameter	PHB	PHBV	PHB4B	PHBHx	PP
Melting temperature (°C)	177	145	150	127	176
Glass transition temperature (°C)	2	−1	−7	−1	−10
Crystallinity (%)	60	56	45	34	50–70
Tensile strength (MPa)	43	20	26	21	38
Extension to break (%)	5	50	444	400	400

(PHBV: containing 20% 3 HV-monomers; PHB4B: containing 16% 4HB-monomers; and PHBHx containing 10% 3HHx-monomers (Tsuge 2002))

15.3.3 Polyhydroxyalkanoates: Biodegradation

One of the major issues about conventional plastics that concern environmentalists all over the world is their safe disposal, thus, making biodegradability of PHAs their most attractive feature. PHAs can be biodegraded by several naturally occurring microorganisms in the environment. Several factors such as polymer composition, enzymology of the process, exposed surface area, and different environmental factors such as moisture, temperature, molecular weight, and pH are responsible for biodegradation of PHAs (Lee 1996a; Boopathy 2000). Under suitable environmental conditions such as moisture, temperature, and oxygen availability, bacteria, fungi, and actinomycetes can depolymerize and utilize the polymer material as a source of nutrient. These microorganisms found in soil, water, and other natural habitats inhabit the surface of PHA polymers, and synthesize extracellular enzymes, such as the polyhydroxybutyrate depolymerases and hydrolases, that hydrolyse the ester bonds of the polymer into water soluble monomers and oligomers. The microbes finally degrade these compounds into water and carbon dioxide (Jendrossek and Handrick 2002; Choi et al. 2004; Chanprateep 2010). Figure 15.2 shows a schematic representation of PHA biodegradation. While biodegradation of PHA has been reported to be fastest in anaerobic sewage sludge, it is decomposed at the slowest rate in seawater. Lee (1996a) has reported the time taken for complete degradation of P (3HB-co-3 HV) in anaerobic sewage, soil, and sea water to be 6, 75, and 350 weeks, respectively. However, in the soil, the rate of bioplastic degradation has been reported to be enhanced with an increase of the bacterial biomass in soil. 16S rDNA PCR-DGGE analysis indicates that the bacterial diversity in the soil is not affected by the degradation of bioplastics. Moreover, the degradation of bioplastic also does not affect the nitrogen circulation activity in the soil (Adhikari et al. 2016). A wide variety of scl- and mcl-PHA degrading bacteria have been isolated and phylogenetically identified in detail (Nam et al. 2002). The bacterial community related to PHB decomposition in the environment consists of both gram-positive and gram-negative bacteria, mainly, *Bacillus* sp., *Streptomyces* sp., *Acidovorax facilis,* and *Variovorax paradoxus* (Mergaert et al. 1993).

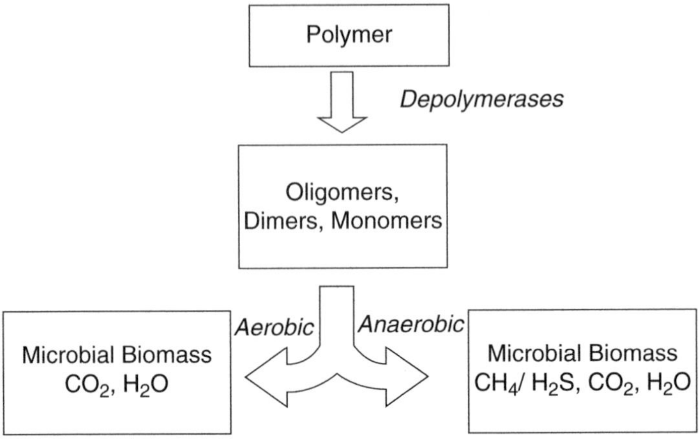

Fig. 15.2 Degradation of biopolymers under aerobic and anaerobic conditions (Gu 2003)

15.4 Polyhydroxybutyrate (PHB)

Poly (3-hydroxybutyrate) [P (3HB)], first described by Lemoigne, a French scientist in the year 1925 (Lemoigne 1926), is presently one of the most commonly used PHAs (Lu et al. 2009). PHB, a homopolymer of 3-hydroxybutyric acid, having methyl group as the side chain, is comparable to polypropylene (PP) mechanically, but is truly biodegradable in the environment (Lee 1996a). PHB granules have been found in the cytoplasm and subcellular compartments, such as the vacuoles, of several gram-positive and gram-negative bacteria including, archaebacteria, cyanobacteria, and photosynthetic bacteria (Chee et al. 2010). PHB granules are chiefly synthesized and accumulated under unbalanced conditions of growth and nutrition. When the nutritive environment of a cell becomes excessively rich in carbon and lacks other essential nutrients, such as nitrogen, phosphorus, magnesium, or oxygen, bacterial cells become stressed. These conditions induce PHB mobilization into the cells, which then acts as an energy store and aids in bacterial survival during stress conditions (Verlinden et al. 2007). These granules can be readily isolated from the microbial cells.

15.4.1 PHB Biosynthesis

Microorganisms respond very quickly to changing environments; they have a natural tendency of accumulating essential nutrients which assist survival during various stress conditions. PHB producing microorganisms accumulate carbon sources during unfavorable growth conditions. Carbon is assimilated and converted to hydroxyalkanoate (HA) compounds. These HA compounds are then polymerized to yield PHB, which are stored as water insoluble granules inside the cell. A cell can accumulate large quantities of PHB as it does not affect the osmotic balance or physiological fitness of the cell (Steinbuchel 2001).

Several pathways are involved in the synthesis of PHA. Complete elucidation of these pathways has been accomplished for *R. eutropha*, *Zoogloea ramigera*, and *Azotobacter beijerinckii* (Anderson and Dawes 1990). The scl-PHA producing bacteria, biosynthesize polyhydroxybutyrate (P (HB)), polyhydroxyvalerate (P (HV)), and their copolymers (P (HB-co-HV)). PHB, which is one of the common scl-PHA produced by bacteria, is synthesized through a three-step pathway, wherein, the excess sugars present in the cell enter into the PHB biosynthetic pathway by getting condensed to acetyl-coenzyme A (acetyl-CoA). The pathway is initiated by the combination of two moles of acetyl-CoA to give rise to acetoacetyl-CoA, which then undergoes a reduction reaction to form 3-hydroxybutyryl-CoA; this in turn polymerizes to give rise to one mole of PHB. Figure 15.3 shows a schematic representation of the PHB biosynthetic pathway (Verlinden et al. 2007).

The synthesis of P (HV) also follows the same pathway, except for the variation in the substrate used for the polymer production. In P (HV) synthesis, the initial substrates are acetyl-CoA and propionyl CoA, which are first condensed into 3-ketovaleryl CoA and then reduced to 3-hydroxyvaleryl CoA. This molecule is then further polymerized into the growing chain of PHA by PHA synthase enzyme. Similarly, if propionic acid is used as a substrate, it leads to the formation of PHB-PHV copolymer. Herein, removal of a carbonyl carbon from propionyl CoA leads to the formation of acetyl-CoA. The HB unit of the copolymer is formed by the combination of two moles of acetyl-CoA, whereas, the HV unit is formed by the reaction of one mole of acetyl-CoA with one mole of propionyl CoA.

15.4.2 Enzymes and Gene(s) Involved in PHB Production

The PHB biosynthetic pathway revolves around three main enzymes, namely, β-ketothiolase, acetoacetyl-co-reductase, and PHA synthase. β-ketothiolase catalyzes the first step in PHB formation and is a member of the family of enzymes involved in the thiolytic cleavage of substrate into acyl-CoA and acetyl Co-A. It catalyzes the Claisen condensation of two molecules of acetyl-CoA to give rise to acetoacetyl-CoA. This enzyme is found throughout nature, from higher eukaryotes to yeasts to prokaryotes and is found to be present in soluble form in in vivo conditions. The second step in the PHB biosynthetic pathway is catalyzed by the enzyme acetoacetyl-CoA-reductase, which is also a soluble protein, and is involved in a stereospecific reduction reaction, resulting into the formation of 3-hydroxybutyryl-CoA from acetoacetyl-CoA. The last reaction in the pathway is catalyzed by the enzyme PHB synthase which performs an ester bonded polymerization of 3-hydroxybutyryl moieties. It is the key enzyme in the PHB biosynthetic pathway which determines the type of PHA synthesized. PHB synthase is also a soluble enzyme; however, it exists in soluble form only in the absence of PHB synthesis. After PHB biosynthesis, it gets associated with the PHB granules and remains so during the entire PHB storage period. Table 15.2 summarizes the function and the genes responsible for encoding all the three enzymes of the PHB biosynthetic pathway.

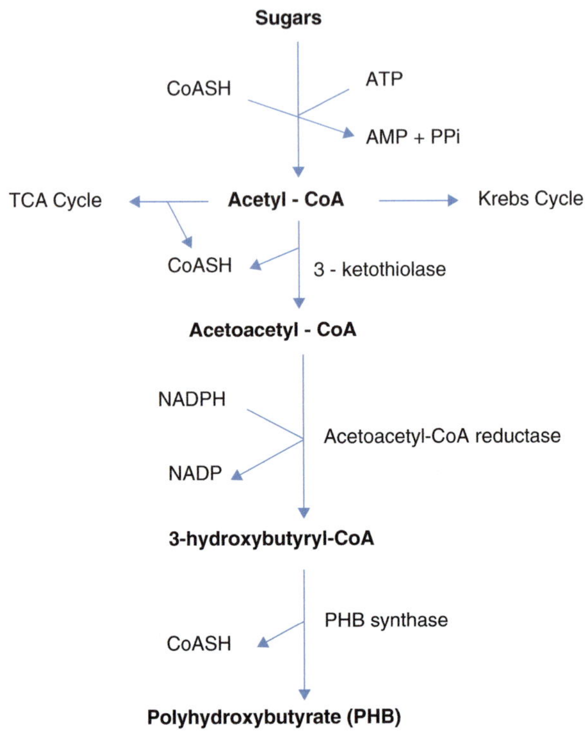

Fig. 15.3 Biosynthetic pathway of PHB in *Ralstonia eutropha* (Verlinden et al. 2007)

Table 15.2 Enzymes involved in the biosynthesis of PHB

Enzyme	Coding gene	Function
3-Ketothiolase	*phb*A	Combines two acetyl-CoA molecules to yield acetoacetyl-CoA
Acetoacetyl-CoA reductase	*phb*B	Reduces acetoacetyl-CoA to D(-)3-hydroxybutyryl-CoA
PHA synthase	*phb*C	Polymerizes 3-hydroxybutyryl-CoA to long chain molecules of PHB

(Compiled from: Steinbuchel 1991; Byrom 1994; Nawrath et al. 1994)

The genes involved in PHA biosynthetic pathways have been studied and characterized from several different species (Madison and Huisman 1999; Nehra et al. 2015a), and it has been observed that the organization of genes involved in PHB biosynthesis is highly varied. In *Ralstonia eutropha* and many other scl-PHA producing bacteria, three genes, namely, *phaA*, *phaB,* and *phaC* are essential for PHB synthesis and accumulation, and are present in tandem as a single operon, namely, *phb*CAB (Steinbuchel 2001). In *Alcaligenes eutrophus* and *Zoogloea ramigera*, *phaA* and *phaB* genes have been found to be closely linked, and are present

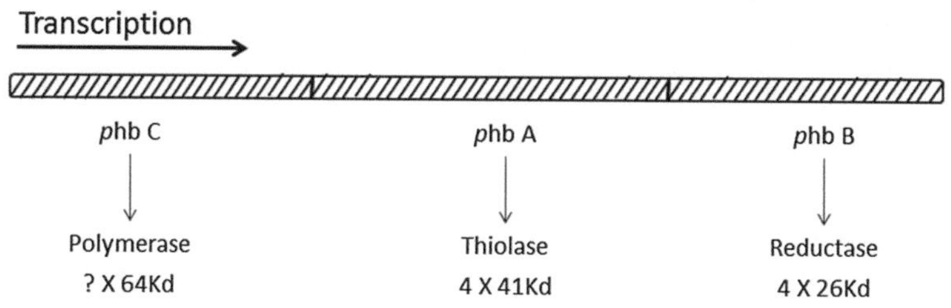

Fig. 15.4 The *phbC-phbA-phbB* locus in *A. eutrophus* H16 (Peoples and Sinskey 1989)

downstream of *phbC* (Peoples and Sinskey 1989). The structure of *phbC*, *phbA*, and *phbB* locus in *Alcaligenes eutrophus* H16 is shown in Fig.15.4.

However, in many mcl-PHA producing bacteria like Pseudomonads, the *pha* loci contain two *pha*C genes (Huisman et al. 1991) separated by the *phaZ* gene encoding for an intracellular depolymerase. In some other bacteria like *Paracoccus denitrificans* and *Rhizobium meliloti*, the loci for *phb*AB are not linked to the loci for *phb*C. This implies that biosynthetic pathway for PHB differs among bacteria in terms of loci clustering and orientation.

15.5 Production of Biopolymer by Biological Entities

15.5.1 Biopolymer Production by Naturally Occurring Microbes

Several bacteria synthesize PHA as inclusion bodies to serve as electron sink and a carbon source (Shah 2014; Lathwal et al. 2015; Nehra et al. 2015b). Under electron microscope these granules appear as highly refractive inclusion bodies. *Bacillus megaterium* was the first bacterium that was found to accumulate PHA (Lemoigne 1926) as early as in the beginning of the twentieth century. By the end of this century, more than 200 different bacteria belonging to the gram-positive and gram-negative groups had been reported to accumulate PHA intracellularly under defined environmental conditions. A few researchers have observed that maximum density of PHB granules is recorded from the clayey soil strains (Hawas et al. 2016); however, the size and number of these granules in the cell varies depending on the type of species. *Ralstonia eutropha* has been found to accumulate PHA as high as 80% of the cell biomass. In *Alcaligenes eutrophus*, 8–13 granules of PHA per cell with a diameter in the range of 0.2–0.5 μm have been reported (Byrom 1994). A few naturally occurring microorganisms such as *Cupriavidus necator*, *Bacillus megaterium*, *Pseudomonas oleovorans,* and *Alcaligenes latus* (Bharti and Swetha 2016) have been found to be capable of utilizing several different carbon sources including industrial wastes and plant oils for PHA production. Among the different microorganisms, *C. necator* is one of the best-known and cost-effective PHA producing bacteria. Besides these commonly available bacteria, several other strains such as *Sinorhizobium meliloti*, *Pseudomonas putida*, *Methylobacterium* sp., and *Azotobacter chroococcum* have also been reported to be PHB producers (Kunasundari and Sudesh 2011).

15.5.2 Biopolymer Production in Genetically Modified Microorganisms

PHAs (Polyhydoxyalkanoates) are better alternative in comparison to chemically synthesized petroleum-based plastics. However, PHA production in natural producers is affected by several factors, mainly due to the mobilizing enzymes in the form of depolymerases and esterases present in vivo. Moreover, many native organisms grow slowly and the production cost of PHB by bacterial fermentation is very high. An alternative to increase the PHB content in native organisms is to use a mutant devoid of PHA degradation pathways, or heterologously express the genes in fast-growing organisms lacking PHB biosynthesis and degrading genes. Hence, for lowering the production cost, there is a need not only to develop better fermentation processes but also to select new genetically modified microorganisms.

PHA biosynthesis genes have been expressed in various heterologous systems. Organisms growing at a faster rate and accumulating high amounts of PHB using cheaper carbon sources and having easy downstream processing are most suitable for metabolic engineering. Among the various heterologous systems that have been developed for mass production of PHB, recombinant *E. coli* and *R. eutropha* have been found to be of commercial importance (Lee and Choi 2001). There are different reasons for the popularity of recombinant *E. coli* as PHA producer; as it produces high level of PHA, and there is no need for providing the limiting concentration of nutrients as PHA production is not natural. Genes of the PHB biosynthetic pathway from several microorganisms have been cloned into *E. coli* for the development of recombinant *E. coli* expression system. However, constitutive expression of PHB biosynthetic genes from *Alcaligenes eutrophus* into *E. coli* have resulted into very efficient PHB production by the recombinant *E. coli* (Nomura et al. 2004), wherein, the accumulation of PHB has been reported to be as high as 80–90% of dry cell weight.

15.5.3 Biopolymer Production by Genetically Modified Eukaryotic Organisms

Among the eukaryotic organisms, only *Saccharomyces cerevisiae* has been transformed to accumulate PHB. This has been achieved by expressing the *pha*C in the *S. cerevisiae* cytoplasm. In contrast to *E. coli*, yeast does not require the expression of ketothiolase and acetoacetyl reductase genes for PHA accumulation, as two native enzymes of the β-oxidation pathway in yeast, namely, Erg10 and Fox2 proteins, in addition to their normal function, can also function as β-ketothiolase and acetoacetyl-CoA reductase (Leaf et al. 1996).

15.5.4 Biopolymer Production by Genetically Modified Plants

More recently, in addition to the recombinant microorganisms, plants are also proving themselves as an attractive source for production of PHA. By transferring of relevant genes, plants can be engineered to produce PHB. To establish PHA biosynthesis in plants, *pha*C and *pha*B alone or together with *pha*A have been used. In many cases *pha*A expression is not required as it is inherently present in plants and is available for PHB synthesis. Replacement of *pha*A by *bkt*B in plants enables them to accumulate the copolymers (Valentin et al. 1999). PHB biosynthesis in transgenic plants has been recognized in different organelles of the plant cell. Acetyl-CoA, the precursor required for production of PHB is present in copious amounts in plants, particularly, the movement of acetyl-CoA is higher in chloroplast because fatty acid synthesis occurs in this organelle and it also requires acetyl-CoA as a precursor (Dalton et al. 2012). Pilot experiments have been performed for exploring the feasibility of PHB production in plants by incorporation of genes encoding the enzymes that are required for synthesis of PHB in *Alcaligenes eutrophus* and other bacteria. The PHB biosynthesis genes from microorganisms have been successfully expressed in *Arabidopsis thaliana* (Nawrath et al. 1994; Poirier et al. 1995), *Brassica napus* (Valentin et al. 1999; Liu and Steinbuchel 2000). However, till date, the PHB accumulated in most of these transgenic plants has not been found to be as high as has been reported in transgenic bacteria (Steinbuchel and Hein 2001). In *Arabidopsis thaliana*, an accumulation of 14% of dry weight of PHB was observed, and the produced PHB was found to be of high molecular weight and similar in physical properties to the bacterial PHB. In *B. napus*, PHA accumulation was observed to occur in leucoplasts without affecting the overall growth of the plant (Valentin et al. 1999). Thus, plants form a new attractive platform for the production of PHB.

15.6 Applications of PHAs

PHAs have found applications in major fields of life owing to their valuable properties such as their crystalline and optically active nature, piezoelectricity, and insolubility in water. PHAs are nontoxic, biodegradable, and biocompatible biopolymers and these properties make them competitive with polypropylene. They find applications in five major areas including medical, agricultural, environmental, packaging, and the electronics field.

15.6.1 Medicine

Because of their unique physical and chemical characteristics, the past decade has witnessed an increase in the usage of the biopolymers in the medical field. Most of the members of the biodegradable plastics are biocompatible; they do not cause any allergic reactions to humans. Their major applications in medicine include their use as surgical implants and implant matrices for controlled drug delivery. Implant matrices are basically drug delivery devices in which rods of copolymer are incorporated with sufficient quantity of drug and implanted into the body. These implants provide drug release at targeted sites, or, if desired, also help in sustained release of the drugs inside the body (Brigham and Sinskey 2012).

Due to the biodegradability and biocompatibility action of PHAs, they can be used as scaffold material in tissue engineering. PHAs have been reportedly used as heart valve scaffold in sheep (Sodian et al. 1999; Hassan et al. 2016). Copolymer of PHB and PGA (polyglycolic acid) has been used as pulmonary artery scaffold and pulmonary valve leaflets. Due to their high tensile strength property, PHAs are being used in tissue repair as sutures for healing of muscle-fascial wounds, and also as bone, ligament, and tendon repair devices. Poly-4-hydroxybutyrate has also been used in the fabrication of surgical material.

15.6.2 Agricultural Applications

Biomaterials are used for agricultural applications such as greenhouse coverings, fumigation, mulching, and the like. Besides this, they are also used as agricultural mulches and as planting containers (Arun et al. 2006). Biodegradability of these polymers offers additional advantage, since these are finally broken down and converted into compost along with other biodegradable materials.

15.6.3 Environmental Applications

PHAs can be used as solid substrate for denitrification of water. Due to the PHAs property of being insoluble in water, they can be used constantly as a source of reducing power for denitrification and also as matrices for microbial growth. According to a few reports, denitrification process that has been carried out by using PHAs is providing good results for removal of nitrogen (Hirashi and Khan 2003).

15.6.4 Packaging

PHAs can be used to make foils, packaging films, paper coatings, bottles, and bags. Latex of PHA can be used to cover paper or cardboard to make water resistant surfaces. Further, as PHAs possess good oil absorbing properties, they have been used as oil indicator in cosmetic oil-blotting films (Saharan et al. 2012).

15.6.5 Electronics

Due to the piezoelectric property of PHAs, they have been used in the making of different articles such as headphones, loudspeakers, microphone, ultrasonic detector, shock wave sensors, and pressure sensors for keyboards (Yogesh et al. 2012).

The applications of PHAs thus seem to have permeated all the major sectors, and almost no aspect of human life remains untouched with it.

15.7 Conclusion

Due to environmental concerns that have been created by usage of synthetic plastics, researchers have been forced to search for environment-friendly plastics which are fully biodegradable. These biodegradable plastics commonly referred to as "bioplastics" have become an important and exciting new field having the potential to help save the environment as well as to slow down the depletion of nonrenewable resources. Although PHAs, the most common form of bioplastics exhibiting materialistic properties very similar to the conventional plastics, seem to be quite promising, their large-scale production is still limited because of the high expense of substrates involved in the production. However, the intervention of biotechnology in this field in the form of development of recombinant bacterial systems for the heterologous expression of PHB biosynthetic pathway has given a new impetus to the biopolymer research. The present-day research in this area is focused on exploring various bacterial systems for developing a process which is commercially feasible and economical. But, this is a highly daunting task; several tests and experiments need to be carried out to bring out this technology from its infant stages, and to make bioplastics actually feasible as a replacement for the incredibly versatile polypropylene. However, it can still be concluded that PHAs have a high potential to be used as an alternative to synthetic plastics and have an extremely promising future.

References

Adhikari D, Mukai M, Kubota K, Kai K et al (2016) Degradation of bioplastics in soil and their degradation effects on environmental microorganisms. J Agri Chem Environ 5:23–34

Albuquerque MGE, Eiroa M, Torres C, Nunes BR, Reis MAM (2007) Strategies for the development of a side stream process for polyhydroxyalkanoate (PHA) production from sugar cane molasses. J Biotechnol 130:411–421

Anderson AJ, Dawes EA (1990) Occurrence, metabolism, metabolic role, and industrial uses of bacterial Polyhydroxyalkanoates. Microbiol Rev 54(4):450–472

Anderson AJ, Williams DR, Taidi B, Dawes EA, Ewing DF (1992) Studies on copolyester synthesis by *Rhodococcus ruber* and factors influencing the molecular mass of polyhydroxybutyrate accumulated by *Methylobacterium extroquens* and *Alcaligenes eutrophus*. FEMS Microbiol Rev 103:93–102

Antipov EM, Dubinsky VA, Rebrov AV et al (2006) Strain-induced mesophase and hard-elastic behaviour of biodegradable polyhydroxyalkanoates fibers. Polymer 47:5678–5690

Arun A, Murrugappan RM, David Ravindran AD, Veeramanikandan V, Balaji S (2006) Utilization of various industrial wastes for the production of poly-β-hydroxy butyrate (PHB) by *Alcaligenes eutrophus*. Afr J Biotechnol 5(17):1524–1527

Bharti SN, Swetha G (2016) Need for bioplastics and role of biopolymer PHB: a short review. J Pet Environ Biotechnol 7:272

Boopathy R (2000) Factor limiting bioremediation technologies. Bioresour Technol 74:63–67

Brigham CJ, Sinskey AJ (2012) Applications of Polyhydroxyalkanoates in the medical industry international. J Biotechnol Wellness Industries 1:53–60

Byrom D (1994) Polyhydroxyalkanoates. In: Mobey DP (ed) Plastics from microbes: microbial synthesis of polymers and polymer precursors. Hanser, Munich, pp 5–33

Casini E, de Rijk TC, de Waard P, Eggink G (1997) Synthesis of poly (hydroxyalkanoate) from hydrolyzed linseed oil. J Environ Polym Degr 5:153–158

Chanprateep S (2010) Current trends in biodegradable polyhydroxyalkanoates. J Biosci Bioeng 110(6):621–632
Chee JY, Yoga SS, Lau NS et al (2010) Bacterially produced Polyhydroxyalkanoate (PHA): converting renewable resources into Bioplastics. Curr Res, Tech Educ Topics in Appl Microbiol Microbial Biotechnol 2:1395–1404
Choi GG, Kim HW, Rhee YH (2004) Enzymatic and non-enzymatic degradation of poly (3-hydroxybutyrate-co-3-hydroxyvalerate) copolyesters produced by Alcaligenes sp MT-16. J Microbiol 42:346–352
Dalton DA, Ma C, Murthy GS, Strauss SH (2012) Bioplastic production by transgenic poplar. Information Systems Biotechnol:7–10
De Smet MJ, Eggink G, Witholt B et al (1983) Characterization of intracellular inclusions formed by *Pseudomonas oleovorans* during growth on octane. J Bacteriol 154:870–878
Endres HJ, Siebert-Raths A (2011) Basics of PHA. Bioplastics Magazine 6(5):43–45
Gu Ji D (2003) Microbiological deterioration and degradation of synthetic polymeric materials: recent research advances. Int Biodeter Biodegr 52:69–91
Hassan M, Bakhiet E, Ali E, Hussien H (2016) Production and characterization of polyhydroxybutyrate (PHB) produced by *Bacillus sp.* isolated from Egypt. J Applied Pharmaceutical Sci 6(04):046–051
Hawas J, Banna T, Belal E, Aziz A (2016) Production of bioplastic from some selected bacterial strains. Intern J Curr Microbiol App Sci 5(1):10–22
Hirashi A, Khan ST (2003) Application of polyhydroxyalkanoates for denitrification in water and wastewater treatment. Appl Microbiol Biotechnol 61:103–109
Hori K, Soga K, Doi Y (1994) Effects of culture conditions on molecular weights of poly (3-hydroxyalkanoates) produced by *Pseudomonas putida* from octanoate. Biotechnol Lett 16:709–714
Huisman GW, Wonink EW, Meima R et al (1991) Metabolism of poly (3 hydroxyalkanoates) (PHAs) by *Pseudomonas oleovorans*. J Biol Chem 266:2191–2198
Ishizaki A, Tanaka K, Taga N (2001) Microbial production of poly-D-3-hydroxybutyrate from CO_2. Appl Microbiol Biotechnol 57:6–12
Jendrossek D, Handrick R (2002) Microbial degradation of polyhydroxyalkanoates. Annual Rev Microbiol 56:403–432
Kang HO, Chung WC, Kim HW et al (2001) Cometabolic production of copolyesters consisting of 3-hydroxyvalerate and medium-chain-length 3-hydroxyalkanoates by *Pseudomonas* sp. DSY-82. Antonie Van Leeuwenhoek 80:185–191
Kunasundri B, Sudesh K (2011) Isolation and recovery of microbial polyhydroxyalkanoates. *eXPRESS* Polymer Lett 5(7):620–634
Lathwal P, Nehra K, Singh M, Jamdagni P, Rana JS (2015) Optimization of culture parameters for maximum polyhydroxybutyrate production by selected bacteria isolated from rhizospheric soils. Polish J Microbiol 64(3):227–239
Leaf TA, Peterson MS, Stoup SK, Somers D, Srienc F (1996) *Saccharomyces cerevisiae* expressing bacterial polyhydroxybutyrate synthase produces poly-3-hydroxybutyrate. Microbiol 142:1169–1180
Lee PC, Lee WG, Lee SY, Chang HN (2000) Succinic acid production with reduced by-product formation in the fermentation of *Anaerobiospirillum succiniciproducens* using glycerol as a carbon source. Biotechnol Bioeng 72:41–48
Lee SY (1996a) Bacterial Polyhydroxyalkanoates. Biotechnol Bioeng 49(1):1–14
Lee SY (1996b) Plastic bacteria? Progress and prospects for polyhydroxyalkanoates production in bacteria. TIBTECH 14:431–438
Lee SY, Choi JI (2001) Production of microbial polyester by fermentation of recombinant microorganisms. Adv Biochem Eng Biotechnol 71:183–207
Lemoigne M (1926) Produits de dehydration et de polymerisation de l'acide ß-oxobutyrique. Bull Soc Chim Biol 8:770–782
Liu SJ, Steinbuchel A (2000) A novel genetically engineered pathway for synthesis of poly (hydroxyalkanoic acids) in *Escherichia coli*. Appl Environ Microbiol 66:739–743

Lu J, Tappel RC, Nimura CT (2009) Mini-review: biosynthesis of poly (hydroxyalkanoates). Polymer Rev 49(3):226–248

Macrae RM, Wilkinson JF (1958a) Poly-f1-hydroxybutyrate metabolism in washed suspensions of *Bacillus cereus* and *Bacillus megaterium*. J Gen Microbiol 19:210–222

Macrae RM, Wilkinson JF (1958b) The influence of cultural conditions on poly-β-hydroxybutyrate synthesis in *Bacillus megaterium*. Proc R Phys Soc Edin 27:73–78

Madison LL, Huisman GW (1999) Metabolic engineering of poly (3-hydroxyalkanoate): from DNA to plastic. Microbiol Mol Biol Rev 63:21–53

Marchessault RH, Yu GE (2004) Crystallization and material properties of polyhydroxyalkanoates. In: Biopoly-mersed. Steinbuchel A, pp 157–248

Matsusaki H, Abe H, Doi Y (2000) Biosynthesis and properties of poly3-hydroxybutyrate-co-3-hydroxyalkanoates by recombinant strains of *Pseudomonas* species. Biomacromolecules 1:17–22

Mergaert J, Webb A, Anderson C, Wouters A, Swings J (1993) Microbial degradation of poly (3-hydroxybutyrate) and poly (3-hydroxybutyrate-co-3-hydroxyvalerate) in soils. Appl Environ Microbiol 59:3233–3238

Nam JS, Kim HC, Kim DY, Rhee YH (2002) Distribution and diversity of microbial communities relating to biodegradation of medium-chain-length poly(3-hydroxyalkanoates) in soils. Proceedings of 9th international symposium on the genetics of industrial microorganisms. Gyeongju, Korea, p 192

Nawrath C, Poirier Y, Somerville C (1994) Targeting of the polyhydroxybutyrate biosynthetic pathway to the plastids of *Arabidopsis thaliana* results in high levels of polymer accumulation. Proc Natl Acad Sci U S A 91:12760–12764

Nehra K, Chhabra N, Sidhu PK, Lathwal P, Rana JS (2015a) Molecular identification and characterization of poly-b-hydroxybutyrate (PHB) producing bacteria isolated from contaminated soils. Asian J Microbiol Biotech Environ Sci 17(4):281–290

Nehra K, Dahiya N, Lathwal P, Singh M (2015b) Influence of nutritional and environmental parameters on PHB production by bacteria isolated from contaminated soils. Interna J Pharma & Biosci 6(1):676–691

Nomura CT, Tanaka T, Gan Z, Kuwabara K et al (2004) Effective enhancement of short-chain-length-medium-chain length polyhydroxyalkanoate copolymer production by co-expression of genetically engineered 3-ketoacyl-acyl-carrier-protein synthase III (*fab*H) and polyhydroxyalkanaotes synthesis genes. Biomacromolecules 5:1457–1464

Padermshoke A, Sato H, Katsumoto Y et al (2004) Thermally induced phase transition of poly (3-hydroxybutyrate-co-3-hydroxyhexanoate) investigated by two-dimensional infrared correlation spectroscopy. Vib Spectrosc 36:241–249

Peoples OP, Sinskey AJ (1989) Poly-β-hydroxybutyrate (PHB) Biosynthesis in *Alcaligenes eutrophus* H16 Identification and Characterization of the PHB Polymerase Gene (phbC). The J Biological Chem 264(26):1529–15303

Poirier Y, Nawrath C, Somerville C (1995) Production of polyhydroxyalkanoates, a family of biodegradable plastics and elastomers, in bacteria and plants. Biotechnol 13:142–150

Saharan BS, Sharma A, Sharma D (2012) Bioplastics-for sustainable development: a review. Intern J Microbial Resource Technol 1(1):11–23

Schubert P, Steinbuchel A, Schlegel HG (1988) Cloning of the *Alcaligenes eutrophus* genes for the synthesis of poly-β-hydroxybutyrate (PHB) and synthesis of PHB in *Escherichia coli*. J Bacteriol 170:5837–5847

Shah KR (2014) Optimization and production of Polyhydroxybutarate (PHB) by *Bacillus subtilis* G1S1from soil. Intern J Curr Microbiol App Sci 3(5):377–387

Sharma M, Dhingra H (2016) Poly-β-hydroxybutyrate: a biodegradable polyester, biosynthesis and biodegradation. British Microbiol Res J 14(3):1–11

Sharmila T, Meenakshi SA, Kandhymathy K, Bharathidasan R, Mahalingam R (2011) Screening and characterization of Polyhydroxybutyrate producing bacteria from sugar industry effluents. World J Sci and Technol 1(9):22–27

Slater SC, Voige WH, Dennis DE (1988) Cloning and expression in *Escherichia coli* of the *Alcaligenes eutrophus* H16 poly-β-hydroxybutyrate biosynthetic pathway. J Bacteriol 170:4431–4436

Soam A, Singh AK, Singh R, Shahi SK (2012) Optimization of culture conditions for bio-polymer producing *Bacillus mycoides* (WSS2) bacteria from sewage. Curr Discovery 1(1):27–32

Sodian R, Sperling JS, Martin DP et al (1999) Tissue engineering of a trileaflet heart valve-early in vitro experiences with a combined polymer. Tissue Eng 5(5):489–494

Solaiman DKY, Ashby RD (2005) Rapid genetic characterization of poly (hydroxyalkanoate) synthase and its applications. Biomacromolecules 6:532–537

Solaiman DKY, Ashby DR, Foglia TA, Marmer WN (2006) Conversion of agricultural feedstock and co-products into poly (hydroxyalkanoates). Appl Microbiol Biotechnol 71:783–789

Steinbuchel A (1991) Polyhydroxyalkanoic acids. In: Byrom D (ed) Biomaterials: novel materials from biological sources. MacMillan Publishers, New York, pp 123–213

Steinbuchel A (2001) Perspectives for biotechnological production and utilization of biopolymers: metabolic engineering of Polyhydroxyalkanoate biosynthesis pathways as a successful example. Macromolecular Biosci 1:1–24

Steinbuchel A, Hein S (2001) Biochemical and molecular basis of microbial synthesis of polyhydroxyalkanoates in microorganisms. Adv Biochem Engin Biotechnol 71:82–119

Steinbuchel A, Valentin HE (1995) Diversity of bacterial polyhydroxyalkanoic acids. FEMS Microbiol Lett 128:219–228

Sudesh K, Abe H, Doi Y (2000) Synthesis and properties of polyhydroxyalkanoates: biological polyesters. Prog Polym Sci 25:1503–1555

Tappel RC, Pan W, Bergey NS, Wang Q et al (2014) Engineering *Escherichia coli* for improved production of short-chain- length-co-medium-chain-length poly [(R) 3-hydroxyalkanoate] (SCL-co-MCL PHA) copolymers from renewable Nonfatty acid Feedstocks. ACS Sustainable Chem Engin 2:1879–1887

Tsuge T (2002) Metabolic improvements and use of inexpensive carbon sources in microbial production of polyhydroxyalkanoates. J Biosci Bioeng 94:579–584

Tsz-Chun M, Chan PL, Lawford H, Chua H, Lo WH, Yu PHF (2005) Microbial synthesis and characterization of physicochemical properties of polyhydroxyalkanoates (PHAs) produced by bacteria isolated from activated sludge obtained from the municipal wastewater works in Hong Kong. Appl Biochem Biotechnol 122:731–740

Valentin HE, Broyles DL, Casagranade LA, Colburn SM et al (1999) PHA production, from bacteria to plants. Int J Biol Macromol 25:303–306

Verlinden RAJ, Hill DJ, Kenward MA, Williams CD, Radecka I (2007) Bacterial synthesis of biodegradable polyhydroxyalkanoates. J Appl Microbiol 102:1437–1449

Wallen LL, Rohwedder WK (1974) Polymeric beta hydroxyalkanoates from activated sludge. Environ Sci Technol 8:576–579

Wani SJ, Shaikh SS, Sayyed RZ (2016) Microbial biopolymers in biomedical field. MOJ Cell Sci & Report 3:3

Witholt B, Kessler B (1999) Perspectives of medium chain length poly (hydroxyalkanoates), a versatile set of bacterial bioplastics. Curr Opinion Biotechnol 10:279–285

Yogesh C, Pathak B, Fulekar MH (2012) PHA – production application and its bioremediation in environment. Intern Res J Environ Sci 1(2):46–52

Biotechnological Strategies for Remediation of Toxic Metal(loid)s from Environment

16

M. Anju

Abstract

Metal(loid)s are ubiquitous in the environment. They are known to exert toxic effects not only on plants and animals, but also on human beings, and are persistent in the environment; therefore, their remediation is imperative. Remediation of metal(loid)s presents a distinct challenge because they are nonbiodegradable and hence cannot be metabolized like organic pollutants; rather, they accumulate in the environment. Thus, goals of remediation of metal(loid)s pollution are generally to extract them from a large volume and transfer them to a smaller volume, to change their speciation so that their toxicity is reduced, or to increase their volatility. Present review aims to provide a succinct overview of potential biotechnological strategies using a vast array of biological materials, especially bacteria, fungi, algae, yeasts, and higher plants for metal(loid)s remediation. Among the various techniques used for remediation of heavy metal(loid)s, phytoremediation has emerged as one of the safest, innovative, environment-friendly, solar-driven and effective technique. A variety of plants, especially the hyperaccumulators, have been used for the phytoremediation of toxic metal(loid)s. To make phytoremediation more efficient, a number of adjunct approaches such as chelate-assisted phytoremediation and microbe-assisted phytoremediation have been explored. Plant growth-promoting bacteria (PGPB) have the potential to improve the efficacy of phytoremediation of toxic metal(loid)s by transforming metal(loid)s' solubility and bioavailability through the action of biosurfactants, organic acids, siderophores, redox processes, and methylation. Another direct method of enhancing phytoremediation is overexpressing the genes involved in metal(loid)s uptake, transportation, sequestration, and metabolism. Transgenic approaches successfully employed to promote phytoextraction of metal(loids)

M. Anju (✉)
Department of Energy and Environmental Sciences, Chaudhary Devi Lal University, Sirsa 125 055, Haryana, India
e-mail: anjumalik27@yahoo.com

S.K. Gahlawat et al. (eds.), *Plant Biotechnology: Recent Advancements and Developments*, DOI 10.1007/978-981-10-4732-9_16

mainly involve manipulation of transition metal transporters, enhanced production of metal-detoxifying chelators (metallothioneins and phytochelatins) and enzymes of sulfur metabolism, and phytovolatilization of mercury and selenium. Microbes and hyperaccumulators with extraordinary capabilities of metal(loid)s tolerance, accumulation, and detoxification are the most logical species to provide an important reservoir of candidate genes for use in genetic engineering strategies for phytoremediation. Genetic manipulation of fast-growing species by transferring unique genes from natural hyperaccumulators is a promising approach which has been demonstrated with selenocysteine methyltransferase (SMT) transgenics volatilizing selenium. Mercury volatilization and tolerance have been achieved by the introduction of bacterial pathways into plants. The tools of genetic engineering to modify plants for enhanced metal uptake, transport, and sequestration have opened doors to new avenues for creation of "remediation" cultivars.

16.1 Introduction

Heavy metals are the elements (metals and metalloids) that have atomic density > 5 gcm^{-3} and atomic mass greater than 20 (Rascio and Navari-Izzo 2011). Often metalloids such as arsenic and selenium with specific gravities of 5.73 gcm^{-3} and 4.79 gcm^{-3}, respectively, are also included among the heavy metals. In this chapter, the term "heavy metal(loids)" has been used as a convenient shorthand, which refers to both metals and metalloids. Heavy metal(loid)s include both biologically "essential" (Zn, Cu, Mn, Fe, Mo, and Co) and "nonessential" (Cd, Pb, Hg, and As) elements. Both essential and nonessential metal(loid)s exert toxic effects on animals, plants, and/or humans at exorbitant concentrations (Ye et al. 2002; Alloway 1990).

The term trace element has not been defined strictly, rather used loosely in the literature and has differing meanings in different scientific disciplines. Trace elements refer to elements (both metals and metalloids) that occur in natural and perturbed environments in small amounts and that, when present in sufficient bioavailable concentrations, are toxic to living organisms (Adriano 2001). Trace metals and heavy metals are the other terms that have been used synonymously for trace elements.

Trace metals are ubiquitous in the environment, occurring in varying concentrations in air, water, soil, bedrock, and all biological matter (Bowen 1979). Some geological parent materials, including black sedimentary sulfide-bearing shales (alum shales), ultramafic rocks, sedimentary ironstones, metallogenic mineralization, and basic igneous rocks, often contain large amounts of heavy metal(loid)s. These are released into the environment from a broad spectrum of natural as well as anthropogenic/technogenic sources (Alloway 2013; Nriagu 1978; Adriano 2001). Natural sources are weathering of mineral deposits, sea salt sprays, forest fires, emissions from active volcanoes, etc. The main anthropogenic sources are mining and smelting activities, metal processing industries, incineration of wastes and

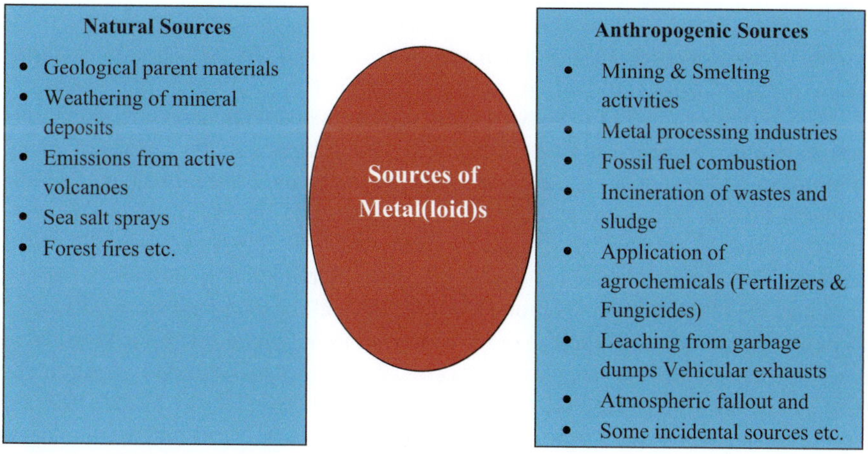

Fig. 16.1 Sources of metal(loid)s in environment

sludge, fossil fuel combustion, leaching from garbage dumps, application of agrochemicals (both inorganic phosphate fertilizers and fungicides), vehicular exhausts, atmospheric deposition and some incidental sources, etc. (Fig. 16.1). Of the two modes of input, anthropogenic sources are of greater environmental threat as metal(loid)s forms from anthropogenic sources are environmentally unstable, thus more soluble and bioavailable than natural forms. Pyrometallurgical nonferrous metal production is the major global source of airborne cadmium, copper, arsenic, lead, and zinc (Nriagu and Pacyna 1988). The metallurgicals are also primary sources of cadmium, lead, and nickel for aquatic ecosystems, whereas mine tailings, smelter wastes, and atmospheric fallout are the most important sources of metals worldwide for soils (Nriagu and Pacyna 1988).

16.2 Why to Remediate Heavy Metal(loid)s?

All heavy metals (e.g., Cd, Co, Cu, Cr, Ni, Pb, Zn, Hg) and metalloid elements (e.g., As, Se, Sb) are toxic to plants and animals when exceeding critical concentrations (Ye et al. 2002). Major health risks associated with heavy metal(loids)s are kidney damage, bone effects, and fractures from cadmium; neurological damage from mercury; neurotoxic effects of lead; risks of skin cancer as well as other skin lesions such as hyperkeratosis and pigmentation changes from arsenic; lung cancer on occupational exposure to arsenic, primarily by inhalation (Järup 2003); cardiovascular disease, chronic anemia, cognitive impairment, nervous system, brain, bones, teeth, skin, and many others (Iqbal 2012; Wuana and Okieimen 2011).

The contamination of the surface of earth by human interventions is a relatively old phenomenon; the evaluation, assessment, and remediation of negative impacts are relatively recent. The identification of a contamination problem is only a small part of the contamination scenario. The decision to remediate implies that the ownership of the contamination problem has been accepted. As metal(loid)s contaminated environment often present an unacceptable risk to human and ecological health, the remediation of the environment contaminated with heavy metal(loid)s is an important problem for many countries and concentrates the efforts of many authorities and scientists. If the remediation measures have been taken, the "ecodisasters" such as Minamata methylmercury incident and Itai-Itai disease – the cadmium poisoning (Jintsu valley, Japan) – would not have found a place in the world history. The cleanup of heavy metal(loids)s contamination is utmost necessary because: (1) metal(loid)s are among the most difficult contaminants to treat as they cannot be destroyed; (2) their toxicity can last long in nature; (3) metal(loid)s can only be transformed in valence and species, but cannot be degraded by any technique including biotreatment; (4) some metal(loid)s even could be transformed from relevant low toxic species to more toxic species under certain environmental conditions, e.g., mercury is transformed to more toxic methylmercury; (5) unlike organic substances, metal(loid)s are essentially nonbiodegradable and therefore accumulate in the environment; (6) the bioaccumulation and bioaugmentation of metal(loid)s by food chain could damage normal physiological activity and endanger human life; and (7) finally, some are lethal in trace concentrations and may have carcinogenic, mutagenic, teratogenic effects and also act as endocrine disruptors.

16.3 Remediation Strategies for Toxic Metal(loid)s

Toxic metal(loid)s pollution is a major environmental problem worldwide. Since, increased metal(loid)s concentrations have adverse effects on most living organisms, hitherto, various cleanup technologies based on physical, chemical, and biological approaches have been developed and employed to remediate contaminated environment. These cleanup technologies can be applied on- or off-site. The physical approaches are thermal desorption or vapor stripping of semivolatile or volatile compounds and electrokinetics. Chemical approaches include all techniques involving reagents or external compounds (Yousfi and Bermond 2001). Established conventional physical and chemical remediation techniques have inherent limitations; for example, they involve intensive labor, are often prohibitively expensive and environmentally invasive, bring irreversible changes in soil properties, cause disturbance of native soil microflora, and also have the potential for introducing secondary contamination (Dermont et al. 2008; Pulford and Watson 2003; Ali et al. 2013). Therefore, a remediation technique that is not only cost-effective and efficient, but also eco-friendly, would be a valuable addition to the current remediation approaches.

Owing to the limitations associated with conventional techniques, there has been an increased interest in alternative bio-based, eco-friendly approaches commonly known as *biological remediation*. This approach is often considered as cost-effective, environment-friendly, and the most effective as it relies on natural processes. Various in situ and ex situ techniques are used for biological remediation.

Among various biological remediation techniques, bioremediation and phytoremediation have been used most successfully.

16.4 Bioremediation

Bioremediation is an alternative which has the possibility to destroy or render harmless various environmental contaminants using natural organisms, from bacteria to plants, or their derivatives. Biodiversity is the precondition for bioremediation. Bioremediation has emerged as a promising alternative for pollutant cleanup. It is often considered an environment-friendly and cost-effective approach and is progressively making inroads for environmental remediation. The basic principle of bioremediation is change in the redox reactions, increasing/decreasing the solubility by changing pH or using different complexation reactions, and adsorption/uptake of a substance from the environment (Smith et al. 1994). Bioremediation, involving the use of microbes to detoxify and degrade various environmental contaminants, has received increasing interest in recent times to clean up a polluted environment (Farhadian et al. 2008; Malik 2004; Gadd 2000).

Bioremediation (i.e., green technologies or phytotechnologies when relied upon plants) mainly deals with biological interventions of biodiversity aimed at environmental contamination assessment and alleviating pollution (Prasad et al. 2010). If the process occurs in the same place which was afflicted by pollution, then it is called in situ bioremediation. On the contrary, deliberate relocation of the contaminated material (soil and water) to a different place to intensify biocatalysis is referred to as ex situ treatment (Prasad et al. 2010). Advances in bioremediation harness tools of genetic, molecular, microbial, and protein engineering.

16.4.1 Biosorption and Bioaccumulation

Biosorption and bioaccumulation are processes by which the microorganisms, or biomass, bind to and concentrate contaminants and heavy metals from the environment (Joutey et al. 2015). However, both biosorption and bioaccumulation function in different manners.

Bioaccumulation is a double-stage, slow, partially reversible, active metabolic transport in living biomass, whereas biosorption by dead biomass (or by some molecules and/or their active groups) is single-stage, quick, reversible, passive, and based mainly on the "affinity" between the (bio-)sorbent and sorbate (Chojnacka 2010b; Volesky 2007).

The adsorption, ion-exchange, complexation, and coordination mechanisms have been found to play a dominant role in biosorption (He and Chen 2014). A broad range of key functional groups, such as hydroxyl, carboxyl, sulfhydryl, phosphate (P), and amine groups, are known to play a dominant role in biosorption of trace elements (He and Chen 2014), and the magnitude of biosorption is directly related to these functional groups (Bolan et al. 2013).

Because of the simplicity, apparent efficiency, analogy to ion-exchange technology, and abundant availability of inactive and dead biomass, biosorption has been heralded as an innovative and promising technology for removal/recovery of metal(loid)s from aqueous solutions. A vast array of biological materials, especially bacteria, fungi, algae, yeasts, and industrial and agricultural wastes, have received increasing attention for heavy metal removal and recovery due to their good performance, low cost, and large available quantities (Wang and Chen 2009).

Several living organisms investigated as potential biosorbents include bacteria such as *Bacillus subtilis*, *Magnetospirillum gryphiswaldense*, and *Pseudomonas* (Vijayaraghavan and Yun 2008; Zhou et al. 2012); fungi such as *Rhizopus arrhizus* (Wang and Chen 2006) and *Trichoderma reesei* (Kim et al. 2003); yeast such as *Saccharomyces cerevisiae* (Wang and Chen 2006); and algae (e.g., brown, red and green algae, modified algae and the derivatives) such as *Sargassum* spp. and *Chaetomorpha linum* (He and Chen 2014; Romera et al. 2006) and microalgae both freshwater and marine such as *Chlorella* spp., *Spirulina* spp., *Chlamydomonas* spp., *Scenedesmus* spp., and *Spirogyra* spp. (Kumar et al. 2015).

Bioaccumulation is the intracellular accumulation of sorbate, which takes place in two steps: (1) the first is the quick sorption, identical with biosorption of metals by microbial biomass and its by-products, and (2) the second is slower and includes physiological transport of sorbate into the insides of cells by the metabolically active transport system (Chojnacka 2010a; Adriano et al. 2004). The pollutants transported across cell wall and membrane are bound to intracellular structures (Kujan et al. 1995). For bioaccumulation, the species which are resistant to high loads of pollutants and do not have mechanisms which protect them from excessive accumulation inside the cell should be selected (Kocberber and Donmez 2007; Deng and Wilson 2001). Rather, bioaccumulating organisms should have a mechanism of intracellular binding, such as special proteins rich in thiol groups – phytochelatins (PCs) and metallothioneins (MTs) – which are synthesized as the response to the presence of toxic metal ions in their living environment, and complex those pollutants, thus excluding them from normal metabolic processes (Chojnacka 2010a).

Bioaccumulation of toxic metal ions by growing yeast such as *S. cerevisiae*, *Schizosaccharomyces pombe*, *Kluyveromyces marxianus*, and *Candida* sp. (Dönmez and Aksu 1999); aquatic plants – free-floating, emergent, or submerged macrophytes (Kara 2004); fungi (Almeida et al. 2015); algae (Jaiswar et al. 2015); marine bacteria such as *Vibrio harveyi* (Abd-Elnaby et al. 2011); and genetically engineered *Escherichia coli* (Deng and Wilson 2001) has been widely discussed in literature. In line with that, certain purple nonsulfur marine bacterial isolates, e.g., *Rhodobium marinum* and *Rhodobacter sphaeroides*, have also been found to possess the potential of removing heavy metals like Cu, Zn, Pb, and Cd from the contaminated environments either by biosorption or by biotransformation (Panwichian et al. 2011).

16.5 Phytoremediation

Phytoremediation is defined as the use of certain plants and trees to remove, contain, or render harmless environmental contaminants (Salt et al. 1998), such as metals and/or organic contaminants such as solvents, crude oil, polyaromatic hydrocarbons (PAHs), etc. Phytoremediation can be applied to both organic and inorganic pollutants present in solid (e.g., soil), liquid (e.g., water), and gaseous substrates (Salt et al. 1998). The discovery of "hyperaccumulators," defined as plants – often endemic to naturally mineralized soils – that accumulate high concentrations of metals in their foliage (Baker and Brooks 1989), has led to the revolutionary idea of using plants to remediate metal pollution.

Phytoremediation has been described to be a novel, efficient, nonintrusive, inexpensive, solar energy driven, carbon neutral, environment- and eco-friendly, in situ applicable, aesthetically pleasing, and socially acceptable remediation strategy (Prasad et al. 2010; Garbisu et al. 2002). It utilizes plants and associated microorganisms to remove, transform, or stabilize metal(loid)s in both water and soil. Soil stabilization, CO_2 sequestration, biodiversity improvement, and watershed management are ancillary to phytoremediation (Dickinson et al. 2009).

16.5.1 Basic Strategies of Phytoremediation of Metal(loids)

Phytoremediation covers several different strategies, classified based on the fate of the contaminant: phytoextraction/phytoaccumulation, phytodegradation, rhizodegradation /phytotransformation, phytofiltration (rhizofiltration, blastofiltration), phytostabilization (phytoimmobilization), phytovolatilization/phytoevaporation, and phytodesalination (Ali et al. 2013; Mahar et al. 2016). Out of these, four strategies, each having a distinct mechanism of action, employed for cleanup of metal(loid)s pollution are phytoextraction, phytofiltration, phytostabilization, and phytovolatilization/phytoevaporation.

16.5.1.1 Phytoextraction/Phytoaccumulation

The terms phytoextraction, phytoaccumulation, phytoabsorption, and phytosequestration are used synonymously. Phytoextraction is the uptake of contaminants from soil or water by a particular group of metallophytes called hyperaccumulators to in situ cleanup of metals and metalloids from polluted substrates (Ginocchio et al. 2004). Pollutant-accumulating plants uptake large amounts of trace elements from substrates through roots which are translocated to and accumulated in aerial/aboveground harvestable biomass, i.e., shoots, leaves, etc. (Salt et al. 1998; Ginocchio et al. 2004).

16.5.1.2 Phytofiltration (Rhizofiltration and Blastofiltration)

Phytofiltration refers to the removal of pollutants from contaminated waters or aqueous waste streams by plants (Dushenkov et al. 1995; Salt et al. 1998). Phytofiltration may use plant roots (rhizofiltration), seedlings (blastofiltration), or excised plant shoots (caulofiltration, Latin *caulis* = shoot) (Rozas et al. 2006; Mesjasz-Przyby et al. 2004). In rhizofiltration, plant roots absorb, precipitate, and concentrate toxic metals from polluted effluents (Dushenkov et al. 1995). An ideal plant for rhizofiltration should have rapidly growing roots with the ability to remove toxic metals from solutions over extended periods of time (Salt et al. 1995).

16.5.1.3 Phytostabilization/Phytoimmobilization

Unlike other phytoremediation techniques, phytostabilization is not intended to remove metal contaminants from a site, but rather to stabilize them by accumulation in roots or precipitation within root zones, minimizing the risk to human health and the environment (Padmavathiamma and Li 2007). Metal-tolerant plant roots and associated microbes can arrest the mobility of heavy metals in soils through sorption by roots, complexation, or metal valence reduction in rhizosphere and precipitation. Heavy-metal-tolerant plants are used to reduce the mobility and bioavailability of heavy metals in the environment either by immobilization or by prevention of migration to groundwater or their entry into the food chain, thereby reducing the risk of further environmental degradation (Pulford and Watson 2003; Salt et al. 1998; Erakhrumen 2007).

Phytostabilization involves the use of excluder metallophytes for in situ immobilization of metal(loid)s in polluted soils and mining passives. Trace elements are taken up by the roots where they are biologically sequestered and accumulated in innocuous forms; they are not translocated into aerial tissues of plants and thus aerial metal concentrations of excluder metallophytes are same as normal plant. Phytostabilization thus reduces the trace element availability and toxicity to other organisms in the substrate and into food webs, and avoids their leaching to groundwater (Ginocchio et al. 2004). Although this strategy is well acknowledged, it is not a permanent solution as the metals are not removed from soil; only their mobility is restricted.

16.5.1.4 Phytovolatilization/Phytoevaporation

Phytovolatilization or phytoevaporation refers to the volatilization of pollutants into the atmosphere via plants (Burken and Schnoor 1999). This technique can be used for organic pollutants and some metal(loid)s like Hg, Se, and As (Mahar et al. 2016). Toxic metals such as Se, As, and Hg can be biomethylated to form volatile molecules that can be lost to the atmosphere. Application of this technique is restricted as the contaminants are only transferred from one environmental sphere to another and not removed completely. Phytovolatilization of Se and Hg is controversial as there is skepticism about whether the volatilization of these elements into the atmosphere is desirable or safe (Padmavathiamma and Li 2007).

16.5.1.5 Phytodegradation

Phytodegradation is aimed at degradation of organic pollutants by plants and associated microbes (Salt et al. 1998; Burken and Schnoor 1997). Application of this strategy is restricted to the cleanup of organic pollutants only as metal(loid)s cannot be degraded through the metabolic activities of plants and associated microbes.

16.5.1.6 Rhizodegradation

Rhizodegradation refers to the breakdown of organic contaminants through the activity existing in the rhizosphere due to the presence of nutrients-containing exudates and enzymes produced by plants or soil microbes such as bacteria and fungi.

16.5.1.7 Phytodesalination

Phytodesalination relies on the ability of Na-hyperaccumulating/halophytic plants for the removal of sodium/salts from salt-affected soils (Rabhi et al. 2010; Sakai et al. 2012). It is based on the capacity of some halophytes to accumulate enormous sodium quantities in their shoots (Rabhi et al. 2010). Phytodesalination is an emerging technique and is of great importance particularly in arid and semiarid regions where insufficient precipitations and inappropriate irrigation systems are unable to reduce the salt burden in the rhizosphere of plants and suitable physic-chemical methods are too expensive (Zorrig et al. 2012).

16.6 Metallophytes: The Plants Evolved in Metalliferous Soils

Metallophytes are unique plants that have evolved biological mechanisms to resist, tolerate, or thrive on the toxic metalliferous soils, and are typically endemic to their native metalliferous soils (Baker et al. 2010). Metallophytes are botanical curiosities (Alford et al. 2010). True metallophytes have evolved on substrates derived from weathered mineral deposits for many thousands or millions of years, and consequently have highly specialized mechanisms of resistance or tolerance. The evolution of metal tolerance is a result of natural selection. Over countless years, the selective pressure resulted in the evolution of mechanisms enabling them to tolerate or detoxify the metals (Whiting et al. 2002). Indeed, metallophytes have often diverged genetically and morphologically to form new taxa endemic to their individual metalliferous area.

Metallophytes are the plants that have evolved to survive on metal-rich soils. The ability of metallophytes to tolerate extreme metal concentrations commends them for site restoration at mine closure and for rehabilitation and revegetation of mines and metal-contaminated sites. They can also be used to remediate contaminated land by exploiting their metal-accumulating properties to scavenge metals from metalliferous soils (phytoextraction). There are many thousands of species of metal-tolerant, nonaccumulating plants that might be considered for phytostabilization. Metallophytes can also be exploited for the development of environmental phytotechnologies such as phytoextraction, phytostabilization, and phytomining.

Metallophytes can be categorized into three basic types on the basis of strategies developed for growing on metalliferous soils: metal excluders, metal indicators, and metal hyperaccumulators (Baker 1981).

16.6.1 Metal Excluders

The majority of metallophyte taxa are able to tolerate specific metals in the substrate by physiologically restricting the entry of metals into the root and/or their transport to the shoot and were termed "excluders" by Baker (1981).

16.6.2 Metal Indicators

As the name indicates, internal concentration of metals in these plants reflects heavy metal concentration in the substrate (McGrath et al. 2002).

16.6.3 Metal Hyperaccumulators

To describe plants with Ni concentrations more than 1000 mg/kg dry weight (0.1%), Brooks et al. (1977) coined the term "hyperaccumulator." Hyperaccumulators are defined as plants, often endemic to naturally mineralized soils, that accumulate high concentrations of metals in their foliage (Baker and Brooks 1989). Hyperaccumulators make up a subset of the category known as metallophytes – plants that occur on metal-enriched soils. Hyperaccumulators have an innate capacity to absorb metals at levels 50–500 times greater than average plants (Lasat 2000). Although "hyperaccumulators" belong to distantly related families, they have common characteristics like the ability to grow on metalliferous soils and to accumulate extraordinarily high amounts of heavy metals in aerial plant parts, far in excess of the levels found in the majority of species, without suffering phytotoxic effects (Rascio and Navari-Izzo 2011). Basic distinguishing features of hyperaccumulators from nonhyperaccumulators taxa are: (i) greater potential of metal uptake, (ii) faster root-to-shoot translocation of metals, and (iii) a greater efficiency in metal sequestration and detoxification in aboveground plant parts (Rascio and Navari-Izzo 2011).

Although no standard has been defined scientifically for hyperaccumulators (Nazir et al. 2011), many researchers have defined their own criterion for hyperaccumulators. The criterion defined by Baker and Brooks (1989) is the most cited one. The higher plants capable of accumulating >1 mg Au and Ag; >100 mg Se, Cd, and Ta; >1000 mg Co, Cu, Cr, Ni, Pb, U, and As; and >10,000 mg Mn and Zn per 1 kg of the dry mass of shoots on soils rich in heavy metal(loid)s are considered as "hyperaccumulators" (Baker and Brooks 1989).

van der Ent et al. (2013) have tried to outline considerations for establishing metal(loid)s hyperaccumulation and (re)defined some of the associated terminologies. A hyperaccumulator can be defined as a plant whose leaves contain a metallic element (metal and metalloid) at a concentration exceeding a specified threshold, when growing in nature (not in experimental cultivation). The threshold concentration should be two to three orders of magnitude higher than in leaves of most species on normal soils, and at least one order of magnitude greater than the usual range found in plants from metalliferous soils. Based on these concepts, the proposed nominal threshold criteria for different metals and metalloids in dried foliage (all in units of μg metal per g of dry leaf tissue) are: 100 for Cd, Se, and Tl; 300 for Co, Cr, and Cu; 1000 for As, Ni, and Pb; 3000 for Zn; and 10,000 for Mn with plants growing in their natural habitat.

Approximately 500 taxa are known to hyperaccumulate one or more heavy metals and metalloids; thus, hyperaccumulation is a rare phenomenon known in less than 0.2% of the world's inventory of vascular plants (van der Ent et al. 2013; Rascio and Navari-Izzo 2011; Pollard et al. 2014) and mainly includes *Asteraceae*, *Brassicaceae*, *Caryophyllaceae*, *Cyperaceae*, *Cunoniaceae*, *Fabaceae*, *Flacourtiaceae*, *Lamiaceae*, *Poaceae*, *Violaceae*, *Scrophulariaceae*, and *Euphorbiaceae* families (Bolan et al. 2014; Padmavathiamma and Li 2007; Prasad 2003). The majority of plants that hyperaccumulate heavy metals are obligate metallophytes that are endemic to metalliferous soils (Rascio and Navari-Izzo 2011; Pollard et al. 2014). Approximate numbers of taxa hyperaccumulating various elements are as follows: Ni (450), Cu (32), Co (30), Se (20), Pb (14), Zn (12), Mn (12), As (5), Cd (2), and Tl (2). Hyperaccumulators have been recorded and experimentally confirmed for elements such as Ni, Zn, Cd, Mn, As, and Se, but remained largely unsubstantiated for Pb, Cu, Co, Cr, and Tl (van der Ent et al. 2013). Of the known hyperaccumulators, a huge majority (>450 taxa) are hyperaccumulators of nickel, generally occurring on serpentine (ultramafic) soils (Pollard et al. 2014).

16.7 Phytoextraction of Heavy Metalloids

Phytoextraction refers to the removal of metal(loid)s from solid (soil and sediments) or liquid substrates (water) by plants and their accumulation into harvestable plant parts. It appears to be the most promising and commercially viable alternative to existing remediation strategies. The efficiency of phytoextraction depends on many factors like mobility and bioavailability of the heavy metals in soil particularly in rhizosphere, speciation of the heavy metals, soil properties, and the ability of concerned plant species to absorb and uptake metals.

Hyperaccumulators can be used for phytoremediation of toxic metal(loid)s as well as for phytomining of precious heavy metals (such as Au, Pd, and Pt). The phytoremediation potential of most hyperaccumulating species is, however, limited because most of them are metal selective, have not been found for all elements of interest, often are in tight association with a specific habitat, and, above all, have slow growth rates, low biomass, and shallow root systems, which limit the speed of metal removal (Salt et al. 1995; Cunningham et al. 1995; Ebbs et al. 1997). In contrast, an ideal plant for phytoextraction, besides having an innate capacity to (hyper) accumulate metal(loid)s, should also possess a rapid growth rate, produce reasonably high biomass, be ideally repulsive to herbivores, have profusely branched deep root system, be easy to cultivate, have the ability to grow on nutrient-poor soil and a wide geographic distribution, and be relatively easily harvestable (Kärenlampi et al. 2000; Garbisu et al. 2002; Punshon et al. 1996).

Phytoextraction of heavy metal(loid)s has been tested through two different approaches: (1) the use of hyperaccumulators species, such as *Thlaspi caerulescens* (Pb, Zn, Cd, Ni), *Alyssum* spp. (Ni, Co), and *Pteris vittata* (As), which produce comparatively less aboveground biomass but accumulate target heavy metals to a greater extent; (2) the application of other fast-growing plants, such as *Brassica juncea* (Indian mustard), *Salix* and *Populus* spp., which produce relatively large aboveground biomass but accumulate target metal(loid)s to a lesser extent, so that overall accumulation is comparable to that of hyperaccumulators (Dickinson et al. 2009; Rozas et al. 2006).

Although no plant is presently known that fulfills all criteria of an ideal plant for phytoextraction, a rapidly growing nonaccumulator plant could be genetically modified and/or engineered so that it attains most of the novel traits of an ideal plant for phytoextraction.

16.8 Phytoextraction Efficiency

Bioconcentration factor (BCF) and translocation factor (TF) are the two indices used commonly for quantification of phytoextraction efficiency. BCF indicates the metal accumulation efficiency of a plant species in its organs from its ambience. BCF is calculated as the ratio of metal concentration in plant tissues at harvest to initial concentration of metals in external environment (Zhuang et al. 2007; Zayed et al. 1998).

$$\text{Bioconcentration Factor}\left(BCF\right)=\mathrm{C}_{\text{harvested tissue}}/\mathrm{C}_{\text{soil}}$$

$\mathrm{C}_{\text{harvested tissue}}$ – target metal concentration in the harvested plant tissue.
C_{soil} – target metal concentration in the soil (substrate).

Translocation factor (TF) indicates roots to shoots metal translocation efficiency of a plant. It is calculated as the ratio of metal concentration in plant's aerial parts to metal concentration in plant's root (Padmavathiamma and Li 2007).

$$\text{Translocation Factor}\left(TF\right)=\mathrm{C}_{\text{shoot}}/\mathrm{C}_{\text{root}}$$

$\mathrm{C}_{\text{shoot}}$ – target metal concentration in plant shoots.
C_{root} – target metal concentration in plant roots.

Bioconcentration factor or accumulation factor (A) and translocation factor (TF) can also be reported in percentage by multiplying the above factors by 100, respectively (Wilson and Pyatt 2007; Zacchini et al. 2009). Both bioconcentration factor (BCF) and translocation factor (TF) values are strong determinants in evaluation, screening, and selection of plants for phytoextraction purposes (Wu et al. 2011). If the BCF is > 1, then the plants can be accumulators; if the BCF is = 1, then there are no influences; and if the BCF is < 1, then the plants can be excluders (Radulescu et al. 2013). Hyperaccumulators have BCF greater than one, sometimes reaching 50–100 (Cluis 2004).

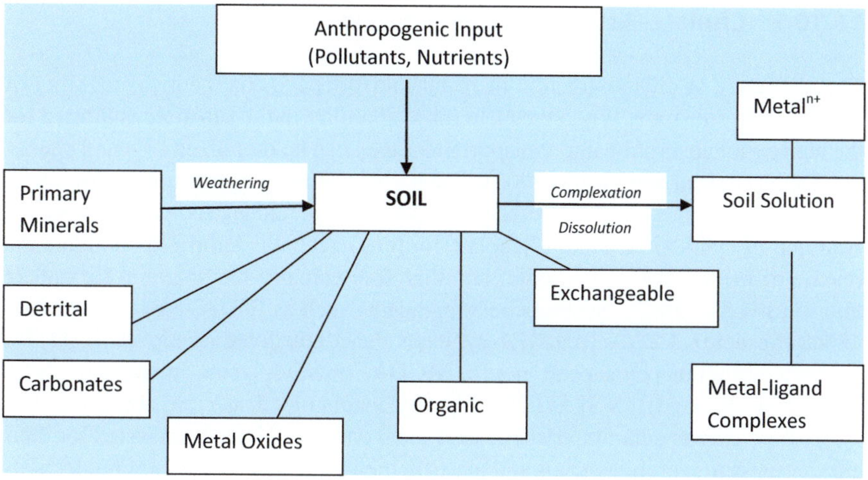

Fig. 16.2 Fate of pollutants and nutrient ions in soil ecosystem (Naidu et al. 2001)

16.9 Bioavailability of Metal(loid)s

The effectiveness of phytoextraction for remediating heavy metal contamination is highly dependent on the availability of metals for plant uptake. Metals in soils exist in different fractions: (1) free metal ions and soluble metal complexes in the soil solution; (2) metal ions occupying ion exchangeable sites and adsorbed on inorganic soil constituents; (3) organically bound metals; (4) precipitated or insoluble compounds particularly oxides, carbonates, and hydroxides; and (5) metals in the crystal lattices of primary minerals (Fig. 16.2). Of these five fractions, metals present only in fraction (1) and possibly in fraction (2) are readily available to plants. Chemical speciation plays an important role in solubility and potential availability of metals on soils. The fate of the metal(loid)s in the soil depends on soil properties as well as environmental factors. The phytoremediation efficiency for metals is often limited by the bioavailability of metals. Manipulation of soil environment to enhance the availability of metals is critical for effective phytoremediation.

16.10 Approaches for Phytoextraction

By far, phytoextraction is the preferred way of dealing with metal(loid)s pollution in the environment. A variety of plants have been tested for the purpose of phytoremediation of toxic metal(loid)s. To make this technology more efficacious, a number of adjunct strategies have been explored worldwide. Different strategies being taken to enhance the metal availability are: (1) chelate-assisted enhanced phytoextraction (also termed as induced phytoextraction), (2) long-term continuous phytoextraction, (3) microbe-assisted phytoremediation, and (4) genetically engineered plants for enhanced phytoremediation.

16.10.1 Chelate-Assisted Phytoextraction

The efficiency of phytoextraction of metal(loid)s depends on the bioavailability of metals in the substrate. When metal(loid)s availability in the substrate is limited for the active root absorption and transportation, they can be mobilized by using chelating agents (Prasad and Freitas 2003). The use of chelating agents to increase metal availability to plants is called "assisted," "induced," or "enhanced" phytoextraction. Addition of synthetic chelating (chelate forming) agents is known to stimulate the release of metals into soil solution and therefore enhance the potential for uptake into roots (Lasat 2002). Several chelating agents such as EDTA (ethylenediaminetetraacetic acid), CDTA (trans-1,2-cyclohexylenedinitrilotetraacetic acid), DTPA (diethylenetriaminepentaacetic acid), MGDA (methylglycine diacetate), NTA (nitrilotriacetic acid), [S,S]-EDDS (S,S-ethylenediaminedisuccinic acid), HEDTA (N-hydroxyethylenediaminetriacetic acid), and citric acid have been tested for their effectiveness to mobilize metals and increase their phytoextraction. Low-molecular-weight organic acids (LMWOAs) such as citric, malic, oxalic, acetic, and malonic acids, elemental sulfur, and ammonium fertilizers are other potential metal mobilizing agents.

EDTA is the most frequently cited chelating agent used in phytoextraction research and has been successfully utilized to remediate heavy metal-contaminated soils because of its efficiency, availability, and relatively low cost (Lestan et al. 2008). Chelating agents enhance the phytoextraction of Cu, Cd, Pb, Zn, and Ni (Evangelou et al. 2007). The effectiveness of chelators to mobilize metals depends on various factors: metal-to-ligand ratio, formation constants of metal-ligand complexes, metal species and distribution among soils factions, soil pH, presence of competing cations, etc.

16.10.2 Continuous Phytoextraction

Continuous phytoextraction relies on the application of specialized plants known as hyperaccumulators, which have the physiological capacity to accumulate metals during their normal growth cycle (Salt and Baker 2000). These plants have the capacity to naturally accumulate metals between 0.1% and 3% of their shoot dry biomass (Baker and Brooks 1989; Baker et al. 1999). This unique ability of hyperaccumulators to accumulate high foliar metal concentrations makes these plants suitable for continuous phytoextraction. The major limitations of most metal phytoextraction processes are (1) the bioavailability of the target metal(s) and (2) the ability of various plants to accumulate metals within their aboveground biomass (Raskin and Ensley 2000). Metal phytoextraction (as well as plant growth) can sometimes be facilitated by soil microorganisms living in intimate association with plant roots (Shilev et al. 2001).

16.10.3 Microbe-Assisted Phytoremediation

The phytoremediation of metals is technically challenging as they cannot be broken down and are persistent in the environment. The success of phytoremediation depends upon accumulation of high concentration of metals, while yielding a large plant biomass. The bioavailability of the target metals in the substrate is one of the major limitations of most metal phytoextraction processes. Chelate-assisted and microbe-assisted phytoextractions are the two approaches which have been tested to enhance metal phytoextraction. The synergistic use of microbes has several advantages over the use of chelating agents, as the microbial metabolites are biodegradable, less toxic, and may be produced in situ at rhizosphere soils (Rajkumar et al. 2012; Ma et al. 2011).

Microorganisms and plants have evolved an arsenal of resistance mechanisms, both active and incidental, to prevent heavy metal poisoning. The mechanisms evolved are active metal efflux (predominantly in eubacteria), synthesis of metal-binding peptides like metallothioneins (MTs) (blue-green algae, fungi, and plants) and phytochelatins (plants and some fungi), vacuolar sequestration (fungi and plants), redox transformations (bacteria), and several others, including extracellular precipitation or chelation of free metal ions (Silver and Phung 1996; Silver 1996; Gadd 2010).

Plant growth-promoting bacteria (PGPB) may be free-living, in symbiotic associations, or endophytic (Glick 2012). PGPB facilitate plant growth either by binding to the plant's outer surface such as the roots (rhizosphere), leaves, or stem (phyllosphere), or by inhabiting the interior surfaces of plants forming an endophytic relationship (Glick 2014). Plant-associated bacteria isolated from rhizoplane and phylloplane surfaces are known as epiphytes, whereas those inhabiting the interior of tissues within cells, in the intercellular space, or in the vascular system of plant hosts, without causing harm and disease symptoms to the host plants, are called endophytes (Azevedo et al. 2000; Kuklinsky-Sobral et al. 2004). Recently, four plant growth-promoting bacteria (PGPB), among them two heavy-metal-tolerant rhizosphere strains SrN1(*Arthrobacter* sp.) and SrN9 (*Bacillus altitudinis*) isolated from rhizosphere soil, while two endophytic strains SaN1(*Bacillus megaterium*) and SaMR12 (*Sphingomonas*) identified from the roots of the Cd/Zn hyperaccumulator *Sedum alfredii* Hance, enhanced Cd extraction by oilseed rape (*Brassica napus*) (Pan et al. 2016). Thus, the use of PGPB as adjuncts can enhance the phytoextraction efficiency.

Although microorganisms are omnipresent in the environment, the rhizosphere soil is an important habitat and ecosystem for various microorganisms, viz., protozoa, algae, fungi, and bacteria, presumably because of the high levels of nutrients such as amino acids, sugars, and organic acids exuded from plant roots. The rhizospheric microbial activities can enhance the phytoremediation efficiency either directly or indirectly.

Direct processes assist directly in heavy metal solubilization, bioavailability, and final accumulation by plants (Gadd 2004; Glick 2010). PGP rhizobacteria can directly improve the heavy metal availability to plants and hence the phytoremediation process by transforming the metal solubility, bioavailability, and transport by lowering soil pH, through the action of chelators (viz. organic acids, siderophores, biosurfactants), biomethylation, bioaccumulation, biosorption, and redox processes (Rajkumar et al. 2009, 2012; Ma et al. 2011; Ullah et al. 2015).

In contrast, through indirect processes, the microbes confer plant tolerance to metal stress and/or improve plant growth, increase plant biomass, and prevent phytopathogens from inhibiting plant growth, which in turn assists phytoremediation.

Microorganisms play an imperative role in metal(loid)s transformation, thereby influencing their solubility, mobility, bioavailability, and toxicity. Metal mobilization can arise from a variety of leaching mechanisms (chemolithotrophic and chemo-organotrophic), complexation by metabolites and siderophores, methylation (results in volatilization) and demethylation, and redox reactions. Immobilization can result from sorption to biomass or exopolymers, transport and intracellular sequestration or precipitation as organic and inorganic compounds, e.g., oxalates (fungi) and sulfides, accumulation, reduction, etc. Reduction of higher valency species may effect mobilization/immobilization (Gadd 2004, 2010). From toxicological or environmental viewpoints, these processes are important for three reasons. They may alter (i) the toxicity, (ii) the water solubility, and/or (iii) the mobility of the element (Alexander 1999). Altered metal solubility and mobility by microbes can be exploited to bioremediate different forms of elements. The ubiquity and importance of microbes in metal(loid)s transformation make microbe-assisted phytoremediation an important approach in metal(loid)s remediation.

16.10.3.1 Bacterial Metal Resistance

Metal(loid)s are present everywhere in the environment, and so are bacteria. Metals exhibit a range of toxicities toward bacteria. In response to metal toxicities, bacteria have developed several resistance mechanisms, both active and incidental, by which they can immobilize/mobilize or transform metals, thus reducing their toxicity to tolerate heavy metal ions. Mechanisms of bacterial metal resistance are efflux pumps, impermeability to metals, volatilization, extracellular polymeric substances (EPS)/bioaccumulation, metal complexation, redox transformations, and enzymatic detoxification (Gadd 2010; Silver and Phung 1996, 2009). Bacteria use both chromosomal and plasmid-encoded mechanisms for sequestration and detoxification of metals. Bacteria have plasmid-encoded genes for resistances to many toxic ions of heavy metal(loids)s, including Ag^{+}, Cd^{2+}, Co^{2+}, Cu^{2+}, Hg^{2+}, Ni^{2+}, Tl^{+}, Zn^{2+}, and Sb^{3+} (Silver and Phung 1996, 2009). Although bacterial metal(loid)s resistance systems have been found predominantly on plasmids, frequently related resistance systems are also determined by bacterial chromosomal genes, e.g., Cd^{2+} efflux mediated by P-type ATPase in *Bacillus*, Hg^{2+} resistance in *Bacillus*, and arsenic efflux in *E. coli* (Silver and Phung 1996).

16.10.3.2 Siderophores

Generally, under aerobic conditions, iron exists mainly as ferric ion (Fe^{3+}) forming insoluble hydroxides and oxyhydroxides, which are not available for biological uptake. Iron is an essential element for virtually all forms of life, with the possible exception of certain *Lactobacilli*. To solubilize Fe for its efficient uptake, the microbes have evolved the most common strategies of siderophore production (Neilands 1983).

Siderophores are low-molecular-mass (400–1000 Daltons) coordination molecules secreted by a large variety of organisms such as fungi, yeasts, bacteria, and plants in response to low iron levels and to assist Fe assimilation. This is the largest class of compounds known to bind, transport, or shuttle iron. Apart from trivalent iron (Fe^{3+}), siderophores can also form stable complexes with Al, Cd, Cu, Cr, Ga, In, Pb, Zn, as well as with actinides and radionuclides including U and Np (Neubauer et al. 2000; Gadd 2010; Glick and Bashan 1997; Schalk et al. 2011).

On the basis of the chemical structure of the metal-binding functionality, siderophores are broadly classified into three main groups, namely, the catecholates, hydroxamates, and (α-hydroxy-) carboxylates (Raymond and Dertz 2004). Rhizospheric PGPB most commonly produce siderophore. Under environmental stress such as the limited supply of nutrients, elevated heavy metals levels, etc., rhizospheric PGPB exhibit their optimum growth and siderophore production activity, which might thus be of particular use for phytoremediation purposes (Rajkumar et al. 2010).

Several siderophore-producing bacteria (SPB) associated with rhizosphere of plants, such as *B. juncea* (*Variovorax paradoxus*, *Flavobacterium* sp., *Rhodococcus* sp., *Ralstonia* sp., *Arthrobacter* sp., *Stenotrophomonas* sp., *Pseudomonas* sp.) (Belimov et al. 2005), *Thlaspi goesingense* (*Methylobacterium* sp., *Burkholderia terricola*, *Okibacterium fritillariae*, *Rhodococcus* sp., *Microbacterium* sp.) (Idris et al. 2004), and endophytic bacteria from *Alyssum bertolonii* (*Staphylococcus* sp., *Microbacterium* sp., *Pseudomonas* sp., *Curtobacterium* sp., *Bacillus* sp., *Arthrobacter* sp., *Paenibacillus* sp., *Leifsonia* sp.) (Barzanti et al. 2007), have been isolated and characterized from metal-rich soils.

The effects of SPB inoculation on metal uptake by hyperaccumulators have been the focus of current research (Dimkpa et al. 2008; Braud et al. 2009; Dimkpa et al. 2009a, b; Carrillo-Castañeda et al. 2003). Rhizospheric siderophore-producing bacteria *Pseudomonas aeruginosa* enhanced the bioavailable concentrations of Cr and Pb in the rhizosphere, making them available for maize uptake (Braud et al. 2009). However, the studies of Kuffner et al. (2008) on Willows (*Salix caprea*) and Tripathi et al. (2005) on *Phaseolus vulgaris* show that siderophore-producing bacteria have either no effect or negatively affected the metal uptake by plants. More recently, rhizoremediation of cadmium-contaminated soil associated with hydroxamate siderophores isolated from Cd-resistant plant growth-promoting *Dietzia Maris* and *Lysinibacillus* strains has been observed by Gusain et al. (2016).

Siderophores not only solubilize unavailable forms of iron but also form complexes with divalent metal ions which can be assimilated by root-mediated processes. Siderophores play a major role in metal mobilization and accumulation. So, rhizospheric siderophore-producing microbes are believed to play a vital role in heavy metal phytoextraction (Ma et al. 2011; Rajkumar et al. 2010; Dimkpa et al. 2009a, b; Braud et al. 2009). Also, metal-resistant SPB play an important role in the successful survival and growth of plants in contaminated soils by alleviating the metal toxicity and supplying the plant with nutrients, particularly iron (Rajkumar et al. 2010).

16.10.3.3 Low-Molecular-Weight Organic Acids (LMWOAs)

LMWOAs such as gluconic, oxalic, malic, succinic, citric and 2-ketogluconic acids produced by plant-associated microbes, play an important role in enhancing the mobility and solubility of heavy metals in the rhizosphere by complexation reactions (Ma et al. 2011; Rajkumar et al. 2012; Ullah et al. 2015). Studies by Li et al. (2010) with rhizosphere bacteria of *Sedum alfredii* (Cd-Zn hyperaccumulating plant) and Saravanan et al. (2007) with endophytic diazotroph *Gluconacetobacter diazotrophicus* revealed that Cd/Zn and Zn solubilities were increased, respectively, due to the release of low-molecular-mass organic acids.

16.10.3.4 The Biosurfactants

Biosurfactants – the surface-active substances – are another important group of metabolites produced by living organisms such as microbes. These are amphiphilic molecules consisting of a nonpolar (hydrophobic) tail and a polar/ionic (hydrophilic) head. These are generally classified into two types: (1) high-molecular-weight polymers that bind tightly to surfaces and (2) low-molecular-weight molecules efficiently lowering the surface and interfacial tensions (Ron and Rosenberg 2001). Microbial biosurfactants enhance metal solubility and bioavailability in soil solution by forming complexes with heavy metals at the soil interface and desorbing metals from the soil matrix (Rajkumar et al. 2012). In vitro study of Juwarkar et al. (2007) on metal mobilization potential of *P. aeruginosa* BS2 and the study of Venkatesh and Vedaraman (2012) on mobilization of Cu in contaminated soils by *P. aeruginosa* demonstrated that the biosurfactants produced by microorganisms increased the heavy metal mobilization in polluted soils. Similarly, a study by Sheng et al. (2008) showed that the inoculation of soils with biosurfactant-producing *Bacillus sp.* J119 significantly enhanced biomass of tomato plants and Cd uptake in plant tissue. Thus, microbial biosurfactants have the potential to improve metal mobilization and enhance metal uptake.

16.10.3.5 Redox Reactions

Redox transformation of metal(loid)s is one of the survival/resistance mechanisms in microorganisms. Microorganisms can mobilize and immobilize metal(loid)s and organometallic compounds by reduction and oxidation processes (Gadd 2010; Lovley 2000). Metal(loid)s such as Hg, Se, As, and Cr most commonly undergo microbial oxidation/reduction reactions. Redox reactions change the trace elemental speciation which leads to decreased or increased mobility. Generally, most of the metals are less soluble in their higher oxidation state. For example, reduction of higher valency Fe(III) and Mn(IV) to Fe(II) and Mn(II), respectively, increases metal solubilities (Gadd 2010; Lovley 2000).

Microorganisms can remove a number of metals and metalloids from the environment or waste streams by reducing them to a lower redox state (Lovley 1995). Microbial reduction of metals to lower redox states may decrease the mobility and toxicity for several elements (Wall and Krumholz 2006; Lovley 2001), e.g., upon reduction of U(VI) to U(IV) and Cr(VI) to Cr(III), their solubility decreases, leading to immobilization (Smith and Gadd 2000). Aerobic and anaerobic microbial reduction of Cr^{6+} to Cr^{3+} has been one of the most widely studied forms of metal bioremediation (Lovley 1995).

Mercury undergoes a biological reduction process in soils, sediments, and water. Bacteria and fungi can enzymatically reduce reactive mercuric ion, Hg^{2+}, to nonreactive metallic mercury Hg(0), which serves as a resistance and detoxification mechanism as Hg^0 is subjected to volatilization losses (Barkay and Wagner-Dobler 2005; Gadd 2010). Bacteria are more important than eukaryotic phytoplankton in the reduction of Hg(II). Mercury resistance mechanisms in bacteria are detailed in Sect. 16.10.4.3.1. Also, dissimilatory metal reducing bacterium (DMRB) *Shewanella oneidensis* MR-1 in the presence of electron donors constitutively reduced ionic mercury (Hg^{2+}) to elemental mercury (Hg[0]) (Wiatrowski et al. 2006). Microorganisms that reduce Hg^{2+} to volatile Hg^0 as a mechanism for mercury resistance naturally contribute to the volatilization of mercury from contaminated environments, and there is the possibility that the stimulation of this metabolism might enhance mercury remediation (Lovley 1995).

Plant-associated microbes can also immobilize the heavy metals in the rhizosphere through metal reduction reactions. Chatterjee et al. (2009) reported that the inoculation of Cr-resistant bacteria *Cellulosimicrobium cellulans* to the seeds of green chili grown in Cr(VI)-contaminated soils decreased Cr uptake into the shoot and root as compared with uninoculated controls, indicating that bacteria reduced the mobile and toxic Cr(VI) to nontoxic and immobile Cr(III) in the soil.

Microbial reduction of the highly soluble oxidized form of selenium, Se^{6+}, to insoluble elemental selenium, Se(0), by microorganisms is a natural mechanism for the removal of selenium from contaminated surface and groundwater (Oremland 1994). Di Gregorio et al. (2005) demonstrated that the bacterium *Stenotrophomonas maltophilia* isolated from the rhizosphere of *Astragalus bisulcatus* significantly reduced soluble and harmful Se(IV) to insoluble and unavailable Se(0) and thereby reduced the plant Se uptake. Microbes can mobilize as well as immobilize heavy metals. Above studies exemplify mechanisms by which microbes immobilize metals within the rhizosphere soil, reflecting that microbes can be explored for phytostabilization applications.

16.10.3.6 Biomethylation

Biomethylation refers to the enzymatic transfer of a previously formed methyl group from some donor atom to some acceptor atom within a living organism. Metals such as Hg, Sn, and Pb are methylated primarily by bacteria and metalloids As, Se, and Te principally by fungi, yeasts, and algae (Thayer 2002). Methylcobalamin (CH_3COB_{12}) and S-adenosylmethionine (SAM) are the major methylating agents involved in the methylation of Hg and As/Se, respectively (Bolan et al. 2013).

Methylation markedly alters the solubility, volatility, and toxicity of metal compounds formed (Gadd 2004, 2010; Bolan et al. 2014). Methylation is considered to be the major process by which volatilization of As, Hg, and Se occurs in soils and sediments (Adriano et al. 2004). Biomethylation transforms the trace metals to their methyl derivatives, most of which are volatile and removed consequently through volatilization. It has been proposed as a biological detoxification mechanism for the removal of trace elements (Frankenberger and Losi 1995).

Selenium has reduced toxicity when methylated (Ranjard et al. 2003). Se biomethylation is a potential mechanism for removing Se from contaminated environments, as methylated Se compounds, such as dimethylselenide (DMSe) and dimethyl selenenyl sulfide (CH_3SeSCH_3), are believed to be less toxic than dissolved Se oxyanions. Bacteria and fungi predominate among the Se-methylating microbes in soils and sediments, while bacteria are active Se-methylaters in aquatic environments (Dungan and Frankenberger 1999; Adriano et al. 2004). Volatile Se-methylated species, viz., dimethyl selenide [$(CH_3)_2Se$; DMSe] and dimethyl diselenide [$(CH_3)_2Se_2$; DMDSe], are often lost from the soil (Dungan and Frankenberger 1999; Karlson and Frankenberger 1993). Se volatilization can be enhanced by optimizing many environmental and soil factors, e.g., organic amendments and frequent tillage, which lead to an increase in diffusive transport through soil (Zhang and Frankenberger 1999; Stork et al. 1999).

Several aerobic and anaerobic microorganisms (e.g., bacteria and microscopic fungi) can methylate arsenic compounds such as arsenate (As(V), AsO_4^{3-}), arsenite (As(III), AsO_2^-), and methylarsonic acid ($CH_3H_2AsO_3$) to volatile arsines, mono-, dimethyl-[$(CH_3)_2HAs$], or trimethylarsine [$(CH_3)_3As$], which can be easily released into the atmosphere (Rahman et al. 2014). Arsenic (As) and antimony (Sb), both metalloids, are subject to biomethylation under oxic as well as anoxic conditions (Bentley and Chasteen 2002; Thayer 2002). This methylation process has been found to be conducted by a broad spectrum of both microorganisms and higher organisms (Thayer 2002), and can fundamentally change mobility, availability, and toxicity of metalloids.

Methylation of Hg occurs under both aerobic and anaerobic conditions (Boening, 2000). Under anaerobic conditions, Hg^{2+} ions can be methylated biologically to form either monomethyl or dimethyl Hg. Methylated mercury is more toxic (neurotoxic) and more biologically mobile than the other forms of mercury (Barkay and Wagner-Dobler 2005; Adriano et al. 2004). Both microbial and abiotic transformations control methylmercury production in environment. Mercuric ion (Hg^{2+}) may be methylated by bacteria or fungi to the more toxic methylmercury [$(CH_3)Hg^+$] (Barkay and Wagner-Dobler 2005). Methylmercury can be further methylated to form volatile dimethylmercury by several bacteria. Some bacteria can enzymatically reduce [$(CH_3)Hg^+$] and phenylmercury to volatile elemental mercury (Barkay and Wagner-Dobler 2005; Gadd 2010).

16.10.3.7 Biosorption and Cytosolic Sequestration (Bioaccumulation)

Both biosorption and bioaccumulation have been discussed in Sect. 16.4.1. All microbial materials can be effective biosorbents for metals except for mobile alkali metal cations like Na^+ and K^+, and biosorption can be an important passive process in living and dead organisms (Gadd 2009; Wang and Chen 2009).

Several bacteria have developed a cytosolic sequestration as a resistance mechanism for metals. Once inside the cell, metal ions might also become compartmentalized or converted into more innocuous forms, e.g., metal detoxification through metal-binding proteins, termed metallothioneins produced by microorganisms, plants, and animals. This process constitutes an effective detoxification mechanism and the respective microbes might be able to accumulate metals in high intracellular concentrations (Haferburg and Kothe 2007).

16.10.4 Genetically Engineered Plants for Enhanced Phytoremediation

Ideally a plant suitable for phytoremediation should possess an inherent capacity to hyperaccumulate and tolerate metal(loid)s in aboveground tissues, be ideally repulsive to herbivores, adapt to the local environmental conditions, be easy to cultivate and have a wide geographic distribution, possess a profusely distributed root system, be able to grow fast and produce large biomass, possess resistance to diseases and pests (Kärenlampi et al. 2000; Garbisu et al. 2002; Punshon et al. 1996; Pilon-Smits and Pilon 2002; Mahar et al. 2016). At present, no plant fulfills all these criteria. Transgenic approach is an important potential strategy to merge the high metal(loid) s accumulation capacity with such novel traits. Some of the high biomass metallophytes with well-established genetic manipulation procedures, which appear to be good candidates as engineered phytoremediation plants, include *B. juncea*, sunflower *Helianthus annuus*, yellow poplar *Liriodendron tulipifera*, and shrub tobacco *Nicotiana glauca* (Eapen and D'Souza 2005).

At the cellular level, the plant metal homeostasis network involves the modification of the solubility of extracellular metal ions by electron transfer, chelation or acidification of the apoplast, the uptake of metal ions, chelation and/or the trafficking within the cell, delivery into cellular compartments and organelles, and storage or efflux of metal ions under conditions of excess (Clemens et al. 2002). Thus, the main biotechnological strategies envisioned to develop genetically engineered plants suitable for phytoremediation are: (1) manipulating metal(loid)s uptake and transport across biomembranes mediated by metal transporter genes; (2) enhancing efficiency of intracellular chelation (with glutathione, phytochelatins, metallothioneins, etc.) and sequestration into subcellular compartments/vacuoles, both to detoxify accumulated metal(loid)s; and (3) biotransformation of metal(loid)s to less toxic and volatile forms, i.e., phytovolatilization.

16.10.4.1 Metal Transporters

Transition metal transporter genes are of pivotal importance in metal(loid)s uptake, partitioning to plant organs and cells, metal delivery to metal requiring proteins, as well as metal sequestration into specialized compartments (viz., vacuoles, chloroplast, mitochondria) (Clemens et al. 2002; Krämer et al. 2007). Metal transporters are grouped into two categories: metal uptake and metal efflux transporters. Metal transporter genes belonging to heavy metal ATPases transporter, natural resistant-associated macrophage proteins (NRAMPs), yellow stripe-like protein transporters

(YSL), zinc-regulated transporter, iron-regulated transporter-related protein (ZIP), plant cation diffusion facilitators (CDF) also known as metal tolerance proteins (MTPs), oligopeptide transporter (OPT), ATP-binding cassette transporter superfamily (ABC), multidrug resistance-associated proteins (MRPs) efflux transporter family belonging to ATP-binding cassette transporter superfamily (ABC), and nicotianamine synthase (NAS) and S-adenosyl-methionine synthetase (SAMs) families have been identified and manipulated in plant genome (Kramer 2010; Tomar et al. 2016). Recent studies have shown that members of the aquaporins transporter family transport arsenite in plants (Zhao et al. 2009; Bienert et al. 2008; Mosa et al. 2012). As(III) transporter plant aquaporins and bacterial, yeast, and mammalian cells aquaglyceroporins belong to the major intrinsic protein (MIP) superfamily.

Genetic manipulation of metal transporters for phytoremediation purposes need to be focused at three levels: (1) metal uptake by roots from soil, (2) metal translocation from root to shoot, and (3) metal sequestration into cell wall or vacuoles. The overexpression of metal transporter genes may lead to enhanced metal uptake, translocation, and/or sequestration, depending on the tissues where the gene is expressed (root, shoot, vascular tissue), and on the intracellular targeting (cell membrane, vacuolar membrane). Genetic manipulation of metal transporters has led to enhanced metal tolerance and accumulation in many plant species (Table 16.1). For example, *Arabidopsis thaliana* overexpressing yeast *YCF1* (Yeast Cadmium Factor 1) resulted in enhanced tolerance and accumulation of Pb(II) and Cd(II) (Song et al. 2003). Arsenic is naturally present in soil in the form of arsenate (AsO_4^{-3}). Due to chemical similarity between arsenate and inorganic phosphate, Pi(PO_4^{-3}), As(V) is usually taken up into the plants via phosphate transporters. Recently, transgenic *Arabidopsis* overexpressing high-affinity phosphate transporters PHT1 or PHT7 from *Arabidopsis*, in combination with or without YCF1, leads to improved As accumulation and tolerance in *Arabidopsis* (LeBlanc et al. 2013).

16.10.4.2 Metal-Binding Proteins

Chelation and sequestration of metals by particular ligands are important mechanisms used by most plants and microorganisms to deal with heavy metal stress. Metallothioneins (MTs) and phytochelatins (PCs) are the two best-characterized thiol-rich peptides which bind to a variety of metals to form stable complexes in cytosol and reduce their toxicity by sequestration into vacuole in plant cells (Cobbett and Goldsbrough 2002; Goldsbrough 2000). These ligands are widely distributed in plants, and many physiological and genetic studies indicate that PCs and MTs are critical for metal tolerance and accumulation in plants (van Hoof et al. 2001; Hartley-Whitaker et al. 2001; Cobbett and Goldsbrough 2002). These peptides have been extensively used to enhance the metal accumulation ability of microorganism by producing them in different subcellular locations. In one such effort, the *A. thaliana* phytochelatin synthase (AtPCS) responsible for enzymatic PC synthesis was overexpressed in *E. coli*, leading to 20-fold higher heavy metal accumulation (Sauge-Merle et al. 2003).

Table 16.1 Transgenic plants with manipulated metal transporters genes

Gene	Source	Target plant	Product	Effects	References
YCF1	Yeast	*Arabidopsis*, *Poplar*	Yeast vacuolar bis(glutathionato) Cd conjugate transporter of ATP-binding cassette superfamily	Increased biomass and tolerance to Cd and Pb	Song et al. (2003)
AtCAX2	*A. thaliana*	Tobacco	Calcium vacuolar transporter	Enhanced metal accumulation in shoots and higher root tonoplast transport in transgenics	Hirschi et al. (2000)
ZAT	*A. thaliana*	*A. thaliana*	Putative vacuolar Zn^{2+} transporter	Increased the Zn^{2+} tolerance and twofold higher root Zn accumulation	van der Zaal et al. (1999)
PgIREG1	*Psychotria gabriellae*	Yeast and *Arabidopsis*	Vacuolar transporter	Confers Ni tolerance in yeast and transgenic plants	Merlot et al. (2014)
HMT1	*S. pombe*	*A. thaliana*	Transporter	Enhanced tolerance to and accumulation of As, Cd, Cu, and Zn	Huang et al. (2012)
AtHMA3	*A. thaliana*	*Arabidopsis thaliana*	Vacuolar transporter	Increased tolerance to and accumulation of Cd, Co, Pb, and Zn	Morel et al. (2009)
ABCC1	*A. thaliana*	*A. thaliana*	Vacuolar transporter	Increased resistance to Cd	Park et al. (2012)
AtABCC1 & *AtPCS1*	*A. thaliana*	*A. thaliana*	Vacuolar PC transporters	Greater As tolerance	Song et al. (2010)
*mer*C	*A. ferrooxidans*	*A. thaliana*	Bacterial Hg^{2+} importer	Enhanced mercury accumulation	Sasaki et al. (2006)
*Mer*E	*E. coli*	*A. thaliana*	Mercury transporter	Significantly more accumulation of methyl-Hg and mercuric ions	Sone et al. (2013)
AtNramp1		*Arabidopsis*	Fe transporter	Only the transgenic plants survived for longer periods at 600 μM Fe concentration	Curie et al. (2000)

(continued)

Table 16.1 (continued)

Gene	Source	Target plant	Product	Effects	References
NtCBP4	*N. tabacum*	*N. tabacum*	Tobacco calmodulin-binding channel protein	Conferred Ni^{2+} tolerance and Pb^{2+} hypersensitivity in transgenic plants	Arazi et al. (1999)
Truncated tobacco NtCBP4	*N. tabacum*	*N. tabacum*	Tobacco calmodulin-binding channel protein	Improved tolerance to and less accumulation of Pb^{2+}	Sunkar et al. (2000)
Disrupted CNGC1 gene	*Arabidopsis*	*Arabidopsis*	Cyclic nucleotide-gated ion channel 1	Conferred Pb^{2+} tolerance	Sunkar et al. (2000)
PHT1 or PHT7 with/without YCF1	*Arabidopsis*	*Arabidopsis*	High-affinity phosphate transporters	Increased As accumulation and tolerance	LeBlanc et al. (2013)
AtABCC3	*Yeast*	*A. thaliana*	ABC-type transporter	Enhanced Cd tolerance	Brunetti et al. (2015)

16.10.4.2.1 Metallothioneins (MTs)

Metallothioneins (MTs) are low-molecular-weight (5–10 kDa), gene-encoded, sulfhydryl-rich proteins which bind with metal ions in metal-thiolate clusters. MTs are encoded by a family of genes present ubiquitously in cyanobacteria, eukaryotic microorganisms, some prokaryotes, fungi, animals, and higher plants (Cobbett and Goldsbrough 2002). On the basis of their cysteine content and structure, MTs are grouped into three classes: Cys-Cys, Cys-X-Cys, and Cys-X-X-Cys motifs (where X is any amino acid). In plants, MTs are thought to be involved in metal tolerance or homeostasis, detoxification, and distribution as they bind with metal ions by the formation of mercaptide bonds (Jagtap and Bapat 2015). Metallothioneins (MTs) are capable of forming high-affinity coordination compounds of heavy metal ions via cysteine residues shared along the peptide sequence in Cys-X-Cys or Cys-Cys motifs. MTs were originally isolated as Cu, Zn, and Cd binding proteins in animals (Yang et al. 2009). A wide range of MT genes from various organisms, including mouse, hamster, yeast, humans, plants, and cyanobacteria, have been overexpressed in bacteria and plants (Table 16.2).

Table 16.2 Metallothionein genes in transgenic bacteria and plants

MT gene	Source	Genetically modified plant species	Effect	Reference(s)
mt-1	Mouse (*Mus musculus*)	*Lycopersicon esculentum*	1.6 times higher average zinc content in transgenic plant leaves than in WT	Sheng et al. (2007)
mt-IA	Mouse (*Mus musculus*)	*Nicotiana tabacum*	Tolerated 200 mM $CdCl_2$ at the seedling level	Pan et al. (1994)
mt-β-glucuronidase fusion	Chinese hamster (*Cricetulus griseus*)	*N. tabacum*	In transgenic tobacco, the levels of 109Cd accumulation in leaves were reduced to about 70% of those in untransformed control plants	Hattori et al. (1994)
cup1	Yeast (*S. cerevisiae*)	*N. tabacum*	Two to three times higher Cu content than the control but no Cd tolerance	Thomas et al. (2003)
ScMTII	Yeast (*S. cerevisiae*)	*N. tabacum*	The transgenic tobacco plant accumulated 3.5–4.5-fold more Cd above the threshold level of 100 mg Cd/kg	Daghan et al. (2013)
MTL4	Humans (*Homo sapiens*)	PGP *Mesorhizobium huakuii* subsp. *rengei* associated with plant *A. sinicus*	Twofold increase in Cd^{2+} uptake	Sriprang et al. (2002)
smtA	Cyanobacteria (*Synechococcus* sp.)	*A. thaliana*	Transgenic plants overexpressing SmtA showed higher survival rate than the WT and enhanced tolerance to high zinc stress	Xu et al. (2010)

(continued)

Table 16.2 (continued)

MT gene	Source	Genetically modified plant species	Effect	Reference(s)
psmtA	Pea (*Pisum sativum*)	*Populus alba*	Transgenic poplar lines showed enhanced ability to translocate and accumulate Zn and Cu compared with NT and AL22 to heavy metal stress	Turchi et al. (2012)
SaMT2	*Sedum alfredii*	*Yeast* (*S. cerevisiae*) *Tobacco*	Improved Cd tolerance and accumulation in transgenic yeast and tobacco	Zhang et al. (2014)

16.10.4.2.2 Phytochelatins (PCs)

Phytochelatins (PCs) are low-molecular-weight, glutathione (GSH)-derived, cysteine-rich metal-binding peptides, synthesized enzymatically in some fungi, photosynthetic organisms, and plants (angiosperms, gymnosperms, and bryophytes) in response to heavy metal stress (Cobbett 2000; Gadd 2010; Clemens 2006a, b). These peptides play a key role in metal detoxification in plants by efficient sequestration of multiple metal(loid)s ions in metal(loid)-thiolate complexes which are sequestered into vacuoles (Clemens 2006a, b; Cobbett and Goldsbrough 2002). Overproduction of PCs appears to be an inducible rather than a constitutive mechanism, observed especially in metal nontolerant plants as part of their defense mechanism against metals (Freeman et al. 2005). PCs – a family of thiol-rich peptides – are synthesized nontranslationally (nonribosomally) from GSH by the enzyme phytochelatin synthase to form molecules of $(\gamma\text{-EC})_nG$ (n ~ 2–11) (Kramer 2010). Different metals, including Cd, Hg, Ag, Cu, Ni, Au, Pb, and Zn, are found to induce PC synthesis in a wide variety of plant species; however, Cd is by far the most potent activator (Mejare and Bülow 2001). Transgenic plants overexpressing phytochelatin synthase resulted in enhanced phytochelatin levels and heavy metal tolerance (Table 16.3). Phytochelatins are also reported to be involved in arsenic detoxification (Schmoger et al. 2000). Heavy metals (viz., Cu, Cd, Pb, and Hg) stress can induce phytochelatin (PC) synthase in some engineered bacteria (Gadd 2010). In one such case, on exposure to Cd or As, the cellular metal accumulation increased by 20- and 50-fold, respectively, in *E. coli* expressing PC synthase gene (AtPCS) from *A. thaliana* (Sauge-Merle et al. 2003).

Table 16.3 Transgenic plants overproducing phytochelatin synthase (PCS)

Phytochelatin synthase gene	Gene origin	Genetically modified plant species	Effect	Reference(s)
TaPCS1	*Triticum aestivum*	*Nicotiana glauca*	1.6 times longer roots on media with 800 μM Pb^{2+} or 50 μM Cd^{2+}. Shoots of transformed line NgTP1 accumulated more Cd, Cu, Ni, Pb, and Zn from polluted soil	Martinez et al. (2006)
CdPCS1	*Ceratophyllum demersum*	*Arabidopsis*	Significantly enhanced metal accumulation in aerial parts without significant difference in growth parameters in comparison to WT	Shukla et al. (2013)
AtPCS1	*A. thaliana*	*N. tabacum*	Longer roots and higher Cd^{2+} accumulation in roots and shoots	Pomponi et al. (2006)
AtPCS1	*A. thaliana*	*B. juncea*	1.9 and 1.4 times longer roots on media with 100 μM Cd^{2+} and 500 μM AsO_4^{3-}, respectively	Gasic and Korban (2007)

16.10.4.2.3 Glutathione (GSH)

Glutathione (GSH) is not only a precursor of phytochelatin and iso-PC synthesis, but also directly eliminates reactive oxygen radicals (RORs) induced by heavy metal stress in cells and acts as a fundamental antioxidant molecule (Schutzendubel and Polle 2002). Chelation with GSH or its oligomers [(γ Glu-Cys)$_n$-Gly] phytochelatins (PCs) is a well-recognized detoxification mechanism of thiol-reactive heavy metal(loid)s and other xenobiotics (Cobbett et al. 1998). The toxicity of metals to plants gets reduced once the metals are bound to thiol peptides, and these metal-thiol complexes are then sequestered into vacuoles by glutathione-conjugating pumps, GCPs, protecting cellular components from the reactive metal(loid)s (Dhankher et al. 2002; Song et al. 2010). GSH is synthesized by two ATP-dependent reactions catalyzed by γ-Glu-Cys synthase (GSH1, i.e., γ-ECS) and GSH synthetase (GSH2). An unusual γ-glutamyl peptide bond is formed between the side-chain carboxyl group of Glu and Cys and then Gly is added to form GSH (Meister 1988). GSH homeostasis in plants is maintained by the γ-glutamyl cycle, which involves GSH synthesis and degradation and the recycling of component amino acids (Paulose et al. 2013).

Indian mustard (*B. juncea*) plants overexpressing bacterial GSH1, i.e., (γ-ECS) showed increased tolerance and accumulation of Cd in the shoot (Zhu et al. 1999a). Also, Indian mustard (*B. juncea*) overexpressing *E. coli gsh*II gene encoding GSH synthetase (GS) in the cytosol exhibited improved tolerance to, and accumulated significantly more, Cd than wild type (WT) (Zhu et al. 1999b). Further, Bennett et al. (2003) demonstrated that overexpression of *E. coli gsh*I and *gsh*II can indeed multiply the natural phytoextraction potential of *B. juncea*. In a similar approach, Bittsanszky et al. (2005) genetically transformed *Penicillium canescens* poplar lines overexpressing the *E. coli gsh*I gene encoding γ-ECS in the cytosol (11ggs) or in the chloroplasts (6LgI). Enhanced Cd, Cr, and particularly Cu uptake in 11ggs poplar clones found in the study were related to the higher GSH and γ-EC concentrations in this clone. The low inducibility of glutathione S-transferase (GST) isoenzymes in the transgenic clones indicated the marked stress tolerance of these clones to Zn exposure.

Similarly, arsenic (As) tolerance in plants can also be improved by modifying GSH and PCs. Double transgenic *Arabidopsis*, coexpressing two bacterial genes: *E. coli* arsenate reductase (*ars*C) and γ-glutamylcysteine synthetase (γ-ECS), grown in the presence of 125 μM sodium arsenate showed threefold more As accumulation in the aboveground biomass and almost 17-fold higher biomass than wild type plants (Dhankher et al. 2002). The transgenic *A. thaliana*, engineered to express the bacterial γ-*ECS* gene under the control of a strong constitutive actin regulatory sequence (*A2*), the levels of γ-EC and its derivatives, glutathione (GSH) and PCs, were increased to 3- to 20-fold higher concentrations than the increases that occurred in wild type (WT). The *A2::ECS* transgenic plants were highly resistant to arsenic and weakly resistant to mercury (Li et al. 2005).

Furthermore, genetically engineered Eastern cottonwood plants expressing bacterial ECS had elevated thiol-group levels, enhanced growth on levels of arsenate toxic to control plants in vitro. Moreover, roots of transgenic plants accumulated approximately twofold more As than control roots, while shoots accumulated approximately two-thirds less than control shoots (LeBlanc et al. 2011).

These studies demonstrated that overexpression of the bacterial gamma-glutamylcysteine synthetase (γ-ECS) gene resulted in a higher tolerance and accumulation of metal(loid)s such as Cd, Hg, and As in plants.

16.10.4.3 Genetic Modifications for Phytovolatilization

Phytovolatilization of metal(loid)s refers to the accumulation of metal(loid)s species in plant cells and their conversion to less toxic and volatile forms, which are subsequently liberated into the atmosphere (Kotrba et al. 2009). For metal(loid)s such as mercury (Hg), selenium (Se), and arsenic (As), phytovolatilization seems to be an alternative promising and viable strategy as the pollutants are removed from a site without the need of harvesting plant biomass and further disposal.

16.10.4.3.1 Phytovolatilization of Mercury

All chemical forms of mercury, i.e., Hg^0, Hg^{2+}, Hg^+, and R-Hg^+, are toxic, but the extent of toxicity varies from one form to another; organic being the most and

Table 16.4 Some important genes of *mer operon*

Function	Gene	Reference(s)
Transport	*mer* T	Nascimento and Chartone-Souza (2003)
	mer P	Nascimento and Chartone-Souza (2003)
	mer C	Sasaki et al. (2005)
	mer E	Kiyono et al. (2009)
	mer F	Wilson et al. (2000)
	mer H	Schue et al. (2009)
Regulation	*mer* R	Lund and Brown (1989)
	mer D	Kiyono and Pan-Hou (1999)
Reduction	*mer* A	Ruiz and Daniell (2009)
	mer B	Ruiz and Daniell (2009)
Conferring resistance to phenyl mercury	*mer* G	Kiyono and Pan-Hou (1999)

elemental as the least toxic (Jan et al. 2009). Organic mercury is considered more toxic because of its high stability combined with lipid solubility, enabling the diffusion across the placental and blood–brain barrier (Nascimento and Chartone-Souza 2003; Beijer and Jernelov 1979).

Due to persistence and toxicity of various forms of mercury in natural environments, large numbers of microorganisms, in particular bacteria, have developed a surprising array of resistance mechanisms based on clustering genes in a single operon referred as "*mer operon*." The *mer operon* usually consists of structural genes encoding functional proteins associated with various functions such as transport, regulation, reduction, and conferring resistance to phenyl mercury (Table 16.4).

On the basis of organization of *mer* operon genes, two modes of mercury resistance are encountered in bacteria: narrow spectrum (systems with *merA* but not *merB*, occur mostly in gram-negative bacteria conferring resistance to inorganic mercury and limited range of resistance to organomercurials) and broad spectrum (systems with both *merA* and *merB*, occur mostly in gram-positive bacteria conferring resistance to both organic and inorganic forms of mercury) (Silver and Phung 1996). Bacteria with broad spectrum bacterial resistance have developed a two-stage, two-enzyme detoxification strategy for recycling of organomercurials to less toxic, volatile elemental form (Hg^0) (Ogunseitan 2002). In broad-spectrum mercury resistance, *merB* gene-encoded organomercurial lyase enzyme catalyzes the removal of Hg^{2+} from organic mercury compounds (R-Hg) by protonolytic cleavage of carbon–mercury bond of organic mercury compounds (R-Hg) (Silver and Phung 1996; Summers 1986). The second enzyme, i.e., mercuric reductase encoded by *mer*A, reduces inorganic (Hg^{2+}) to volatile elemental mercury (Hg^0) using NADPH as an electron donor (Summers 1986).

$$\text{R-Hg}^{+} \underset{}{\overset{\text{Organomercurial lyase } (\textit{merB} \text{ encoded})}{\rightleftarrows}} \text{Hg}^{2+} \xrightarrow[\text{NADPH}]{\text{Mercuric reductase } (\textit{merA} \text{ encoded})} \text{Hg}^{0}$$

The potential of these biotransformation reactions has been recognized as a promising strategy for phytovolatilization of mercury. The focus of enhanced phytoremediation of mercury contamination is to transform toxic organic mercury to less toxic, volatile mercury by introducing an entirely new pathway from another organism such as bacteria. *merA*, *merB*, or a combination of both genes from bacteria have been introduced into plants that together convert methylmercury to volatile elemental mercury.

Various plant species such as *A. thaliana* (Rugh et al. 1996), rice *Oryza sativa* (Heaton et al. 2003), cottonwood *Populus deltoides* (Che et al. 2003), yellow poplar *L. tulipifera* (Rugh et al. 1998), and tobacco *N. tabacum* (Heaton et al. 2005; He et al. 2001) constitutively expressing modified *merA* were resistant to at least ten times greater concentrations of Hg(II) than those that killed nontransgenic controls and showed significant levels of Hg(0) volatilization. Transgenic *A. thaliana* overexpressing a modified bacterial *merB* gene grew vigorously on a wide range of concentrations of monomethylmercuric chloride and phenylmercuric acetate as compared to nontransformed plants, which were severely inhibited or died at the same organomercurial concentrations (Bizily et al. 1999). In an attempt to more efficiently detoxify methylmercury, *Arabidopsis* (Bizily et al. 2000), Eastern cottonwood (Lyyra et al. 2007), and common wetland plant saltmarsh cordgrass *Spartina alterniflora* (Czakó et al. 2006) were engineered to coexpress *merA* and *merB*. Transgenic plants detoxified organic mercury by converting it to volatile and significantly less toxic elemental mercury (Bizily et al. 2000). Transformation of *Spartina alterniflora* resulted in increased tolerance to phenylmercuric acetate and mercuric chloride (Czakó et al. 2006).

Ruiz et al. (2003) used chloroplast engineering by integrating a native operon containing the *merA* and *merB* genes into tobacco (*Nicotiana tabacum*) chloroplast genome. Transgenic tobacco showed high levels of tolerance to the organomercurial compound phenylmercuric acetate (PMA). Chloroplast engineering to enhance Hg phytoremediation is particularly beneficial as it prevents the escape of transgenes via pollens to other related plants and there is no need for codon optimization to improve transgene expression.

16.10.4.3.2 Phytovolatilization of Selenium

True Se hyperaccumulation occurs in four to five genera in the *Brassicaceae*, *Fabaceae*, and *Asteraceae* families (Pilon-Smits 2015). Se hyperaccumulators accumulate organic forms like methyl-SeCys, γ-glutamyl-Se-methylselenocysteine, and selenocystathionine, while most plants accumulate inorganic Se (Anderson 1993; Terry et al. 2000). Like other plants, hyperaccumulators can volatilize Se, but mostly in the form of dimethyldiselenide (DMDSe), which originates from methyl-SeCys (Terry et al. 2000). In hyperaccumulating plant species, the biosynthesis of

MetSeCys is catalyzed by selenocysteine methyltransferase (SMT). SMT methylates SeCys specifically into MetSeCys (Neuhierl and Boeck 1996). MetSeCys is further methylated to volatile dimethyldiselenide (DMDSe; Terry et al. 2000).

Selenium is toxic at high concentrations because it replaces sulfur in proteins. Various transgenic approaches have been used to enhance plant selenium accumulation, tolerance, and volatilization, primarily in *B. juncea*. Van Huysen et al. (2003) showed that overexpression of an *A. thaliana* cystathionine-γ-synthase (CγS) enzyme in *B. juncea* resulted in threefold enhanced Se volatilization from either selenate or selenite, showing that this is the first enzyme in the conversion of selenocysteine (SeCys) to selenomethionine (SeMet) and is a limiting factor for Se volatilization.

In another transgenic approach, fast-growing plants *B. juncea* and *A. thaliana* were genetically engineered to overexpress a selenocysteine methyltransferase (SMT) gene from a Se hyperaccumulator (*Astragalus bisulcatus*) in order to increase phytoremediation potential (LeDuc et al. 2004; Ellis et al. 2004). In addition to an improved Se accumulation in SMT *B. juncea*, this plant showed a higher dimethyl selenide (DMSe) production than control wild type (WT) plants and an acquired ability to produce dimethyldiselenide (DMDSe) (LeDuc et al. 2004). The SMT transgenics showed enhanced Se accumulation in the form of nontoxic methyl-SeCys, as well as enhanced Se tolerance, and greater production of volatile Se, most likely as DMDSe (Kubachka et al. 2007; LeDuc et al. 2004; Montes-Bayon et al. 2002; Meija et al. 2002). Volatile forms were efficiently evaporated from the leaves of transgenes grown in culture media and polluted soil (Banuelos et al. 2007; LeDuc et al. 2004). SMT plants had increased Se tolerance, accumulation, and volatilization ability, all the traits necessary for Se phytoremediation.

Other strategies used are to overexpress ATP sulfurylase (Pilon-Smits et al. 1999), selenocysteine lyase (SL) (Garifullina et al. 2003; Banuelos et al. 2007), and double transgenic ATP sulfurylase(APS)xSMT plants (LeDuc et al. 2006) in Indian mustard (*B. juncea*). Indian mustard plants overexpressing APS were found to have increased tolerance and accumulation of selenium, but did not volatilize more Se than wild type (WT) (Van Huysen et al. 2004; LeDuc et al. 2006; Banuelos et al. 2005, 2007). Selenium accumulation was significantly enhanced in the shoots of transgenic Indian mustard (*B. juncea*) overexpressing genes encoding the enzymes selenocysteine lyase (*cpSL*) and selenocysteine methyltransferase (*SMT*) in comparison to wild type Indian mustard (Banuelos et al. 2007).

The above-mentioned studies on genetically engineered plants bear the testimony to the fact that with the use of novel techniques of genetic engineering, it is possible to manipulate metal(loid)s tolerance, accumulation, and/or metabolization in plants, and thus to create the ideal plant for environmental remediation.

16.11 Challenges for Phytoremediation

Although phytoremediation seems to be an attractive option for heavy metal(loid)s removal, the applicability of the technique is debatable since practical implications are not so evident (Whiting et al. 2004; do Nascimento and Xing 2006; Dickinson et al. 2009; Karami and Shamsuddin 2010; Bhargava et al. 2012; Pollard et al. 2014; Clemens 2001; Mahar et al. 2016).

- Long period (usually several years) required for remediation.
- Phytoextraction efficacy of most hyperaccumulators is generally limited by their slow growth rate and low biomass.
- Correct disposal of the metal(loid)s containing biomass following phytoextraction.
- Limited depth that can be accessed by roots.
- Decline in phytoextraction efficiency under increasing metal concentration.
- Lack of knowledge on the agronomic practices and management.
- Pest and disease attack in climate-affected tropical and subtropical regions may affect the metal accumulation capacity of some plants.
- Limited solubility and bioavailability of metal ions in soil.
- Invasion of hyperaccumulators may affect the indigenous floral diversity.
- Sustainability of phytoremediation depends mainly on the climatic conditions.
- Difficulty in establishing and maintaining vegetation at some sites with high toxic levels.
- Risk of transfer of accumulated metals into the food chain in case of mishandling of biomass.
- The vast majority of the hyperaccumulator species discovered so far are Ni hyperaccumulators, while species accumulating Cu, Pb, Cd, Zn, Co, and As are much less numerous.
- Finding the right combination of high levels of trace element uptake and high productivity has been largely elusive.
- Majority of plants known to hyperaccumulate metal(loid)s are obligate metallophytes that have restricted geographic distribution only in metalliferous soils.
- Metallophytes, the plants that provide much of the genetic resource base for phytoremediation, are increasingly threatened by extinction due to mining.
- Techno-economic perspective and environmental safety of transgenic plants are always questioned.
- Currently most research on phytoremediation is restricted to laboratory (controlled conditions), rather than in the actual field.

16.12 Conclusion

Remediation of heavy metal(loid)s is crucial. Phytoremediation, and especially phytoextraction, is becoming more popular as a method of remediating heavy metal(loid)s contamination as it has many advantageous features that make it an appropriate and successful technology. Considerable attention has been given to the possibility of using hyperaccumulators for phytoremediation/phytomining of contaminated or natural metal-rich environment. To make this technology more efficient, a number of adjunct approaches such as chelate-assisted phytoremediation and microbe-assisted phytoremediation have been explored. Microbes especially the plant growth-promoting (PGP) bacteria-assisted phytoremediation is an emerging technology that appears to solve heavy metal(loid)s problems without any negative effects. These bacteria enhance the solubility and bioavailability of heavy metals and facilitate phytoremediation directly through microbial biosurfactants, organic acids, siderophores, redox processes, and biomethylation. Additionally, PGPB indirectly facilitate phytoremediation by producing growth-promoting agents. Transgenic plants (over)expressing desirable genes from microbes, plants, and animals are being developed successfully to enhance the ability of plants to tolerate, accumulate, or volatilize metal(loid)s. A vast amount of knowledge regarding various approaches of phytoremediation of metal(loid)s has been generated through emerging scientific tools and techniques; however, currently most research is limited to laboratory and greenhouse-scale studies and only a few studies have been conducted to test the efficiency of phytoremediation in actual field conditions. Now, the emphasis should shift from "know-how" to "do-how."

References

Abd-Elnaby H, Abou-Elela GM, El-Sersy NA (2011) Cadmium resisting bacteria in Alexandria Eastern Harbor (Egypt) and optimization of cadmium bioaccumulation by *Vibrio harveyi*. Afr J Biotechnol 10:3412–3423

Adriano DC (2001) Trace elements in terrestrial environments: biogeochemistry, bioavailability and risks of metals, 2nd edn. Springer, New York

Adriano DC, Wenzel WW, Vangronsveld J, Bolan NS (2004) Role of assisted natural remediation in environmental cleanup. Geoderma 122:121–142

Alexander M (1999) Biodegradation and bioremediation, 2nd edn. Academic, San Diego

Alford ER, Pilon-Smits EAH, Paschke MW (2010) Metallophytes – a view from the rhizosphere. Plant Soil 337:33–50

Ali H, Khan E, Sajad MA (2013) Phytoremediation of heavy metals—concepts and applications. Chemosphere 91:869–881

Alloway BJ (1990) Soil processes and the behaviour of metals. In: Alloway BJ (ed) Heavy metals in soils. Blackie and Son Ltd, Glasgow

Alloway BJ (2013) Sources of heavy metals and metalloids in soils. In: Alloway BJ (ed) Heavy metals in soils. Springer, Dordrecht, pp 11–50

Almeida SM, Umeo SH, Marcante RC, Yokota ME, Valle JS, Dragunski DC, Colauto NB, Linde GA (2015) Iron bioaccumulation in mycelium of *Pleurotus ostreatus*. Braz J Microbiol 46:195–200

Anderson JW (1993) Selenium interactions in sulfur metabolism. In: De Kok LJ (ed) Sulfur nutrition and assimilation in higher plants-regulatory, agricultural and environmental aspects. SPB Academic, The Hague, pp 49–60

Arazi T, Sunkar R, Kaplan B, Fromm H (1999) A tobacco plasma membrane calmodulin-binding transporter confers Ni^{2+} tolerance and Pb^{2+} hypersensitivity in transgenic plants. Plant J 20:171–182

Azevedo JL, Maccheroni JW, Pereira JO, de Araújo WL (2000) Endophytic microorganisms: a review on insect control and recent advances on tropical plants. Electron J Biotechnol 3:40–65

Baker AJM (1981) Accumulators and excluders—strategies in the response of plants to heavy metals. J Plant Nutr 3:643–654

Baker AJM, Brooks RR (1989) Terrestrial higher plants which hyperaccumulate metallic elements – a review of their distribution, ecology and phytochemistry. Biorecovery 1:81–126

Baker AJM, Ernst WHO, van der Ent A, Malaisse F, Ginocchio R (2010) Metallophytes: the unique biological resource, its ecology and conservational status in Europe, central Africa and

Latin America. In: Batty L, Hallberg KB (eds) Ecology of industrial pollution. Cambridge University Press/British Ecological Society, Cambridge, UK, pp 7–40
Baker AJM, McGrath SP, Reeves RD, Smith JAC (1999) Metal hyperaccumulator plants: a review of the biological resource for possible exploitation in the phytoremediation of metal- polluted soils. In: Terry N, Banuelos GS (eds) Phytoremediation of contaminated soil and water. Lewis Publishers/CRC Press, Boca Raton, pp 85–107
Banuelos G, Terry N, LeDuc DL, Pilon-Smits EAH, Mackey B (2005) Field trial of transgenic Indian mustard plants shows enhanced phytoremediation of selenium contaminated sediment. Environ Sci Technol 39:1771–1777
Banuelos G, LeDuc DL, Pilon-Smits EAH, Tagmount A, Terry N (2007) Transgenic Indian mustard overexpressing selenocysteine lyase, selenocysteine methyltransferase, or methionine methyltransferase exhibit enhanced potential for selenium phytoremediation under field conditions. Environ Sci Technol 41:599–605
Barkay T, Wagner-Dobler I (2005) Microbial transformations of mercury: potentials, challenges, and achievements in controlling mercury toxicity in the environment. Adv Appl Microbiol 57:1–52
Barzanti R, Ozino F, Bazzicalupo M, Gabbrielli R, Galardi F, Gonnelli C, Mengoni A (2007) Isolation and characterization of endophytic bacteria from the nickel hyperaccumulator plant *Alyssum bertolonii*. Microb Ecol 53:306–316
Beijer K, Jernelov A (1979) Methylation of mercury in aquatic environments. In: Nriagu JO (ed) The biogeochemistry of mercury in the environment. Elsevier/North Holland Biomedical Press, Amsterdam, pp 203–230
Belimov AA, Hontzeas N, Safronova VI, Demchinskaya SV, Piluzza G, Bullitta S, Glick BR (2005) Cadmium-tolerant plant growth-promoting bacteria associated with the roots of Indian mustard (*Brassica juncea* L. Czern.) Soil Biol Biochem 37:241–250
Bennett LE, Burkhead JL, Hale KL, Terry N, Pilon M, Pilon-Smits EAH (2003) Analysis of transgenic Indian mustard plants for phytoremediation of metal-contaminated mine tailings. J Environ Qual 32:432–440
Bentley R, Chasteen TG (2002) Microbial methylation of metalloids: arsenic, antimony, and bismuth. Microbiol Mol Biol R 66:250–271
Bhargava A, Carmona FF, Bhargava M, Srivastava S (2012) Approaches for enhanced phytoextraction of heavy metals. J Environ Manag 105:103–120
Bienert GP, Thorsen M, Schüssler MD, Nilsson HR, Wagner A, Tamás MJ et al (2008) A subgroup of plant aquaporins facilitate the bi-directional diffusion of $As(OH)_3$ and $Sb(OH)_3$ across membranes. BMC Biol 6:26. doi:10.1186/1741-7007-6-26
Bittsanszky A, Komives T, Gullner G, Gyulai G, Kiss J, Heszky L, Radimszky L, Rennenberg H (2005) Ability of transgenic poplars with elevated glutathione content to tolerate zinc(2+) stress. Environ Int 31:251–254
Bizily SP, Rugh CL, Summers AO, Meagher RB (1999) Phytoremediation of methylmercury pollution: merB expression in *Arabidopsis thaliana* confers resistance to organomercurials. Proc Natl Acad Sci U S A 96:6808–6813
Bizily SP, Rugh CL, Meagher RB (2000) Phytodetoxification of hazardous organomercurials by genetically engineered plants. Nat Biotechnol 18:213–217
Boening DW (2000) Ecological effects, transport, and fate of mercury: a general review. Chemosphere 40:1335–1351
Bolan NS, Choppala G, Kunhikrishnan A, Park J, Naidu R (2013) Microbial transformation of trace elements in soils in relation to bioavailability and remediation. In: Whitacre DM (ed) Reviews of environmental contamination and toxicology, vol 225. Springer, New York, pp 1–43
Bolan N, Kunhikrishnan A, Thangarajan R, Kumpiene J, Park J, Makino T, Kirkham MB, Scheckel K (2014) Remediation of heavy metal(loid)s contaminated soils to mobilize or to immobilize? J Hazard Mater 266:141–166
Bowen HJM (1979) Environmental chemistry of the elements. Academic press, New York

Braud A, Jézéquel K, Bazot S, Lebeau T (2009) Enhanced phytoextraction of an agricultural Cr, Hg and Pb contaminated soil by bioaugmentation with siderophore producing bacteria. Chemosphere 74:280–286

Brooks RR, Lee J, Reeves RD, Jaffrre T (1977) Detection of nickeliferous rocks by analysis of herbarium specimens of indicator plants. J Geochem Explor 7:49–57

Brunetti P, Zanella L, De Paolis A, Di Litta D, Cecchetti V, Falasca G, Cardarelli M (2015) Cadmium inducible expression of the ABC-type transporter AtABCC3 increases phytochelatin-mediated cadmium tolerance in Arabidopsis. J Exp Bot 66(13):3815–3829

Burken JG, Schnoor JL (1997) Uptake and metabolism of atrazine by poplar trees. Environ Sci Technol 31:1399–1406

Burken JG, Schnoor JL (1999) Distribution and volatilisation of organic compounds following uptake by hybrid poplar trees. Int J Phytoremediation 1:139–151

Carrillo-Castañeda G, Muñoz JJ, Peralta-Videa JR, Gomez E, Gardea-Torresdey JL (2003) Plant growth-promoting bacteria promote copper and iron translocation from root to shoot in *Alfalfa* seedlings. J Plant Nutr 26:1801–1814

Chatterjee S, Sau GB, Mukherjee SK (2009) Plant growth promotion by a hexavalent chromium reducing bacterial strain, *Cellulosimicrobium cellulans* KUCr3. World J Microbiol Biotechnol 25:1829–1836

Che D, Meagher RB, Heaton ACP, Lima A, Rugh CL, Merkle SA (2003) Expression of mercuric ion reductase in eastern cottonwood (*Populus deltoides*) confers mercuric ion reduction and resistance. Plant Biotechnol J 1:311–319

Chojnacka K (2010a) Biosorption and bioaccumulation – the prospects for practical applications. Environ Int 36:299–307

Chojnacka K (2010b) Biosorption and bioaccumulation in practice. Nova Science Publishers

Clemens S (2001) Developing tools for phytoremediation: towards a molecular understanding of plant metal tolerance and accumulation. Int J Occup Med Environ Health 14:235–239

Clemens S (2006a) Toxic metal accumulation, responses to exposure and mechanisms of tolerance in plants. Biochimie 88:1707–1719

Clemens S (2006b) Evolution and function of phytochelatin synthases. J Plant Physiol 163:319–332

Clemens S, Palmgren M, Krämer U (2002) A long way ahead: understanding and engineering plant metal accumulation. Trends Plant Sci 7:309–315

Cluis C (2004) Junk-greedy greens: phytoremediation as a new option for soil decontamination. BioTeach J 2:61–67

Cobbett CS (2000) Phytochelatins and their roles in heavy metal detoxification. Plant Physiol 123:825–832

Cobbett C, Goldsbrough P (2002) Phytochelatins and metallothioneins: roles in heavy metal detoxification and homeostasis. Annu Rev Plant Biol 53:159–182

Cobbett CS, May MJ, Howden R, Rolls B (1998) The glutathione-deficient, cadmium-sensitive mutant, *cad2-1*, of *Arabidopsis thaliana* is deficient in glutamylcysteine synthetase. Plant J 16:73–78

Cunningham SD, Berti WR, Huang JW (1995) Phytoremediation of contaminated soils. Trends Biotechnol 13:393–397

Curie C, Alonso JM, Le Jean M, Ecker JR, Briat JF (2000) Involvement of *Nramp1* from *Arabidopsis thaliana* in iron transport. Biochem J 347:749–755

Czakó M, Feng X, He Y, Liang D, Márton L (2006) Transgenic *Spartina alterniflora* for phytoremediation. Environ Geochem Health 28:103–110

Daghan H, Arslan M, Uygur V, Koleli N (2013) Transformation of tobacco with ScMTII gene enhanced cadmium and zinc accumulation. Clean Soil Air Water 41:503–509

Deng X, Wilson DB (2001) Bioaccumulation of mercury from wastewater by genetically engineered *Escherichia coli*. Appl Microbiol Biotechnol 56:276–279

Dermont G, Bergeron M, Mercier G, Richer-Laflèche M (2008) Soil washing for metal removal: a review of physical/chemical technologies and field applications. J Hazard Mater 152:1–31

Dhankher OP, Li Y, Rosen BP, Shi J, Salt D, Senecoff JF, Sashti NA, Meagher RB (2002) Engineering tolerance and hyperaccumulation of arsenic in plants by combining arsenate reductase and gamma-glutamylcysteine synthetase expression. Nat Biotechnol 20:1140–1145

Di Gregorio S, Lampis S, Vallini G (2005) Selenite precipitation by a rhizospheric strain of Stenotrophomonas sp isolated from the root system of *Astragalus bisulcatus*: a biotechnological perspective. Environ Int 31:233–241

Dickinson NM, Baker AJM, Doronila A, Laidlaw S, Reeves RD (2009) Phytoremediation of inorganics: realism and synergies. Int J Phytoremediation 11:97–114

Dimkpa CO, Svatos A, Merten D, Büchel G, Kothe E (2008) Hydroxamate siderophores produced by *Streptomyces acidiscabies* E13 bind nickel and promote growth in cowpea (*Vigna unguiculata* L.) under nickel stress. Can J Microbiol 54:163–172

Dimkpa CO, Merten D, Svatoš A, Büchel G, Kothe E (2009a) Siderophores mediate reduced and increased uptake of cadmium by *Streptomyces tendae* F4 and sunflower (*Helianthus annuus*), respectively. J Appl Microbiol 107:1687–1696

Dimkpa CO, Merten D, Svatoš A, Büchel G, Kothe E (2009b) Metal-induced oxidative stress impacting plant growth in contaminated soil is alleviated by microbial siderophores. Soil Biol Biochem 41:154–162

do Nascimento CWA, Xing B (2006) Phytoextraction: a review on enhanced metal availability and plant accumulation. Sci Agric 63:299–311

Dönmez G, Aksu Z (1999) The effect of copper(II) ions on the growth and bioaccumulation properties of some yeasts. Process Biochem 35:135–142

Dungan RS, Frankenberger WT (1999) Microbial transformations of selenium and the bioremediation of seleniferous environments. BioRem 3:171–188

Dushenkov V, Kumar PBAN, Motto H, Raskin I (1995) Rhizofiltration: the use of plants to remove heavy metals from aqueous streams. Environ Sci Technol 29:1239–1245

Eapen S, D'Souza SF (2005) Prospects of genetic engineering of plants for phytoremediation of toxic metals. Biotechnol Adv 23:97–114

Ebbs SD, Lasat MM, Brady DJ, Cornish J, Gordon R, Kochian LV (1997) Heavy metals in the environment: phytoextraction of cadmium and zinc from a contaminated soil. J Environ Qual 26:1424–1430

Ellis DR, Sors TG, Brunk DG, Albrecht C, Orser C, Lahner B, Wood KV, Harris HH, Pickering IJ, Salt DE (2004) Production of Se-methylselenocysteine in transgenic plants expressing selenocysteine methyltransferase. BMC Plant Biol 4:1–11

Erakhrumen AA (2007) Phytoremediation: an environmentally sound technology for pollution prevention, control and remediation in developing countries. Educ Res Rev 2:151–156

Evangelou MW, Ebel M, Schaeffer A (2007) Chelate assisted phytoextraction of heavy metals from soil, effect, mechanism, toxicity, and fate of chelating agents. Chemosphere 68(6):989–1003

Farhadian M, Vachelard C, Duchez D, Larroche C (2008) In situ bioremediation of monoaromatic pollutants in groundwater: a review. Bioresour Technol 99:5296–5308

Frankenberger WT, Losi ME (1995) Application of bioremediation in the cleanup of heavy elements and metalloids. In: Skipper HD, Turco RF (eds) Bioremediation: science and applications, Soil science special publication No, vol 43. Soil Science Society of America Inc, Madison, pp 173–210

Freeman JL, Garcia D, Kim D, Hopf A, Salt DE (2005) Constitutively elevated salicylic acid signals glutathione-mediated Ni tolerance in Thlaspi Ni hyperaccumulators. Plant Physiol 137:1082–1091

Gadd GM (2000) Bioremedial potential of microbial mechanisms of metal mobilization and immobilization. Curr Opin Biotechnol 11:271–279

Gadd GM (2004) Microbial influence on metal mobility and application for bioremediation. Geoderma 122:109–119

Gadd GM (2009) Biosorption: critical review of scientific rationale, environmental importance and significance for pollution treatment. J Chem Technol Biotechnol 84:13–28

Gadd GM (2010) Metals: minerals and microbes: geomicrobiology and bioremediation. Microbiology 156:609–643

Garbisu C, Hernández-Allica J, Barrutia O, Alkorta I, Becerril JM (2002) Phytoremediation: a technology that uses green plants to remove contaminants from polluted areas. Rev Environ Health 17(3):173–188

Garifullina GF, Owen JD, Lindblom S-D, Tufan H, Pilon M, Pilon-Smits EAH (2003) Expression of a mouse selenocysteine lyase in *Brassica juncea* chloroplasts affects selenium tolerance and accumulation. Physiol Plant 118:538–544

Gasic K, Korban SS (2007) Transgenic Indian mustard (*Brassica juncea*) plants expressing an *Arabidopsis* phytochelatin synthase (AtPCS1) exhibit enhanced As and Cd tolerance. Plant Mol Biol 64:361–369

Ginocchio R, Baker AJM, Cucuzza J (2004) Phytoremediation. J Mining Environ Manag, November 7–10

Glick BR (2010) Using soil bacteria to facilitate phytoremediation. Biotechnol Adv 28:367–374

Glick BR (2012) Plant growth-promoting bacteria: mechanisms and applications. Scientifica 963401, pp 15. doi:10.6064/2012/963401

Glick BR (2014) Bacteria with ACC deaminase can promote plant growth and help to feed the world. Microbiol Res 169:30–39

Glick BR, Bashan Y (1997) Genetic manipulation of plant growth-promoting bacteria to enhance biocontrol of phytopathogens. Biotechnol Adv 15:353–378

Goldsbrough P (2000) Metal tolerance in plants: the role of phytochelatins. In: Terry N, Banuelos G (eds) Phytoremediation of contaminated soil and water. Lewis Publishers, Boca Raton, pp 221–233

Gusain P, Paliwal R, Singh V (2016) Rhizoremediation of cadmium contaminated soil associated with hydroxamate siderophores isolated from Cd resistant plant growth promoting *Dietzia Maris* and *Lysinibacillus* strains. Int J Phytorem. Published online 03 Sept 2016, http://dx.doi.org/10.1080/15226514.2016.1225281

Haferburg G, Kothe E (2007) Microbes and metals: interactions in the environment. J Basic Microbiol 47:453–467

Hartley-Whitaker J, Ainsworth G, Vooijs R, Ten Brookum W, Schat H, Meharg AA (2001) Phytochelatins are involved in differential arsenate tolerance in *Holcus lanatus*. Plant Physiol 126:299–306

Hattori J, Labbe H, Miki BL (1994) Construction and expression of a metallothionein-b-glucuronidase gene fusion. Genome 37:508–512

He J, Chen JP (2014) A comprehensive review on biosorption of heavy metals by algal biomass: materials, performances, chemistry, and modeling simulation tools. Bioresour Technol 160:67–78

He YK, Sun JG, Feng XZ, Czakó M, Márton L (2001) Differential mercury volatilization by tobacco organs expressing a modified bacterial merA gene. Cell Res 11:231–236

Heaton ACP, Rugh CL, Kim T, Wang NJ, Meagher RB (2003) Toward detoxifying mercury polluted aquatic sediments with rice genetically engineered for mercury resistance. Environ Toxicol Chem 22:2940–2947

Heaton ACP, Rugh CL, Wang N-J, Meagher RB (2005) Physiological responses of transgenic *mer*A-TOBACCO (*Nicotiana tabacum*) to foliar and root mercury exposure. Water Air Soil Pollut 161:137–155

Hirschi KD, Korenkov VD, Wilganowski NL, Wagner GJ (2000) Expression of *Arabidopsis* CAX2 in tobacco altered metal accumulation and increased manganese tolerance. Plant Physiol 124:125–133

Huang J, Zhang Y, Peng JS, Zhong C, Yi HY, Ow DW, Gong JM (2012) Fission yeast HMT1 lowers seed cadmium through phytochelatin-dependent vacuolar sequestration in *Arabidopsis*. Plant Physiol 158:1779–1788

Idris R, Trifonova R, Puschenreiter M, Wenzel WW, Sessitsch A (2004) Bacterial communities associated with flowering plants of the Ni hyperaccumulator *Thlaspi goesingense*. Appl Environ Microbiol 70:2667–2677

Iqbal MP (2012) Lead pollution—a risk factor for cardiovascular disease in Asian developing countries. Pak J Pharm Sci 25:289–294

Jagtap UB, Bapat VA (2015) Genetic engineering of plants for heavy metal removal from soil. In: Sherameti I, Varma A (eds) Heavy metal contamination of soils. Soil biology. Springer International Publishing, Cham, pp 433–470

Jaiswar S, Kazi MA, Mehta S (2015) Bioaccumulation of heavy metals by freshwater algal species of Bhavnagar, Gujarat, India. J Environ Biol 36:1361–1366

Jan AT, Murtaza I, Ali A, Haq QMR (2009) Mercury pollution: an emerging problem and potential bacterial remediation strategies. World J Microbiol Biotechnol 25:1529–1537

Järup L (2003) Hazards of heavy metal contamination. Br Med Bull 68:167–182

Joutey NT, Sayel H, Bahafid W, El Ghachtouli N (2015) Mechanisms of hexavalent chromium resistance and removal by microorganisms. Rev Environ Contam Toxicol 233:45–69

Juwarkar AA, Nair A, Dubey KV, Singh SK, Devotta S (2007) Biosurfactant technology for remediation of cadmium and lead contaminated soils. Chemosphere 10:1996–2002

Kara Y (2004) Bioaccumulation of copper from contaminated wastewater by using *Lemna minor*. Bull Environ Contam Toxicol 72:467–471

Karami A, Shamsuddin ZH (2010) Phytoremediation of heavy metals with several efficiency enhancer methods. Afr J Biotechnol 9:3689–3698

Kärenlampi S, Schat H, Vangronsveld J, Verkleij JAC, van der Lelie D, Mergeay M, Tervahauta AI (2000) Genetic engineering in the improvement of plants for phytoremediation of metal polluted soils. Environ Pollut 107:225–231

Karlson U, Frankenberger WT (1993) Biological alkylation of selenium and tellurium. In: Sigel H, Sigel A (eds) Metal ions in biological systems. Marcel Dekker, New York, pp 185–227

Kim SK, Park CB, Koo Y-M, Yun HS (2003) Biosorption of cadmium and copper ions by *Trichoderma reesei* RUT C30. J Ind Eng Chem 9:403–406

Kiyono M, Pan-Hou H (1999) *mer* G gene product is involved in phenyl mercury resistance in *Pseudomonas* strain K-62. J Bacteriol 181:726–730

Kiyono M, Sone Y, Nakamura R, Pan-Hou H, Sakabe K (2009) The *Mer E* protein encoded by transposon Tn21 is a broad mercury transporter in *Escherichia coli*. FEBS Lett 583:1127–1131

Kocberber N, Donmez G (2007) Chromium(VI) Bioaccumulation capacities of adapted mixed cultures isolated from industrial saline wastewaters. Bioresour Technol 98:2178–2183

Kotrba P, Najmanova J, Macek T, Ruml T, Mackova M (2009) Genetically modified plants in phytoremediation of heavy metal and metalloid soil and sediment pollution. Biotechnol Adv 27:799–810

Kramer U (2010) Metal hyperaccumulation in plants. Annu Rev Plant Biol 61:517–534

Krämer U, Talke IN, Hanikenne M (2007) Transition metal transport. FEBS Lett 581:2263–2272

Kubachka KM, Meija J, LeDuc DL, Terry N, Caruso JA (2007) Selenium volatiles as proxy to the metabolic pathways of selenium in genetically modified *Brassica juncea*. Environ Sci Technol 41:1863–1869

Kuffner M, Puschenreiter M, Wieshammer G, Gorfer M, Sessitsch A (2008) Rhizosphere bacteria affect growth and metal uptake of heavy metal accumulating willows. Plant Soil 304:35–44

Kujan P, Votruba J, Kamenik V (1995) Substrate-dependent bioaccumulation of cadmium by growing yeast *Candida utilis*. Folia Microbiol 40(3):288–292

Kuklinsky-Sobral J, Araújo WL, Mendes R, Geraldi IO, Pizzirani-Kleiner AA, Azevedo JL (2004) Isolation and characterization of soybean-associated bacteria and their potential for plant growth promotion. Environ Microbiol 6:1244–1251

Kumar KS, Dahms H-U, Won E-J, Lee J-S, Shin K-H (2015) Microalgae – a promising tool for heavy metal remediation. Ecotoxicol Environ Saf 113:329–352

Lasat MM (2000) Phytoextraction of metals from contaminated soil: a review of plant /soil/metal interaction and assessment of pertinent agronomic issues. J Hazard Subs Res 2:1–25

Lasat MM (2002) Phytoextraction of toxic metals: a review of biological mechanisms. J Environ Qual 31:109–120

LeBlanc MS, Lima A, Montello P, Kim T, Meagher RB, Merkle S (2011) Enhanced arsenic tolerance of transgenic eastern cottonwood plants expressing gamma-glutamylcysteine synthetase. Int J Phytoremediation 13(7):657–673

LeBlanc MS, McKinney EC, Meagher RB, Smith AP (2013) Hijacking membrane transporters for arsenic phytoextraction. J Biotechnol 163:1–9

LeDuc DL, Tarun AS, Montes-Bayon M, Meija J, Malit MF, Wu CP, AbdelSamie M, Chiang C-Y, Tagmount A, deSouza MP, Neuhierl B, Bock A, Caruso JA, Terry N (2004) Overexpression of selenocysteine methyltransferase in *Arabidopsis* and Indian mustard increases selenium tolerance and accumulation. Plant Physiol 135:377–383

LeDuc DL, AbdelSamie M, Montes-Bayon M, Wu CP, Reisinger SJ, Terry N (2006) Overexpressing both ATP sulfurylase and selenocysteine methyltransferase enhances selenium phytoremediation traits in Indian mustard. Environ Pollut 144:70–76

Lestan D, Luo C, Li X (2008) The use of chelating agents in the remediation of metal-contaminated soils: a review. Environ Pollut 153:3–13

Li Y, Dhankher OP, Carreira L, Balish RS, Meagher RB (2005) Arsenic and mercury tolerance and cadmium sensitivity in *Arabidopsis* plants expressing bacterial gamma-glutamylcysteine synthetase. Environ Toxicol Chem 24:1376–1386

Li WC, Ye ZH, Wong MH (2010) Metal mobilization and production of short-chain organic acids by rhizosphere bacteria associated with a Cd/Zn hyperaccumulating plant *Sedum alfredii*. Plant Soil 326:453–467

Lovley DR (1995) Bioremediation of organic and metal contaminants with dissimilatory metal reduction. J Ind Microbiol 14:85–93

Lovley DR (2000) Fe(III) and Mn(IV) reduction. In: Lovley DR (ed) Environmental microbe – metal interactions. American Society of Microbiology, Washington, pp 3–30

Lovley DR (2001) Anaerobes to the rescue. Science 293:1444–1446

Lund PA, Brown NL (1989) Regulation of transcription in *Escherichia coli* from mer promoters in the transposon Tn501. J Mol Biol 205:343–353

Lyyra S, Meagher RB, Kim T, Heaton A, Montello P, Balish RS, Merkle SA (2007) Coupling two mercury resistance genes in eastern cottonwood enhances the processing of organomercury. Plant Biotechnol J 5:254–262

Ma Y, Prasad MNV, Rajkumar M, Freitas H (2011) Plant growth promoting rhizobacteria and endophytes accelerate phytoremediation of metalliferous soils. Biotechnol Adv 29:248–258

Mahar A, Wang P, Ali A, Awasthi MK, Lahori AH, Wang Q, Li R, Zhang Z (2016) Challenges and opportunities in the phytoremediation of heavy metals contaminated soils: a review. Ecotoxicol Environ Saf 126:111–121

Malik A (2004) Metal bioremediation through growing cells. Environ Int 30:261–278

Martinez M, Bernal P, Almela C, Vélez D, García-Agustín P, Serrano R, Navarro-Aviñó J (2006) An engineered plant that accumulates higher levels of heavy metals than *Thlaspi caerulescens*, with yields of 100 times more biomass in mine soils. Chemosphere 64:478–485

McGrath SP, Zhao FJ, Lombi E (2002) Phytoremediation of metals, metalloids and radionuclides. Adv Agron 75:1–56

Meija J, Montes-Bayon M, LeDuc DL, Terry N, Caruso J (2002) Simultaneous monitoring of volatile selenium and sulfur species from se accumulating plants (wild-type and genetically modified) by GC-MS and GC-ICP-MS using SPME for sample introduction. Anal Chem 74:5837–5844

Meister A (1988) Glutathione metabolism and its selective modification. J Biol Chem 263:17205–17208

Mejare M, Bülow L (2001) Metal binding proteins and peptides in bioremediation and phytoremediation of heavy metals. Trends Biotechnol 19:67–73

Merlot S, Hannibal L, Martins S, Martinelli L, Amir H, Lebrun M, Thomine S (2014) The metal transporter PgIREG1 from the hyperaccumulator *Psychotria gabriellae* is a candidate gene for nickel tolerance and accumulation. J Exp Bot 65:1551–1564

Mesjasz-Przyby OJ, Nakonieczny M, Migula P, Augustyniak M, Tarnawska MM, Reimold WU, Koeberl C, Przyby OW, Owacka EG (2004) Uptake of cadmium, lead, nickel and zinc from soil and water solutions by the nickel hyperaccumulator *Berkheya coddii*. Acta Biol Cracov Ser Bot 46:75–85

Montes-Bayon M, LeDuc DL, Terry N, Caruso J (2002) Selenium speciation in wild-type and genetically modified Se accumulating plants with HPLC separation and ICP-MS/ES-MS detection. J Anal At Spectrom 17:872–879

Morel M, Crouzet J, Gravot A, Auroy P, Leonhardt N, Vavasseur A, Richaud P (2009) AtHMA3, a P1B-ATPase allowing Cd/Zn/Co/Pb vacuolar storage in *Arabidopsis*. Plant Physiol 149:894–904

Mosa KA, Kumar K, Chhikara S, Mcdermott J, Liu Z, Musante C (2012) Members of rice plasma membrane intrinsic proteins subfamily are involved in arsenite permeability and tolerance in plants. Transgenic Res 21:1265–1277

Naidu R, Krishnamurti GSR, Wenzel W, Megharaj BNS (2001) Heavy metal interactions in soils and implications for soil microbial biodiversity. In: Prasad MNV (ed) Metals in the environment: analysis by biodiversity. Mercel Dekker Inc, New York, pp 401–432

Nascimento AMA, Chartone-Souza E (2003) Operon *mer*: bacterial resistance to mercury and potential for bioremediation of contaminated environments. Genet Mol Res 2(1):92–101

Nazir A, Malik RN, Ajaib M, Khan N, Siddiqui MF (2011) Hyperaccumulators of heavy metals of industrial areas of Islamabad and Rawalpindi. Pak J Bot 43:1925–1933

Neilands JB (1983) Siderophores. In: Eichhorn L, Marzilla LG (ed.) Advances in inorganic biochemistry. Elsevier, New York, pp 137–166

Neubauer U, Furrer G, Kayser A, Schulin R (2000) Siderophores, NTA, and citrate: potential soil amendments to enhance heavy metal mobility in phytoremediation. Int J Phytoremediation 2:353–368

Neuhierl B, Boeck A (1996) On the mechanism of selenium tolerance in selenium accumulating plants: purification and characterization of a specific selenocysteine methyltransferase from cultured cells of *Astragalus bisulcatus*. Eur J Biochem 239:235–238

Nriagu JO (ed) (1978) Biogeochemistry of lead in the environment, vol I & II. Elsevier, Amsterdam

Nriagu JO, Pacyna JM (1988) Quantitative assessment of worldwide contamination of air, water and soils by trace elements. Nature 333:134–139

Ogunseitan AO (2002) Episodic bioavailability of environmental mercury: implications for biotechnological control of mercury pollution. Afr J Biotechnol 1(1):1–9

Oremland RS (1994) Biogeochemical transformations of selenium in anoxic environments. In: Frankenberger WTJ, Benson SN (eds) Selenium in the environment. Marcel Dekker, New York, pp 389–419

Padmavathiamma PK, Li LY (2007) Phytoremediation technology: hyperaccumulation metals in plants. Water Air Soil Pollut 184:105–126

Pan AH, Yang M, Tie F, Li L, Che Z, Ru B (1994) Expression of mouse metallothionein-I gene confers Cd resistance in transgenic tobacco plants. Plant Mol Biol 24:341–351

Pan F, Meng Q, Luo S, Shen J, Chen B, Khan KY, Japenga J, Ma X, Yang X, Feng Y (2016) Enhanced Cd extraction of oilseed rape (*Brassica napus*) by plant growth promoting bacteria isolated from Cd Hyperaccumulator *Sedum alfredii* Hance. Int J Phytorem Published online 03 Sep 2016, http://dx.doi.org/10.1080/15226514.2016.1225280

Panwichian S, Kantachote D, Wittayaweerasa B, Mallavarapu M (2011) Removal of heavy metals by exopolymeric substances produced by resistant purple non sulphur bacteria isolated from contaminated shrimp ponds. Electron J Biotechnol 14: ISSN: 0717–3458

Park J, Song W-Y, Ko D, Eom Y, Hansen TH, Schiller M, Lee TG, Martinoia E, Lee Y (2012) The phytochelatin transporters AtABCC1 and AtABCC2 mediate tolerance to cadmium and mercury. Plant J 69:278–288

Paulose B, Chhikara S, Coomey J, Jung HI, Vatamaniuk O, Dhankher OP (2013) A γ-glutamyl cyclotransferase protects *Arabidopsis* plants from heavy metal toxicity by recycling glutamate to maintain glutathione homeostasis. Plant Cell 25:4580–4595

Pilon-Smits EAH (2015) Selenium in plants. In: Lüttge U, Beyschlag W (eds) Progress in botany, vol 76. Springer International Publishing, Cham, pp 93–107

Pilon-Smits E, Pilon M (2002) Phytoremediation of metals using transgenic plants. Crit Rev Plant Sci 21:439–456

Pilon-Smits EAH, Hwang S, Lytle CM, Zhu Y, Tai JC, Bravo RC, Chen Y, Leustek T, Terry N (1999) Overexpression of ATP sulfurylase in Indian mustard leads to increased selenate uptake, reduction, and tolerance. Plant Physiol 119:123–132

Pollard AJ, Reeves RD, Baker AJM (2014) Facultative hyperaccumulation of heavy metals and metalloids. Plant Sci 217–218:8–17

Pomponi M, Censi V, DiGirolamo V, DePaolis A, di Toppi LS, Aromolo R, Costantino P, Cardarelli M (2006) Overexpression of *Arabidopsis* phytochelatin synthase in tobacco plants enhances Cd^{2+} tolerance and accumulation but not translocation to the shoot. Planta 223:180–190

Prasad MNV (2003) Phytoremediation of metal polluted ecosystems: hype for commercialization. Russ J Plant Physiol 50:686–700

Prasad MNV, Freitas HMO (2003) Metal hyperaccumulation in plants biodiversity prospecting for phytoremediation technology. Electron J Biotechnol 6(3):285–321

Prasad MNV, Freitas H, Fraenzle S, Wuenschmann S, Markert B (2010) Knowledge explosion in phytotechnologies for environmental solutions. Environ Pollut 158:18–23

Pulford ID, Watson C (2003) Phytoremediation of heavy metal contaminated land by trees-a review. Environ Int 29:529–540

Punshon T, Dickinson NM, Lepp NW (1996) The potential of Salix clones for bioremediating metal polluted soil. In: Glimmerveen I (ed) Heavy metals and trees. Proceedings of a discussion meeting, Glasgow. Institute of Chartered Foresters, Edinburgh, pp 93–104

Rabhi M, Ferchichi S, Jouini J, Hamrouni MH, Koyro H-W, Ranieri A, Abdelly C, Smaoui (2010) Phytodesalination of a salt-affected soil with the halophyte *Sesuvium portulacastrum* L. to arrange in advance the requirements for the successful growth of a glycophytic crop. Bioresour Technol 101: 6822–6828

Radulescu C, Stihi C, Popescu IV, Dulama ID, Chelarescu ED, Chilian A (2013) Heavy metal accumulation and translocation in different parts of *Brassica oleracea L.* Rom J Physiol 58:1337–1354

Rahman S, Kim K-H, Saha SK, Swaraz AM, Paul DK (2014) Review of remediation techniques for arsenic (As) contamination: a novel approach utilizing bio-organisms. J Environ Manag 134:175–185

Rajkumar M, Ae N, Freitas H (2009) Endophytic bacteria and their potential to enhance heavy metal phytoextraction. Chemosphere 77:153–160

Rajkumar M, Ae N, Prasad MNV, Freitas H (2010) Potential of siderophore- producing bacteria for improving heavy metal phytoextraction. Trends Biotechnol 28:142–149

Rajkumar M, Sandhya S, Prasad MNV, Freitas H (2012) Perspectives of plant-associated microbes in heavy metal phytoremediation. Biotechnol Adv 30:1562–1574

Ranjard L, Nazaret S, Cournoyer B (2003) Freshwater bacteria can methylate selenium through the thiopurine methyltransferase pathway. Appl Environ Microbiol 69:3784–3790

Rascio N, Navari-Izzo F (2011) Heavy metal accumulating plants: how and why do they do it? And what makes them so interesting? Plant Science 180:169–181

Raskin I, Ensley BD (2000) Phytoremediation of toxic metals: using plants to clean up the environment. Wiley, New York, p 304

Raymond KM, Dertz E (2004) Biochemical and physical properties of siderophores. In: Crosa JH, May AR, Payne SM (eds) Iron transport in bacteria. ASM Press, Washington, DC, pp 3–17

Romera E, González F, Ballester A, Blázquez ML, Muñoz JA (2006) Biosorption with algae: a statistical review. Crit Rev Biotechnol 26:223–235

Ron EZ, Rosenberg E (2001) Natural roles of biosurfactants. Environ Microbiol 3:229–236. doi:10.1046/j.1462-2920.2001.00190.x

Rozas MA, Alkorta I, Garbisu C (2006) Phytoextraction and phytofilteration of arsenic. Rev Environ Health 21(1):43–56

Rugh CL, Wilde HD, Stack NM, Thompson DM, Summers AO, Meagher RB (1996) Mercuric ion reduction and resistance in transgenic *Arabidopsis thaliana* plants expressing a modified bacterial *mer*A gene. Proc Natl Acad Sci U S A 93:3182–3187

Rugh CL, Senecoff JF, Meagher RB, Merkle SA (1998) Development of transgenic yellow poplar for mercury phytoremediation. Nat Biotechnol 16:925–928

Ruiz NO, Daniell H (2009) Genetic engineering to enhance mercury phytoremediation. Curr Opin Biotechnol 20:1–7

Ruiz ON, Hussein HS, Terry N, Daniell H (2003) Phytoremediation of organomercurial compounds via chloroplast genetic engineering. Plant Physiol 132:1344–1352

Sakai Y, Ma Y, Xu C, Wu H, Zhu W, Yang J (2012) Phytodesalination of a salt affected soil with four halophytes in China. J Arid Land Stud 22:17–20

Salt DE, Baker AJM (2000) Phytoremediation of metals. In: Rehm HJ, Reeds G (eds) Biotechnology: environmental processes II, vol 11b, 2nd edn. Wiley-VCH Verlag Gmbh, Weinheim, pp 385–397

Salt DE, Blaylock M, Kumar NPBA, Viatcheslav D, Ensley BD, Chet I, Raskin I (1995) Phytoremediation: a novel strategy for the removal of toxic metals from the environment using plants. Nat Biotechnol 13:468–474

Salt DE, Smith RD, Raskin I (1998) Phytoremediation. Annu Rev Plant Physiol 49:643–668

Saravanan VS, Madhaiyan M, Thangaraju M (2007) Solubilization of zinc compounds by the diazotrophic, plant growth promoting bacterium *Gluconacetobacter diazotrophicus*. Chemosphere 66:1794–1798

Sasaki Y, Minakawa T, Miyazaki A, Silver S, Kusano T (2005) Functional dissection of a mercuric ion transporter *Mer* C from *Acidithiobacillus ferrooxidans*. Biosci Biotechnol Biochem 69(7):1394–1402

Sasaki Y, Hayakawa T, Inoue C, Miyazaki A, Silver S, Kusano T (2006) Generation of mercury hyperaccumulating plants through transgenic expression of the bacterial mercury membrane transport protein *MerC*. Transgenic Res 15:615–625

Sauge-Merle S, Cuine S, Carrier P, Lecomte-Pradines C, Luu DT, Peltier G (2003) Enhanced toxic metal accumulation in engineered bacterial cells expressing *Arabidopsis thaliana* phytochelatin synthase. Appl Environ Microbiol 69:490–494

Schalk IJ, Hannauer M, Braud A (2011) New roles for bacterial siderophores in metal transport and tolerance. Environ Microbiol 13:2844–2854

Schmoger MEV, Oven M, Grill E (2000) Detoxification of arsenic by phytochelatins in plants. Plant Physiol 122:793–801

Schue M, Dover LG, Besra GS, Parkhill J, Brown NL (2009) Sequence and analysis of a plasmid encoded mercury resistance operon from *Mycobacterium marinum* identifies *Mer*H, a new mercuric ion transporter. J Bacteriol 191(1):439–444

Schutzendubel A, Polle A (2002) Plant responses to abiotic stresses: heavy metal-induced oxidative stress and protection by mycorrhization. J Exp Bot 53:1351–1365

Sheng J, Liu K, Fan B, Yuan Y, Shen L, Ru B (2007) Improving zinc content and antioxidant activity in transgenic tomato plants with expression of mouse metallothionein-I by mt-1 gene. J Agric Food Chem 55:9846–9849

Sheng XF, He LY, Wang QY, Ye HS, Jiang C (2008) Effects of inoculation of biosurfactant producing *Bacillus sp.* J119 on plant growth and cadmium uptake in a cadmium amended soil. J Hazard Mater 155:17–22

Shilev SI, Ruso J, Puig A, Benlloch M, Jorrin J, Sancho E (2001) Rhizospheric bacteria promote sunflower (*Helianthus annus* L.) plant growth and tolerance to heavy metals. Minerva Biotechnologica 13:37–39

Shukla D, Kesari R, Tiwari M, Dwivedi S, Tripathi RD, Nath P, Trivedi PK (2013) Expression of *Ceratophyllum demersum* phytochelatin synthase CdPCS1 in *Escherichia coli* and *Arabidopsis* enhances heavy metal(loid)s accumulation. Protoplasma 250:1263–1272

Silver S (1996) Bacterial resistances to toxic metal ions – a review. Gene 179:9–19

Silver S, Phung LT (1996) Bacterial heavy metal resistances: new surprises. Annu Rev Microbiol 50:753–789

Silver S, Phung LT (2009) Heavy metals, bacterial resistance. In: Schaechter M (ed) Encyclopedia of microbiology. Elsevier, Oxford, pp 220–227

Smith WL, Gadd GM (2000) Reduction and precipitation of chromate by mixed culture sulphate-reducing bacterial biofilms. J Appl Microbiol 88:983–991

Smith LA, Alleman BC, Copley-Graves L (1994) Biological treatment options. In: Means JL, Hinchee RE (ed) Emerging technology for bioremediation of metals. Lewis Publishers, Boca Raton, pp1–12

Sone Y, Nakamura R, Pan-Hou H, Sato MH, Itoh T, Kiyono M (2013) Increase methylmercury accumulation in *Arabidopsis thaliana* expressing bacterial broad-spectrum mercury transporter *MerE*. AMB Express 3:52

Song W-Y, Sohn EJ, Martinoia E, Lee YJ, Yang Y-Y, Jasinski M, Forestier C, Hwang I, Lee Y (2003) Engineering tolerance and accumulation of lead and cadmium in transgenic plants. Nat Biotechnol 21:914–919

Song WY, Park J, Mendoza-Cózatl DG, Suter-Grotemeyer M, Shim D, Hörtensteiner S, Geisler M, Weder B, Rea PA, Rentsch D, Schroederc JI, Youngsook L, Enrico M (2010) Arsenic tolerance in *Arabidopsis* is mediated by two ABCC-type phytochelatin transporters. Proc Natl Acad Sci U S A 107:21187–21192

Sriprang R, Hayashi M, Yamashita M, Ono H, Saeki K, Murooka Y (2002) A novel bioremediation system for heavy metals using the symbiosis between leguminous plant and genetically engineered rhizobia. J Biotechnol 99:279–293

Stork A, Jury WA, Frankenberger WT (1999) Accelerated volatilization rates of selenium from different soils. Biol Trace Elem Res 69:217–234

Summers AO (1986) Organization, expression, and evolution of genes for mercury resistance. Annu Rev Microbiol 40:607–634

Sunkar R, Kaplan B, Bouché N, Arazi T, Dolev D, Talke IN, Maathuis FJM, Sanders D, Bouchez D, Fromm H (2000) Expression of a truncated tobacco NtCBP4 channel in transgenic plants and disruption of the homologous *Arabidopsis* CNGC1 gene confer Pb2Ctolerance. Plant J 24:533–542

Terry N, Zayed AM, De Souza MP, Tarun AS (2000) Selenium in higher plants. Annu Rev Plant Physiol Plant Mol Biol 51:401–432

Thayer JS (2002) Biological methylation of less-studied elements. Appl Organomet Chem 16:677–691

Thomas JC, Davies EC, Malick FK, Endreszl C, Williams CR, Abbas M, Petrella S, Swisher K, Perron M, Edwards R, Ostenkowski P, Urbanczyk N, Wiesend WN, Murray KS (2003) Yeast metallothionein in transgenic tobacco promotes Cu uptake from contaminated soils. Biotechnol Prog 19:273–280

Tomar PR, Dixit AR, Jaiwal PK, Dhankher OP (2016) Engineered plants for heavy metals and metalloids tolerance. In: Jaiwal PK et al (eds) Genetic manipulation in plants for mitigation of climate change. Springer, New Delhi, pp 143–168

Tripathi M et al (2005) Isolation and functional characterization of siderophore-producing lead- and cadmium-resistant *Pseudomonas putida* KNP9. Curr Microbiol 50:233–237

Turchi A, Tamantini I, Camussi AM, Racchi ML (2012) Expression of a metallothionein A1 gene of *Pisum sativum* in white poplar enhances tolerance and accumulation of zinc and copper. Plant Sci 183:50–56

Ullah A, Heng S, Munis MFH, Fahad S, Yang X (2015) Phytoremediation of heavy metals assisted by plant growth promoting (PGP) bacteria: a review. Environ Exp Bot 117:28–40

van der Ent A, Baker AJM, Reeves RD, Pollard AJ, Schat H (2013) Hyperaccumulators of metal and metalloid trace elements: facts and fiction. Plant Soil 362:319–334

van der Zaal BJ, Neuteboom LW, Pinas JE, Chardonnens AN, Schat H, Verkleij JA, Hooykaas PJ (1999) Overexpression of a novel *Arabidopsis* gene related to putative zinc-transporter genes from animals can lead to enhanced zinc resistance and accumulation. Plant Physiol 119:1047–1055 van Hoof NALM, Hassinen VH, Hakvoort HWJ, Ballintijn KF, Schat H, Verkleij JAC, Ernst WHO, Karenlampi SO, Tervahauta AI (2001) Enhanced copper tolerance in *Silene vulgaris* (Moench) Garcke populations from copper mines is associated with increased transcript levels of a 2b-type metallothionein gene. Plant Physiol 126:1519–1526

Van Huysen T, Abdel-Ghany S, Hale KL, LeDuc D, Terry N, Pilon-Smits EAH (2003) Overexpression of cystathionine-γ-synthase in Indian mustard enhances selenium volatilization. Planta 218:71–78

Van Huysen T, Terry N, Pilon-Smits EAH (2004) Exploring the selenium phytoremediation potential of transgenic *Brassica juncea* overexpressing ATP sulfurylase or cystathionine γ-synthase. Int J Phytoremediation 6:111–118

Venkatesh NM, Vedaraman N (2012) Remediation of soil contaminated with copper using rhamnolipids produced from *Pseudomonas aeruginosa* MTCC 2297 using waste frying rice bran oil. Ann Microbiol 62:85–91

Vijayaraghavan K, Yun Y-S (2008) Bacterial biosorbents and biosorption. Biotechnol Adv 26:266–291

Volesky B (2007) Biosorption and me. Water Res 41:4017–4029

Wall JD, Krumholz LR (2006) Uranium reduction. Annu Rev Microbiol 60:149–166

Wang J, Chen C (2006) Biosorption of heavy metals by *Saccharomyces cerevisiae*: a review. Biotechnol Adv 24:427–451

Wang J, Chen C (2009) Biosorbents for heavy metals removal and their future. Biotechnol Adv 27:195–226

Whiting SN, Reeves RD, Baker AJM (2002) Mining, metallophytes and land reclamation. Mining Environ Manag 10:11–16

Whiting SN, Reeves RD, Richards D et al (2004) Research priorities for conservation of metallophyte biodiversity and their potential for restoration and site remediation. Restor Ecol 12:106–116

Wiatrowski HA, Ward PM, Barkay T (2006) Novel reduction of mercury (II) by mercury-sensitive dissimilatory metal reducing bacteria. Environ Sci Technol 40:6690–6696

Wilson B, Pyatt FB (2007) Heavy metal bioaccumulation by the important food plant, *Olea europaea* L. in an ancient metalliferous polluted area of Cyprus. Bull Environ Contam Toxicol 78:390–394

Wilson JR, Leang C, Morby AP, Hobman JL, Brown NL (2000) *Mer* F, a mercury transport protein: different structure but a common mechanism for mercuric ion transport. FEBS Lett 472:78–82

Wu Q, Wang S, Thangavel P, Li Q, Zhong H, Bai J, Qiu R (2011) Phytostabilization potential of *Jatropha curcas* L. in polymetallic acid mine tailings. Int J Phytoremediation 13:788–804

Wuana RA, Okieimen FE (2011) Heavy metals in contaminated soil: a review of sources, chemistry, risks and best available strategies for bioremediation. ISRN Ecol 20. doi:10.5402/2011/402647

Xu J, Tian YS, Peng RH, Xiong AS, Zhu B, Hou XL, Yao QH (2010) Cyanobacteria MT gene SmtA enhances zinc tolerance in *Arabidopsis*. Mol Biol Rep 37:1105–1110

Yang Z, Wu Y, Li Y, Ling HQ, Chu C (2009) OsMT1a, a type 1 metallothionein, plays the pivotal role in zinc homeostasis and drought tolerance in rice. Plant Mol Biol 70:219–229

Ye ZH, Baker AJM, Wong MH (2002) Problems of toxicities. In: Wong MH, Bradshaw AD (eds) The restoration and management of derelict land: modern approaches. World Scientific, Hackensack, pp 66–79

Yousfi I, Bermond A (2001) Physical and chemical approach to assess the effectiveness of several amendments used for in situ remediation of trace metals-contaminated soils by adding solid phases. In: Iskandar IK (ed) Environmental restoration of metals-contaminated soils. Lewis Publishers, Baca Raton, pp 3–20

Zacchini M, Pietrini F, Mugnozza GS, Iori V, Pietrosanti L, Massacci A (2009) Metal tolerance, accumulation and translocation in poplar and willow clones treated with cadmium in hydroponics. Water Air Soil Pollut 197:23–34

Zayed A, Gowthaman S, Terry N (1998) Phytoaccumulation of trace elements by wetland plants: I. Duckweed. J Environ Qual 27:715–721

Zhang YQ, Frankenberger WT (1999) Effects of soil moisture, depth, and organic amendments on selenium volatilization. J Environ Qual 28:1321–1326

Zhang J, Zhang M, Tian S, Lu L, Shohag MJI, Yang X (2014) Metallothionein 2 (*SaMT2*) from *Sedum alfredii* Hance confers increased Cd tolerance and accumulation in yeast and tobacco. PLoS One 9(7):e102750

Zhao FJ, Ma JF, Meharg AA, McGrath SP (2009) Arsenic uptake and metabolism in plants. New Phytol 181:777–794

Zhou W, Zhang Y, Ding X, Liu Y, Shen F, Zhang X, Deng S, Xiao H, Yang G, Peng H (2012) Magnetotactic bacteria: promising biosorbents for heavy metals. Appl Microbiol Biotechnol 95:1097–1104

Zhu YL, Pilon-Smits EAH, Tarun AS, Weber SU, Jouanin L, Terry N (1999a) Cadmium tolerance and accumulation in Indian mustard is enhanced by overexpressing gamma-glutamylcysteine synthetase. Plant Physiol 121:1169–1178

Zhu YL, Pilon-Smits E, Jouanin L, Terry N (1999b) Overexpression of glutathione synthetase in Indian mustard enhances cadmium accumulation and tolerance. Plant Physiol 119:73–80

Zhuang P, Yang Q, Wang H, Shu W (2007) Phytoextraction of heavy metals by eight plant species in the field. Water Air Soil Pollut 184:235–242

Zorrig W, Rabhi M, Ferchichi S, Smaoui A, Abdelly C (2012) Phytodesalination: a solution for salt-affected soils in arid and semi-arid regions. J Arid Land Stud 22:299–302

Genetic Engineering of Poplar: Current Achievements and Future Goals

17

Rakesh Yadav, Neha Yadav, Umesh Goutam, Sandeep Kumar, and Ashok Chaudhury

Abstract

Global biomass demand for industrial applications is ever increasing especially in biofuels and pulp industries. Poplar is likely to have great biological advantages over the other forest trees which include small genome size, large number of species, rapid juvenile growth, ease of clonal propagation, easy recovery of genetic transformants, and available genome draft. Wide and sustainable farming of rapidly growing trees such as poplars may supplement to attain the requirement of renewable resources. This chapter covers the progress in both basic and applied studies in poplar genetic tailoring. Certain advices are given for future direction of the research in poplar genetic tailoring so as to achieve the needs of environmental cleansing system and the timber industries. Emerging new thoughts for designing wood improvement approaches are discussed.

R. Yadav (✉) • A. Chaudhury
Department of Bio and Nano Technology, Guru Jambheshwar University of Science and Technology, Hisar 125 001, Haryana, India
e-mail: ry.yadav01@gmail.com

N. Yadav
Department of Bio and Nano Technology, Guru Jambheshwar University of Science and Technology, Hisar 125 001, Haryana, India

Central Instrumentation Laboratory, Central University of Punjab, Bathinda 151 001, Punjab, India

U. Goutam
Department of Biotechnology, Lovely Professional University, Phagwara 144 402, Punjab, India

S. Kumar
Germplasm Evaluation Division, Biochemistry Laboratory, National Bureau of Plant Genetic Resources (ICAR), New Delhi 110 012, India

S.K. Gahlawat et al. (eds.), *Plant Biotechnology: Recent Advancements and Developments*, DOI 10.1007/978-981-10-4732-9_17

17.1 Introduction

Poplar is a model tree species for being the first woody plant that has been transformed and sequenced. It belongs to *Salicacae* family comprising deciduous forest trees with ~20–35 randomly scattered species especially in the Northern Hemisphere. Various species were largely categorized into three main groups including aspens, poplars, and cottonwoods. The two main species of poplars, which are grown in India, are *Populus deltoides*, *Populus ciliata*, and now recently, hybrids of poplar are gaining importance. The *P. deltoides* is grown in the plains of India, whereas *P. ciliata* is grown mainly in hilly regions as it is adapted to higher altitude of the Himalayan belt and is endemic to this region. *P. deltoides* was first brought to eastern Uttar Pradesh, and now it has become a well-known woody tree in Tarai region and states of Haryana and Punjab. P. are rapidly growing deciduous trees that produce 10–30 m^3/ha/year of timber in a brief tenure of 6–8 years (Yadav et al. 2010).

Poplar species have been the matter of numerous biotechnological research and molecular biology studies owing to their easiness of sexual and asexual reproduction, the competence for interspecific crossability and the accessibility of the reference genome sequence (Kang et al. 2009). In vitro propagation facilitates culturing of clonal trees that hold the desired traits for testing. Numerous in vitro micropropagation systems have been given for poplar and its hybrids, for improving biomass production of these fast growing, better qualities, fiber-containing trees, by screening of elite cultivars or by genetic tailoring.

Recently, transgenic technology gained great advancement in the controlled improvement of plant characteristics and has emerged out to be the most important tool in contemporary molecular breeding. Many reports have included transgenosis and genetic manipulations especially stress-tolerance genes, insect-resistant transgenosis, downregulation of lignin biosynthesis pathway gene, and phytoremediation genes in the poplar genome. This chapter reveals recent progress in genetic engineering of poplar and future directions of biotechnological intervention.

17.2 Current Achievements

17.2.1 Production of Transgenic Poplars

Genetic tailoring has transformed improvement of silviculturally important traits of many economically important forest crops. Unlike the traditional breeding approaches, the transgenic method permit introduction of novel genes from entirely divergent source and broaden the gene pool of the germplasm. It plays a pivotal role in forestry, where, the traditional breeding program requires a long time to select for desired traits. Several methods including *Agrobacterium*-mediated transformation, electroporation, and biolistics can be employed to transmit non-self DNA into the poplar genome (Han et al. 1996). Poplar genetic engineering include, but are not restricted to enhance disease resistance, herbicide tolerance, insect resistance, improving wood quality and fast growth, enhancing the use of poplars in phytoremediation, and tailoring wood quality by revising the lignin biosynthetic pathway.

17.2.1.1 Reporter and Marker Genes Introduced Through Genetic Engineering in Poplar

The significance of *Agrobacterium tumefaciens* as a gene transfer approach was first known when it was discovered that the crown galls were in fact formed as a consequence of the successful transfer and integration of genes from Ti-plasmid of *Agrobacterium* species into the nuclear genome of the plant cells (Chilton et al. 1977). An effective technique for introducing cloned genes into plant cells was given by Herrera-Estrella et al. (1983) and Zambryski et al. (1983). It was first used for the production of morphologically normal plants that contained and expressed foreign gene(s).

Poplar genetic engineering has been reported back way by Parsons et al. in 1986. They have invented the reproducible transformation system in hybrid *Populus* (*P. trichocarpa* x *P. deltoides*) using two wild-type *A. tumefaciens* strains, A281 and A348, whereas Fillaiti et al. (1987) first reported the efficient retrieval of genetically tailored poplar plants coding a bacterial *aroA* gene for expression of herbicide tolerance. Subsequently, many reports including the easy transformation of poplar by means of *A. tumefaciens* have been published. Brasileiro et al. (1991) developed a gene transfer protocol in hybrid poplar by inoculating intermodal sections with *Agrobacterium* suspension having a disarmed strain that harbor the suitable genes *neo* and *gus* in a binary vector pBI 121 in addition to an oncogenic strain that stimulates shoot differentiation. Poplar plants screened for kanamycin resistance did not harbor oncogenic T-DNA were expressing *neo* and *gus* genes. Leple et al. (1992) performed transformation on the leaf and stem explants of a hybrid poplar (*P. tremula* x *P. alba*) using the nopaline disarmed strain C58/plasmid pMP90 with binary vector plasmid pBI 121 to test three other vectors. One possessing *npt-II* gene fused to the *CaMV19S* promoter for kanamycin resistance, whereas, the other two vectors carried an additional *npt-II* gene under the control of *CaMV35S* promoter. Devantier et al. (1993) described a transient gene expression system for poplar using microprojectile-mediated DNA delivery in cell suspension cultures. *GUS* assay revealed 58.9% of cells coding the β-glucuronidase gene positioned in the nuclear genome or its vicinity and the remaining cells had microparticles in the cytoplasm.

Transformation of leaf segments of hybrid poplar, *P. kitakarniensis* (*P. sieboldii* x *P. grandidentata*) by employing the disarmed *A. tumefaciens* strain LBA 4404, harboring the binary vector pBI 121, has been reported which carries the *npt-II* and *gus* genes under the control of the NOS and *CaMV35S* promoters. Expression of a peroxidase isoenzyme was altered by incorporating the *prx Al* gene, encoding the anionic peroxidase isoenzyme prx Al. Transgenic poplars exhibit a 10-fold increase in total peroxidase activity as compared to the control plants (Kajita et al. 1994). Similarly, transgenic quaking aspen (*P. tremuloides*) has given amplification of *npt-II* and *gus* gene segments from genome of regenerated transgenic plantlets using PCR; further, incorporation of these genes into the nuclear genome was verified by

Southern hybridization technique (Tsai et al. 1994). Genetically modified *P. nigra* for expressing human proinsulin (PINS) gene into leaf explants has been verified by *npt-II* assay, PCR using '*npt-II*' specific primers and Southern hybridization (Kim et al. 1995).

Biolistic transformation of callus of hybrid poplar (*P. trichocarpa* x *P. deltoides*) has shown higher level of *gus* expression that is driven by the maize ubiquitin promoter instead of the *CaMV35S* promoter (Wang et al. 1995). Mijnsbrugge et al. (1996) produced transgenic poplar (*P.tremula* x *P. alba*) harboring *gus* gene having *Arabidopsis* S-adenosyl-L- methionine synthetase (sam-s) gene promoter P-sam-1. Histochemical analyses of the transgenics have shown *GUS* activity basically in cortex and phloem tissues all over the plant and in root tips. Fluorometric assays exhibited high *GUS* activity in the tissues outside (cortex, phloem and cork) as compared to those inside (xylem and pith) of the cambial layer mRNA gel blot analysis confirmed a high level of transcript of the endogenous sam-s gene in tissues inside and outside the cambial layer.

In addition to *A. tumefaciens*, transformation using wild-type *A. rhizogenes* has also been reported in cottonwoods (*Populus* spp.). Hairy roots were formed in diverse clones including *P. trichocarpa*, *P. deltoides*, *P. trichocarpa* x *P. deltoides* hybrids and *P. alba* x *P. grandidentata*, else some of these were obstinate to plant regeneration or genetic modification. The frequency of hairy roots emergence ranging from 17 to 92%, and nearly all hairy roots could be regenerated into transgenic plants (Han et al. 1997a). Further, Han et al. (1997b) introduced matrix attachment region ('MAR') fragments derived from a tobacco gene to increase the frequency of *Agrobacterium*-mediated modification in some poplar hybrids. A binary vector harboring a *gus* reporter gene (uid A) having an intron and *npt-II* gene was altered to carry flanking 'MAR' elements in between the T-DNA borders. Vectors with or without 'MARs' were then used to transform poplar clones, *P. tremula* x *P. alba* and *P. trichocarpa* x *P. deltoides*. 'MARs' enhanced '*gus*' gene expression by about tenfold in the two hybrid poplar clones. Frequency of kanamycin-resistant poplar shoots recovered was also higher using vectors containing 'MARs'.

A receptive, precise and rapid method to verify transgene expression by using RT-PCR in poplar tree (*P. alba*) protoplasts has been designed by employing total RNA as template. Expression of cDNA encoding goat growth hormone (somatotropin) was verified at the mRNA level about a month after electroporation event meant for DNA transfer (JingBo et al. 1997). *Populus* hybrid transformed with *A. tumefaciens* LBA 4404 strain harboring the construct pm-gfp-ER5, which included *gfp*, a green fluorescent protein gene from the jelly fish *Aequorea Victoria*, has shown an increase of at least 10 times fluorescence of the transformed shoot extract compared to the untransformed controls, indicating the introduction and stable expression of pm-gfp-ER5 in the poplar genome (Main et al. 1998). Expression of the gfp gene in biolistically transformed *P. tremula* x *P. alba* hybrid confirmed its stable integration into plant genomes (Tian et al. 1999).

Son and Hyun (1998) gave the protocol for genetic transformation of poplar hybrid (*P. nigra* x *P. maximowiczii*) by *A. tumefaciens* containing *gus* gene. Fluorometric and histochemical *GUS* assays verified that vector pEHA 101 is more efficient in regeneration of transformants than LBA 4404/ pBI 121. Moreover, a treatment with methylation inhibitor 5-azacytidine promises a hike in transgene expression level from 5.7 to 26.7%. In same year, a highly efficient protocol for genetic transformation and plant regeneration through shoot-bud induction in root culture in aspen (*P. tremula*) had been reported. Aspen stem fragments were transformed by employing wild-type *A. rhizogenes* (LBA 9402) with the binary p35S*gus*INT plasmid, witnessed high levels of transitory *GUS* expression and emerged out numerous adventitious roots (Tzfira et al. 1998).

A developmentally regulated pattern of expression which was driven by poplar PAL (phenylalanine ammonia-lyase) promoters in transgenic poplar has been thoroughly studied by Gray et al. (1999). Two 'PAL' genes, 'PAL 1' and 'PAL 2', from a *P. trichocarpa* x *P. deltoides* F1 hybrid, have been cloned previously. In transgenic poplar, 'PAL2-*gus*' fusion regulated the highest degree of expression in young leaves, roots and stems. In young stems and leaves, high *GUS* activity was found in subepidermal or epidermal cells and in primary phloem and xylem fibers. *GUS* activity was poor in woody stems and was low or absent in developing secondary xylem. Transgenic Lombardy poplar (*P. nigra*), having the rice gene for a homeodomain protein (*oshl*) and a gene for neomycin phosphotransferase-II (*npt-II*), triggered morphological anomalies in the stems and leaves of the newly regenerated plants which advocates that '*oshl*' can function as a regulator of morphogenesis in transgenic poplar, as it does in *Arabidopsis*, transgenic rice and tobacco plants (Mohri et al. 1999).

Agrobacterium-mediated transformation procedure effective on diverse cottonwood hybrids (genus *Populus*) was optimized by Han et al. (2000). An attempt was made for *Agrobacterium*-mediated transformation of *npt-II* and *gus* genes into 11 hybrid cottonwood cultivars and one *P. deltoides* cultivar using kanamycin as selectable marker. Shani et al. (2000) transformed aspen plants (*P. tremula*) with a bacterial β-glucuronidase gene (*uidA*) under the regulation of a promoter derived from *A. thaliana* elongation-specific endo-1,4-β-glucanase (cel1). The transformants have expressed stable, high level of β-glucuronidase (*GUS*) activity in young, fast growing tissues. A transformation protocol was optimized for two elite aspen hybrid clones (*P. canescens* x *P. grandidentata* and *P. tremuloides* x *P. davidiana*) by Dai et al. (2003). Two hybrid poplar clones were co-cultivated by *A. tumefaciens* strain EHA 105 harboring the binary Ti-plasmid pB1121 having the *uidA* coding for β-glucuronidase (*GUS*) and the *npt-II*. Introduction of transgene was verified by *GUS* assays, polymerase chain reaction and Southern blot analysis. The black cottonwood namely Nisqually-1 (*P.trichocarpa*), whose genome was sequenced first, was successfully transformed by *Agrobacterium* strain carrying a binary plasmid vector harboring genes for both and *npt-II gus*. Histochemical assays analyzed expression of the *gus* gene in stem, leaf and root parts of transgenic plants. PCR verified the integration of both reporter genes and selectable marker in all lines which were showing for β-glucuronidase activity (Ma et al. 2004).

The capability of double-stranded RNA interference (RNAi) tools for suppression of *gus* gene expression in poplar has been described by Meyer et al. (2004). The functionality of vectors producing RNAs those were capable of duplex formation with specific sequences for *gus* gene was tested, both in transitory assays by tailoring protoplasts with the RNAi structure and in stably modified poplar plants showing *GUS* expression. *Agrobacterium*-mediated genetic transformation of the *GUS* expressing plants with a *gus*-specific RNAi construct exhibited a proficient down regulation of the reporter gene. White poplar (*P.alba* L.) transformed via disarmed *A. tumefaciens* LBA 4404/p35S*gus*TNT/pCAMBIA 2301 strain has shown *gus* gene expression 2 months after the co-cultivation experiment (Sanchez et al. 2004). Transformed petiolar tissue of Himalayan poplar (*P. ciliata*) has stable expression of *gus* gene (β-glucuronidase) in binary vector pBI 121 along with *npt-II* marker (Thakur et al. 2005). Affluent plastids transformation with spectinomycin resistance gene and *gfp* gene in *P.alba* result in site-specific incorporation of *gfp* genes and attaining of a high homoplastomic condition (Okumura et al. 2006). Heterologous promoters in transgenic *P. tremula* x *P. alba* plants have been extensively described by Studart et al. (2006). The profile and extent of expression of *gus* reporter gene controlled by six heterologous promoters has been confirmed by histochemical staining of stem, root and leaf tissues. Yevtushenko and Misra (2010) described the stable transformation of a hybrid poplar *P. nigra* L. x *P. maximowiczii* A. Henry (genotype NM6; recalcitrant to genetic transformation) to express the β-glucuronidase (*GUS*) gene. Besides this, a novel method for in situ bud tailoring of *P. cathayana* Rehd via *Agrobacterium* was optimized by Yang et al. 2010. Later on, an effective regeneration and transformation protocol was developed for the production of two hybrid aspen clones, namely *P. alba* x *P. berolinensis* and *P. davidiana* x *P. bolleana*. A transformation vector tailored to encode the reporter gene *uidA*, expressing *GUS*, by the help of *CaMV35S* promoter, was employed to find transformation result at initial phase of plant regeneration and to ameliorate the factors that may influence plant transformation efficiency. Transgenic plants were generally produced by employing this system in less than 4 weeks post co-cultivation. The morphology and growth features of transgenic plants were similar to the control plants, and the transgenes could be transmissible by vegetative propagation, as verified by PCR, RT–PCR, Southern blotting and β-glucuronidase staining analysis (Wang et al. 2011).

Zheng et al. (2012) have reported an effective approach for transient *GUS* gene expression that works better on diverse plant species, including herbaceous plants (such as *Arabidopsis* and tobacco) and trees (such as birch, cork, poplar, tamarisk, willow, and aralia), and support that it may be applicable widely for gene transient expression studies in plants. Various factors, viz. co-cultivation time, acetosyringone concentration and *Agrobacterium* cell density, which considerably affected transient expression rate were described thoroughly. Transgenic poplar lines harboring constructs carrying the *GUS* gene driven by truncated Pt-*RbcS* promoters revealed distinctive expression patterns for five different promoter constructs. The Pt-*RbcS* promoter was expressed preferentially in photosynthetic tissues such as leaves and stems (Wang et al. 2013). Thus, so far, success in genetic transformation of poplar has been limited to some of hybrids and several species of poplar other than *P. deltoides*.

17.2.1.2 Isolation and Functional Characterization of New Genes via Genetic Engineering

Enhancer trapping is a beneficial approach in functional characterization of promoters and isolation of novel genes which are driving tissue-specific expression in tree species. Employing this method, a new gene ET304 has been isolated from *Populus* (aspens and cottonwoods). Both putative ET304 promoter and ET304 enhancer trap line fused to a *GUS* reporter gene exhibited a strong root-predominant expression profile and drove the expression to root-specific tissue and cell types. *GUS* activity was found in root apical meristem, lateral root primordia, elongation zone and cortex. ET304 promoter sequence harbors a canonic auxin response element (*AuxRE*) present upstream to the enhancer trap insertion site. In a coordinated lateral root triggering system ET304 promoter showed an auxin-responsive expression in fresh emerging lateral roots of both *Arabidopsis* and poplar (Filichkin et al. 2006).

17.2.1.3 Genetic Engineering for Disease Resistance

Populus spp. is vulnerable to a number of fungal, viral and bacterial diseases which cause tremendous yield loss and deterioration of wood quality. Various antifungal, antibacterial and antiviral genes have been incorporated into poplar genome to develop transgenic plants resistant to theses pathogens.

Chalcone synthase (*chsA*) gene, *CaMV35S* promoter with a double enhancer and the *gus* and *npt-II* marker genes from *Petunia naringenin* has been introduced in hybrid poplar (*P. tremula* x *P. alba*) for improving resistance to pathogens by increasing flavonoid content. Several transgenic lines furnished strong *chsA* overexpression and were characterized by flavonoids found in both peripheral and cortical tissues of the stem (Nicolescu et al. 1996). Transgenic hybrid poplar (*P. deltoides* x *P. nigra* OGY) containing two genes encoding antimicrobial peptides *ESF12* and *AcAMP12* has been produced by using *A. tumefaciens* LBA 4404 strain (Catranis et al. 1997). Peptide AcAMP12 is an analogue of Ac-AMP1 which is among the smallest chitin binding proteins. Synthetic peptide ESF12 resembles the amphipathic alpha-helix found in magainins. Transformed aspen (*P. tremula*) with octopine, nopaline and succinaminopine-type *A. tumefaciens* carrying an '*ita*' (inhibition of tumorigenicity of *A. tumefaciens*) gene were produced by Ovadis et al. (1998). Resistance against poplar fungal pathogen, *Septoria musiva*, in hybrid poplar clones has been developed by transferring genes, *AcAMP12* and *ESF12*, encoding antimicrobial peptides (Liang et al. 2002). White poplar (*P. alba*) has shown enhancement in disease resistance by incorporating a cDNA insert expressing stilbene synthase gene from grapevine (*Vitis vinifera*). Expression of this gene led to synthesis of two new compounds as the trans-and cis-isomers of resveratrol-3-glucoside (piceid) as detected by HPLC, UV-spectrometry, enzymatic hydrolysis and electrospray mass spectrometry (Giorcelli et al. 2004).

An endochitinase gene (*ech42*) from the biocontrol fungus *T. harzianum* was incorporated into hybrid poplar (*P. nigra* x *P. maximowiczii*) and black spruce (*Picea mariana*) via *Agrobacterium*-mediated genetic transformation approach. Six transgenic poplar lines and fifteen black spruce lines were recovered and reported to show an enhanced accumulation of the transcript expressing the recombinant endochitinase gene in all the transgenic plants tested. In vitro assays of transgenic lines using inoculated leaf disks method support an increased resistance to the leaf rust pathogen *Melampsora medusae* (Noel et al. 2005).

17.2.1.4 Genetic Engineering for Insect Resistance

Populus species are critically injured by *Coleopteran* and *Lepidopteran* insect pests that cause defoliation along with remarkable reduction in tree growth. These insect pests include the *Lepidopteran* such as *Hyphantria cunea, Clostera anastomosis* and *Lymantria dispar*, and the *coleopteran* cotton wood leaf beetle viz. *Chrysomela tremulae*, *Chrysomela scripta* and willow leaf beetle – *Plagiodera versicolora.*

Among earlier reports, stable modification of *Populus* species and incorporation of pest resistance gene by electric arc discharge particle acceleration method was reported by McCown et al. (1991). They took three diverse target tissues-protoplast derived cells, nodes and stems and two correlated hybrid genotypes of *Populus* (*P. alba* x *P. grandidentata* 'Crandon' and *P. nigra* 'Betulifolia' x *P. trichocarpa*) for transformation. Four transformed plants of one hybrid genotype, NC5339, containing the gene was very resistant to feeding by two lepidopteran pests (the forest caterpillar *Malacosoma disstria* and the gypsy moth, *Lymantria dispar*). Suspension cultures of *P. alba* x *P. grandidentata* cv. 'Crandon' were transformed with vectors containing the maize *Ac* transposable element and a truncated insect toxin gene from *Bacillus thuringiensis* (*Bt*). Transgenic plants were regenerated by culturing the transformed callus on medium containing 0.11–27.0 μM TDZ showing insect resistance (Howe et al. 1994).

Tailored *P. deltoides* has been produced which retain '*Bt* toxin' gene. PCR analysis of kanamycin-resistant transformant revealed that the *Bt* gene had been stably integrated into the nuclear genome (Chen et al. 1995). Transgenic plants of poplar hybrid 741{*P. alba* x (*P. tremula* var. davidiana x *P. simoii*) x *P. tomentosa*} has also been produced by incorporating *Bt.* gene. Transgenic plantlets containing the modified *Bt.* toxin gene were identified by PCR testing (Zheng et al. 1995). Transgenic plants of *P. nigra* have shown an expression of *Bt.* gene (Chen et al. 1996). Genetically modified hybrid poplar (*P. tremula* x *P. tremuloides*) has conquered expression of a proteinase inhibitor (*OCI*) and *cry3A* transgene to improve resistance against *Chrysomela tremulae*. Transgenic poplar leaves were fed on by the insects which has shown decreased larval growth, enhanced mortality and modified development as compared to the control (Cornu et al. 1996). Insect-resistant poplar (*P. nigra*) plants have been bioassayed with *Apocheima cinerarium* and *Lymantria dispar* on the leaves of transformed clones illustrated high levels of insecticidal activity by Wang et al. (1996) whereas bioassayed for field resistance to *Apocheima cinerarium* by Sala et al. (2000). The molecular analysis demonstrated integration and expression of the *Bt* gene. Insect resistance could also be acquired by introducing the potato proteinase inhibitor-II (*pin2*) transgene in hybrid poplar clone, *Populus* x *euramericana* 'Ogy' (Heuchelin et al. 1997) and the clone 'Hansen' (*P. alba* x *P. grandidentata*; Klopfenstein et al. 1997). Rao et al. (2000) transformed hybrid poplar (*P. deltoides* x *P. simonii*) plants and regenerated insect-resistant transgenic plants which have integrated *Bt* gene and *npt-II* gene.

Delledonne et al. (2001) produced transgenic *P. alba* L. plants encoding a new *A. thaliana* cysteine proteinase inhibitor (*Atcys*) gene. The incorporation and expression of the cysteine proteinase inhibitor (Atcys) gene into tree genome was verified by Southern and northern blot techniques. Papain inhibitory activity was found in transgenic poplar lines by means of a particular in vitro assay. These activities were found adequate to restrain most of the digestive proteinase activity of chrysomelidae beetle (*Chrysomela populi* L.) and conferred resistance to *C. populi* larvae on screened transgenic plants. A near correlation between the resistance to poplar leaf beetle and inhibition of papain was found in all confirmed transgenic lines. Results indicated that *Atcys* could be used in breeding programs aimed at the screening of new poplar cultivars resistant to major insect pest.

Transgenic hybrid aspen (*P. tremula* x *P. alba*) plants have been shown for overexpressing a *Ptd PPO1* gene from hybrid poplar (*P. trichocarpa* x *P. deltoides*) for incorporating resistance against forest tent caterpillar (*Malacosoma disstria*). Transgenic plants have revealed a higher poly phenol oxidase enzyme activity, *Ptd PPO1* mRNA levels and PPO protein accumulation. In leaf disc bioassays, forest tent caterpillar larvae feeding on polyphenol oxidase overexpressing transgenics have imparted significantly higher mortality and have reduced average weight gain in contrast to larvae feeding on control leaves (Wang and Constabel 2004).

The efficacy of hybrid triploid poplars (*P. tomentosa* x *P. tomentosa* x *P. bolleana*), genetically tailored with cowpea trypsin inhibitor (*CPTI*) gene in field for 2 years against *Malacosoma disstria*, *Stilpnotia candida* and *Lymantria dispar* have been tested. Foliage of transgenic poplars educed a rise in larval mortality rate and a fall in foliage consumption, feces excretion, wet weight gains, deposited pupa number and pupae weight. PCR and Southern blotting analysis verified the stable integration of *CPTI* gene, while proteinase inhibitory assays revealed its elevated expression in the two seasons of transgenic trees (Zhang et al. 2005c).

Introduction of chimeric gene for spider insecticidal peptide and *Bt* gene in *P. xeuramericana* cv. "114/69" have led to gain sensitivity of cefazolin sodium cefamezin and kanamycin, along with insecticidal activity (Zhang et al. 2005b). Spider neurotoxin gene also has prospective as insecticidal properties into hybrid poplar *P. nigra* x *P. simonii*. The fusion of *BGT* gene from the spider, *Atrax robustus*, and the C terminal of *cry1A(b)* gene from *Bacillus thuringiensis* integrated in the genome of hybrid poplar have affected negatively to the growth rate of *L. dispar* fed on the transgenics leaves (Lin et al. 2006). Similarly, a fusion of insect-resistant genes *Bt. cryl Ac* gene and arrowhead proteinase inhibitor (*API*) gene could effectively control insects to hybrid triploid of Chinese white poplar (Yang et al. 2006). Transgenic poplars (*P. alba* x *P. glandulosa*) containing the *cry3A* gene were found to be toxic against a stem borer larvae, *Anoplophora glabripennis* (Zhang et al. 2006a).The content of Cry1Ac toxic protein dynamically altered in transgenic poplar. The content initially rise during the annual growth cycle, then fall in the long and the short branches of the crown and in the root system, peaking in August, and eventually the protein did not accumulate. At peak occurrence of the *Hyphantria cunea* insects, its density of in the transgenic forest was radically lower than in the wild poplar forest (Zhang et al. 2016).

17.2.1.5 Genetic Engineering for Herbicide Tolerance in Poplar

Poplars are cultivated as agro-forestry crops and the major reduction in their productivity is because of the weed growth that competes for nutrition and other resources. Thus, protocol has been designed to particularly hinder the weed growth without influencing the crop plant. One of the techniques is to isolate genes that confer resistance to such herbicides and incorporate them into plants so that herbicide application can suppress the weeds without affecting the growth of crop plant. A tailoring of poplar species using *Agrobacterium*-mediated gene transfer technique was made by Fillaiti et al. (1987) by introducing marker gene (*npt-II*) and herbicide resistance gene (*aroA*).

Devillard (1991) produced transgenic aspen [*P. tremula* x *P. alba* (*P. canescens*)] hybrids and poplar [*P. trichocarpa* x *P. deltoides* (*P. interamericana*)] hybrids for integrating a gene which confers resistance to Basta (glufosinate). Brasileiro et al. (1992) transformed *P. tremula* x *P. alba* hybrid with mutant acetolactate synthase gene '*crs-1*', that confers resistance to sulfonylurea/chlorsulfuron. Transgenic were produced by electroporation of protoplasts with a decay-wave electroporation machine. Transformants were selected on phosphinothricin medium with the phosphinothricin acetyl transferase (*PAT*) gene and on chlorsulfuron medium with the acetolactate synthase (*ALS*) gene (Chupeau et al. 1994). Genetically modified hybrid poplar clone NC-5339 (*P. alba* x *P. grandidentata* cv. Crandon) for glyphosate tolerance with mutant, *aroA* gene along with *npt-II* gene have acquired glyphosate tolerance by a combinatorial impact both high level expression of the *aroA* gene under the control of the *CaMV35S* promoter and transport of mutant EPSP synthase into the chloroplasts (Donahue et al. 1994).

Transgenic plants of poplar (*P. tremula* x *P. alba* and *P. trichocarpa* x *P. deltoides*) hybrids containing a herbicide tolerance gene were tested under field condition for the herbicide Basta (glufosinate ammonium) resistance (Paques et al. 1995). The *bar* gene from *Streptomyces hygroscopicus* transferred to white poplar plants (*P. alba*) have been revealed for complete resistant against herbicide phosphinothricin-Basta; (Confalonieri et al. 2000). Glyphosate tolerance gene (*aroA*) introduced into hybrid poplars (*P. tremula* x *P. alba* and *P. balsamifera* x *P. deltoides*) have indicated little adverse effects on growth along with high levels of tolerance to 'Roundup' (glyphosate) herbicide (Meilan et al. 2000).

Genetically transformed *P. alba* plants have been regenerated after co-cultivation of stem-segments-derived calli with *A. tumefaciens* strain GV3101 harboring genes for resistance to the herbicide bialophos (*bar*) and β-glucuronidase. No chimeric transgenics were obtained in this transformation experiment and the morphology of regenerated plants resembled that of the original parental genotype (Igasaki et al. 2002). A combined integration of *CP4* and *gox* gene into hybrid poplars prevent foliar damage or reduced growth upon higher level of Round up® spray beyond normal commercial rates. Damage control was linked with expression of the *CP4* gene. Growth of the transformant lines integrating only *CP4* gene was significantly higher than double transgenics and conferred low damage upon glyphosate treatment (Meilan et al. 2002).

17.2.1.6 Genes for Abiotic Stress Tolerance Transferred in Poplar

Genetic engineering of poplars also steers toward eternal solution against global problem of various abiotic stresses such as drought tolerance, salinity tolerance and various oxidative stresses, etc. Overexpression of glutathione reductase (*GR*) or glutathione synthetase (*GS*) in the cytosol in transgenic hybrid poplar, *P. tremula* x *P. alba*, has no effect on glutathione levels, but chloroplast *GS* expression lead to doubling of leaf glutathione and an increase in reduction state. The transgenics conferring high level of chloroplast *GR* expression also exhibit high photo-inhibition resistance along with low oxidative stress (Foyer et al. 1995). A water stress responsive protein (*BspA*) expressing in transgenic aspen (*P. tremula*) shows low osmotic potential than controls which make them tolerant against drought and salt-stress. Similarly, transgenic '*BspA*' aspen plants grow normally under stress because of upregulation of nutrient transport and restoration of osmotic balance (Wang et al. 1999). An exogenous gene *BetA* integrated in *P. simonii* x *P. nigra* hybrid enhances the salt resistance in transgenic lines (Yang et al. 2001). The transformation of *mtlD*/*gutD* divalent genes to *P. deltoides* and *P. cathayana* and its stable integration into the genome led to coding their characteristic proteins. Salt tolerance level of such transgenics was enhanced many-fold (Fan et al. 2002). Likewise, tailored *P. tomentosa* for containing antisense phospholipase D gamma (*PLD* gamma) gene bear high salt strength of the soil (Liu et al. 2002).

In hydroponic and in vitro culture, transgenic *P. tomentosa* having the *mtlD* gene (coding mannitol-1-phosphate dehydrogenase) conferred high salt tolerance. The transgenic buds rooted usually on MS medium containing 50 mM NaCl; however, wild-type buds failed to root. Although, height was reduced by about 50% in the transgenic plants compared with the wild-type plants in the absence of salt, yet, relative growth rate was much less influenced by salt stress in transgenic plants (Hu et al. 2005). Transgenic *P. deltoides* for containing anti-*PLD* gamma gene and chitinase harbor high salt tolerance and disease resistance (Zou et al. 2006). Liu et al. (2008) incorporated the apple *SPDS* gene (spermidine synthase) into *P. tomentosa*, and four transgenic lines were verified by PCR, single copy integration verified by Southern blotting, and the comparative expression levels (one to three) was revealed by quantitative real-time PCR (qPCR). Liu et al. (2008) introduced the *PLD*/*AtNHX1* gene (phospholipase D/*Arabidopsis thaliana* Na^+/H^+ antiporter gene) into *P. tomentosa*, and transgenic lines were having enhanced salt tolerance. Zhang et al. (2008) increased expression of the *AtPLDa* gene in *P. tomentosa* to stably improve salt and drought tolerance. A capable approach to enhance salinity tolerance is the cloning and characterization of transcription factor genes which control a number of upstream or downstream genes, thus triggering several genes that function together in improving tolerance to multiple stresses. *DREB1C* gene (a dehydration-responsive element binding transcription factor) integrated into an elite hybrid *P.* x *euramericana* cv. Nanlin 895 has a higher tolerance to salt and drought than control plants (Yang et al. 2009).Variation in the transcriptomes of the transgenic poplar line conferring five exogenous, stress-related genes show improved resistance to multiple stresses including salt, drought, waterlogging, and insect feeding. The non-transgenic lines were analyzed employing high-throughput transcriptome sequencing methods to reveal the utility of the differentially expressed genes by employing numerous functional annotation approaches (Zhang et al. 2014).

Recently, a *GSK3*/shaggy-like kinase (*AtGSK1*) was known to link with various signal transduction pathways particularly in the control of salt and drought tolerance. *AtGSK1* gene, from *A. thaliana* introduced to a hybrid poplar (*P. alba* x *P. tremula* var. grandulosa), conferred salt and drought tolerance in the transgenic plants. While culturing on media, the transgenic plantlets emerged vigorously proliferating roots even on media containing 125 mM NaCl and callus on media containing 150 mM NaCl. The transgenic poplars cultured in pots with NaCl grow much faster than the non-transgenic poplars because of stomatal conductance and higher photosynthetic rates and evaporation rates under stress. The entire concentration of leaf Na^+ enhanced radically in transgenic poplars in critical salt stress (150 mM NaCl) but that of leaf K^+ dropped under the same saline conditions. The relative electrical conductivity and chloride content of the transgenic poplars under saline treatment were weaker than control plants (Han et al. 2013). Overexpression of PtSOS2, salt overly sensitive 2 (SOS2) gene, performs an important function in enhancing the salinity tolerance of poplars, lowering the harm to membrane structures and increasing antioxidative enzyme regulation and osmotic regulation under saline conditions (Zhou et al. 2014).

Low temperature is amongst the major abiotic stresses liable for lowered contention, production and biomass growth in various crops, including forest trees. Integration of *ThCAP* gene is responsive to several stress conditions along with organ-specific expression. Transgenic *Populus* (*P. davidiana* x *P. bolleana*) hybrid for *ThCAP* translates under control of the *CaMV35S* promoter, conferring higher tolerance to frosting temperature than control seedlings. Therefore, *ThCAP* plays a pivotal role in frost tolerance (Guo et al. 2009). C-repeat binding factors (CBFs) perform an important function during abiotic stresses. The PeCBF4a, a member of CBF gene family, was isolated from *P. Euphratica* after induced expression of the gene by dehydration, salinity and low temperature. The gene was introduced to overexpress in *Populus tomentosa*. The photosynthetic rate (Pn) of 2 months old transgenic lines was enhanced 34.7–165.7% under abiotic stresses along with the instantaneous water use efficiency (iWUE) which was increased 48.9–103.7%, whereas, the non-photochemical quenching coefficient (NPQ) mean was dropped by 12.02–23.64%. The photochemical quenching (qP) index rise 8.75–22.31% along with increased superoxide dismutase (SOD) activities and lowered malondialdehyde (MDA) levels. Higher accumulation levels of proline and sugars were found in transgenics over controls. Overexpression of PeCBF4a gene also caused dwarfed phenotypes (Tian et al. 2016).

17.2.1.7 Genetic Engineering for Modifying Wood Quality and Fast Growth in Poplars

Fast development in transgenic technology promises to devise genetic tailoring tools that alter economically desirable traits related to yield and growth in *Populus* sp. In early days, root induction on cuttings of hybrid poplar lines (*P. deltoides* x *P. nigra* and *P. nigra* x *P. maximowiczii*) by transforming with T-DNA of *A. rhizogenes* was attempted with the objective for enhancing biomass accumulation by genetically tailoring hybrid poplar root systems but this failed (Charest et al. 1992). Later, Schwartzenberg et al. (1994) developed transgenic poplar (*P. tremula* x *P. alba*) with the agrobacterial *ipt* gene for rise in innate cytokinin level. The *ipt*-transformed stem segments regenerate calli which emerged out many shoot buds without the supply of external cytokinins. Microcuttings of the *ipt* transformants germinated buds that gave rise to extensively branching shoots with tiny internodes and failed to root. Wood traits and growth profile may be altered in transgenic hybrid aspen (*P. tremula* x *P. trernuloides*) expressing *A. tumefaciens* T-DNA-IAA-biosynthetic '*iaaM*' and '*iaaH*' -genes. Out of several lines regenerated, only few lines were altered in their growth profile such as modification in height, internode elongation, stem diameter growth, degree of apical dominance and leaf enlargement. Two transgenic lines exhibited the noticeable trait difference along with alteration in wood anatomy, remarkable decrease in vessel size, increase in vessel density and alteration in ray development pattern. Results advocate the possibility of influencing wood properties in tree species by regulation of IAA concentration and supply.

Transgenic aspen (*P. tremula*) and its hybrid (*P. tremula* x *P. tremuloides*) with various gene construct using two types of promoters help to determine the influence of the reporter gene '*rolC*', controlled under viral or plant origin promoters, on genetic and morphological expression of different transgenic aspen clones. Transgenic aspen plants harboring the '*rolC*' gene of *A. rhizogenes* under driving of the *CaMV35S* promoter showed a decrease in size and bore smaller leaves, but under the regulation of the light-inducible *rbcS* promoter of potato, there was only a slight reduction in size compared with the untransformed controls (Fladung et al. 1997).

Bacterial glutathione synthetase (*GSS*) or gamma-glutamylcysteine synthetase (gamma-*ECS*) integrated into *P. tremula* x *P. alba* hybrid led to overexpression of bacterial gamma-*ECS*, not of *GSS*, significantly enhanced cysteine, gamma-EC and *GSH* in phloem exudates (Herschbach et al. 1998). A molecular method was designed to enhance glutamine formation in transgenic poplar tree because of overexpression of a pine (*Pinus*) glutamine synthetase (*GS*) gene. A chimeric construct consisting of the *CaMV35S* promoter joined with pine cytosolic 'GS-cDNA' and the nopaline synthetase polyadenylation fragment was cloned into pBin19 for genetic modification of a hybrid poplar clone (INRA 7171-B4, *P. tremula* x *P. alba*) through *A. tumefaciens*. Pine 'GS1' gene expression in poplar was found to be associated with rise in the concentration of total soluble protein and chlorophyll content in foliage of transformed plants. Moreover, the mean net increase in height of *GS*-overexpressing clones was considerably higher than that of control clones. The result advised that the proficiency of nitrogen consumption may be tailored in trees by genetic modification of glutamine biosynthesis gene (Gallardo et al. 1999). A pine cytosolic glutamine synthetase (*GS9*) cDNA under the control of *CaMV35S* promoter stably integrated into hybrid poplar (*P. tremula* x *P. alba*) via *A. tumefaciens* ectopically expressed to increase in total soluble protein levels, GS activity, and chlorophyll level in the leaf tissue (Kirby et al. 1999).

Li et al. (1999a) produced transgenic cottonwoods (*P. deltoides*) containing 'ACC oxidase' (1-amino cyclopropane carboxylic acid, an ethylene releaser). The ethylene release rate was reduced significantly to ~28% of than that of the control plants. Simultaneously, they have also generated transgenic *P. deltoides* plants with antisense *ACC* synthase gene for inhibition of ethylene biosynthesis. The fraction of ethylene released from transgenic poplar was reduced to 22% than non-transformed control plants (Li et al. 1999b). Four phenotypically different selected clones of *P. tremula* were transformed with '*rol*' gene from *A. rhizogenes* to correlate several known phenotypic changes with '*rol*' gene expression, viz. removal of stem apical dominance consequent to development and branching of around four axillary buds per explants, in comparison of lack of axillary sprouting in a '*uidA*' transgenic aspen line and control plants. The *rol* transgenic tree exhibits a greater cumulative stem height and increase in growth rate and hence a greater stem formation index (Tzfira et al. 1999). Growth and cellulose deposition in transgenic *P. alba* was enhanced by incorporating an *Aspergillus xyloglucanase* gene. The expression of transgene has enlarged the stem size even in the sucrose media. The elongation in internodes corresponded to rise in cellulose level and specific gravity, exhibiting that the elimination of xyloglucan might enhance cellulose density in the secondary xylem (Park et al. 2004). Shani et al. (2004) produced transgenic poplar (*P. tremula*) plant containing *A. thaliana* cell gene for growth enhancement. The transgenic plants harbor significant phenotypic modification, viz. taller plants, larger leaves, increased stem diameter, dry weight, wood volume index and a greater percentage of cellulose and hemicellulose, as compared to the control plants. The hardwood cuttings of transgenic *P. tomentosa* containing *rolB* gene enhance rooting ability of the plant (Xiong et al. 2005). Transgenic poplar (*P. alba* x *P. glandulosa*) expressing levansucrase from *Bacillus subtilis* were transferred along with vacuolar targeting signal sequences and constitutive promoters (Zhang et al. 2005a).

Transgenic poplar (*P. tremula* L.) delivered with large fragments of T-DNA as a cluster having *ugt* and *acb* genes exhibit enhanced growth. The transgenic poplar conferred increased growth energy, linked to higher levels of IAA as compared to the non-transgenic plants (Salyaev et al. 2006). The *vhb* gene introduced into white poplar (*P.alba* L.) has enhanced cell growth and protein synthesis and modified the oxygen dependent product biosynthesis and the susceptibility to nitrosactive and oxidative stresses. One amongst six selected transgenic lines exhibited considerably higher values for stem biomass and plant height in greenhouse conditions. Under in vitro and in vivo submergence conditions, growth parameters and total protein level of transgenic *vhb* poplars were similar to control plants (Zelasco et al. 2006). The transgenic *vgb* poplar (*P. alba* x *P. glandulosa*) shows increasing growth rate which is especially important for low-quality wood applications, so this has become an important goal in poplar breeding. Expression of *Vitreoscilla* hemoglobin gene (*vgb*) was under the control of constitutive promoters. These transgenic were morphologically similar to control mother plant but exhibit noticeably higher growth rates in terms of height and diameter (Zhang et al. 2006b).

17.2.1.8 Genetic Engineering for Enhancement of Phytoremediation Abilities

Phytoremediation is an extremely useful, in situ pollutant amputation system for extraction of contaminants from ecosystem commodities such as soil, sediments, air and water. It implies very sensitive and capable area of bioresource tools. It is noticeably inexpensive and operative for removal of metals, solvents, pesticides, crude oil and explosive pollutants. Eco-intoxication with xenobiotics and heavy metals is a global threat arising from industrial, mining, military and agricultural practices. Poplar has been known for being an excellent contender for phytoremediation and also for reducing environmental pollutants by its high rate of photosynthesis (Soudek et al. 2004; Yadav et al. 2010).

Amusingly, for proper growth and reproduction, plants take up not only macronutrients, N, P, K^+, S, Ca^{2+} and Mg^{2+}, but vital micronutrients, including Fe^{2+}, Zn^{2+}, Mn^{2+}, Ni^{2+}, Cu^{2+} and Mo^{2+}, also. Plants have unique mechanisms to take up, translocate and store these nutrients (Yadav et al. 2010). The molecular organization and properties of membrane transporter proteins regulate ion uptake selectivity. Metal ions are transported across the membrane by precise membrane transporter proteins which develop low molecular weight complex structures along with phytochelatins (PCs). Glutathione (GSH, γ -L-glutamyl-L-cysteinylglycine) cater metal-phytochelator complex (M.PC) formation (Li et al. 1997; Grill et al. 1989). M.PC complex is detoxified by turning out into high molecular weight complex after actively transported to vacuole which finally accumulated at high concentration (Shah and Nongkynrih 2007). Many pathways, including octadecanoid and phenylpropanoid pathways, are stimulated simultaneously by activated oxygen group. The stimulated kinases of several cascades mediate triggering the transcription factor that sequentially stimulates various tolerance and defense genes. Cellular cytoplamic mechanisms upregulate intracellular intensity of metal ions within the physiologically tolerable limit. Metals ions, such as Zn^{+2}, Ni^{+2}, Mn^{+2} and Cu^{+2}, are vital micronutrients which act as co-factors for many enzymes mediating significant cellular metabolic processes. Some metal superaccumulator species take up substantial extent of non-essential metals. Occasionally, plant may take up heavy metal by misidentification between the ion and its chemical analogue such as cadmium absorption instead of zinc because metal accumulation is an energy powered process (Chaney et al. 1994), while, metal accumulation in leaves permit hyper-accumulator species to eliminate predators, including fungi, bacteria and caterpillars (Boyd and Martens 1994; Pollard and Baker 1997). Certain over-contaminated zones are designated as Superfund sites which are polluted to terrifyingly elevated levels. Every year, huge expenditure is made for remediating such polluted sites. Poplar species can be used for industrial waste management and Superfund site remediation such as phytodegradation, phytotransformation, detoxification, phyto-accumulation and phytovolatilization.

Genetic modification is the most capable tool for tailoring phytoremediation in poplars and other plants (Doty 2008; Abhilash et al. 2009; Yadav et al. 2010). Plenty of reports are available for improvement of phytoremediation capability in plants through genetic transformation. This technology has two key advantages over conventional breeding. Firstly, cross kingdom transfer of gene and secondly selected commercial genotypes can be genetically tailored for specific characters by putting the bits and pieces of the genome together.

Upregulation of the metabolic genes, uptake, accumulation or transport of certain pollutants in transgenic plants is supposed to be adroit technique for improving the phytoremediation ability. Recombinant DNA technology is a swift mechanism for genetic tailoring in various plants including woody tree species. γ-glutamylcysteine synthetase (γ-ECS) is the key regulatory enzyme in the synthesis of the ubiquitous tripeptide thiol compound γ-L-glutamyl-L-cysteinyl-glycine (GSH) protein expressed by microbial gene. Incorporation and overexpression of this gene play a vital role in the antioxidative defense of plant cells against oxidative stress mediated by native phytoremediation cascades (Noctor et al. 1998). Overexpression of γ-ECS in hybrid poplars accumulate elevated amounts of phytochelatins along with cadmium. Although, the overexpression of γ-ECS permits higher accumulation of cadmium in shoot tissues, it has only a supplementary consequence on cadmium tolerance (Koprivova et al. 2002). Expression of *merA* gene (mercuric reductase) in transgenic yellow poplar has drastically elevated phytovolatilization efficiency for elemental mercury (Meagher 2000). An incorporation of mammalian cytochrome P450 enzyme identified to regulate the rate-limiting stage in the metabolism of certain pollutants, viz. trichloroethylene (TCE), vinyl chloride, carbon tetrachloride, chloroform, benzene and ethylene dibromide, also hike the phytoremediation potential. Newman et al. (1997) produced transgenic tumor cell line of hybrid poplar (*P. trichocarpa* x *P. deltoides* clone H11-11) which was efficient to degrade TCE to trichloroethanol, trichloroacetic acid and dichioroacetic acid. Such transgenic plants can transpire TCE in measurable amounts. TCE uptake and biotransformation event already exist in poplar trees which favor its potential for of in situ remediation of TCE. Besides this, the mammalian enzymes were fully functional in plants and do not require other mammalian enzymes or genes such as cytochrome b5 and oxidoreductase, identified for complete function of mammalian P450s. Doty et al. (2007, 2008a) were successful to enhance phytoremediation potential about hundred-fold by enhancing the expression of *P450 2E1* gene of rabbit in hybrid poplar. An exciting improvement in the phytovolatilization potential of poplar for TCE and benzene up to 79% could be attained by the short exposure of contaminant for 7 days. Calligari et al. (2008) were introduced useful multi-traits including enhanced phytoremediation abilities, diseases and pests resistance, tailored composition and wood properties, herbicide tolerance and growth rate into interspecific *Populus euramericana* hybrids.

Genes for increasing proliferative root system, biomass production and enhancing growth rate favor superfluous accumulation of pollutants while the others for enhancing chelation make plants with hyperaccumulation potential. The genes coding for biodegradative enzymes, metallothioneins, metal transporters and transportation, improving rhizosphere activity, detoxifying enzyme, sequestering enzyme, phytovolatilization, and changing oxidation state of metal ions contribute for enhancement of phytoremediation potential of a plant through modified routes. Genetic tailoring of fast-growing poplar trees for improved potential to metabolize or chelate toxic metals may be achieved by commercial plantation in future (Liu et al. 2000). Transgenic for *gshl P.* x *canescens* clones could activated by Zn to be upshoted for elevated heavy metal (Cu) uptake (Gyulai et al. 2005). Overexpressed gene (i.e., γ-glutamylcysteine synthetase; gshI) in transgenic gray poplar could efficiently diminish cadmium toxicity on the root and aerial part of the plant (Dai et al. 2011). The YCF1-expressing transgenic poplar plants show high growth, reduced toxicity symptoms and increased Cd content in the aerial tissue (Shim et al. 2013).

17.2.1.9 Genetic Engineering for Modification of Lignin Biosynthesis Pathway

Lignin is the major phenolic sink in tree trunk which account for 18–25% of dry woody biomass. Lignin reduce forage digestibility by ruminants and interfere with cellulosic biomass conversion into bioenergy and pulp; however, it is a chief structural constituent of the cell wall. For making the process of paper manufacturing economical and eco-friendly, the lignin content of wood demand modification. Two factors including the lignin content and composition play a vital role in determining the extent of use of chemicals during the pulping process. Especially in paper and pulp industries, both these factors have a direct impact on the quality of the pulp which may cause discoloration and yellowing of paper. Removal of the lignin from the complexed cellulose and hemicellulose is made by cooking the tree wood in large volumes of sodium sulfide (Na_2S) and sodium hydroxide (NaOH). The residual lignin is eventually separated by strong bleaching agents such as chlorine dioxide (ClO_2), hydrogen peroxide (H_2O_2), sodium hypochlorite (NaOCl). These extensive chemical delignification and bleaching processes are very expensive, environmental polluting, energy intensive that results in low biomass consumption. Therefore, genetic modification of lignin biosynthesis pathway for reduction or alteration in the composition of lignin is a promising technique.

Biological synthesis of lignin include oxidative polymerization of three monomers, namely coumaryl, coniferyl and sinapyl alcohols (monolignols) which mediates formation of hydroxyphenyl (H), guaiacyl (G) and syringyl (S) residues respectively. Lignin exhibits a wide range of structural variability and monomeric composition depending upon the plant species, tissues and cellular locations, stage of development, etc. Various key enzymes linked to the lignin biosynthetic pathway such as PAL, C4H, COMT, CAD, etc., have been cloned and characterized by Baucher et al. (1996). This have now made possible for global researchers to modify both the lignin quality and quantity in various plant species through genetic transformation tool. Phenylalanine ammonia-lyase (PAL) is a major regulatory enzyme in the lignin biosynthesis cascade which controls carbon flux in plants. Downregulation of PAL was first reported in transgenic tobacco by Elkind et al. (1990). However, along with lowered lignin in the transgenic tobacco plants show drastic effects on the growth and development, with localized lesions, altered leaf shape, stunted growth, altered flower morphology and pigmentation. These findings advocate not targeting PAL for modulating the lignin content in transgenic plants.

The cinnamyl alcohol dehydrogenase (CAD) enzyme activity in xylem was reduced up to 70% in transgenic poplar hybrid *P. tremula* x *P. alba* transformed with antisense construct (Leple et al. 1992). Antisense RNA technology also emerged as an alternative tool to downregulate CAD in poplar (Tollier et al. 1995). Cinnamyl alcohol dehydrogenase (*CAD*) is involved in the last step of reduction of cinnamyl aldehyde to cinnamyl alcohol in the lignin biosynthetic pathway. Downregulation of *CAD* in transgenic poplar (*P. tremula* x *P. alba*) by any of the approach among antisense and co-suppression can cause 70% reduction of CAD activity and development of red coloration of the xylem cells due to incorporation of increased amounts of aldehydes in the lignin. However, there is no change in lignin content or the composition of syringyl/guaiacyl monomers, still wood of suppressed *CAD* transgenic poplar plants can be easily delignified with lowered Kappa number and without affecting the extent of cellulose degradation after chemical kraft pulping (Baucher et al. 1996). Genetic modification for reduction in caffeic acid-o-methyltransferase (COMT) activity in transgenic poplar lowered enzyme activity up to 95% (Van-Doorsselaere et al. 1995). Sense suppression of COMT activity up to 78% with no abnormality was a great achievement by Tsai et al. (1994) in transgenic Aspen; however, no modification in Klason lignin content was observed. Further research showed a reduction of 17% lignin content in genetically tailored poplar; however, a compensatory rise in 6% cellulose content was likely to restore plant strength along with 10% increase in pulp yield. The xylem tissue in the transgenic lines was of red-brown color because of an increase in the amount of coniferaldehyde which is integrated into the lignin structures. Such woods from transgenic plants can either be useful or even poor for kraft pulping was highly decided by the extent of inhibition of COMT activity. The color change in transgenic poplar can be correlated with COMT deficient maize brown midrib mutants (Jouanin et al. 2000). Later on, Franke et al. (2000) have reported overexpression of F5H gene from *Arabidopsis* under the control of cinnamate 4-hydroxylase (*C4H*) promoter in transgenic poplars and tobacco. These transgenic plants begin accumulating the S units which can be estimated by nitrobenzene oxidation and derivatization method, followed by reductive cleavage. The significant increase in S units has a good impact on making the process of delignification simple, which is very promising for paper and pulp industry.

Four different coummarate: CoA ligase (4CL) catalyzes the activation of hydroxylated cinnamic acids to their relative thioesters. Transgenic tobacco plants having reduced 4CL activity develop brown xylem tissues in the trunk (Kajita et al. 1997). The alkaline nitrobenzene oxidation diagnosed high extent of condensed units in such brown xylem cells. A compartmentalized expression of two structurally and functionally distinct *4CL* genes, *Pt4CL1* and *Pt4CL2*, are expressed differentially in poplar. Northern hybridization study reveal that *Pt4CL1* mRNA expressed specifically in lignified xylem tissues, on the other hand, *Pt4CL2* rRNA expressed exclusively in epidermal layers of leaf and stem. *Pt4CL1* was known to be involved in lignin biosynthesis in developing xylem tissues; however, *Pt4CL2* was involved in biosynthesis of flavanoids and other phenolic compounds in epidermal cells (Hu et al. 1998). A downregulation of lignin biosynthesis by lowering *Pt4CL1* gene expression through antisense approach in transgenic poplar lead to 45% reduction of lignin along with a 15% compensatory rise in cellulose content. The leaf, root and stem growths were substantially enhanced. The structural integrity was restored at cellular level as well as whole plant level in the transgenic plants. The composition of monomeric lignin does not change. However, there is an increase in the non-lignin cell wall constituents, viz. p-coumaric, sinapic and ferulic acid. The transgenic poplar plants bear thicker stems and longer internodes (Hu et al. 1999). Thus, antisense technology could efficiently regulate the level of 4CL expression in poplars and apparently provide better alternative approach for production of low-lignin plants (Yadav et al. 2008, 2009).

O-methyltransferase (*pOMT8*) and peroxidase (*Shpx6*) antisense cDNA fragments from *Stylosanthes humilis* used for modification of hybrid poplar (*P. tremula* x *P. alba*) plants result in constitutive expression of *pOMT8* and *Shpx6* driven by *CaMV35S* promoter, eventually, lead to modification in total lignin content and peroxidase activity. Such approach worth for restructuring plant cell wall for optimal utilization in paper and pulp industry. Transgenic hybrid poplars under expressing *COMT* and *CAD* had a significant lowering in lignin content with an increase in cellulose content. Besides, a dramatic change in lignin structure was made by COMT downregulation which supports changes in paper pulp quality. The alteration substantially reduces lignin level even in younger poplar trees upto17%. Lignin structure change in such transgenic trees with about double condensed bonds, an entire lack of syringyl units, and the integration of 5-hydroxyguaiacyl units to the most significant extent (Jouanin 1997; Jouanin et al. 1999; Jouanin et al. 2000). Baucher et al. (2000) produced genetically modified poplar with multi-transgene constituted of antisense of caffeoyl-CoA-O-methyltransferase, cinnamyl alcohol dehydrogenase, cinnamoyl-CoA reductase and peroxidase, which resulted in significant changes in lignin structure. Therefore, multiple transgene introduction technique could be more promising toward wood quality improvement for pulp and paper industry.

Genetic transformation through antisense CCoAOMT approach in American poplar, *P. trichocarpa*, produce 12% decrease Klason lignin level, an 11% enhanced syringyl/guaiacyl ratio in the non-condensed lignin fraction along with a rise in lignin-attached p-hydroxybenzoate of xylem change in its composition (Meyermans et al. 2000), whereas, in hybrid poplar *P. tremula* x *P. alba* transformation result to a decrease in both guaiacyl and syringyl lignins (Zhong et al. 2000). Fourier transform infrared spectroscopy reveals the reduction in lignin content with a less condensed and less cross-linked lignin structure in wood.

Transgenic poplars under expressing caffeic acid/5-hydroxyferulic acid-O-methyltransferase and cinnamyl alcohol dehydrogenase made the lignin polymer less amenable to alkaline delignification, while the lignin characteristics of the CAD antisense poplars favor a reduction in the volume of alkali required for paper pulp making using the kraft procedure (Jouanin et al. 2004). A study of anatomical and chemical characteristics of transgenic hybrid poplar (*P. nigra* x *P. maximowiczii*) downregulated with antisense O-methyltransferase (OMT) gene didn't found any visibly abnormal phenotype in the fibers and vessel elements. The antisense OMT gene transferred to poplar is not enough to reduce the overall content of Klason lignin the expression of transformation remain confined to vessel walls only (Wi et al. 2004).The antisense 4CL (4-coumarate: CoA ligase) gene transformed into triploid Chinese white poplar (*P. tomentosa*) stably repressed the gene expression along with a remarkable reduction of lignin content up to 41.73% without significant difference in holocellulose content (Caihong et al. 2004).

Recently, genetic tailoring of lignin biosynthesis pathway for repressing the lignin expression level by post-transcriptional silencing approach has been reported in Himalayan poplar (*P. ciliata* Wall.) plants through antisense cinnamyl alcohol dehydrogenase (*CAD*) gene transfer. The lignin content of transgenic poplar was slightly reduced by a fraction of 3–4% (Thakur et al. 2012). Function for microRNAs (miRNAs) is well known in developmental biology of model annual plant species and also in perennials trees. The overexpression of a distinct Corngrass1 (*Cg1*) miRNA gene cloned from maize in poplar expressed to produce considerably larger axillary meristem branching, smaller internode length, and up to a 30% decrease in stem lignin level. The miRNA overexpression is a new method for modifying lignin level and composition in poplar. These transgenic plants offer an approach for exploring the major role of *MIR156* in controlling developmental activities in poplar (Rubinelli et al. 2012). A R2R3-MYB transcription factor that was involved in control of lignin biosynthesis at the time of secondary cell wall formation, named PtrMYB152, was isolated from *P. trichocarpa* and expressed a protein of 321 amino acids having conserved R2R3-MYB domain after overexpression in the transgenic *Arabidopsis* plants which resulted in specific triggering of lignin biosynthetic genes, and caused ectopic deposition of lignin in stem and petiole of transgenic plants (Li et al. 2014)

17.3 Future Goals in Tailoring the Poplar Genome

17.3.1 Stress Tolerance

Plants are exposed to harsh environmental conditions of drought, acidity, salinity, alkalinity, elevated temperature and extreme cold during growth and development. In reaction to stress, plants produce stress-response genes and alter their physiology to resist and adapt to adversities. Genetic modification and breeding technologies can enhance stress tolerance in hardwood trees through transformation of stress-response genes. Poplar is amongst most economically important hardwood trees in temperate regions of the world (Zhao et al. 2013) and is quiet responsive to improvement through genetic engineering.

17.3.2 Pest Resistance

Insect attack not only kills the trees, but also can cause them to be more vulnerable to other health problems. Foliar spray of pesticides is costly and targets diverse species of insects, sometime friendly insects also. Genetically engineered annual crops such as corn and cotton already are widely grown with insect resistance genes. Trees, although, have to grow and thrive for many years until harvest and struggle with several generations of insect pest attacks. Accordingly, tailored insect protection promise even higher commercial value; however, comprehensive tests are essential to reveal that the resistance genes would certainly be expressed for prolonged life of the tree. Challengingly, the genetically modified poplars are likely to need extended studies of gene flow and their consequences on forest ecosystems, which are hard to achieve.

17.3.3 Biofuel Production

A big challenge to price-competitive biofuel production has been the processing cost required to obtain the sugar contained in plant fiber. Biochemically, sugar to ethanol conversion is easy, but getting to the sugar is tough job. A feasible solution would be to come up with better chemistry, but a unique approach is producing better plants. For inventing plants with easy sugar, researchers are making best efforts in plant fiber coming from differences in genetics, environment and growing conditions. Plant fiber is constituted of a tough matrix of cell walls that support the plant to stand erect. The cell walls are made up of lignin, cellulose and hemicellulose. Cellulose and hemicellulose are the sugars needed to make biofuels, but entangled in a hard-to-process lignin matrix.

Even though, many biological processes to produce lignin are known, yet to uncover the genes that regulate lignin synthesis. To understand more about function of such genes, we need to analyze the complete genome of various poplar genotypes which are broadly used for biofuel production. High throughput computational biology to look for small differences among the billions of nucleotides of DNA constituting each genome is needed. This approach is highly promising to find plants with desired set of genes and the traits. This approach is already being used to discover a number of genes that regulate the formation of lignin, cellulose and hemicellulose. Another alternative approach of gene discovery in practice is transformation. Scientifically, one may be able to tailor poplar species that yield high sugar content by altering the way they produce lignin and cellulose. High level of sugar in tailored plants converts is directly linked to higher biofuel production which is an economic and sustainable approach.

17.3.4 Phytoremediation

Poplars are best candidates for detoxifying and cleaning up many organic contaminants, such as herbicide atrazine, chlorinated hydrocarbon solvents, and a range of explosives (Kõmives and Gullner 2000). It also deliberate extremely high levels of the endogenous detoxifying and reducing agent GSH (Noctor et al. 1998) and an efficient phase II detoxification construct made of the enzyme system GST and the endogenous co-substrate GSH that catalyzes the metabolism of toxic electrophilic species mediated conjugation reaction (Kõmives et al. 2003).

Genetic modification approach could enhance phytoextraction and phytoaccumulation abilities of poplar by introducing new essential traits. These recombinant DNA technology based tools will surely advocates phytoremediation industries in future and poplar will emerge as a model candidate for helping the environment and the ecosystem. Photosynthesis powered inexpensive tools of environmental cleansing strongly support commercialization of phytoremediation process. Overexpression of YCF1 in *Arabidopsis thaliana* proved that vacuoles of YCF1-transgenic plants have a great uptake of elemental cadmium (Song et al. 2006). Hence, understanding of translocation and metal uptake events in higher plants at molecular and genetic levels could give a breakthrough for genetically engineering the model poplar phytoremediator species. The merger of interdisciplinary technologies may produce useful multi-traits in poplar and emerge it as an essential component of future forests. Non-availability of direct remediation technology at commercial scale make transgenic poplars for various desirable tailored traits, more promising for the forests, ecology, environment and finally of the phytoremediation industries (Giachetti and Sebastiani 2006).

17.3.5 Lignin Engineering

Past efforts to lower lignin content by genetic transformation by downregulation of genes encoding *COMT*, *CAD* in tree species were not highly successful. Few reports were with modified the lignin structure. Whereas, downregulation of *PAL* enzyme restricted overall phenylpropanoid biosynthetic pathway leaded to wide abnormalities in transgenic plants with collapsed cell wall and stunted plant growth. Thus there arises an important question whether the lignin level can actually be lowered without compromising with or harming the structural integrity and normal growth of trees. However, excellent report of Hu et al. (1999) by employing *Pt4CL1* gene in antisense strategy significantly surface the way for future genetic manipulation of lignin and will serve as an effective strategy for generating crop plants and tree species with improved digestibility and lowered lignin content suitable for paper and pulp industry.

17.4 Conclusion

The poplars are among the finest success stories to date in the genetic transformation of forest trees, although, less advanced than genetically modified annual crops such as cotton, corn and soybeans. The advancement is useful in the pulp and paper industries, and in an emerging biofuel industry. Considering plantation health, wood yield and productivity, these genetically tailored trees could be very important. A number of good reasons rush toward engineering of forest trees via various approaches including stress tolerance, insect and pest resistance, lignin regulation, and phytoremediation. Several approaches promise to enhance wood and biomass yield of the poplar tree; however, its perennial nature is likely to spread the characteristics to other trees which drive us for wisely addressing the forest biosafety issues. Keeping in view the idea of gene flow and its consequences on forest ecosystems, some mechanism to control gene flow such as engineering sterility genes must be taken into consideration to save the rest of forests from unidentified or harmful consequences.

References

Abhilash PC, Jamil S, Singh N (2009) Transgenic plants for enhanced biodegradation and phytoremediation of organic xenobiotics. Biotechnol Adv 27:474–488

Baucher M, Chabbert B, Pilate G, Doorsselaere JV et al (1996) Red xylem and higher lignin extractability by down regulating a cinnamyl alcohol dehydrogenase in poplar. Plant Physiol 112:1479–1490

Baucher M, Montagu MV, Boerjan W, Arencibia AD (2000) Improvement of wood quality for the pulp and paper industry by genetic modification of lignin biosynthesis in poplar. In. Proceedings of the International Symposium on Plant Genetic Engineering, Havana, Cuba, pp 215–221

Boyd RS, Martens SN (1994) Nickel hyperaccumulated by *Thlaspi montanum var. montanum* is acutely toxic to an insect herbivore. Oikos 70:21–25

Brasileiro ACM, Leple JC, Muzzin J, Ounnoughi D et al (1991) An alternate approach for gene transfer in trees using wild type *Agrobacterium* strains. Plant Mol Biol 17:441–452

Brasileiro ACM, Tourneur C, Leple JC, Combes V, Jouanin L (1992) Expression of the mutant *Arabidopsis thaliana* acetolactate synthase gene confers chlorsulfuron resistance to transgenic poplar plants. Transgenic Res 1:133–141

Caihong J, Huayan Z, Hongzhi W, Zhifeng X et al (2004) Obtaining the transgenic poplars with low lignin content through down-regulation of 4CL. Chin Sci Bull 49(9):905–909

Calligari P, Collot T, Zelasco S, Nervo G (2008) Genetic transformation of elite poplar clones for useful traits. In: 23rd session of international poplar commission on poplars, Willows and People's Wellbeing. Beijing, China, p 28

Catranis CM, Maynard CA, Powell WA (1997) Stable transformation of hybrid poplar. Phytopathology 87(6):515–516

Chaney RL, Green CE, Filcheva E, Brown SL (1994) Effect of iron, manganese, and zinc enriched biosolids compost on uptake of cadmium by lettuce from cadmium-contaminated soils. In: Clapp CE, Larson WE, Dowdy RH (eds) Sewage sludge: land utilization and the environment. SSSA Miscellaneous Publication, Madison, pp 205–207

Charest PJ, Steward D, Budicky PL (1992) Root induction in hybrid poplar by *Agrobacterium* genetic transformation. Can J For Res 22:1832–1837

Chen V, Han Y, Tian Y, Li L, Nie S (1995) Study on plant regeneration from *Populus deltoides* explants transformed with Bt. toxin gene. Sci Silvae Sin 31:97–103

Chen Y, Li O, Li L, Han YF, Tian YC (1996) Western blot analysis of transgenic *Populus nigra* plants transformed with *Bacillus thuringiensis* toxin gene. Sci Silvae Sin 37(3):274–276

Chilton MD, Drummond MH, Merlo DJ, Saiky D et al (1977) Stable incorporation of plasmid DNA into higher plant cells, the molecular basis of crown gall tumorigenesis. Cell 11:263–271

Chupeau MC, Pautot V, Chupeau Y (1994) Recovery of transgenic trees after electroporation of poplar protoplasts. Transgenic Res 3:13–19

Confalonieri M, Belenghi B, Balestrazzi A, Negri S et al (2000) Transformation of elite white poplar (*Populus alba* L.) cv. 'Villafrance' and evaluation of herbicide resistance. Plant Cell Rep 19:978–982

Cornu D, Leple JC, Bonade-Bottino M, Ross A et al (1996) Expression of a proteinase inhibitor and a *Bacillus thuringiensis* delta-endotoxin in transgenic poplars. In: Boerjan W, Neale DB (eds) Somatic cell genetics and molecular genetics of trees. Kluwer Academic Publishers, Dordrecht, pp 131–136

Dai WH, Cheng ZM, Sargent W (2003) Plant regeneration and *Agrobacterium*-mediated transformation of two elite aspen hybrid clones from *in vitro* leaf tissues. In Vitro Cell Dev Biol Plant 39:6–11

Dai H, Jia G, Feng S, Wei A et al (2011) Phytoremediation with transgenic poplar. J Food Agric Environ 9(3&4):710–713

Delledonne M, Allegro G, Belenghi B, Balestrazzi A et al (2001) Transformation of white poplar (*Populus alba* L.) with a noval *Arabidopsis thaliana* cysteine proteinase inhibitor gene and analysis of insect pest resistance. Mol Breed 7:35–42

Devantier YA, Moffatt B, Jones C, Charest PJ (1993) Microprojectile-mediated DNA delivery to the Salicaceae family. Can J Bot 71:1458–1466

Devillard C (1991) Obtaining genetically transformed roots of *Populus tremula* x *P. alba* and *P. trichocarpa* x *P. deltoides* after inoculation with *Agrobacteirum rhizogenes*. Annales de Researches Sylvicoles 5:22

Donahue RA, Davis TD, Michler CM, Riemenschneider CF et al (1994) Growth, photosynthesis and herbicide tolerance of genetically modified hybrid poplar. Can J For Res 24(12):2377–2383

Doty SL (2008) Enhancing phytoremediation through the use of transgenics and endophytes. New Phytol 179:318–333

Doty SL, James CA, Moore AL, Vajzovic A et al (2007) Enhanced phytoremediation of volatile environmental pollutants with transgenic trees. Proc Natl Acad Sci U S A 104:16816–16821

Doty SL, Kang JW, James A, Vajzovic A et al (2008a) Enhancing phytoremediation and plant growth in poplar and willow. In: 23rd session of international poplar commission on poplars, Willows and People's Wellbeing. Beijing, China, p 55

Elkind Y, Edwards R, Movandad M, Hedrick SA et al (1990) Abnormal plant development and down-regulation of phenylpropanoid biosynthesis in transgenic tobacco containing a heterologus phenylalanin ammonia-lyase gene. PNAS 87:9057–9061

Fan JF, Han Y, Li L, Peng X, Li JR (2002) Studies on transformation of *mtlD/gutD* divalent genes to *Populus deltoids* x *P. cathayana*. Biotechnol Lett 24:281–286

Filichkin SA, Wu Q, Busov V, Meilan R et al (2006) Enhancer trapping in woody plants: isolation of the ET304 gene encoding a putative AT-hook motif transcription factor and characterization of the expression patterns conferred by its promoter in transgenic *Populus* and *Arabidopsis*. Plant Sci 171:206–216

Fillaiti JJ, Sellmer J, McCown B, Haissig B, Comai L (1987) *Agrobacterium*-mediated transformation and regeneration of *Populus*. Mol Gen Genomics 206:192–199

Fladung M, Kumar S, Ahuja MR (1997) Genetic transformation of *Populus* genotypes with different chimeric gene constructs: transformation efficiency and molecular analysis. Transgenic Res 6:111–121

Foyer CH, Souriau N, Perret S, Lelandais M et al (1995) Over-expression of glutathione reductase but not glutathione synthetase leads to increases in antioxidant capacity and resistance to photoinhibition in poplar trees. Plant Physiol 109(3):1047–1057

Franke R, McMichael CM, Meyer K, Shirley AM et al (2000) Modified lignin in tobacco and poplar plants over expressing the *Arabidopsis* gene encoding ferulate 5-hydroxylase. Plant J 22(3):223–224

Gallardo F, Fu JM, Canton FR, Garciagutierrez A et al (1999) Expression of a conifer glutamine synthetase gene in transgenic poplar. Planta 210:19–26

Giachetti G, Sebastiani L (2006) Metal accumulation in poplar plant grown with industrial wastes. Chemosphere 64:446–454

Giorcelli A, Spar-voli F, Mattivi F, Tava A et al (2004) Expression of the stilbene synthase (stsy) gene from grapevine in transgenic white poplar results in high accumulation of the antioxidant resveratrol glucosides. Transgenic Res 13(3):203–214

Gray M, Moutor EK, Cukovic D, Carlson JE, Dounglas CJ (1999) Developmentally regulated patterns of expression directed by poplar PAL promoters in transgenic tobacco and poplar. Plant Mol Biol 39:657–669

Grill E, Loffler S, Winnacker EL, Zenk MH (1989) Phytochelatins, the heavy metal binding proteins are synthesized from glutathione by a specific γ-glutamylcysteine dipeptidyl transpeptidase (phytochelatin synthase). Proc Natl Acad Sci U S A 86:6838–6842

Guo XH, Jiang J, Lin SJ, Wang BC et al (2009) A *ThCAP* gene from *Tamarix hispida* confers cold tolerance in transgenic *Populus* (*P. davidiana* x *P. bolleana*). Biotechnol Lett 31:1079–1087

Gyulai G, Humphreys MO, Bittsanszky A, Skot KP et al (2005) AFLP analysis and improved phytoextraction capacity of transgenic gshl-poplar clones (*Populus* x *canescens*) for copper *in vitro*. In: Proceedings of an international OECD workshop, Matrahaza, Hungary, 2004

Han KH, Gordon MP, Strauss SH (1996) Cellular and molecular biology of *Agrobacterium*-mediated transformation of plant and its application to genetic transformation of *Populus*. In: Stettler RF, Bradshaw HD, Heilman PE, Hinckley TM (eds) Biology of *Populus* and its implications for management and conservation. NRC Research Press, Ottawa/Ontario, pp 201–202

Han KH, Gordon MP, Strauss SH (1997a) High frequency transformation of cotton woods (genus *Populus*) by *Agrobacterium tumefaciens*. Can J For Res 24:464–470

Han KH, Ma CP, Strauss SH (1997b) Matrix attachment regions (MARs) enhance transformation frequency and transgene expression in poplar. Transgenic Res 6:415–420

Han KH, Meilan R, Ma C, Strauss SH (2000) An *Agrobacterium tumefaciens* transformation protocol effective on a variety of cotton wood hybrids (genus *Populus*). Plant Cell Rep 19:315–320

Han MS, Noh EW, Han SH (2013) Enhanced drought and salt tolerance by expression of AtGSK1 gene in poplar. Plant Biotechnol Rep 7:39–47

Herrera-Estrella L, Depicker A, Montagu MV, Schell J (1983) Expression of chimeric genes transferred into plant cell using a Ti plasmid vector. Nature 303:209–213

Herschbach C, Jouanin L, Rennenberg H (1998) Over-expression of gamma-glutamylcysteine synthetase, but not of glutathione synthetase, elevates glutathione allocation in the phloem of transgenic poplar trees. Plant Cell Physiol 39(4):447–451

Heuchelin SA, McNabb MS, Klopfenstein NR (1997) *Agrobacterium* -mediated transformation of *Populus* x *euramericana* 'Ogy' using the chimeric CaMV 35S pin2 gene fusion. Can J For Res 27(7):1041–1048

Howe GT, Goldfarb B, Strauss SH (1994) *Agrobacterium*-mediated transformation of hybrid poplar suspension cultures and regeneration of transformed plants. Plant Cell Tissue Organ Cult 36:59–71

Hu WJ, Kawaoka A, Tsai CJ, Lung J et al (1998) Compartmentalized expression of two structurally and functionally distinct 4-coumarate: CoA ligase genes in aspen (*Populus tremuloides*). PNAS 95(9):5407–5412

Hu WJ, Harding SA, Lung J, Popko JL et al (1999) Repression of lignobiosynthesis promotes cellulose accumulation and growth in transgenic trees. Nat Biotechnol 17:808–812

Hu L, Lu H, Liu QL, Chen XM, Jiang XN (2005) Over expression of mtlD gene in transgenic *Populus tomentosa* improves salt tolerance through accumulation of mannitol. Tree Physiol 25(10):1273–1281

Igasaki T, Ishida Y, Mohri T, Ishikawa H, Shinohara K (2002) Transformation of *Populus alba* and direct selection of Transformants with the herbicide bialaphos. Bulletin FFPRI 1:235–240

JingBo Q, Ishihara Y, Kuroda H, Sakai F, Sakai H, Komano T (1997) Transient expression of goat growth hormone gene in poplar (*Populus alba* L.) protoplasts: a quick method for detection of foreign gene expression in mRNA level. Biosci Biotechnol Biochem 61:1580–1581

Jouanin L (1997) Modification of lignin composition in trees by genetic engineering. C R Seances Soc Biol Fil 191(2):155–159

Jouanin L, Brasiliero AM, Leple JC, Pilate G, Cornu D (1999) Genetic transformation: a short review of methods and their applications, results and perspectives for forest trees. Ann For Sci 50:325–336

Jouanin L, Goujon T, Nadai V, Martin MT et al (2000) Lignification in transgenic poplars with extremely reduced caffeic acid O-methyl transferase activity. Plant Physiol 123:1363–1374
Jouanin L, Goujon T, Sibout R, Pollet B et al (2004) Comparison of the consequences on lignin content and structure of COMT and CAD down-regulation in poplar and *Arabidopsis thaliana*. Plantation Forest Biotechnology for the 21st Century, pp 219–229
Kajita S, Osakabe K, Katayamsa Y et al (1994) *Agrobacterium*-mediated transformation of poplar using a disarmed binary vector and the overexpression of a specific member of a family of poplar peroxidase genes in transgenic poplar cell. Plant Sci 103:231–239
Kajita S, Hishiyama S, Tomimura Y, Katayama Y, Omori S (1997) Structural characterization of modified lignin in transgenic tobacco plants in which the activity of 4-Coumarate: Coenzyme A ligase is depressed. Plant Physiol 114:871–879
Kang B, Osburn L, Kopsell D, Tuskan GA, Cheng ZM (2009) Micropropagation of *Populus trichocarpa* 'Nisqually-1': the genotype deriving the *Populus* reference genome. Plant Cell Tissue Organ Cult 99:251–257
Kim YW, Noh EW, Youn Y, Noh ER (1995) Genetic transformation of *Populus nigra* using *Agrobacterium tumefaciens* LBA 4404/pBKS-1. Research Report of the Forest Genetics Research Institute (Kyonggido) 3:160–166
Kirby EG, Wu D, Fu J, Gallard F, Canovas FM (1999) Expression of a pine cytosolic glutamine synthetase gene in transgenic poplar. In: Forestry biotechnology-99: proceedings of a conference, Keble College, University of Oxford, UK, p 43
Klopfenstein NB, Allen KK, Avila FJ, Heuchelin SA, Martinez J, Carman RC, Hall RB, Hart ER, McNabb HS Jr (1997) Proteinase inhibitor II gene in transgenic poplar: chemical and biological assays. Biomass Bioenergy 12(4):299–311
Kõmives T, Gullner G (2000) Phytoremediation. In: Wilkinson RE (ed) Plant-environment interactions, 2nd edn. Marcel Dekker, New York, pp 437–452
Kõmives T, Gullner G, Rennenberg H (2003) Roles of glutathione and glutathione-related enzymes in remediation of polluted soils by transgenic poplars. In: Davidian JC, Grill D, De Kok LJ et al (eds) Sulfur transport and assimilation in plants. Regulation, interaction and signaling. Backhuys Publishers, Leiden, pp 101–109
Koprivova A, Kopriva S, Jager D, Will B et al (2002) Evaluation of transgenic poplars over-expressing enzymes of glutathione synthesis for phytoremediation of cadmium. Plant Biol 4:664–670
Leple JC, Brasileiro ACM, Michel MF, Delmotte F, Jouanin L (1992) Transgenic poplars: expression of chimeric genes using four different constructs. Plant Cell Rep 11:137–141
Li ZS, Lu YP, Zhen RG, Szezypka M et al (1997) A new pathway for vacuolar cadmium sequestration in *Saccharomyces cerevisiae*: YCF1-catalyzed transport of bis (glutathionato) cadmium. Proc Natl Acad Sci U S A 94:42–47
Li H, Yifan L, Liang T, Yingchuan L et al (1999a) Cloning of ACC oxidase cDNA and its inhibition of ethylene synthesis by its antisense RNA in transgenic *P. deltoides*. For Res 12:223–228
Li ML, Han YF, Qiu DY, Li N, Tian YC (1999b) Cloning of ACC synthase cDNA and its inhibition of ethylene synthesis by its antisense RNA in transgenic *Populus deltoides*. Sci Silvae Sin 35(3):10–15
Li C, Wang X, Lu W, Tian QR, Sun Y, Luo K (2014) A poplar R2R3-MYB transcription factor, PtrMYB152, is involved in regulation of lignin biosynthesis during secondary cell wall formation. Plant Cell Tissue Organ Cult 119(3):553–563
Liang H, Catranis CM, Maynard CA, Powell WA (2002) Enhanced resistance to the poplar fungal pathogen, *Septoria musiva* in hybrid poplar clones transformed with genes encoding antimicrobial peptides. Biotechnol Lett 24:383–389
Lin T, Wang ZY, Liu KY, Jing TZ, Zhang CX (2006) Transformation of spider neurotoxin gene with prospective insecticidal properties into hybrid poplar *Populus simonii* x *P. nigra*. Acta Entomol Sin 49(4):593–598
Liu JR, Suh MC, Choi D (2000) Phytoremediation of cadmium contamination: overexpression of metallothionin in transgenic tobacco plants. Bundesgesundheitsbl-Gesundheitsforsch-Gesundheitsschutz 43:126–130

Liu B, Li HS, Wang QH, Cui DC (2002) Transformation of *Populus tomentosa* with anti-*PLD* gene. Hereditas (Beijing) 24:40–44

Liu J, Wang C, Wang B, Liu J, Zhao J et al (2008) Study on the salt tolerance of genetically modified triploid Chinese white poplar. For Res Beijing 21:379–385

Ma C, Strauss SH, Meilan R (2004) *Agrobacterium*-mediated transformation of the genome-sequenced poplar clone, Nisqually-1 (*Populus trichocarpa*). Plant Mol Biol Rep 22:1–9

Main GD, Williamson A, Trvine RJ, Gartland JS et al (1998) The use of green fluorescent protein (*gfp*) as a reporter gene in tree genetic manipulations. Tree Biotechnology: Towards the Millennium. Nottingham University Press, UK, pp 315–320

McCown BH, McCabe DE, Russel DR, Robinson DJ et al (1991) Transformation of *Populus* spp. and incorporation of pest resistance by electric discharge particle acceleration (direct gene transfer) method. Plant Cell Rep 9:590–594

Meagher RB (2000) Phytoremediation of toxic elemental and organic pollutants. Curr Opin Plant Biol 3:153–162

Meilan R, Han KYH, Ma CP, James RR et al (2000) Development of glyphosate-tolerant hybrid cottonwoods. TAPPI J 83:164–166

Meilan R, Han KH, Ma C, DiFazio SP et al (2002) The CP4 transgene provides high levels of tolerance of round up Reg. herbicide in field-grown hybrid poplars. Can J For Res 32(6):967–976

Meyer S, Nowak K, Sharma VK, Schulze J et al (2004) Vectors for RNAi technology in poplar. Plant Biol 6(1):100–103

Meyermans H, Morreel K, Lapierre C, Pollet B (2000) Modificaition in lignin and accumulation of phenolic glucosides in poplar xylem upon down regulation of caffeoyl-coenzymeA-o-methyltransferase, an enzyme involved in lignin biosynthesis. J Biol Chem 275(47):36899–36909

Mijnsbrugge K, Montagu M, Inze D, Boerjan W (1996) Tissue-specific expression conferred by the S-adenosyl-L-methionine synthetase promoter of *Arabidopsis thaliana* in transgenic poplar. Plant Cell Physiol 31(8):1108–1115

Mohri T, Igasaki T, Futamura N, Shinohara K (1999) Morphological changes in transgenic poplar induced by expression of the rice homeobox gene *OSHI*. Plant Cell Rep 18:816–819

Newman LA, Strand SE, Choe N, Duffy J et al (1997) Uptake and biotransformation of trichioroethylene by hybrid poplars. Environ Sci Technol 31(4):1062–1067

Nicolescu C, Sandre C, Jouanin L, Chriqui D (1996) Genetic engineering of phenolic metabolism in poplar in relation with resistance against pathogens. Acta Botanica Gallica 143(6):539–546

Noctor G, Arisi AM, Jouanin L, Kunert KJ et al (1998) Glutathione: biosynthesis, metabolism and relationship to stress tolerance explored in transformed plants. J Exp Bot 49:249–279

Noel A, Levasseur C, Van QL, Seguin A (2005) Enhanced resistance to fungal pathogens in forest trees by genetic transformation of black spruce and hybrid poplar with a *Trichoderma harzianum* endochitinase gene. PIMP Physiol Mol Plant Pathol 67(2):92–99

Okumura S, Sawada M, Park YW, Hayashi T et al (2006) Transformation of poplar (*Populus alba*) plastids and expression of foreign proteins in tree chloroplasts. Transgenic Res 15(5):637–646

Ovadis M, Chernin L, Tzfira T, Canaan V et al (1998) Transformation of tobacco and aspen plants with the 'ita' locus of IncQ plasmid confers resistance to *Agrobacterium tumefaciens*. In: Plant biotechnology and *in vitro* biology in the 21st century. Proceedings of the IXth international congress of the International Association of the Plant Tissue Culture and Biotechnolgy 36: 286–288

Paques M, Bercetche J, Bruneau G, Bregeon JM et al (1995) Genetic engineering for poplar improvement. Estat et perspectives de la populicutture. Colloque Organise 29 et 30 mars 1995. Ecolc superieure du Bois, Nantis, France. Comptes Rendus de ('Academic d' Agriculture de France) 81(3):153–162

Park YW, Baba K, Furuta Y, Tida T et al (2004) Enhancement of growth and cellulose accumulation by overexpression of xyloglucanase in poplar. FFBS Lett 564(1/2):183–187

Parsons TJ, Sinkar VP, Steller RF, Nester EW, Garden MP (1986) Transformation of poplar by *Agrobacterium tumefaciens*. Biotechnology 4:533–536
Pollard JA, Baker AJM (1997) Deterrence of herbivory by zinc hyperaccumulation in *Thlaspi caerulescens* (Brassicacea). New Phytol 135:655–658
Rao HY, Chen Y, Yuang MR, Wang MX et al (2000) Genetic transformation of poplar NL-80106 transferred by '*Bt*.' gene and its insect-resistance. J Plant Res Environ 9:1–5
Rubinelli PM, Chuck G, Li X, Meilan R (2012) Constitutive expression of the Corngrass1 microRNA in poplar affects plant architecture and stem lignin content and composition. Biomass Bioenergy 54:312–321
Sala F, Castiglione S, Jianjun H, Zheng Y, Han Y (2000) Field and molecular evaluation of insect-resistant transgenic poplar (*Populus nigra* L.) trees. In: Proceeding of the international symposium on plant genetic engineering, Havana, Cuba, p 137–147
Salyaev R, Rekoslavskaya N, Chepinoga A et al (2006) Transgenic poplar with enhanced growth by introduction of the ugt and acb genes. New For 32(2):211–229
Sanchez N, Manzanera JA, Bueno MA (2004) *Agrobacterium*-mediated transformation of white poplar (*Populus alba* L.). Int J Exp Bot 73:123–130
Schwartzenberg K, Doumas P, Jouanin L, Pilate G (1994) Enhancement of the endogenous cytokinin concentration in poplar by transformation with *Agrobacterium* T-DNA gene *ipt*. Tree Physiol 14(l):27–35
Shah K, Nongkynrih JM (2007) Metal hyperaccumulation and bioremediation. Biol Plant 51(4):618–634
Shani Z, Dekel M, Jensen CJ, Tzfira T et al (2000) *Arabidopsis thaliana* Endo-1, 4-β-glucanase (cell) promoter mediates 'uida' expression in elongating tissues of aspen (*Populus tremula*). J Plant Physiol 156:118–120
Shani Z, Dekel M, Tsabary G, Goren R, Shoseyov O (2004) Growth enhancement of transgenic poplar plants by overexpression of *Arabidopsis thaliana* endo-1,5-β-glucanase (cell). Mol Breed 14(3):321–330
Shim D, Kim S, Choi YI, Song WY et al (2013) Transgenic poplar trees expressing yeast cadmium factor 1 exhibit the characteristics necessary for the phytoremediation of mine tailing soil. Chemosphere 90(4):1478–1486
Son S, Hyun J (1998) Transformation of *Populus nigra* × *P. maximowiczii* using *Agrobacterium tumefaciens* vector. J Korean For Soc 87:164–172
Song J, Lu S, Chen ZZ, Lourenco R, Chiang VL (2006) Genetic transformation of *Populus trichocarpa* genotype Nisqually 1: a functional genomic tool for woody plants. Plant Cell Physiol 47(11):1582–1589
Soudek P, Tykva R, Vaněk T (2004) Laboratory analyses of 137Cs uptake by sunflower, reed and poplar. Chemosphere 55:1081–1087
Studart GC, Lacorte C, Brasileiro ACM (2006) Evaluation of heterologous promoters in transgenic *Populus tremula* x *P. alba* plants. Biol Plant 50(1):15–20
Thakur AK, Sharma S, Srivastava DK (2005) Plant regeneration and genetic transformation studies in petiole tissue of Himalayan poplar (*Populus ciliata* Wall.) Curr Sci 89:664–668
Thakur AK, Aggarwal G, Srivastava DK (2012) Genetic modification of lignin biosynthetic pathway in *Populus ciliata* Wall.via *Agrobacterium*-mediated antisense CAD Gene transfer for quality paper production. Nat Acad Sci Lett 35(2):79–84
Tian LN, Levee V, Mentag R, Charest PJ, Seguin A (1999) Green fluorescent protein as a tool for monitoring transgene expression in forest tree species. Tree Physiol 19(8):541–546
Tian Q, Chen J, Wang D, Wang HL et al (2016) Overexpression of a *Populus euphratica* CBF4 gene in poplar confers tolerance to multiple stresses. Plant Cell Tissue Organ Cult:1–17. doi:10.1007/s11240-016-1118-y
Tollier MT, Chabbert B, Lapierre C, Monties B et al (1995) Lignin composition in transgenic poplar plants with modified cinnamyl alcohol dehdrogenase activity with reference to dehydropolymer models of lignin. In: Proceedings of 17th International Conference, Palma de Mallorca, Spain, p 339–340

Tsai CJ, Podila GK, Chiang VL (1994) *Agrobacterium* -mediated transformation of quaking aspen (*Populus tremuloides*) and regeneration of transgenic plants. Plant Cell Rep 14:94–97

Tzfira T, Jensen CS, Wang W, Zuker A et al (1998) Transgenic *Populus tremula*: a step-by-step protocol for its *Agrobacterium*-mediated transformation. Plant Mol Biol Rep 15:219–235

Tzfira T, Vainstein A, Altman A (1999) *rol*-gene expression in transgenic aspen (*Populus trernula*) plants results in accelerated growth and improved stem production index. Trees Struct Funct 14:49–54

Van-Doorsselaere J, Baucher M, Chognot E et al (1995) A novel lignin in poplar trees with a reduced caffeic acid/5 hydroxyferulic acid O-methyl transferase activity. Plant J 5:855–864

Wang JH, Constabel CP (2004) Polyphenol oxidase over-expression in transgenic *Populus* enhances resistance to herbivory by forest tent caterpillar (*Malacosoma disstria*). Planta 220(l):87–96

Wang YC, Emerick RM, Denchev PD, Conger RV, Tuskan GA (1995) A biolistic approach for the transient expression of a *GUS* reporter gene in callus cultures of hybrid poplar. In Vitro Cell Dev Biol Plant 31(4):226

Wang GJ, Castiglione S, Cheri Y, Li L et al (1996) Poplar (*Populus nigra* L.) plants transformed with a *Bacillus thuringiensis* toxin gene: insecticidal activity and genomic analysis. Transgenic Res 5:289–301

Wang WX, Tzfira T, Levin N, Shoseyov O, Altman A (1999) Plant tolerance to water and salt stress: the expression pattern of a water stress responsive protein (Bsp A) in transgenic aspen plants. In: Plant biotechnology and *in vitro* biology in the 21st century. Proceedings of the IXth international congress of the International Association of Plant Tissue Culture and Biotechnology 36:66–68

Wang H, Wang C, Liu H, Tang R, Zhang H (2011) An efficient *Agrobacterium*-mediated transformation and regeneration system for leaf explants of two elite aspen hybrid clones *Populus alba* x *P. berolinensis* and *Populus davidiana* x *P. bolleana*. Plant Cell Rep 30:2037–2044

Wang L, Li L, Xu L, Zhou J et al (2013) Isolation and functional analysis of the poplar *RbcS* Gene promoter. Plant Mol Biol Rep 31:120–127

Wi SG, Lee KH, Park BD, Park YG, Kim YS (2004) Anatomical, chemical and topochemical characteristics of transgenic poplar down-regulated with O-methyltransferase. J Kor Wood Sci Technol 32(3):15–24

Xiong J, Liang J, Chen XY, Li W, Li H, Liu Y (2005) The rooting ability of rolB transformed clones of *Populus tomentosa*. J Beijing For Univ 27(5):54–58

Yadav R, Kumar D, Arora P, Raghuvanshi S, Dilbaghi N, Chaudhury A (2008) Genetic modification in *Populus deltoides*: An antisense approach for lignin repression. Published in Proceedings "International Conference on Poplars, willows and people's wellbeing" Beijing, China, October 26–30

Yadav R, Arora P, Kumar D, Katyal D, Dilbaghi N, Chaudhury A (2009) High frequency direct plant regeneration from leaf, internode, and root segments of Eastern Cottonwood (*Populus deltoides*). Plant Biotechnol Rep 3:175–182

Yadav R, Arora P, Kumar S, Chaudhury A (2010) Perspectives for genetic engineering of poplars for enhanced phytoremediation abilities. Ecotoxicology 19:1574–1588

Yang CP, Liu GF, Liang HW, Zhang H (2001) Study on the transformation of *Populus simonii* x *P. nigra* with salt resistance gene *Bet*-A. Sci Silvae Sin 37(6):34–38

Yang MS, Li ZL, Wang Y, Wang JM, Liang HY (2006) Transformation and expression of two insect-resistant genes to hybrid triploid of Chinese White Poplar. Sci Silvae Sin 42(9):61–68

Yang C, Li H, Cheng Q, Chen Y (2009) Transformation of drought and salt resistant gene (*DREB1C*) in *Populus* x *euramericana* cv. Nanlin 895. Sci Silvae Sin 45:17–21

Yang L, Sun Y, Xie L, Liang A (2010) A novel approach for in situ bud transformation of *Populus* by *Agrobacterium*. Scandinavian J For Res 25:3–9

Yevtushenko DP, Misra S (2010) Efficient *Agrobacterium*-mediated transformation of commercial hybrid poplar *Populus nigra* L. x *P. maximowiczii* A. Henry. Plant Cell Rep 29:211–221

Zambryski P, Joos H, Genetello C, Leemans J et al (1983) Ti plasmid vector for the introduction of DNA into plant cells without alteration of their normal regeneration capacity. EMBOJ 2:2143–2150
Zelasco S, Reggi S, Calligari P, Balestrazzi A et al (2006) Expression of the Vitreoscilla haemoglobin (VHb) – encoding gene in transgenic white poplar: plant growth and biomass production, biochemical characterization and cell survival under submergence, oxidative and nitrosative tress conditions. Mol Breed 17(3):201–216
Zhang BY, Su XH, Huang QJ, Zhang XH, Hu ZM (2005a) Regeneration of transgenic poplar (*Populus alba* x *P. glandulosa*) expressing levansucrase from *Bacillus subtilis*. Sci Silvae Sin 41(3):48–53
Zhang GC, Zou CS, Wang ZY (2005b) Transformation system of chimeric gene for spider insecticidal peptide and *Bt* of *Populus euramericana* cv."114/69". J Northeast For Univ 54(6):43–44
Zhang Q, Zhang ZY, Lin SZ, Lin YZ (2005c) Resistance of transgenic hybrid triploids in *Populus tomentosa* Carr. Against 3 species of Lepidopterans following two winter dormancies conferred by high level expression of cowpea trypsin inhibitor gene. Silvae Genetica 54(3):108–116
Zhang BY, Su XH, Li YL, Zhang YA et al (2006a) Production of *Populus alba* x *P. glandulosa* with a coleopterous insect resistant gene and analysis of insect resistance. J Beijing For Univ 28(2):102–105
Zhang BY, Su XH, Li YL, Huang QJ et al (2006b) Regeneration of vgb-transgenic poplar (*Populus alba* x *P. glandulosa*) and the primary observation of growth. Chin J Agric Biotechnol 3(1):59–64
Zhang TT, Song YZ, Liu YD et al (2008) Overexpression of phospholipase Da gene enhances drought and salt tolerance of *Populus tomentosa*. Chin Sci Bull 53:3656–3665
Zhang W, Chu Y, Ding C, Zhang B et al (2014) Transcriptome sequencing of transgenic poplar (*Populus* x *euramericana* 'Guariento') expressing multiple resistance genes. BMC Genet 15(1):1–17
Zhang Y, Jun Z, Jinping L, Jinmao W et al (2016) Temporal and spatial changes in Bt toxin expression in Bt-transgenic poplar and insect resistance in field tests. J For Res 27(6):1249–1256
Zhao XY, Zheng HQ, Li SW, Yang CP et al (2013) The rooting of poplar cuttings: a review. New For 45(1):21–34
Zheng JB, Zhang YM, Yang WZ, Pei D et al (1995) Plant regeneration from excised leaves of poplar hybrid 741 and transformation with insect resistant Bt. toxin gene. J Hebei Agric Univ 18:20–25
Zheng L, Guifeng L, Xiangnan M, Yanbang L, Yucheng W (2012) A versatile *Agrobacterium*-mediated transient Gene expression system for herbaceous plants and trees. Biochem Genet 50:761–769
Zhong RQ, Morrison WH, Himmlsbach OS, Poole FL II et al (2000) Essential role of caffeoyl-coenzyme A-O-methyltransferase in lignin biosynthesis in woody poplar plants. Plant Physiol 124(2):563–577
Zhou J, Wang J, Bi Y, Wang L et al (2014) Overexpression of PtSOS2 enhances salt tolerance in transgenic poplars. Plant Mol Biol Rep 32(1):185–197
Zou WH, Zhao OA, Cui OC, Wang B (2006) Transformation of *Populus deltoides* with anti-PLD gamma gene and chitinase gene. Sci Silvae Sin 42(1):37–42

MIX
Papier aus verantwortungsvollen Quellen
Paper from responsible sources
FSC® C105338

If you have any concerns about our products,
you can contact us on
ProductSafety@springernature.com

In case Publisher is established outside the EU,
the EU authorized representative is:
Springer Nature Customer Service Center GmbH
Europaplatz 3, 69115 Heidelberg, Germany

Printed by Libri Plureos GmbH
in Hamburg, Germany